1973

W. B. SAUNDERS COMPANY PHILADELPHIA LONDON TORONTO

Introductory Astronomy and Astrophysics

Elske v.P. Smith, Ph.D.
Astronomy Program
University of Maryland
College Park, Maryland

Kenneth C. Jacobs, Ph.D.
Department of Astronomy
Leander McCormick Observatory
University of Virginia
Charlottesville, Virginia

W. B. Saunders Company: West Washington Square
                          Philadelphia, Pa.   19105

                          12 Dyott Street
                          London, WC1A   1DB

                          833 Oxford Street
                          Toronto 18, Ontario

Introduction to Astronomy   ISBN 0-7216-8387-8

Print No.:   9   8   7   6   5   4   3   2   1

Preface

Throughout the ages, mankind has been intrigued by the phenomena of the starry heavens and has continued to wonder about the nature of the Universe in which he lives. It is not surprising, therefore, that astronomy is the oldest observational science. But, together with the complementary theoretical discipline of astrophysics, astronomy is also as young as today, for both sciences have experienced a spectacular renaissance in the past few years. This book is an attempt to convey the feeling and content of *modern* astronomy and astrophysics to the serious scientific student.

By "serious scientific student" we mean the *undergraduate* (at the advanced freshman level or sophomore) who is willing to combine basic mathematical techniques (trigonometry and some calculus) with an elementary knowledge of physics to broaden his understanding of the content and structure of our Universe. Hence, this text is intended to fill a serious *gap*, for there exist no good textbooks more advanced than the excellent survey texts aimed at the nonscientist interested in general astronomy, yet more basic than the high-level works on astronomy and astrophysics available to the graduate student or upper-level undergraduate. Here we are interested in providing a solid foundation in up-to-date astronomy, as well as an elementary introduction to current astrophysics.

For pedagogical purposes, this book follows the usual chronological-proximity ordering: we begin close to home (the planet Earth) and move steadily outward (to the stars, our Galaxy, and the Universe of galaxies). Taken in its entirety, the text should be ideal for a one-year course in "introductory scientific astronomy." However, it may also be used to advantage in a one-semester course covering one of the following: (a) an intensive study of our Solar System (Chapters 2 through 9); (b) an in-depth investigation of "stars and stellar systems" (Chapters 8 to 20, perhaps excluding all but sections 1 and 2 of Chapter 9); or (c) an overall survey of the makeup of our Universe (selecting readings from most chapters). Each instructor will have his own preferences, but the following comments may be useful for those considering a one-semester survey.

> Omit all small-type sections. Bypass the mathematics of vectors and the calculus by omitting section 5 of Chapter 3 [but *not* subsection 5(b)] and section 2 of Chapter 4. Do not include the material on binary stars (Chapter 11), since it is not needed in later chapters.

Furthermore, it is advantageous to delete (or at best, to skim) the following specialized material:

On celestial mechanics, section 1 of Chapter 3 and section 2 of Chapter 6.

On the Solar System, sections 1(b) through 1(e) of Chapter 5, section 6 of Chapter 6, and parts of Chapter 7 at the instructor's discretion.

On radiation theory, section 1(a, ii and iii) and section 4 of Chapter 8, sections 2(b), 2(c), and 3(a) of Chapter 9, and section 1(c) of Chapter 12.

On the Sun, sections 3(b), 5, and 6 of Chapter 9.

On stellar motions, sections 2, 3, and parts of section 4 in Chapter 14.

On variable stars, sections 2(b), 3(b), 3(c), 4(b), 4(c), 4(d), and 5(b) of Chapter 16 and perhaps other sections.

On our Galaxy, sections 1(b), 1(c), 1(d), 2(d), 2(e), and 3 of Chapter 18.

On other galaxies, sections 1(b), 2, and 6(d) of Chapter 19.

Finally, additional time may be saved without irreparable loss by simply skimming Chapters 1, 2, 4, 7, and 17.

Some readers may lament the lack of a chapter on telescopes and astronomical instrumentation. After careful consideration, we decided that such a chapter was inappropriate in this book, which concerns itself with astronomical phenomena and their physics. The study of *how* the data are obtained is better undertaken in a course in practical astronomy. Moreover, the subject of telescopes and their accessories is well covered in several standard astronomy texts. Specific references are given in Appendix A1–3.

We have tried to make this book as up-to-date as possible at the time of its printing, but both astronomy and astrophysics are changing so rapidly that soon new phenomena will be uncovered and better theories will appear to supersede some of those which we have mentioned. Therefore, the student should be constantly aware of the state of flux of astronomical "knowledge," and should strive to learn in particular those time-tested fundamentals which will survive into the future.

No book can be written without help from others, and this text is no exception. We wish to acknowledge the following for their contributions.

Bart J. Bok, Priscilla F. Bok, and James E. Gunn made extensive and most useful comments and criticisms; we believe that the book has greatly benefited from their careful readings. We frequently sought advice on matters of fact, interpretation, or presentation from many of our colleagues in our respective departments. Several colleagues also read parts of the manuscript. Special thanks are due Philip A. Ianna, Alan H. Karp, Frank J. Kerr, G. Siegfried Kutter, Thomas A. Matthews, William K. Rose, Jack B. Saba, and S. Christian Simonson. Parts of an earlier version were read and commented upon by A. G. W. Cameron and Woodruff T. Sullivan III, to whom we express our thanks. Alan Karp's assistance with the proofreading was of enormous help. Many of the line drawings were done by Jay Wilbur. For monumental work in typing, as well as moral support to both of us, thanks go to Bonnie Jacobs. Mary King also did much of the typing. As Chairman of the University of Maryland Astronomy Program, Gart Westerhout lent his greatly valued support during the years the manuscript was in preparation. E. S. says "thanks" to Henry, Geoffrey, and Kenneth for their tolerance and patience during these same years. Photographs and figures were supplied by many individuals and institutions whom we acknowledge here with thanks. Appropriate credits appear with the legends.

In a book of this size, there will undoubtedly remain some undiscovered errors, for which the authors naturally take responsibility. We would appreciate learning of these mistakes—either typographical errors or errors of fact—as soon as possible.

ELSKE V. P. SMITH
KENNETH C. JACOBS

Contents

Part 1
Introduction

Chapter 1
Astronomy and Astrophysics

1-1 DEFINING ASTRONOMY AND ASTROPHYSICS

Astronomy and astrophysics are the scientific disciplines in which we collect, correlate, and interpret data pertinent to our entire observable Universe, from our own planet to the farthest reaches of extragalactic space. Astronomy has its basis in observations of the sky, while astrophysics is more theoretical, applying and extrapolating laboratory physics to astronomical phenomena.

The word "astronomy" derives from two Greek words: *astron*, meaning star, and *nomos*, which has two forms, *nomós*—district or abode, and *nómos*, law or custom, both of which are related to *nemein*, which means to distribute, allot. One could define astronomy, therefore, as a study of the distribution of the stars in space; broadly interpreted, this includes the laws that govern the distribution of stars.

"Astrophysics" similarly consists of two roots: *astron* and *physic* or *physis*, Greek for nature. Hence, astrophysics studies the nature or physics of the stars and star systems.

But by these definitions, most astronomers are also astrophysicists and vice versa. The two fields are different, mutually dependent approaches to the same subject.

In contrasting astronomy and astrophysics, consider the fable of the blind men who first encounter an elephant. Each man describes what is before him. The first grasps the elephant's trunk and says, "Verily, it is a snake." Another, **3** feeling the massive leg, likens it to a tree. Yet another, at the bristly tail, compares that to a broom. The examination and discourse continue but an understanding of the whole is not yet attained. Astronomers are like the blind men of the fable; they stumble upon something and try to describe it. But astronomy goes further; it interprets the observations. An astrophysicist in this fable, having heard tell of this wondrous creature, would try to build a model of the animal by using his knowledge of biology, chemistry, and physics. Astrophysics provides the theoretical framework for interpreting and understanding the observations. We have over-simplified the distinctions between astronomy and astrophysics in our portrayal. Astronomy not only catalogs data; every observation invokes some interpretation. Similarly, astrophysics would be sterile if it were purely theoretical and not grounded in observations.

Sometimes astrophysics anticipates astronomy, predicting phenomena not yet observed or recognized. For example, astrophysicists calculated and built models of neutron stars long before astronomers observed any such objects. Neutron stars, stars so dense that all matter within is reduced to neutrons, are now believed to be the phenomenon responsible for pulsars—objects that emit short radio pulses at very frequent intervals. Astrophysicists interpreted pulsar observations in terms of theoretical models and decided that they could be best explained in terms of neutron stars.

In brief, then, astronomy works from the observations toward an interpretation, while astrophysics works toward the observations from models based on physics. Yet even this distinction is somewhat artificial, for these two sciences are never far apart.

1–2 PERSPECTIVE AND PREVIEW

Astronomy is as old as recorded history; it appeared in the fourth millennium B.C. with the rise of the great river civilizations: the Tigris-Euphrates (Mesopotamia), the Nile (Egypt), the Indus (India), and the Hwang-Ho (China). Early observations were interpreted astrologically. Astrology (from *astron* plus *logos*, meaning discourse) is based on the idea that the heavenly bodies exert a strong influence upon events on Earth; therefore, men can foretell possible events from the positions of the stars and planets. Astronomy first became important to these early agricultural civilizations as a means to forecast the seasons. For example, the rising of the Dog Star, Sirius, was used to predict the annual flooding of the Nile. Early astronomers went beyond this, however, devising calendars, predicting eclipses, preparing lunar and planetary tables, and even evolving cosmologies to interpret their observations. We will not delve into the details of the history of astronomy in this book. The reader who wishes to go further is referred to the reading list at the end of this chapter.

Since astronomy studies the distribution of the stars, the astronomer's first step is to set up a system describing this distribution. Many cultures grouped visible stars into constellations, depicting mythological heroes or familiar creatures. Today we still use the constellations delineated by the Greeks, although we also describe the position of a star by a coordinate system which gives the angular distance from a set of reference points and circles.

The planets wander back and forth among the "fixed" stars of the constellations, and these wanderings have intrigued astronomers through the ages. Some of the greatest thinkers have felt challenged to explain and predict these wanderings (Plato, 427–347 B.C.; Aristotle, 384–322 B.C.; Heracleides, 375–310 B.C.; Ptolemy, 137 A.D.; Copernicus, 1473–1543 A.D.; Galileo, 1564–1642 A.D.; Kepler, 1571–1630 A.D.; and others). Isaac Newton (1642–1727) ushered in the modern study of the motions of planetary bodies (celestial mechanics). He enunciated the three basic laws of mechanics and the law of universal gravitation, and invented differential and integral calculus so that these laws could be applied to the motions of bodies. This grand synthesis embraced all astronomy of the past and initiated the discipline of astrophysics.

The planets themselves are fascinating. Some are dense and rocky, like the Earth, while others, such as Jupiter, have densities comparable to that of water. The explanation involves an understanding of Solar System evolution. The Moon's surface represents a record of the last four billion years of that evolution, and thus is our keystone to the puzzle. With radar technology we have begun to map the surface of Venus, previously hidden from us by an opaque cloud cover. Mars, long thought to be a twin of Earth, was shown by space probes to be more analogous to the Moon. Planetology relies more heavily on geology and atmospheric physics than upon astrophysics, yet the latter does play a role in constructing models of planetary interiors and atmospheres.

Astrophysics truly comes into its own when we turn again to the stars and consider their physical structure. The atmosphere of a star lends itself to detailed analysis with the spectrograph. Astrophysicists use knowledge gained in the laboratory to derive surface temperatures and chemical abundances from stellar spectra. The absorption lines in the solar spectrum were first described in 1817 by J. Fraunhofer. Their identification and interpretation started with the work of G.R. Kirchhoff and W. Huggins in the 1860s. But real progress did not begin until early in this century, with the advent of modern atomic physics. Refinement of interpretations continues to the present.

In the early 1910s, E. Hertzsprung and H. N. Russell independently discovered a relation between stellar temperatures and luminosities. These astronomers laid the groundwork for future astrophysical studies of stellar structure and evolution. Observed quantities, such as stellar luminosities, surface temperatures, sizes, and masses, supply the boundary conditions or "specifications" that must be met by models of a star's interior. These properties are deduced from a plethora of observational data. The results give us not only stellar models, but also a stupendous evolutionary scheme depicting the life history of a star—from its birth in an interstellar cloud of dust and gas; through the long-lived stable stage (where the Sun is at present); through a giant stage (in which the star swells immensely and cools off); through a stage in which mass is lost (either explosively as in a supernova or more sedately as in a planetary nebula); to the end of the star's life as a neutron star or a white dwarf. It was not until the 1930s, when it became clear that nuclear energy was the only energy source that could account for the radiation of the Sun and other stars, that our understanding of stellar structure and evolution was put on a firm footing. H. Bethe received the Nobel Prize in physics for "his discoveries concerning energy production of stars," in particular, the carbon-nitrogen cycle in which hydrogen is converted to helium with carbon and nitrogen acting as "catalysts." A more direct hydrogen-to-helium conversion, the proton-proton chain, is operative in relatively low-mass stars like our Sun.

The Sun is only one of a hundred billion stars in the Milky Way Galaxy—one of the wonders of the night sky that stretches overhead like a river of light and resolves into stars upon inspection with a telescope or even a pair of binoculars. Dark patches in the Milky Way are clouds of dust obstructing the view of stars beyond. The presence of dust was an obstacle and a challenge to the astronomers, and caused confusion in the attempt to delineate the form and size of our Galaxy. W. Herschel (1738–1822) and J. Kapteyn (1851–1922) deduced models of the Universe that did not recognize the presence of dust, while H. Shapley and R. J. Trumpler were instrumental in clarifying its significance. Yet, despite the dust, astronomers discerned that ours is a spiral galaxy, dominated by a central mass and rotating differentially, as first determined by J. H. Oort and B. Lindblad.

Gas, intermingled with the dust and even more prevalent, is detected because of its emission of radio waves. Abundant hydrogen emits a spectral line at 21 cm, and examination of this radiation enables us to study the structure of our Galaxy well beyond the reach of optical telescopes. Intriguing, too, are the interstellar molecules—complex molecules, such as water, ammonia, and even formaldehyde and formic acid—that may be considered the building blocks of life. These complex molecules are even more closely associated with the dust than the hydrogen is. We are only beginning to understand the circumstances under which they exist, radiate, and absorb (at radio frequencies).

Apparently it is clouds like these in which stars are born. The dust and gas, and their associated young stars, are confined to the spiral arms of our Galaxy, but older stars and globular clusters are found in a spherical system around the center. W. Baade proposed the scheme of two stellar populations as represented by these extremes.

Despite the transparency of interstellar dust to radio waves, the nucleus or center of our Galaxy still remains somewhat of a mystery to us. Among the phenomena discovered there by astronomers but still to be explained by the astrophysicists are: the expanding spiral arm, the anomalous chemical abundance of member stars, the high-velocity gas clouds that are either falling into the center or being ejected from it, the presence of strong infrared and radio sources, and the possibility that the center of the Galaxy is the emitter of gravitational waves. Some have proposed that in the center of the Galaxy is a black hole, an object so dense and massive that light cannot escape from it.

Many details concerning our Galaxy were deduced by comparisons with other galaxies, such as the Andromeda galaxy in the Local Group of galaxies. The debate in the 1920s as to whether such galaxies were merely members of the Milky Way or "island universes" was settled when E. P. Hubble identified individual stars in the Andromeda galaxy. It was also Hubble who, in 1929, followed V. M. Slipher's earlier work and showed that the farther a galaxy is from us, the faster it moves away from us; he thereby deduced that the Universe is expanding. Einstein's general theory of relativity can explain this expansion, but we are still seeking to determine just what kind of Universe we live in. Did it start with a big bang? Will it one day return to that infinitely dense primordial "atom" as it oscillates from expansion to contraction? To answer these and similar questions, observations are being pushed as far as possible. Quasars, recently discovered and initially thought to yield clues concerning cosmology, have proved more enigmatic than first realized, answering no questions but raising many new ones. Some astrophysicists have invoked "black holes" to explain the apparently extraordinary energies of these objects.

This has been a quick preview of what we shall cover in depth in the following pages. We also recognize that the division between astronomy and astrophysics is a blurred one, and the difference relatively artificial. Having set the stage, let the play begin!

Reading List

Berry, Arthur: *A Short History of Astronomy*. New York, Dover Publications, 1960.

Boas, M.: *The Scientific Renaissance: 1450–1630*. New York, Harper & Row, 1962.

Dreyer, J. L. E.: *A History of Astronomy From Thales to Kepler*. Second Ed. New York, Dover Publications, 1953.

Durant, Will and Ariel: *The Story of Civilization*. Vols. I–X. New York, Simon and Schuster, 1935–1967.

Encyclopaedia Britannica. Vol. 2, pp, 643 ff., 656 ff. Chicago, Encyclopaedia Britannica Inc., 1967. (Additional material of interest may be found in *Encyclopaedia Britannica* under specific topics.)

Hawkins, Gerald S.: *Stonehenge Decoded*. Garden City, New York, Doubleday & Company, 1965.

Hoyle, Fred: *Astronomy*. Garden City, New York, Doubleday & Company, 1962.

Koestler, Arthur: *The Sleepwalkers*. New York, The Macmillan Company, 1959.

————: *The Watershed: A Biography of Johannes Kepler*. New York, Doubleday & Company, 1960.

Pannekoek, Antonie: *A History of Astronomy*. New York, Interscience Publishers, 1961.

Rapport, S., and Wright, H. (eds.): *Astronomy*. New York, Washington Square Press, 1965.

Shapley, Harlow, and Howarth, Helen E.: *A Source Book in Astronomy*. New York, McGraw-Hill Book Company, 1929.

Shapley, Harlow (ed.): *Source Book in Astronomy: 1900–1950*. Cambridge, Massachusetts, Harvard University Press, 1960.

Struve, Otto, and Zebergs, Velta: *Astronomy of the Twentieth Century*. New York, The Macmillan Company, 1962.

Thompson, Eric. S.: *The Rise and Fall of Mayan Civilizations*. Norman, Oklahoma, University of Oklahoma Press, 1954.

Vaucouleurs, Gérard de: *Discovery of the Universe*. New York, The Macmillan Company, 1957.

Whitney, Charles A.: *The Discovery of Our Galaxy*. New York, Alfred Knopf, 1971.

Chapter 2
Aspects of the
Celestial Sphere

The heavens are the fundamental arena of astronomy and astrophysics. The earthbound observer, however, receives no impression of depth as he views the celestial scene; he senses only relative angular positions. As did the ancient Greeks, we also see a hollow celestial sphere apparently centered upon the Earth's center. The fixed stars appear to be borne across the sky in concert as the sphere turns once each day, and the wandering planets (the ancients counted seven: the Sun, the Moon, Mercury, Venus, Mars, Jupiter, and Saturn) pursue their individual meanderings within the confines of the band of the zodiac.

To make sense of this spectacle we must map the sky, assigning locations to each of the celestial phenomena we plan to study. In Chapter 3 we will specify the three-dimensional spatial position of an event by its Cartesian (rectangular), polar, and spherical coordinates. Since angular positions are our primary interest in positional astronomy, we shall devote the present discussion almost exclusively to *spherical coordinate systems*. On the surface of a sphere the circumference and radius of a circle are not related by 2π and the sum of the interior angles of a triangle is always greater than $180°$. Hence, the familiar plane geometry and trigonometry are no longer applicable; they must be replaced with spherical geometry and trigonometry, for which the most important formulae are given in the Mathematical Appendix at the end of this book. The chapter ends with a presentation of the convenient and historically important systems of planetary configurations and stellar constellations.

To begin, let us consider the surface of a sphere of arbitrary radius. Any plane passing through the center of the sphere intersects the surface in a *great circle*.

We select one plane—usually that plane perpendicular to an axis of rotation—and designate its great circle the *primary circle*. All great circles intersecting the primary circle perpendicularly are called *secondary circles;* all the secondary circles meet at only two points, the *poles*. We define one intersection point of the primary circle and a given secondary circle (the reference circle) as the *point of origin*. A coordinate system may now be set up on the spherical surface as follows: the position of a point A is specified by (a) the angular distance in a conventional direction along the primary circle from the point of origin to the point of intersection closest to A of the secondary circle passing through A, and (b) the shortest angular distance along this secondary circle from the primary circle to point A. Before proceeding to celestial coordinates, let us illustrate these ideas using the surface of the Earth.

2–1 LONGITUDE AND LATITUDE ON THE EARTH

Figure 2–1 shows the familiar longitude-latitude system of terrestrial coordinates. The *equator* is the primary circle, defined by the central plane perpendicular to the Earth's axis of rotation; the rotation axis intersects the surface of the Earth at the north and south poles. Secondary circles pass through the poles, and each semicircle terminating at both poles is termed a *meridian*. The reference semicircle, the *prime meridian*, passes through Greenwich, England, and it meets the equator at the point of origin (0° longitude). *Longitude* is the shortest angular distance along the equator from the prime meridian to a given meridian; it is measured eastward or westward from 0° to 180°. The International Date Line is located essentially at longitude 180° E (or W). *Latitude* is the angular distance north or south from the equator, measured along a meridian in degrees from 0° (the equator) to 90° (the poles). Notice that planes parallel to the equator slice the Earth's surface

Figure 2–1. *Terrestrial Coordinates.* The Earth's surface is shown here, with the primary circle of the equator (E) and the reference secondary circle of the prime meridian (PM). The north pole (NP) and south pole (SP) are indicated, as is a small circle or parallel of latitude (PL). The angular longitude and latitude of the point X are shown by the arrows.

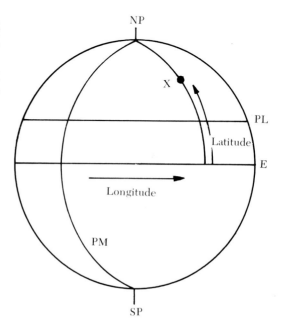

in small circles—the *parallels* of latitude. Some examples of approximate locations specified by this system are: New York City (73°58′ W, 40°40′ N); Sydney, Australia (151°17′ E, 33°55′ S); and Addis Abāba, Ethiopia (38°44′ E, 9°0′ N).

Let us note here that the angular measure which we use today was originally devised by the Babylonians. A circle is subdivided into 360 degrees of arc (°); each degree contains 60 minutes of arc (′); and every minute is composed of 60 seconds of arc (″). As we reiterate in the Mathematical Appendix, there are 3600″ per degree and 1,296,000″ in a complete circle. Since the Earth's circumference is about 40,075 kilometers (24,900 miles), we have the following approximate relations at the equator: one degree is 111 kilometers (69 miles), one arc-minute is 1.9 kilometers (1.15 miles), and one arc-second is 31 meters (101 feet). Note that the distance per unit angle of longitude decreases steadily toward zero as we approach either pole. In Chapter 4 we shall see that the Earth is actually an oblate spheroid (fatter at the equator than at the poles), so that a degree of latitude corresponds to 110.6 kilometers (68.7 miles) at the equator and increases to 111.7 kilometers (69.4 miles) at the poles.

2–2 THE HORIZON SYSTEM

The celestial pageant presents different aspects to different observers on the Earth's surface, but everyone is aware of the celestial hemisphere visible at any given time. Primary observations are clearly location dependent; therefore, let us outline the observer-based *horizon* or *azimuth-altitude* system of coordinates.

On the celestial sphere we shall construct a spherical coordinate system with the observer at its center (see Figure 2–2). The point vertically overhead is termed the *zenith*, while the opposite point (directly underfoot) is the *nadir*. Together these two points define an axis. The plane passing through the observer, perpendicular to this axis, meets the sky at the *celestial horizon*, which is everywhere 90° distant from both zenith and nadir. Because of natural and man-made obstructions, the actual horizon is seldom the celestial horizon; the closest approximation occurs for a sea-level observer in the middle of a calm ocean. Planes parallel to the axis

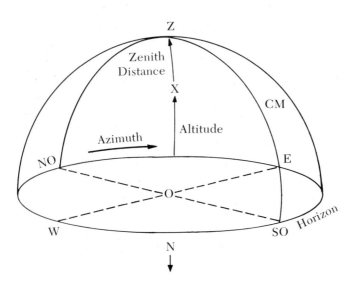

Figure 2–2. *The Horizon System.* The observer (O) sees his zenith (Z) directly overhead, while the nadir (N) is directly underfoot. The celestial horizon contains the four cardinal points: north (NO), east (E), south (SO), and west (W); the observer's celestial meridian (CM) is the great circle containing NO, Z, and SO. The azimuth and altitude of a point X are indicated by the arrows, as is the zenith distance.

and passing through the observer cut the celestial sphere in great circles called *vertical circles*. The reference circle is that vertical circle which includes the observer's zenith and the north and south points on his horizon—this we designate the observer's *celestial meridian*. The north point on the horizon is the point of origin, and east and west lie on the horizon midway between the north and south points.

The position of a celestial phenomenon is specified in the horizon system by giving its instantaneous azimuth and altitude. *Azimuth* is the angular distance along the horizon measured eastward from the north point to the foot of the vertical circle containing the phenomenon; that foot closest to the phenomenon is the one intended, and azimuth ranges from 0° to 360°. *Altitude* is the shortest angular distance upward along this vertical circle from the horizon to the phenomenon, and it ranges from 0° (the horizon) to 90° (the zenith). The complement of a body's altitude is its *zenith distance* (i.e., 90° minus the altitude). Two events on the observer's celestial meridian are of particular interest: a celestial body is said to be in *upper transit* when it crosses the celestial meridian moving westward, and in *lower transit* when it crosses moving eastward. In the case of meteors and artificial satellites, which may rise in the west and set in the east, another criterion for upper transit is necessary: usually upper transit is that crossing of the celestial meridian visible to the observer.

2–3 CELESTIAL EQUATORIAL COORDINATES

We come now to the most important astronomical coordinate system—the *celestial equatorial system*. The full complexities of spherical and historical astronomy are met here, but with a modicum of imagination and applied visualization these coordinates will soon become second nature to us.

Let us recall that the *celestial sphere* is centered upon the Earth's center and that its radius is indefinitely large. The last stipulation follows because we intend to map the entire observable Universe onto the surface of this sphere, and because the lines of sight to any star (other than our Sun) are essentially perfectly parallel for any two earthbound observers. There are many ways to project a spherical coordinate system onto the interior surface of the celestial sphere (see, for example, the ecliptic and galactic systems in section 2–4); the Earth's rotation is the basis of the present method. Though much of the terminology is different, the celestial equatorial coordinate system is almost completely analogous to terrestrial longitude and latitude.

To a given observer, the apparent rotation of the celestial sphere causes all stars to circumnavigate the heavens once each day; hence, the azimuth and altitude of each star are constantly changing with time. By transforming to a spherical coordinate system which rotates with the celestial sphere—the celestial equatorial system—we can obtain positions which are *fixed* (i.e., which do not vary with time) to an accuracy of one part in 10^4 per year. The chief cause of the remaining variations is the Earth's precession (see Chapter 4), which results in the westward precession of the equinoxes by about 50″ per year. The wanderings of the Sun, the Moon, and the planets are not eliminated in our new coordinate system, but they are slowed down considerably.

Let us visualize a stationary celestial sphere, at the center of which the Earth rotates eastward upon its axis once each day. We can imagine halting the Earth's rotation and projecting the terrestrial longitude-latitude coordinate mesh onto the surface of the celestial sphere (see Figure 2–3). The Earth's equatorial plane cuts the celestial sphere in the great circle of the *celestial equator*, and the extensions of the Earth's rotation axis intersect the sphere at the north and south *celestial poles*. Meridians of longitude are mapped into *hour circles* on the celestial sphere, while parallels of latitude appear as small circles concentric to the poles. If we now permit the Earth to resume its rotation, this celestial equatorial coordinate mesh fixed to the celestial sphere appears, to any observer on the Earth's surface, to rotate westward as an entity once each day.

In the celestial equatorial coordinate system, the celestial equator is the primary circle and the hour circles are the secondary circles. The position of a celestial body is specified by its declination (δ) and its right ascension (once designated R.A. but now indicated by α). *Declination*, the analog of terrestrial latitude, is the smallest angular distance (measured in $^{\circ}{}'{}''$) from the celestial equator to the body along the hour circle passing through the body. Positions between the celestial equator and the north celestial pole have positive (+) declination by convention, while those between the celestial equator and the south celestial pole have negative (−) declination; hence, declination ranges from 0° (the celestial equator) to $+90^{\circ}$ (the north celestial pole) or -90° (the south celestial pole).

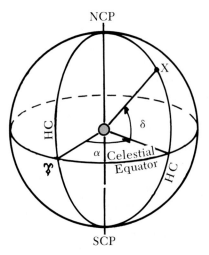

Figure 2–3. *Celestial Equatorial Coordinates.* The Earth lies at the center of the celestial sphere. The Earth's rotation axis extends to the north celestial pole (NCP) and the south celestial pole (SCP), and the Earth's equatorial plane cuts the celestial sphere in the celestial equator. Meridians of longitude project into hour circles (HC), and the direction to the vernal equinox (Υ) is indicated. The arrows show the right ascension (α) and declination (δ) of the point X on the celestial sphere.

Right ascension, the analog of terrestrial longitude, is the angular distance (measured in hours-minutes-seconds of *time*, or ^{h} ^{m} ^{s}) eastward along the celestial equator from the prime hour circle (see below) to the hour circle containing the body. Right ascension ranges from 0^{h} 0^{m} 0^{s} to 23^{h} 59^{m} 59^{s}. To understand how the time units of right ascension come about, consider the prime hour circle to exactly coincide with an observer's local celestial meridian. Since the Earth rotates through 360° in 24 hours, the observer's celestial meridian will lie 15° east of the prime hour circle after one hour. Therefore, we call this right ascension 1^{h} 0^{m} 0^{s}; a rotation of one degree corresponds to four minutes of time ($1^{\circ} = 4^{\text{m}}$), of one arc-minute to four seconds of time ($1' = 4^{\text{s}}$), and of one arc-second to one-fifteenth second of time ($1'' = 1/15^{\text{s}}$). In the celestial equatorial coordinate system, the approximate position of the bright star Sirius is (α, δ) = (6^{h} 43^{m}, $-16^{\circ}39'$).

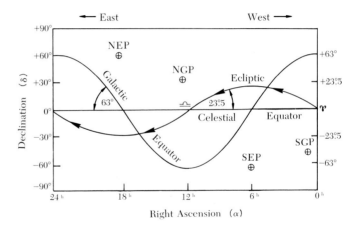

Figure 2-4. *Spherical Coordinates on the Celestial Sphere.* Here we map the celestial sphere onto a rectangle. The coordinates are right ascension (α) and declination (δ). The celestial equator is the horizontal line at declination 0°. The ecliptic oscillates 23°.5 above and below the celestial equator, intersecting it at the vernal equinox (Υ) and the autumnal equinox ($\simeq$); the direction of the Sun's annual motion is indicated by the arrows. The north and south ecliptic poles, (NEP and SEP) are shown, and for completeness we include the galactic equator (inclined 63° to the celestial equator) and the north and south galactic poles (NGP and SGP).

The point of origin of right ascension is the *vernal equinox*, (Υ), historically known as the first point of Aries. This origin, fixed on the celestial sphere, is defined by the celestial equator and the ecliptic. The *ecliptic*, the apparent annual path of the Sun in the sky, is the great circle where the orbit plane of the Earth intersects the celestial sphere. In Figure 2-4 we show a map of the celestial sphere, whereupon are marked the celestial equator and the sinuous ecliptic. These two great circles are inclined at the *obliquity* angle of 23°26'.5 to one another, so that they intersect at only two points—the *equinoxes*. As the Sun progresses eastward along the ecliptic, it crosses the celestial equator moving northward at the vernal equinox (spring) and again six months later moving southward at the *autumnal equinox* (fall). By definition, the right ascension-declination of the vernal equinox is (0^h, $0°$).

For observational purposes we sometimes use the celestial equatorial system, but take as our reference circle the local celestial meridian. Declination measures north-south angular positions, while *hour angle* tells us how far west of the celestial meridian a celestial body is in units of time. The hour angle of an astronomical object depends both on time and on the observer's location, being $0^h\ 0^m\ 0^s$ at upper transit and $2^h\ 30^m\ 0^s$ two and one-half hours later. When we discuss sidereal time in Chapter 4, we shall see that this convenient system permits us to define sidereal time as the hour angle of the vernal equinox.

2-4   ECLIPTIC AND GALACTIC COORDINATES

For describing the motions of bodies within the Solar System, the *ecliptic coordinate system* is extremely useful. Here the ecliptic is the primary circle, and the poles are called the *north ecliptic pole* (that pole closest to the north celestial pole) and the *south ecliptic pole.* *Celestial longitude* (λ) is the angular distance (from 0° to 360°) eastward along the ecliptic from the vernal equinox (the point of origin). *Celestial latitude* (β) is the angle from the ecliptic, measured positively ($+$) toward the north ecliptic pole and negatively ($-$) toward the south ecliptic pole; it ranges from 0° to $\pm90°$. Hence, when the Sun's center lies at the autumnal equinox, its coordinates are (λ, β) = (180°, 0°).

When we discuss phenomena associated with our Galaxy toward the end of this book, we shall find it convenient to employ modern *galactic coordinates*. Here the primary circle is defined by the central plane of the Milky Way, and is called the *galactic equator*. The center of the Galaxy (in Sagittarius), which lies on the galactic equator, is the point of origin. *Galactic longitude* (ℓ or ℓ^{II}) is measured eastward along the galactic equator from the direction to the galactic center, and ranges from $0°$ to $360°$. As viewed from the *north galactic pole*, galactic longitude increases in the counterclockwise direction. *Galactic latitude* (θ or θ^{II}) is the angle from the galactic equator toward the north galactic pole ($+$) or south galactic pole ($-$), and it lies in the range $0°$ to $\pm 90°$. To avoid confusion, it should be noted that prior to August, 1958, a different system of galactic coordinates (ℓ^{I}, θ^{I}) was in use; we shall have no occasion to refer to these obsolete coordinates.

In Figure 2–4 we have also indicated, for completeness, the poles and equators of the ecliptic and galactic coordinate systems. There we see that the great circle of the galactic equator is inclined $63°$ to the celestial equator. Hence, we already know of three different coordinate systems, each of which completely covers the celestial sphere.

2–5   PLANETARY CONFIGURATIONS

Copernicus advanced his heliocentric theory of the Solar System in the 16th century. In Chapter 3 the characteristics and dynamical basis of this heliocentric theory will be discussed in great detail. Important in this discussion, and also of great convenience and historical value, is the positional system of *planetary configurations*. This system of terminology is usually applied to the Moon and the other planets, but it is equally applicable to other planetary satellites, asteroids, and even comets. The origin of this system probably predates Babylonian astronomy (circa 1000 B.C.).

Figure 2–5 illustrates the heliocentric model of our Solar System, wherein the planets all orbit the Sun in the same sense and move in almost-circular orbits lying near the ecliptic plane. In distance from the Sun, the planets range from Mercury (the closest), through Venus, Earth, Mars, Jupiter, Saturn, Uranus, and Neptune, to Pluto (the most distant). Planets closer to the Sun than the Earth are termed *inferior planets;* these are Mercury and Venus. The planets orbiting farther from the Sun than the Earth are called *superior planets;* these are Mars through Pluto. This geocentric nomenclature is very useful, since most astronomical observations are performed on the Earth, and it is very appropriate, since the motions of the inferior planets as seen in our night sky differ markedly from the motions of the superior planets.

Referring again to Figure 2–5, we define *elongation* as the angle seen at the Earth between the direction to the Sun's center and the direction to a given object (e.g., a planet). We speak of eastern or western elongation according to whether the planet lies east or west of the Sun, as seen from the Earth. Elongations of particular geocentric significance are given special names: an elongation of $0°$ is termed *conjunction* (*inferior conjunction* when the planet lies between the Earth and the Sun, and *superior conjunction* when the planet lies on the opposite side of the Sun from the Earth); $180°$ is called *opposition;* $90°$ is *quadrature;* and when an inferior planet attains its maximum elongation, we refer to *greatest elongation*. Only inferior

Figure 2–5. *Planetary Configurations.* Arrows indicate the direction of orbital motion as well as direct rotation. As the Earth rotates toward the east, the Sun rises before a planet at eastern elongation but after one at western elongation or quadrature.

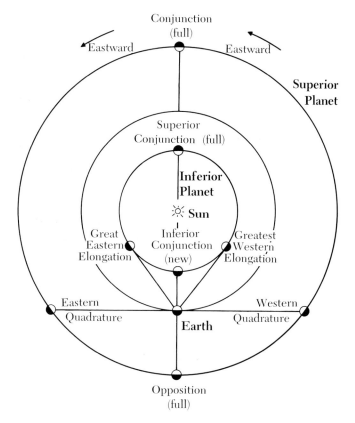

planets may be at inferior conjunction or at greatest elongation (28° for Mercury and 48° for Venus), but they may never be at either quadrature or opposition. For superior planets, inferior conjunction can never occur, and their greatest elongation is 180° (when they are in opposition). Note that our Moon may pass through inferior conjunction, quadrature, and opposition (its greatest elongation) because it is a satellite of the Earth.

2–6   CONSTELLATIONS AND STAR NAMES

Long before the spherical coordinate systems discussed above were constructed, stars were located in the night sky by using constellation designations and relative visual brightnesses. The Babylonians commemorated their gods and heroes by naming certain regions in the starry sky after them, and the ancient Greeks later did the same. Eventually, the entire celestial sphere was arbitrarily subdivided into areas of irregular outline—the *constellations*—with each area generally containing an interesting collection of visible stars. The brightest stars in a given constellation are purported to form the skeletal outline of the mythological figure or beast for which that constellation is named, but an observer must possess a very good imagination to recognize some of the figures. After 1928, the boundaries of the constellations consisted of east-west line segments of constant declination and north-south segments of hour circles; today there are 88 such constellations (see Table A2–1).

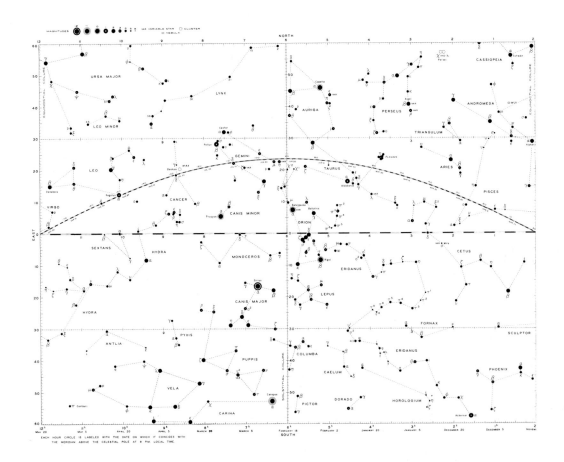

Figure 2–6A. *Constellation and Star Charts.* The celestial equatorial region showing the sinusoidal ecliptic. (Sky Publishing Company)

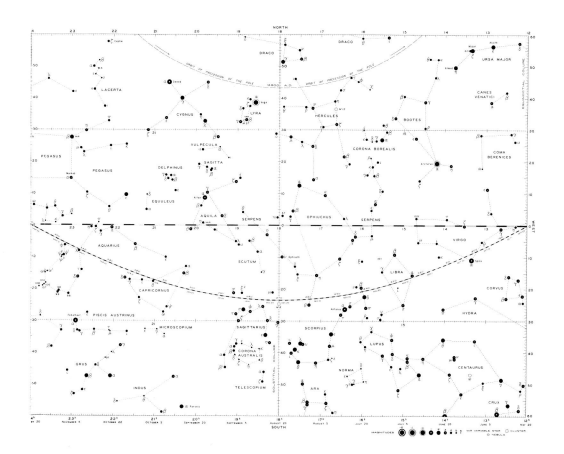

Figure 2–6A, Continued.

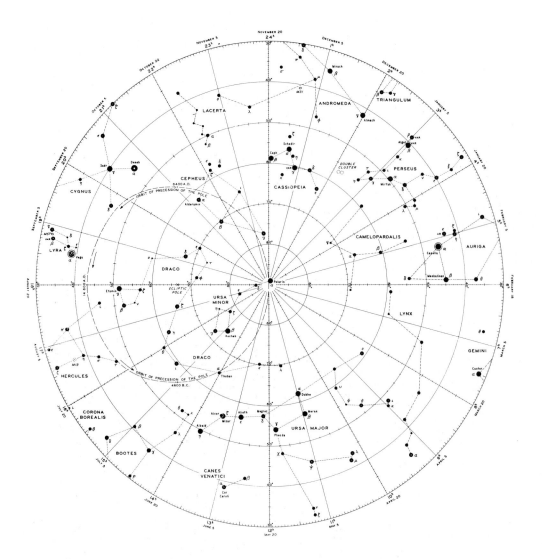

Figure 2–6B.   The north circumpolar region with Polaris near the center. (Sky Publishing Company)

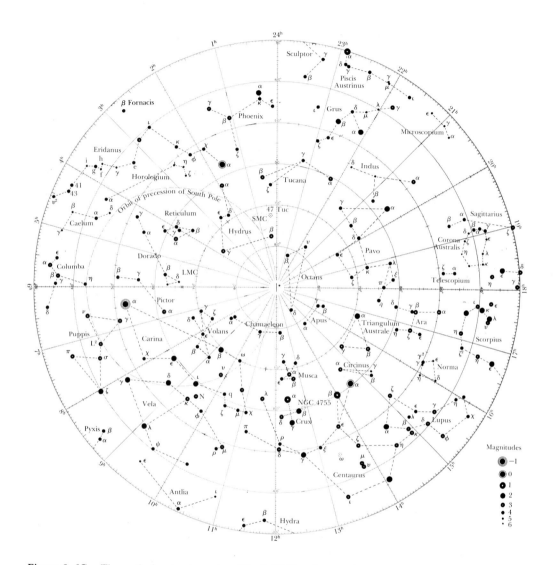

Figure 2–6C.   The south circumpolar region. (Sky Publishing Company)

All constellations have Latin names, but we frequently refer to the anglicized forms, such as the Swan (Cygnus), the Great Bear (Ursa Major), and the Bull (Taurus). The brighter stars within a constellation are ordered by apparent brightness (the brightest is designated α, the next brightest β, etc.) followed by the genitive form of the constellation name. Many stars also have Arabic names, so that we may speak of Betelgeuse (or α Orionis), Denebola (or β Leonis), and Deneb (or α Cygni). In Table A2–1 we list the names, genitive forms, conventional abbreviations, and approximate (α, δ) of the 88 constellations.

The constellation designation of star names serves to approximately locate all of the brighter stars; for the exact locations of these and of the fainter stars we must consult a star catalog (e.g., the *Bonner Durchmusterung*, the *Henry Draper Catalogue*, etc.), where each star is listed by number and by (α, δ). Note that many stars vary in brightness, so that the criterion of relative brightness is not unique; for example, Mira (omicron Ceti) briefly brightens by a factor of 10^3 once every eleven months. To show the forms of the constellations and the relative positions of their stars, we have included the star maps of Figure 2–6.

Among the constellations, you might notice that 14 (12 in ancient times) are traversed by the ecliptic band—the *zodiac*—wherein move the Sun, the Moon, and all of the planets. In order eastward along the ecliptic, with the two recent additions in parentheses, these constellations are: Aries, Taurus, Gemini, Cancer, Leo, Virgo, Libra, (Ophiuchus), Scorpius, Sagittarius, Capricornus, Aquarius, Pisces, and (Cetus). About 3000 years ago the zodiac was divided by the Babylonians into twelve parts of equal width; these astrological *signs of the zodiac* corresponded approximately to the twelve zodiacal constellations at the time of the ancient Greeks, and the vernal equinox marked the beginning of the first sign—the first point in Aries. As a result of the Earth's precession, the vernal equinox is now located in the constellation Pisces. When the vernal equinox moves into Aquarius in the near future, the Age of Aquarius will truly dawn.

Problems

2–1. We have said that the geometry on a spherical surface differs from Euclidean geometry. Consider a sphere of radius R, and note that the "straightest" lines on the surface are arcs of great circles. Now perform the following derivations (the second is optional):

(a) Derive the relation between the arc-radius (s) and the circumference (C) of a circle on the spherical surface; derive the formula for the area of the circle.

(b) Show that as s becomes small (how small?) the familiar Euclidean formulae obtain.

2–2. In its apparent daily rotation, the celestial sphere carries the stars westward in circular paths concentric to the celestial poles. Assume that you are located at (1) the north pole, (2) latitude 38°N, and (3) the equator; for each of these three places answer the following questions:

(a) What are the azimuth and altitude of the north celestial pole?

(b) What is the maximum altitude of the celestial equator?

(c) Over what range of declination do stars always remain visible (these are circumpolar stars)? Invisible (that is, always below the horizon)?

(d) Looking directly east, what is the angle between a rising star path and the horizon?

(e) Give the general formula relating a star's maximum altitude to that star's declination (δ) and your terrestrial latitude.

2–3.   If the hour angle of a star is $16^h\,52^m\,03^s$, and its right ascension is $3^h\,21^m\,17^s$, what is the hour angle of the vernal equinox? Generalize this result by briefly explaining how you would determine the right ascension of a given star, without recourse to star catalogs, charts, or tables.

2–4.   Using graph paper, draw a diagram of the celestial sphere, similar to Figure 2–4, and plot thereon the positions (α, δ) of the twenty brightest stars (see Table A2–6). For your own edification and education, you might also plot the approximate locations of 20 or more of the constellations listed in Table A2–1.

2–5.   For the sake of argument, suppose that Martians exist. Their system of planetary configurations will differ from ours. For the Martians:

(a) Which planets are inferior? Superior?

(b) What is the greatest elongation of Mercury? Venus? Earth? (Use the information provided in Table A2–2).

2–6.   We have defined zenith distance as the complementary angle to altitude.

(a) Show that the altitude of the north (or south) celestial pole is also the latitude of an observer. May we make a similar statement about the minimum zenith distance of the celestial equator; if so, what would that statement be?

(b) To an observer at latitude 38°N, what is the mid-day (noon) zenith distance of the center of the Sun when the Sun's right ascension is (1) 0^h, (2) 6^h, (3) 12^h, and (4) 18^h? If the vernal equinox is in upper transit at noon on March 21, on approximately what days of the year will cases (1)–(4) above occur?

2–7.   In this book we will not discuss lunar and solar eclipses, since an excellent treatment of these phenomena may be found in George Abell's *Exploration of the Universe* (Second Ed., 1969). Do some research on this topic, and answer the following questions:

(a) What were the approximate right ascension and declination of the Moon's center during the total umbral lunar eclipse of February 10, 1971?

(b) Give the same data for the Moon during the total solar eclipse of March 7, 1970.

Reading List

Abell, George: *Exploration of the Universe*. Second Ed. New York, Holt, Rinehart and Winston, 1969.

Allen, Richard H.: *Star Names*. New York, Dover Publications, 1963.

The American Ephemeris and Nautical Almanac. Superintendent of Documents, U.S. Government Printing Office, Washington, D.C. 20402. Published yearly.

Baker, Robert H.: *Introducing the Constellations*. Revised Ed. New York, The Viking Press, 1957.

Baker, Robert H.: *Astronomy*. Seventh Ed. New York, D. Van Nostrand Company, 1959.

Burnham, Robert: *Celestial Handbook*. Flagstaff, Arizona, Northland Press, 1966.

Chauvenet, William: *Spherical and Practical Astronomy*. New York, Dover Publications, 1960.

Davidson, Martin: *Elements of Mathematical Astronomy*. Third Ed. New York, The Macmillan Company, 1962.

Motz, Lloyd, and Duveen, Anneta: *Essentials of Astronomy*. Belmont, California, Wadsworth Publishing Company, 1966.

Mueller, Ivan I.: *Spherical and Practical Astronomy*. New York, Frederick Ungar Publishing Company, 1969.

Norton, Arthur P.: *A Star Atlas*. Fifteenth Ed. Cambridge, Massachusetts, Sky Publishing Corporation, 1966.

Shaw, R. William, and Boothroyd, Samuel L.: *Manual of Astronomy*. Fifth Ed. Dubuque, Iowa, William C. Brown Company, 1967.

Sky and Telescope. Cambridge, Massachusetts, Sky Publishing Corporation. Published monthly.

Smart, William M.: *Spherical Astronomy*. New York, Cambridge University Press, 1962.

Van de Kamp, Peter: *Principles of Astrometry*. San Francisco, W. H. Freeman and Company, 1967.

Woolard, Edgar W., and Clemence, Gerald M.: *Spherical Astronomy*. New York, Academic Press, 1966.

Part 2
The Solar System

Chapter 3
Celestial
Mechanics

3–1 THE HISTORICAL BASIS

While the stars remain fixed on the celestial sphere, and the Sun proceeds ponderously on its annual ecliptic journey among them, the wandering planets perform convoluted gyrations within the zodiac. After superior conjunction (see Figure 2–5), an inferior planet (Mercury or Venus) moves progressively eastward of the Sun, appearing higher and higher in the western sky at sunset as an *evening star*. At greatest eastern elongation, the planet attains its maximum altitude at sunset, then it moves closer to the western horizon at sunset as it sweeps toward inferior conjunction. If conditions are right at inferior conjunction, the planet may be seen to cross the face of the Sun (transit), following which it moves west of the Sun to rise as a predawn *morning star* in the eastern sky. The arching act repeats in the morning twilight as the planet moves to greatest western elongation and on toward superior conjunction.

The superior planets progress steadily eastward across the celestial sphere until they approach opposition. Then, as shown for Mars in Figure 3–1, they slow to a halt, trace a *retrograde* loop toward the west, and finally resume their eastward march. These phenomena both fascinated and frustrated the ancients. Let us briefly discuss the historical efforts to explain these planetary motions—efforts which gave birth to physics and astrophysics, and which culminated in the creation of *celestial mechanics* by Isaac Newton (1687 A.D.).

(a) Ptolemaic Epicycles

The earlier work of Eudoxus and Hipparchus provided the foundation upon which Ptolemy of Alexandria constructed his famous epicyclic theory of the

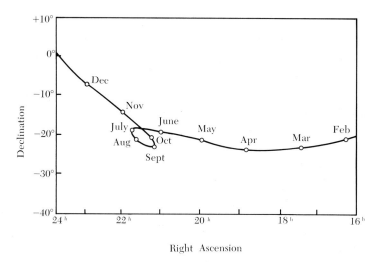

Right Ascension

Figure 3–1. *Retrograde Loop of Mars.* The celestial path of Mars for 1971 is shown; opposition occurred on August 10. Mars proceeded eastward until July 13, retrograded westward near the time of opposition, and resumed its eastward journey on September 11.

Solar System around 140 A.D. Though extremely cumbersome, this theory could account for planetary positions with reasonable accuracy, and it survived for 1400 years.

In Figure 3–2 we illustrate the basic principles of Ptolemy's system. The stationary Earth is placed at the center of the Universe, and about it orbit the planets. To reproduce the observed retrograde loops, Ptolemy set each planet on an *epicycle*, a circle whose center pursues the circular path of the *deferent*. The deferent circle is *eccentric*, that is, not centered upon the Earth. In addition, the center of the epicycle moves at a uniform angular speed about yet another point, the *equant*. This ingenious use of off-set centers permitted Ptolemy to closely mimic the general nonuniform motion of each planet; by adjusting the uniform angular speed of the planet on its epicycle, he could reproduce the position and duration of retrograde motion.

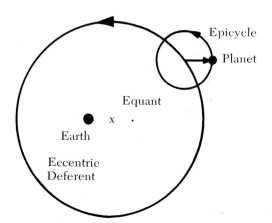

Figure 3–2. *The Epicyclic System of Ptolemy.* The motions explained in the text are: The planet moves uniformly around the circular epicycle, while the center of the epicycle moves along the eccentric deferent circle (not Earth-centered) at a uniform angular rate with respect to the equant point. The planet's angular motion about the Earth is nonuniform and periodically retrograde.

(b) The Heliocentric Theory of Copernicus

During the Dark Ages, astronomical observations improved in the Arabian world, and Ptolemy's theory was augmented to account for these refined data. Eventually, the seven known planets rode upon more than 240 epicycles, and King Alfonso X of Castile remarked that had he been present at the Creation he might have given (the deity) excellent advice. Such was the situation in 1543 A.D. when Copernicus' heliocentric theory of the planetary system appeared.

Copernicus placed the Sun at the center of the Solar System, and made the planets (including the Earth) orbit that central body on eccentric circles. Since he retained the concept of uniform circular motion, Copernicus was forced to keep a few epicycles in his theory. This novel system was slow to be accepted, since its predictions were not significantly better than those of the Ptolemaic theory; nevertheless, the Copernican system was conceptually far simpler than its predecessor. In Figure 3–3 we show the original planetary models put forward by Ptolemy and Copernicus. For convenience, each body is designated by its historically-derived symbol: the Sun ($\odot$), the Moon ($\supset$), Mercury (φ), Venus (φ), the Earth ($\oplus$), Mars ($\circlearrowleft$), Jupiter ($\mathtt{2|}$), Saturn ($\hbar$), and the latecomers Uranus (δ), Neptune (Ψ), and Pluto (P).

Copernicus correctly stated that the farther it lies from the Sun, the slower a planet moves. The arching motions of the inferior planets were easily explained, as were the apparent retrograde loops of the superior planets on the celestial sphere. Figure 3–4 shows how the retrograde gyration arises from the combined motions of the superior planet and the Earth. As we view the moving planet from the

Figure 3–3. *The Planetary Systems of Ptolemy and Copernicus.* **A**, Ptolemy's original geocentric theory of epicycles is displayed, while **B** shows the heliocentric model of Copernicus. Note the relative positions of the Moon ($\supset$), Mercury (φ), and Venus (φ) in Ptolemy's system, and also note that the centers of the epicycles of Mercury and Venus always lie on the line joining the Earth ($\oplus$) and Sun ($\odot$). To simplify the diagrams, equants and eccentric centers are not shown.

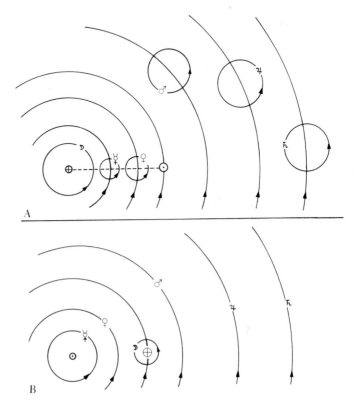

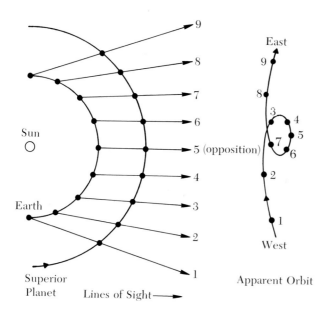

Figure 3–4. *Copernicus Explains Retrograde Motion.* In the Copernican heliocentric theory, the Earth moves faster on its orbit than the superior planet. We show nine lines of sight from the Earth to the superior planet, and display the apparent celestial path of the superior planet with its retrograde loop near the time of opposition.

faster-moving Earth, our line of sight reverses its angular motion twice near opposition, and the loop comes about because the orbits of the two planets are not coplanar.

Copernicus was the first to derive the relationship between the synodic and sidereal periods of a planet. The *synodic period* (*S*) is the apparent time it takes the planet to circumnavigate the celestial sphere as seen from the Earth; for inferior planets (and the Moon) this is the interval between successive inferior conjunctions, while for the superior planets (and the Moon again) it is the interval between successive oppositions. The *sidereal period* (*P*) is the actual time it takes the planet to complete one orbit of the Sun with respect to the fixed stars. Figure 3–5 illustrates the following discussion for circular orbits (an excellent approximation). Designating the Earth's sidereal period by E (365.26 days), we see that the Earth moves at the rate of $360/E$ degrees per day in its orbit, while the planet's rate of angular motion is $360/P$. For a superior planet, the Earth completes one orbit and must then traverse the angle $S(360/P)$ in the time $(S - E)$ to return the superior planet to opposition. Hence,

$$(S - E)(360/E) = S(360/P)$$

or

$$\frac{1}{S} = \frac{1}{E} - \frac{1}{P}$$

For an inferior planet, the Earth is apparently a superior planet, so we merely interchange E and P. We arrive finally at Copernicus' result:

$$\frac{1}{S} = \begin{cases} \dfrac{1}{P} - \dfrac{1}{E} & \text{(inferior)} \\[2ex] \dfrac{1}{E} - \dfrac{1}{P} & \text{(superior)} \end{cases}$$

Figure 3–5. *Synodic and Sidereal Periods.* As the Earth orbits the Sun at the angular rate of 360/E degrees per day, the superior planet moves at the rate 360/P. After one orbit, the Earth is at position 2 and has $(S - E)$ days to reach the second opposition at 3; during the entire synodic period S the superior planet moves from 1 to 3. For an inferior planet, we merely reverse the roles of the Earth and the planet.

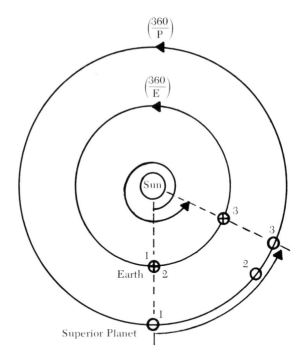

As an example, consider Venus, an inferior planet with the observed synodic period of $S = 583.92$ days. The appropriate formula quickly yields the correct sidereal period of $P = 224.70$ days.

(c)   Galileo's Observations

The telescopic observations which Galileo reported in his *Sidereus Nuncius* (1610 A.D.) strongly supported the heliocentric model of the Solar System. The ancient belief in a perfect and immutable heavenly scene crumbled before his drawings of the wrinkled lunar surface and the moving sunspots on the Sun. Galileo discovered the four largest moons of Jupiter and showed that they orbited Jupiter—not the Earth. This crack in the wall of geocentricism became an irreparable breach with the disclosure of the phases of Venus.

The planets (and the Moon) shine only by the sunlight which they reflect. Half of a planet is always sunlit, while the other half is dark; the *terminator* is the line dividing these two hemispheres. *Geocentric phase* is that fraction of the sunlit hemisphere which can be seen from the Earth. Referring to Figure 3–6 (and back to Figure 2–5), we see that geocentric phase depends upon the elongation of the planet. *New* phase occurs when we see only the dark hemisphere (at inferior conjunction for the Moon, Mercury, and Venus), while *full* phase takes place at opposition when the entire sunlit hemisphere faces us. At *quarter* phase one-half of the hemisphere facing us is sunlit (e.g., at greatest elongation for the inferior planets and at the Moon's quadratures). The superior planets can never be in *crescent* phase (when less than half the observable hemisphere is sunlit), and are practically always in *gibbous* phase (when more than half the planet appears sunlit). In the Ptolemaic model of Figure 3–3A, we see that Venus must always be in crescent phase, since

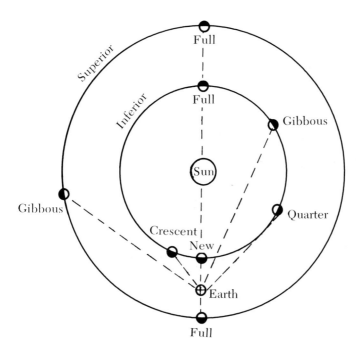

Figure 3–6. *Geocentric Phases.* Superior planets are always in gibbous or full phase. An inferior planet passes through all phases—new at inferior conjunction, then crescent, followed by quarter phase at greatest elongation, and then gibbous phase until it reaches full phase at superior conjunction.

its greatest elongation is rarely greater than 45°; but Galileo observed that Venus shows gibbous phases—hence it must orbit the Sun. This death knell rang out the stationary Earth and ushered in Copernicus, Kepler, and Newton.

(d)   The Methods of Kepler

Using the positional observations of Tycho Brahe and the heliocentric theory of Copernicus, Kepler tediously computed the relative distances of the planets from the Sun. He succeeded in reproducing these distances using elliptical planetary orbits (see section 3–2, Planetary Orbits), and in 1609 and 1619 A.D. he published his three empirical laws of planetary motion. The stage was finally set for Newton's great scheme of dynamics.

Let us examine several of Kepler's methods of distance determination; Copernicus had earlier used the first two methods to find the mean (or average) relative distances from the Sun to the planets. We will use the Sun-Earth distance— the *astronomical unit* (AU)—as our unit of distance. Figure 3–7 illustrates the following discussion.

The Sun-planet distance r (in astronomical units) may be found when an inferior planet reaches greatest elongation (see Figure 3–7A). The angle SEP is observed and the angle EPS is 90°; hence, trigonometry yields $r = \sin(\measuredangle SEP)$. Repeating this procedure over many years gives us the orbit of the inferior planet.

Figure 3–7B shows Copernicus' method for finding the Sun-planet distance of a superior planet. Since both this and the previous method depend upon the planet's synodic period, it takes many years (even centuries) to trace the planet's orbit. The superior planet is in opposition at P when the Earth is at E, and reaches

Figure 3–7. *Distance Determinations.* **A,** When an inferior planet (P) reaches greatest elongation we know the angle SEP; *r* may be found by trigonometry. **B,** A superior planet is in opposition at P. When it reaches quadrature at P′, the Earth is at E′. The time interval between E and E′ gives the angles ESE′ and PSP′. Hence, the angle P′SE′ is known, and trigonometry gives us *r* directly. **C,** A superior planet is at P at the beginning and end of one sidereal period; at these times the Earth is at E and E′. Hence, the angle ESE′ is known, and we can solve for EE′ and the angles SEE′ and SE′E. By observing the angles PES and PE′S we may solve triangle EPE′; *r* then follows by trigonometry.

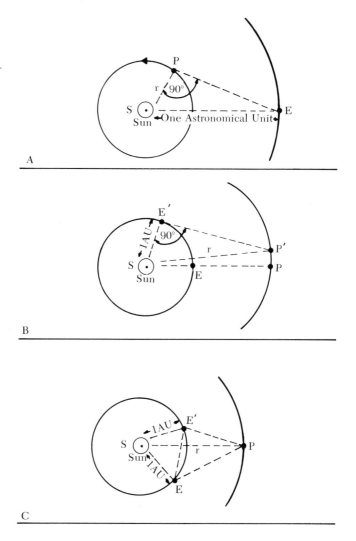

quadrature at P' when the Earth is at E'. The angle $P'E'S$ is clearly 90°, and the angles ESE' and PSP' are found from the time elapsed between E and E'. Since SE' is 1 AU and the angle $P'SE'$ is now known, we may use trigonometry to find $r = 1/\cos{(\measuredangle P'SE')}$.

Kepler's method for finding the distances to a superior planet, while considerably faster than the preceding two procedures, is complicated. In Figure 3–7C the planet is at P at the beginning and end of one sidereal period, and the Earth is at E and E' at these two times. Note that the point P is on the planet's orbit, but is otherwise arbitrary. Since we know the planet's sidereal period, the angle ESE' is also known; we must observe the angles PES and $PE'S$. We can immediately solve the triangle ESE' using the law of cosines and trigonometry (see the Mathematical Appendix) to obtain EE' and the angles SEE' and $SE'E$. Hence, by subtraction, the angles PEE' and $PE'E$ are known, and we may solve the triangle EPE'. Enough information is now available to solve for r, using either triangle SEP or $SE'P$. This was the method by which Kepler first traced out the elliptical orbit of Mars.

3–2   PLANETARY ORBITS

(a)   Kepler's Three Empirical Laws

Using the fine data accumulated by Brahe, Kepler labored for more than twenty years to unravel the mysteries of the orbits of the planets. He tested many models made up of epicycles, equants, and even ovals, but discarded them all—none could reproduce the motions of the planets to the accuracy with which Brahe had observed them. He showed that the orbital planes of the planets pass through the center of the Sun, and after a decade of work discovered the orbit shape to be an *ellipse* [see section 3–2(b) below]. This finding was announced in 1609 as Kepler's First Law—the Law of Ellipses: *The orbit of each planet is an ellipse with the Sun at one of its foci.* (See Figure 3–8A.)

Kepler also investigated the speeds of the planets and found that the closer in its orbit a planet was to the Sun the faster it moved. Drawing a straight line connecting the Sun and the planet (the radius vector), he discovered that he could express this phenomenon quantitatively in Kepler's Second Law—the Law of Areas: *The radius vector to a planet sweeps out equal areas in equal intervals of time.* (See Figure 3–8B.)

Ever seeking a greater harmony in the motions of the planets, Kepler toiled for another decade, and in 1619 put forth Kepler's Third Law—the Harmonic Law: *The squares of the sidereal periods of the planets are proportional to the cubes of the semi-major axes (mean radii) of their orbits.* (See Figure 3–8C.)

The third law may be written as an equation:

$$P^2 = ka^3$$

where P is a planet's sidereal period and a is its average distance from the Sun; the constant k has the same value for every planet. By 1621 Kepler had shown that the four Galilean moons of Jupiter also obey the third law—such were the persistence and genius of this man.

(b)   Geometrical Properties of Elliptical Orbits

An *ellipse* is defined mathematically as the locus of all points, such that the sum of the distances from two foci to any point is a constant; hence, from Figure 3–9 we have

$$r + r' = 2a = \text{constant} \tag{3–1}$$

The line joining the two foci, F and F', intersects the ellipse at the two vertices, A and A'. When r and r' lie along this line we find that a is half the distance between the vertices; we call a the *semi-major axis* of the ellipse. The shape of the ellipse is determined by its *eccentricity*, e, such that the distance from each focus to the center of the ellipse is ae (as shown); when $e = 0$ we have a circle. One-half of the perpendicular bisector of the major axis is termed the *semi-minor axis*, b. Using the dashed lines ($r = r' = a$) in the figure, and the Pythagorean theorem, we find:

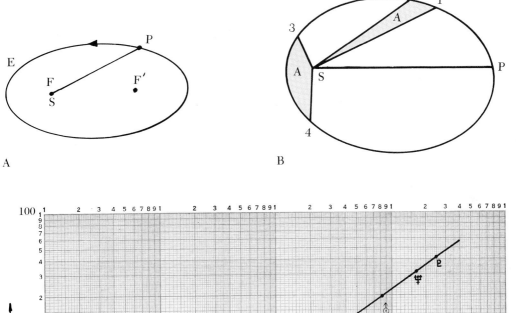

Figure 3–8. *Kepler's Laws Illustrated.* **A**, The planet (P) is in elliptical orbit (E) about the Sun (S), which is at focus F of the ellipse. **B**, The planet (P) spends the same time moving from 1 to 2 as it does moving from 3 to 4. The radius vector (e.g., SP) sweeps out the same area (A) during each of these arbitrary time intervals. **C**, We test Kepler's third law by plotting the semi-major axes (a) of the planets versus their sidereal periods (P) on a log–log graph (see Table A2–2). Such a graph yields a straight line for the equation $P^2 = ka^3$. We see that the planets satisfy this relation extremely well.

$$b^2 = a^2 - a^2e^2 = a^2(1 - e^2) \qquad (3\text{--}2)$$

Following Kepler's first law we place the Sun at focus F. Then vertex A is termed the *perihelion* of the orbit (point nearest the Sun), while vertex A' is called

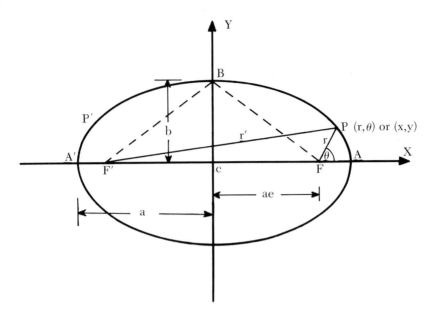

Figure 3–9.   *The Ellipse.* See the text for an explanation of the symbols, coordinates, and properties of the ellipse.

the *aphelion* (point farthest from the Sun). The perihelion distance *AF* is clearly $a - ae = a(1 - e)$, while the aphelion distance *A′F* is $a(1 + e)$. The *mean (average) distance* from the Sun to a planet in elliptical orbit is just the semi-major axis *a*. We prove this by noting that for each point *P* on the ellipse at a distance *r* from focus *F*, there is a symmetrical point *P′* a distance *r′* from *F*; the average of these distances is $(r + r')/2 = a$. Considering any arbitrary but symmetrical pair of points, the same result and our conclusion follow.

It is extremely useful to know the distance from one focus to a point on the ellipse (e.g., the Sun-planet or planet-satellite distance), as a function of the position of that point. Let us center a polar coordinate system (r, θ) at *F*, and let the line *FA* correspond to $\theta = 0$. Now *r* measures the distance *FP*, and θ—the *true anomaly*—measures the counterclockwise angle *AFP*. Using

$$\cos (\pi - \theta) = -\cos \theta$$

and the law of cosines, we have:

$$r'^2 = r^2 + (2ae)^2 + 2r(2ae) \cos \theta$$

but from equation (3–1) $r' = 2a - r$, so that we find:

$$r = \frac{a(1 - e^2)}{1 + e \cos \theta} \tag{3–3}$$

Equation (3–3) is the equation for an ellipse in polar coordinates, as long as we have $0 \le e < 1$.

To derive the *area* of an ellipse, let us find the analog of equation (3–3) in Cartesian coordinates (x, y) positioned at the center of the ellipse. Figure 3–9 and the Pythagorean theorem give us:

$$\left.\begin{array}{l} r'^2 = (x + ae)^2 + y^2 \\[2mm] r^2 = (x - ae)^2 + y^2 \end{array}\right\}$$

Subtracting these two equations and using equation (3–1), we easily find $r' = a + ex$. Substituting back into the first of the above two equations, and employing equation (3–2), we obtain the result:

$$\boxed{(x/a)^2 + (y/b)^2 = 1} \qquad\qquad \textbf{(3–4)}$$

which is the equation for an ellipse in Cartesian coordinates. The area of the ellipse is given by the double integral

$$A = 4 \int_0^b dy \int_0^x dx$$

where, from equation (3–4)

$$x = a[1 - (y/b)^2]^{1/2}$$

The integration is straightforward if we use the substitution $y = b \sin z$ (so that $dy = b \cos z\, dz$), and remember the relation $\sin^2 z + \cos^2 z = 1$; the final answer is:

$$\boxed{A = \pi ab} \qquad\qquad \textbf{(3–5)}$$

The ellipse is one of the curves called *conic sections*. This family of curves, all of which result from slicing a cone, includes the circle, ellipse, parabola, and hyperbola. These are shown in Figure 3–10. From equation (3–3) we see that the ellipse degenerates to a *circle* of radius $r = a$ when $e = 0$. If we increase e, the foci move apart and the ellipse grows longer and narrower. When $e = 1$, one of the foci is at infinity, and we have the *parabola* specified by

$$\boxed{r = 2p/(1 + \cos \theta)} \qquad\qquad \textbf{(3–6)}$$

where p is the distance of closest approach (at $\theta = 0$) to the remaining focus. When the eccentricity is greater than one $(e > 1)$ the open *hyperbola* results:

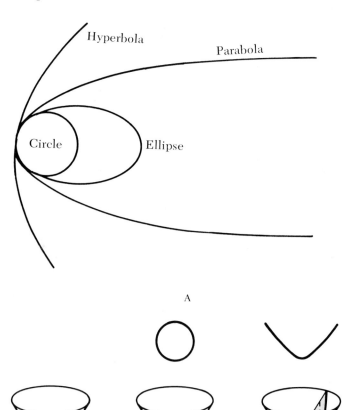

A

Figure 3–10.   *The Conic Sections.*
A, The family of conic section curves includes the circle ($e = 0$), the ellipse ($0 < e < 1$), the parabola ($e = 1$), and the hyperbola ($e > 1$). **B,** When a plane cuts a cone perpendicular to its axis, one obtains a circle; when parallel to a side of the cone, a parabola; and intermediate sections result in an ellipse. A hyperbola is obtained when the plane makes an angle with the side of the cone greater than the opening angle of the cone.

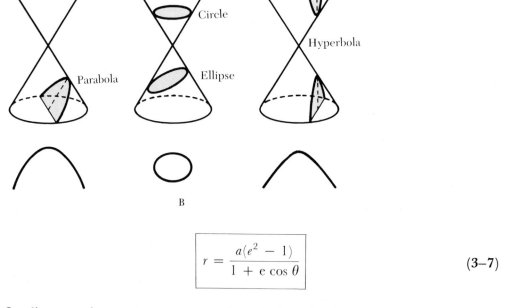

B

$$r = \frac{a(e^2 - 1)}{1 + e \cos \theta}$$

(3–7)

Its distance of nearest approach to the sole focus is $a(e - 1)$. When one body is under the gravitational influence of another, its relative orbit must be one of the conic sections. All planets, satellites, and asteroids describe elliptical orbits; many comets have eccentricities so close to one that they follow essentially parabolic orbits. A few comets have nonperiodic hyperbolic orbits; after one perihelion passage, such a comet leaves the Solar System forever. Artificial space probes have been launched into hyperbolic orbits with respect to the Earth, but they are nearly always captured into elliptical orbits about the Sun. Pioneer 10 is the first space craft with an orbit which, when perturbed by Jupiter, will lead to escape from the Solar System.

3–3 NEWTON'S MECHANICS

Following Kepler's empirical deductions about planetary orbits, Sir Isaac Newton created his awesome scheme of dynamics and gravitation, which he published in the *Principia* in 1687 A.D. Newton's brilliant insight and elegant formulation laid the foundations for the *Newtonian physics* which we know today. In this section we present the first half of his monolithic structure—the theory of mechanics.

Newton assumed that the arena within which motions take place is three-dimensional, Euclidean space—the *absolute space*. These motions are parameterized (i.e., labelled) by *absolute time*, which passes steadily and which is unaffected by any phenomenon in the Universe. The basic entity in the scheme is the *point particle*, which has mass but no extent. In Figure 3–11A we show the trajectory of such a particle. The *position* of the particle of time t, relative to some origin, is indicated by the vector $\mathbf{x}(t)$; the length of this vector is measured in meters (or kilometers, or miles, etc.). At a slightly later time $t + \Delta t$, the particle has moved to $\mathbf{x} + \Delta\mathbf{x}$ at approximately the *velocity*

$$\mathbf{v} \approx \frac{(\mathbf{x} + \Delta\mathbf{x}) - \mathbf{x}}{(t + \Delta t) - t} = \frac{\Delta\mathbf{x}}{\Delta t}$$

As we let $\Delta t \to 0$, the velocity vector becomes parallel to the trajectory at point $\mathbf{x}$, where it is defined by the derivative:

$$\boxed{\mathbf{v} \equiv (d\mathbf{x}/dt)} \tag{3–8}$$

The magnitude of the velocity vector is called the *speed*, and its units are (distance/time, e.g., m sec^{-1} or km sec^{-1}). Noting that the velocity of the particle at t is $\mathbf{v}$, while at $t + \Delta t$ it is $\mathbf{v} + \Delta\mathbf{v}$, we may express the change in velocity by the *acceleration* vector (see Figure 3–11B)

$$\mathbf{a} \approx \frac{(\mathbf{v} + \Delta\mathbf{v}) - \mathbf{v}}{(t + \Delta t) - t} = \frac{\Delta\mathbf{v}}{\Delta t} \Rightarrow \boxed{\mathbf{a} = \frac{d\mathbf{v}}{dt}} \quad \text{(as } \Delta t \to 0) \tag{3–9}$$

The units of acceleration are (speed/time) or (distance/time^2, e.g., m sec^{-2} or km sec^{-2}). Newton also considered the *linear momentum* vector (units of mass $\times$ speed, or kg m sec^{-1}) of the particle, defined by the product

$$\boxed{\mathbf{p} = m\mathbf{v}} \tag{3–10}$$

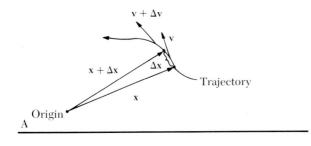

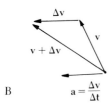

Figure 3–11. *Position, Velocity, and Acceleration.* **A**, At time t the position of a particle in its trajectory is $\mathbf{x}$; at $t + \Delta t$ it is $\mathbf{x} + \Delta\mathbf{x}$. The average velocity is $(\Delta\mathbf{x}/\Delta t)$, and as $\Delta t \to 0$ the instantaneous velocity at $\mathbf{x}$ becomes $\mathbf{v} = (d\mathbf{x}/dt)$. The instantaneous velocity at $\mathbf{x} + \Delta\mathbf{x}$ is $\mathbf{v} + \Delta\mathbf{v}$. **B**, The change in velocity between t and $t + \Delta t$ is $\Delta\mathbf{v}$, from which we define the instantaneous acceleration at $\mathbf{x}$ (as $\Delta t \to 0$) as $\mathbf{a} = (d\mathbf{v}/dt)$.

where m is the particle's mass and $\mathbf{v}$ is its instantaneous velocity. With these kinematical fundamentals firmly in mind, we may proceed directly to Newton's laws of motion.

(a)   The Law of Inertia

In his *Physics*, Aristotle attempted to show that the natural state of a body is one of rest. Our mundane experience seems to verify his observation that all moving objects eventually slow to a halt. Indeed, to explain the flight of an arrow Aristotle said that the air rushing in upon the tail of the arrow "pushed" it along.

Galileo came to a quite different conclusion. He released balls on smooth inclined planes, and observed that they rolled up adjacent inclined planes to approximately the same height from which he had released them. As he made the planes smoother, and inclined the second plane less to the horizontal, he found that the balls rolled farther. Galileo attributed any slowing-down of the balls to *friction*, and conjectured that a perfect ball on a horizontal plane would roll forever at a constant speed.

René Descartes later formulated this principle in the form which Newton adopted as his First Law of Motion (the Law of Inertia): *The velocity of a body remains constant (in both magnitude and direction) unless a force acts upon the body.* For a freely-moving body, the first law may be written as $\mathbf{v}$ = constant; when the constant is zero, we find that a body initially at rest will remain at rest, unless acted upon by a force.

The modern form of Newton's first law is known as the Conservation of Linear Momentum. For a body of constant mass m, we may write $\mathbf{p} = m\mathbf{v}$ = constant, which is equivalent to

$$\boxed{d\mathbf{p}/dt = 0} \quad \text{(force-free)} \tag{3–11}$$

Today we know that equation (3–11) is true even when the body's mass changes.

(b) The Definition of Force

Newton's second law is already implied by the proviso "unless a force acts upon the body" in the first law, for a force causes a change in velocity. Such a change in velocity (either in speed or direction) is conveniently indicated by the acceleration vector [equation (3–9)]. An important special case of acceleration is circular motion, where the speed remains constant while the direction of motion changes.

The concept of *force* was defined in Newton's Second Law of Motion (the Law of Force): *The acceleration imparted to a body is proportional to and in the direction of the force applied and inversely proportional to the mass of the body.* Thus we may write $\mathbf{a} = \mathbf{F}/m$ or more commonly

$$\mathbf{F} = m\mathbf{a} \tag{3–12}$$

Note that force is a vector, with the units (mass × acceleration, e.g., kg m sec^{-2}). If several forces act upon a single body, the resultant acceleration is determined via equation (3–12) using the $\mathbf{F}$ which is the *vector sum* of the individual forces—this is the *principle of superposition*. Figure 3–12 illustrates this principle when the two forces, $\mathbf{F}_1$ and $\mathbf{F}_2$, add to give the resultant $\mathbf{F}$; we may clearly reverse this procedure to decompose the force $\mathbf{F}$ into two or more component forces, indicated here by $\mathbf{F}_x$ and $\mathbf{F}_y$.

In equation (3–12), the body's mass, m, must remain constant. This restriction vanishes in the modern statement of the second law, which is formulated using the body's linear momentum:

$$\mathbf{F} = d\mathbf{p}/dt \tag{3–13}$$

We may recover Newton's form of the second law by using equations (3–9), (3–10), and (3–13) when m is constant. Note that the first law of motion [equation (3–11)] is now a consequence of the second.

Let us inquire further into the concept of *mass*. In dynamics, mass represents the *inertia* of a body, i.e., that body's resistance to any change in its state of motion.

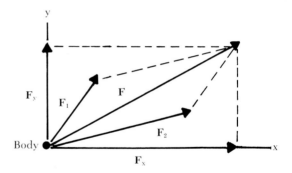

Figure 3–12. *Superposing and Decomposing Forces.* The two forces, $\mathbf{F}_1$ and $\mathbf{F}_2$, act on the body. The resulting motion is determined by $\mathbf{F}$, the vector sum of $\mathbf{F}_1$ and $\mathbf{F}_2$. The orthogonal components $\mathbf{F}_x$ and $\mathbf{F}_y$ will yield the same motion, since we may decompose $\mathbf{F}$ into these two vectors.

If we apply the same force to two bodies, the more massive body will change velocity at a slower rate than the less massive body. In everyday terms, we may think of mass as the amount of material comprising a body; hence, two identical lead billiard balls constitute twice the mass of one such ball. Mass is a scalar quantity characterizing a body, and it depends in no way upon the body's location or state of motion [see section 3–4(b)]. In the mks system of units, a *kilogram* is defined as the mass of one liter (10^3 cubic centimeters) of pure water at a temperature of 4° Celsius.

(c) Action and Reaction

To complete his dynamical theory and make it applicable to a collection of point particles or even to a continuous body, Newton stated his Third Law of Motion (the Law of Action-Reaction): *For every force acting on a body (in a closed system) there is an equal and opposite force exerted by that body.* A simple example will at least illustrate why the third law is necessary. The weight of a book lying on a table must be exactly counterbalanced by the force which the table exerts on the book; otherwise, according to the second law, the book will accelerate off or through the table. The third law is needed to maintain the static situation we see.

The modern version of the third law is the conservation of *total* linear momentum. Though we may treat any number of bodies, let us consider only two bodies here. The total linear momentum of the system is given by $\mathbf{P} = \mathbf{p}_1 + \mathbf{p}_2 = $ constant, when no external force acts upon the system (the first two laws). Considering two instants of time (the latter instant indicated by primes), we have

$$\mathbf{p}_1 + \mathbf{p}_2 = \mathbf{p}_1' + \mathbf{p}_2'$$

If the time interval is Δt, and we call $\Delta\mathbf{p}_1 = \mathbf{p}_1' - \mathbf{p}_1$ and $\Delta\mathbf{p}_2 = \mathbf{p}_2' - \mathbf{p}_2$, the above equation may be rearranged and divided by Δt to give

$$(\Delta\mathbf{p}_1/\Delta t) = -(\Delta\mathbf{p}_2/\Delta t)$$

For arbitrarily small Δt, the deltas become differentials and equation (3–13) yields the third law:

$$\mathbf{F}_1 = -\mathbf{F}_2$$

(d) Summarizing Newtonian Mechanics

As a summary and aid to future studies, let us collect together the modern versions of Newton's three laws of mechanics. Recall that the Newtonian context is absolute space and time, wherein a point particle of mass m describes a trajectory $\mathbf{x}(t)$ with instantaneous velocity $\mathbf{v}(t)$, linear momentum $\mathbf{p} = m\mathbf{v}$, and acceleration $\mathbf{a}(t)$.

First Law (Inertia):

$$\boxed{\mathbf{v} \text{ and } \mathbf{p} = \text{constant}}$$

The velocity and linear momentum of a body remain constant (in both magnitude and direction) unless a force acts upon the body.

Second Law (Force):

$$\boxed{\mathbf{F} = d\mathbf{p}/dt}$$

The time rate of change of the linear momentum of a body (or system of bodies) equals the force acting upon the body (or system).

Third Law (Action-Reaction):

$$\boxed{\begin{array}{c} \mathbf{F}_{\text{exerted}} = -\mathbf{F}_{\text{acting}} \\[6pt] \mathbf{P} = \text{constant} \end{array}}$$

In a closed system, the force exerted by a body is equal and opposite to the force acting upon the body, or, the total linear momentum of a closed system of bodies is constant in time.

3–4 NEWTON'S LAW OF UNIVERSAL GRAVITATION

(a) Centripetal Force and Gravitation

Even before Newton's time it was clear that some force must be acting to keep the planets in orbit about the Sun; Kepler had attributed the elliptical orbits to the force of magnetic attraction. Following a train of thought similar to the simplified derivation which follows, Newton discovered his law of universal gravitation, tested it on the motion of the Moon, and then explained the motions of the planets in detail.

Figure 3–13A shows a body moving in a circular orbit of radius r about a center of force. From symmetry, the speed (v) of the body must be constant, but the direction of the velocity vector is constantly changing. Such a changing velocity represents an acceleration—the *centripetal acceleration* which maintains the circular orbit; from the geometry of the figure we may deduce this acceleration. The time is t at point A when the body's velocity is $\mathbf{v}$. At an infinitesimal time interval (Δt) later, the body has traversed the angle $\Delta\theta$ to B, where the velocity is $\mathbf{v}'$. In Figure 3–13B we show the change in velocity, $\Delta\mathbf{v} = \mathbf{v}' - \mathbf{v}$, by joining the tails of the two velocity vectors; the angle between $\mathbf{v}$ and $\mathbf{v}'$ is clearly $\Delta\theta$. Recalling that the magnitude of both $\mathbf{v}$ and $\mathbf{v}'$ is the speed v, we use trigonometry to deduce (Figure 3–13A)

$$\sin(\Delta\theta/2) = (s/2r) = (v\,\Delta t/2r)$$

and (Figure 3–13B)

$$\sin(\Delta\theta/2) = (\Delta v/2v)$$

where the arc-length s approximates the chord joining points A and B. Therefore, the centripetal acceleration has the magnitude

$$a = (\Delta v/\Delta t) = v^2/r$$

and it points directly toward the center of the circle as Δt vanishes. If m is the mass of the orbiting body, Newton's second law immediately gives us the magnitude of the *centripetal force* as

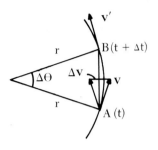

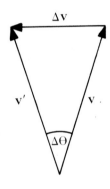

A

Figure 3–13. *Centripetal Acceleration in Circular Orbit.* **A,** A body moves in a circular orbit of radius *r* at speed *v*. In a short time Δt the body moves from *A* to *B* through an angle $\Delta\theta$ and a distance *s*. The velocity at *A* is **v** and at *B* is **v'**. **B,** The change in velocity $\Delta\mathbf{v} = \mathbf{v'} - \mathbf{v}$ during Δt is illustrated. From geometry, the angle between **v** and **v'** is $\Delta\theta$.

B

$$F_{\text{cent}} = ma = mv^2/r \qquad\qquad (3\text{–}14)$$

If *P* is the circular period of the body, then its speed is

$$v = 2\pi r/P$$

but Kepler's third law relates the period and the orbital radius via

$$P^2 = kr^3$$

where *k* is the proportionality constant. Substituting these two results into equation (3–14), we find

$$F = (4\pi^2 m/kr^2)$$

that is, the force maintaining the orbit is *inversely proportional to the square of the radius*. According to Newton's third law, the body (of mass *M*) at the center of the orbit feels an equal, but opposite, force. Since the centripetal force acting on the central body must be proportional to *M*, we see that the *mutual* gravitational force is proportional to the product of the two masses; redefining the constant of proportionality, we therefore have

$$|\mathbf{F}_{\text{grav}}| = GMm/r^2 \qquad (3\text{--}15)$$

Equation (3–15) is Newton's Law of Universal Gravitation; the direction of the gravitational force is along the line joining the two bodies (from the third law of motion). The gravitation constant G has the measured value 6.67×10^{-11} $\text{m}^3 \text{ kg}^{-1} \text{ sec}^{-2}$ ($6.67 \times 10^{-8} \text{ cm}^3 \text{ gm}^{-1} \text{ sec}^{-2}$).

Equation (3–15) expresses the attractive gravitational force between two *point* masses. To find the gravitational attraction due to an extended body, we must sum the vectorial contributions due to each small piece of the body. In general this is rather difficult, but, for a spherically symmetric body acting on a point mass symmetry arguments alone tell us that the gravitational force must act along the line joining the centers of the bodies. By integrating the effects of all parts of the spherical body, it is straightforward to show that *it behaves gravitationally as though its entire mass were concentrated at its center.* This important result will be used many times in the following pages.

(b) Weight and Gravitational Acceleration

Near the Earth's surface, bodies feel a constant downward gravitational acceleration of magnitude [from equation (3–15)]

$$g = GM_{\oplus}/R_{\oplus}^2 \approx 9.807 \text{ m sec}^{-2} \ (32.17 \text{ ft sec}^{-2})$$

where $M_{\oplus}$ is the mass of the Earth and $R_{\oplus}$ is its radius. The Earth's oblate surface and rapid rotation (see Chapter 4) cause variations in the measured value of g from 9.781 m sec^{-2} at the equator to 9.832 m sec^{-2} at the poles.

The *weight* of a body is the force necessary to hold the body motionless in a gravitational field. At the Earth's surface, we comply with this definition by "weighing" the body on a spring-scale, and if the mass of the body is m, we find

$$\text{weight} = mg$$

In contrast to its mass, which is invariant, the weight of a body depends upon the *location* of the body. Consider a man: on the Moon's surface he will weigh approximately one-sixth his normal Earthweight; in a satellite orbit his weight will be zero, since he is falling freely in the gravitational field. Weight and force have the same units, and in the mks system the conventional unit is the *newton* where

$$1 \text{ newton} = 1 \text{ kg m sec}^{-2}$$

(c) Determinations of G and $M_{\oplus}$

Let us describe how Henry Cavendish measured the gravitational constant (G) in 1798 and how Phillip von Jolly established the mass of the Earth $(M_{\oplus})$ in

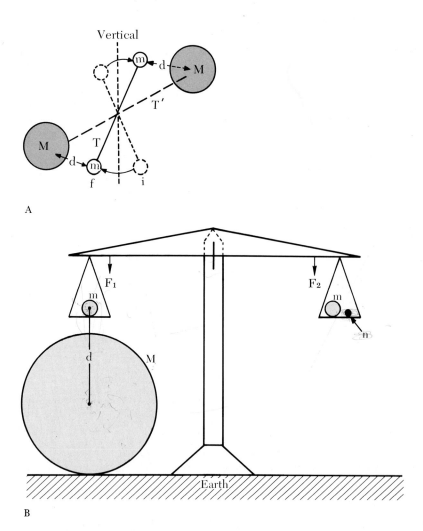

Figure 3–14.   *Measuring G and $M_\oplus$.* **A**, Cavendish used two torsion bars, T and T', free to rotate about the vertical axis, to determine G. After the gravitational force between adjacent M-m balls swings the bars to their equilibrium position, the M-m separations are d. The initial (i) and final (f) states are shown. **B**, The horizontal torsion balance with which von Jolly determined $M_\oplus$ is in equilibrium (horizontal) when masses M and n are absent. When M is placed below the left-hand mass m, n must be added to the right-hand mass m to restore equilibrium ($F_1 = F_2$).

1881. These two methods are representative of the many procedures employed to determine these important constants. Both measurements were performed with a torsion balance.

To find G, Cavendish used the experimental configuration shown schematically in Figure 3–14A. Two small balls of equal mass m hang from a torsion beam, while two larger balls of equal mass M are attached to an independently suspended but coaxially aligned beam. Each adjacent M-m pair is initially placed a distance D apart, but the gravitational force between the balls twists the torsion bars to a static equilibrium distance d between each M-m pair. From symmetry, the gravitational force causing the deflection is $F_{\text{tot}} = 2GMm/d^2$; by directly measuring F_{tot}, M, m, and d, Cavendish found $G = 6.7 \times 10^{-8}$ cm^3 gm^{-1} sec^{-2}.

Phillip von Jolly's apparatus (Figure 3–14B) consisted of a single torsion beam bearing two small masses m; the rotation axis of the beam was aligned horizontally. When he placed a large mass M below one of these small masses, the beam rotated; to restore the original balance he suspended a small mass n from the other mass m. If the equilibrium M-m distance is d, the forces acting on each side of the torsion beam are

$$F_1 = \frac{GMm}{d^2} + \frac{GM_\oplus m}{R_\oplus^2}; \qquad F_2 = \frac{GM_\oplus n}{R_\oplus^2} + \frac{GM_\oplus m}{R_\oplus^2}$$

The beam is horizontal again when $F_1 = F_2$, so we find

$$M_\oplus = \left(\frac{Mm}{n}\right)\left(\frac{R_\oplus}{d}\right)^2 = 5.976 \times 10^{24} \text{ kg}$$

3–5 PHYSICAL INTERPRETATION OF KEPLER'S LAWS

(a) The Law of Areas and Angular Momentum

Newton combined his theory of mechanics and law of gravitation to *derive* all three of Kepler's empirical laws. We too could derive the elliptical orbits of equation (3–3), using equations (3–13) and (3–15), but since this requires a knowledge of vector differential equations, we refer the reader to the fine treatments in the mechanics texts given in this chapter's reading list. We will accept Kepler's first law, and follow Newton's footsteps in deducing the second and third laws.

Let us illustrate Kepler's law of areas for an elliptical orbit (see Figure 3–15). A body orbits the focus F at the position $\mathbf{r}$ with velocity $\mathbf{v}$. During an infinitesimal time interval Δt, the body moves from P to Q, and the radius vector sweeps through the angle $\Delta \theta$. This small angle is given by $\Delta \theta \approx v_t \, \Delta t/r$, where v_t is the component of $\mathbf{v}$ perpendicular to $\mathbf{r}$. During this time the radius vector has swept out the triangle FPQ, the area of which is $\Delta A \approx r v_t \, \Delta t/2$. Therefore, we may write (for $\Delta t \to 0$)

Figure 3–15. *The Law of Areas.* A body at $\mathbf{r}$ moves at velocity $\mathbf{v}$ in elliptical orbit about focus F. In time Δt it passes from P to Q, and the radius vector sweeps out angle $\Delta \theta$ and area ΔA. The component of $\mathbf{v}$ perpendicular to $\mathbf{r}$ is v_t.

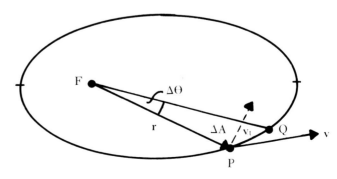

$$(dA/dt) = rv_t/2 = r^2(d\theta/dt)/2 = h/2 \tag{3-16}$$

here the constant h (the angular momentum per unit mass) appears because Kepler's second law states that the rate of change of area with time is a constant.

It is clear that $A/P = h/2$, where $A = \pi ab$ is the total area of the ellipse and P is the orbital period; we may prove this by simply integrating equation (3–16). By combining this result with equation (3–16), and noting that v_t is the total speed at perihelion and aphelion, we can deduce the perihelion and aphelion speeds of a planet orbiting the Sun. For example, at perihelion we have

$$v = h/r = 2A/Pr = 2\pi ab/Pa(1 - e)$$

where equation (3–3) with $\theta = 0$ is used in the last equality. Carrying through the derivation at aphelion, and using equation (3–2), gives:

$$v = \begin{cases} \dfrac{2\pi a}{P}\left[\dfrac{1 + e}{1 - e}\right]^{1/2} & \text{(perihelion)} \\[3mm] \dfrac{2\pi a}{P}\left[\dfrac{1 - e}{1 + e}\right]^{1/2} & \text{(aphelion)} \end{cases} \tag{3-17}$$

For the Earth, a is 1 AU (1.496×10^8 km), P is one year (3.156×10^7 sec), and the orbital eccentricity is $e = 0.0167$; hence, the orbital speed varies from 30.3 km sec^{-1} at perihelion to 29.3 km sec^{-1} at aphelion.

A modern Newtonian derivation of Kepler's second law requires the concept of the orbiting body's *angular momentum*:

$$\boxed{\mathbf{L} = \mathbf{r} \times \mathbf{p} = m(\mathbf{r} \times \mathbf{v})} \tag{3-18}$$

Here m is the body's mass, $\mathbf{r}$ its position vector, and $\mathbf{p}$ its linear momentum [see equation (3–10)]. The *vector cross product* ($\times$) in equation (3–18) is an operation which yields the product of the *perpendicular* components of the two vectors under consideration (see the Mathematical Appendix); hence, if $\mathbf{r}$ and $\mathbf{p}$ are parallel, then $\mathbf{r} \times \mathbf{p} = 0$. Angular momentum is a vector quantity $\mathbf{L}$, with the units kg m^2 sec^{-1}. Differentiating equation (3–18), we have

$$d\mathbf{L}/dt = \mathbf{v} \times \mathbf{p} + \mathbf{r} \times (d\mathbf{p}/dt) = \mathbf{r} \times \mathbf{F} \tag{3-19}$$

since $\mathbf{v}$ is parallel to $\mathbf{p}$ and $(d\mathbf{p}/dt)$ defines force. We call $d\mathbf{L}/dt$ the *torque* (with units kg m^2 sec^{-2}), and see that when $\mathbf{F}$ is collinear with $\mathbf{r}$—a *central force*, such as gravitation—the torque vanishes. Hence, $\mathbf{L}$ is constant in time so that: *angular momentum is conserved for all central forces*. Combining equation (3–18) and Figure 3–15 we immediately find

$$L/m = rv_t = h = \text{constant}$$

just as we said at equation (3–16).

(b)   Newton's Form of Kepler's Third Law

The external forces which act upon the Solar System are essentially negligible; hence, the total linear momentum of the Solar System must remain constant. If the Sun did not move in response to the orbiting planets, this would clearly not be the case. Therefore, the Sun must move about the center of mass (*barycenter*) of the Solar System. Let us apply this idea to an isolated system of two bodies moving in circular orbits because of their mutual gravitational attraction; our final result, Newton's form of Kepler's third law, will also be applicable to elliptical orbits.

In Figure 3–16 we show two masses, m_1 and m_2, orbiting their stationary center of mass at distances r_1 and r_2, respectively. Since the gravitational force acts only along the line joining the centers of the bodies, both bodies must complete one orbit in the same period P (though they move at different speeds, v_1 and v_2). The centripetal forces maintaining the orbits are therefore:

$$F_1 = m_1 v_1^2 / r_1 = 4\pi^2 m_1 r_1 / P^2 \qquad (3\text{–}20\text{a})$$

and

$$F_2 = m_2 v_2^2 / r_2 = 4\pi^2 m_2 r_2 / P^2 \qquad (3\text{–}20\text{b})$$

Newton's third law requires $F_1 = F_2$, so we immediately obtain

$$r_1 / r_2 = m_2 / m_1 \qquad (3\text{–}21)$$

The larger (more massive) body orbits closer to the center of mass than the smaller.

The total separation of the two bodies, $a = r_1 + r_2$, is also the radius of their relative orbits. Now equation (3–21) may be expressed in the form

$$r_1 = m_2 a / (m_1 + m_2) \qquad (3\text{–}22)$$

But the mutual gravitational force, $F_{grav} = F_1 = F_2$, is given by

$$F_{grav} = G m_1 m_2 / a^2 \qquad (3\text{–}23)$$

hence, combining equations (3–20a), (3–22), and (3–23) gives us the result we seek—*Newton's form of Kepler's third law*:

Figure 3–16. *Center of Mass and Kepler's Third Law.* The masses m_1 and m_2 orbit their center of mass (CM) at speeds v_1 and v_2 and distances r_1 and r_2. Their centripetal forces, F_1 and F_2, are both equal to the mutual gravitational force between them.

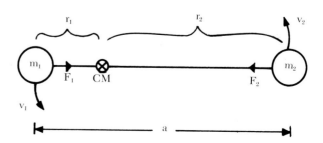

$$P^2 = \left[\frac{4\pi^2}{G(m_1 + m_2)}\right]a^3 \tag{3–24}$$

If body 1 is the Sun and body 2 any planet, then we know that $m_1 \gg m_2$; therefore, $k = (4\pi^2/GM_\odot)$ is the proportionality "constant" (to a good approximation) in Kepler's third law.

(c) Orbital Velocity

To better understand elliptical orbits, let us consider the *orbital velocity* **v**. We may decompose this velocity into two orthogonal components (see Figure 3–17): v_r, the radial speed; and v_θ, the "angular" speed. From equation (3–16) and the discussion following it, we have

$$(d\theta/dt) = (2\pi/P)(a/r)^2(1 - e^2)^{1/2} \tag{3–25}$$

Using the polar equation (3–3) of an ellipse and equation (3–25), we compute the time derivatives indicated below to straightforwardly find:

$$v_r \equiv (dr/dt) = (2\pi a/P)(e\sin\theta)(1 - e^2)^{-1/2} \tag{3–26a}$$

and

$$v_\theta \equiv r(d\theta/dt) = (2\pi a/P)(1 + e\cos\theta)(1 - e^2)^{-1/2} \tag{3–26b}$$

Note that equation (3–26b) reduces to equation (3–17) at perihelion and aphelion. The orbital speed now follows from equations (3–26) as

$$v^2 = v_r{}^2 + v_\theta{}^2 = (2\pi a/P)^2(1 + 2e\cos\theta + e^2)/(1 - e^2) \tag{3–27}$$

Rearranging the polar equation for an ellipse, we may write

$$e\cos\theta = [a(1 - e^2) - r]/r$$

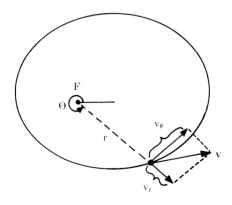

Figure 3–17. *The Components of Orbital Velocity.* At an arbitrary point (r, θ) in an elliptical orbit, the orbital velocity **v** may be decomposed into perpendicular components—the radial speed (v_r) parallel to the radius vector, and the "angular" speed (v_θ).

and, substituting this result into equation (3–27), we finally obtain [with the help of equation (3–24)]:

$$v^2 = G(m_1 + m_2)\left(\frac{2}{r} - \frac{1}{a}\right) \tag{3–28}$$

Therefore, for given masses, *the orbital speed depends only upon the separation and the orbit's semi-major axis*—this useful result will reappear frequently in the following pages.

(d) The Conservation of Total Energy

The concept of energy is extremely useful in celestial mechanics, though its meaning is hard to grasp. We will briefly discuss energy here, but the interested reader should consult the reading list at the end of this chapter for a proper introduction to this subject. We shall begin by speaking obliquely and then give several examples.

Energy is a quantity assigned to one body which indicates that body's ability to change the state of another body. *Heat* is a form of energy, for a hot body will warm a cold body if the two are brought into contact. Electrical energy causes the filament of a light to glow (become hot), so that one type of energy may be converted into another. The energy of a body depends upon the observer of that body. *Kinetic energy* (KE) is a body's energy of *motion*, but if we decide to move along with the body, it has no motion; hence, it has no kinetic energy. *Potential energy* (PE) is due to the *position* of the body; if the body is free to move, this energy may be converted into kinetic energy. In celestial mechanics we sum the kinetic and potential energies to obtain the *total energy*: $TE = KE + PE$.

Assume that a force **F** acts upon a body of mass m which is moving in the trajectory $\mathbf{x}(t)$ about the center of force (see Figure 3–18). In the infinitesimal time dt, the body moves through the vector distance $d\mathbf{x}$. As the body moves from position A to position B, we define the work (units $=$ kg m^2 sec^{-2} $=$ *joules*) done *on* the body by the force as

$$W = \int_A^B \mathbf{F} \cdot d\mathbf{x} \tag{3–29}$$

The *vector dot product* (·) operation in equation (3–29) yields the product of the

Figure 3–18. *The Concept of Work.* The force **F** acts upon the mass m which is orbiting in the trajectory **x**. As the body moves from A to B, its instantaneous displacement is $d\mathbf{x}$.

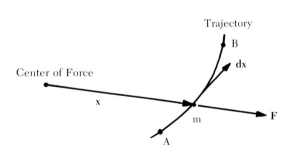

parallel components of **F** and $d\mathbf{x}$ (see the Mathematical Appendix); hence, when **F** and $d\mathbf{x}$ are mutually perpendicular, their dot product vanishes. To evaluate equation (3–29) we note the following:

$$\mathbf{F} \cdot d\mathbf{x} = m(d\mathbf{v}/dt) \cdot \mathbf{v}\, dt = m(\mathbf{v} \cdot d\mathbf{v}) = d(mv^2/2)$$

where Newton's second law and the definitions of velocity and speed have been employed. Hence, equation (3–29) may be integrated directly to give:

$$W = (mv^2/2)_B - (mv^2/2)_A = KE_B - KE_A \tag{3–30}$$

where the kinetic energy is specified by $KE = mv^2/2$. Therefore, *the work done by the force on the body changes the body's kinetic energy.* In a system of two bodies, the kinetic energy is $KE = (m_1 v_1{}^2/2) + (m_2 v_2{}^2/2)$.

If the force is the gravitational force between two bodies, then we have

$$\mathbf{F} \cdot d\mathbf{x} = -(Gm_1 m_2/r^2)\, dr = d(Gm_1 m_2/r)$$

where r is the radial separation of the bodies. Equation (3–29) is again easily integrated:

$$W = (Gm_1 m_2/r)_B - (Gm_1 m_2/r)_A = PE_A - PE_B \tag{3–31}$$

where we have inserted the conventional definition of the mutual gravitational potential energy $PE = -(Gm_1 m_2/r)$. Therefore, the potential energy is the *negative* of the work done by the gravitational force as m_1 moves from $r = \infty$ to $r = r$ with m_2 held fixed. Note that the mutual gravitational potential energy vanishes when the two bodies are infinitely separated.

Since KE and PE have the same units as work, we may combine equations (3–30) and (3–31) to obtain:

$$(KE + PE)_B = (KE + PE)_A$$

or in terms of the total energy of the two-body system:

$$TE_B = TE_A = \text{constant}$$

Therefore, the total energy of our gravitating system is conserved; or,

$$\boxed{TE = (m_1 v_1{}^2/2) + (m_2 v_2{}^2/2) - (Gm_1 m_2/r) = \text{constant}} \tag{3–32}$$

As the bodies move, kinetic and potential energy may be interchanged, but the total energy of the system remains constant.

Let us evaluate the constant TE in equation (3–32) for elliptical orbits. Referring back to section 3–5(b), we recall that the total linear momentum of our isolated system is constant; we choose this constant to be zero, so that $m_1 \mathbf{v}_1 = -m_2 \mathbf{v}_2$ and in terms of the speeds of the bodies:

$$m_1 v_1 = m_2 v_2$$

Since $v = v_1 + v_2$ is the relative speed of either body with respect to the other, we easily find:

$$v_1 = m_2 v/(m_1 + m_2), \qquad v_2 = m_1 v/(m_1 + m_2) \tag{3–33}$$

Substituting this result into equation (3–32) yields

$$TE = m_1 m_2 \left[\frac{v^2}{2(m_1 + m_2)} - \frac{G}{r} \right] \tag{3–34}$$

Using equation (3–24), let us evaluate equation (3–34) at the perihelion of the orbit, where $r = a(1 - e)$ and v is the perihelion speed given by equation (3–17). Our result, $TE = -Gm_1 m_2/2a$, shows that the total energy is negative—the orbit is *bound*; now equation (3–34) takes its final form:

$$v^2 = G(m_1 + m_2)\left(\frac{2}{r} - \frac{1}{a}\right)$$

which is exactly the expression found in equation (3–28). This useful and classic result, which is simply a statement of total energy conservation, is sometimes (archaically) referred to as the *vis viva equation*.

3–6 APPLICATIONS

(a) Using Kepler's Third Law

To fully understand and appreciate the formal manipulations which we have just carried out, the reader should use the results to solve a few physical problems. Let us illustrate this point by considering several simple examples of the application of Newton's form of Kepler's third law.

The correct form of the harmonic law is equation (3–24):

$$P^2 = 4\pi^2 a^3/G(m_1 + m_2)$$

In applying this relation to the Sun ($M_\odot$) and planets (m_p), considerable simplification occurs when we measure the sidereal periods, P, in years and the semi-major axes, a, in AU, for then we have:

$$a^3/P^2 = 1 + (m_p/M_\odot)$$

Note that we have simplified by using Earth units; this clue tells us that we should always use the units most appropriate to the system under consideration. We implement this idea by forming *ratios* of equation (3–24) in the form

$$\left(\frac{m_1 + m_2}{m_1' + m_2'}\right)\left(\frac{P}{P'}\right)^2 = \left(\frac{a}{a'}\right)^3 \qquad (3\text{--}35)$$

where the system m_1 and m_2 (with period P and semi-major axis a) is compared with the *standard system* m_1' and m_2' (with P' and a'). For objects orbiting the Sun or for binary stars (Chapter 11), the standard system is the Sun-Earth system: P is expressed in years, a in AU, and all masses m in solar masses ($M_\odot$). For planetary satellites (either natural moons or artificial satellites), we use the Earth-Moon system: we set $P' = 27.3$ days (or 656 hours), $a' = 3.84 \times 10^5$ km, and $(m_1' + m_2') = M_\oplus$ (or 5.976×10^{24} kg); we then obtain P in days (or hours), a in kilometers, and the masses m in Earth masses (or kilograms).

As our first example, consider a comet with an elliptical period of seven years; we want to find the semi-major axis of its orbit. Since $(M_\odot + m_{\text{comet}}) \approx M_\odot$, equation (3–35) gives $a = (7)^{2/3} = 3.66$ AU. Our final, and more interesting, example is this: find the mass of Uranus (M_U) in terms of the Earth's mass. We observe that Miranda (M_m), a moon of Uranus, orbits the planet in 1.4 days at a mean distance of 128,000 kilometers. Using the Earth-Moon standard system, we have

$$\left(\frac{M_U + M_m}{M_\oplus + M_\text{☽}}\right)\left(\frac{P_m}{P_\text{☽}}\right)^2 \approx \left(\frac{M_U}{M_\oplus}\right)\left(\frac{P_m}{P_\text{☽}}\right)^2 = \left(\frac{a_m}{a_\text{☽}}\right)^3$$

where we have neglected the masses of Miranda and the Moon with respect to the much larger masses of their primary planets, Uranus and the Earth. Substituting the relevant data, we obtain:

$$M_U \approx (27.3/1.4)^2 (128{,}000/384{,}000)^3 M_\oplus \approx 14 M_\oplus$$

(b)   The Launching of Rockets

Let us consider projectiles launched vertically upward from the Earth's surface. Neglecting atmospheric friction, which continuously decreases the projectile's speed, we may use the conservation of total energy to find the height to which the projectile ultimately rises.

Near the Earth's surface, the strength of the downward gravitational force is $F = mg$, where m is the mass of the projectile. Hence, the potential energy at altitude h is $PE = mgh$. If the speed of the body is v at ground level, its kinetic energy there is $KE = mv^2/2$. Conservation of total energy says that

$$TE = (KE + PE)_{\text{ground}} = \text{constant} = (KE + PE)_h$$

so that by evaluating TE at $h = 0$ and at the maximum height (h), we find $mv^2/2 = mgh$ or

$$\boxed{h = v^2/2g} \qquad (3\text{--}36)$$

Since $g = 9.8$ m sec^{-2}, a rock thrown upward with speed $v = 14$ m sec^{-1} will rise to a height of 10 meters before falling back to the ground.

When we consider altitudes greater than about one Earth radius ($h \gtrsim R_\oplus$), our previous approximation breaks down, and we must use the correct formula, equation (3–32). Evaluating this expression at the ground ($R_\oplus$) and at the maximum height ($R_\oplus + h$), we find

$$(mv^2/2) - (GM_\oplus m/R_\oplus) = -GM_\oplus m/(R_\oplus + h)$$

or

$$h = R_\oplus \left[\frac{(v^2 R_\oplus/2GM_\oplus)}{1 - (v^2 R_\oplus/2GM_\oplus)} \right] = (v^2/2g) \left[\frac{R_\oplus}{R_\oplus - (v^2/2g)} \right] \qquad (3\text{–}37)$$

Equation (3–36) follows from equation (3–37) in the limit $(v^2/2g) \ll R_\oplus$, that is, $h \ll R_\oplus$. Note that when $(v^2/2g) = R_\oplus$ (i.e., at the speed $v = (2gR_\oplus)^{1/2} = 11.2$ km sec^{-1}), the projectile escapes to $h = \infty$—this critical speed is called the *escape speed*.

(c)   Orbits of Artificial Satellites and Space Probes

Most major space vehicles are placed into Earth orbit; this is the *parking orbit* from which deep-space probes are accelerated toward the Moon and the planets. Multistage rockets lift the vehicle beyond the Earth's atmosphere, where the final stages accelerate the payload horizontally to the desired orbital speed. The Earth's rotation speed (0.46 km sec^{-1} at the equator) aids launchings toward the east. For a circular orbit at distance r from the center of the Earth ($r = R_\oplus + h$, if h is the altitude of the orbit) the circular speed v_c may be found by balancing the centripetal and gravitational forces:

$$v_c = (GM_\oplus/r)^{1/2} = v_0 \left(\frac{R_\oplus}{R_\oplus + h} \right)^{1/2} \qquad (3\text{–}38)$$

where $v_0 = (GM_\oplus/R_\oplus)^{1/2} = 7.86$ km sec^{-1} is the circular speed at the Earth's surface (neglecting atmospheric friction). Rearranging equation (3–37), we find

$$v = v_0 \left(\frac{2h}{R_\oplus + h} \right)^{1/2}$$

so that $h \to \infty$ at the *escape speed* $v_{\text{escape}} = \sqrt{2}\, v_0$; note that this is the escape speed from the Earth's *surface*, not from a satellite orbit.

The orbit of a space probe depends critically upon the *velocity* at burn-out. Let us consider the simple case of a projectile moving *parallel* to the Earth's surface at this point (point A; see Figure 3–19). The energy equation (3–28) tells us the semi-major axis a of the orbit, if we know the burn-out speed v:

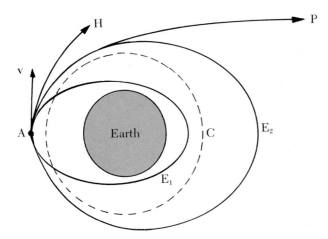

Figure 3–19. *Orbits of Space Vehicles.* At point A, the vehicle has the burn-out velocity **v**. If $v = v_c$, the circular orbit (C) results. When $v = \sqrt{2}\,v_c$, parabolic (P) escape ensues, and when $v > \sqrt{2}\,v_c$, the vehicle escapes on a hyperbolic trajectory (H). When $v_c < v < \sqrt{2}\,v_c$, point A is the perigee of elliptical orbit E_2, while when $0 < v < v_c$, it is the apogee of orbit E_1 (which may intersect the Earth).

$$a = r/[2 - (v/v_c)^2] \qquad\qquad (3\text{–}39a)$$

where r is the distance from the Earth's center to point A; or we may invert equation (3–39a) to find the injection speed needed to attain an orbit of semi-major axis a:

$$v = v_c \left(2 - \frac{r}{a}\right)^{1/2} \qquad\qquad (3\text{–}39b)$$

A circular orbit of radius $r = a$ results when $v = v_c$, as expected. When we have $v = \sqrt{2}\,v_c$, the semi-major axis becomes infinite, and the projectile escapes along a parabolic orbit; hence, this is the escape speed appropriate to radius r. When $v > \sqrt{2}\,v_c$, the space probe escapes on a hyperbolic trajectory. If $v < \sqrt{2}\,v_c$, the projectile enters an elliptical orbit, and point A must clearly be either the *perigee* (point closest to Earth) or *apogee* (point farthest from Earth) of the orbit—the velocity is perpendicular to the radius vector at point A. For $v_c < v < \sqrt{2}\,v_c$, equation (3–39a) tells us that $a > r$ at point A; hence, we insert at perigee, and the orbit never approaches the Earth closer than point A. For $0 < v < v_c$, we have $a < r$, so that point A is the apogee; the satellite will collide with the Earth if the perigee distance is less than $R_\oplus$.

In colliding with molecules of the Earth's tenuous upper atmosphere, a satellite experiences *atmospheric drag* which reduces its total orbital energy. Since the total energy for a circular orbit ($r = a$) is

$$TE = (mv^2/2) - (GM_\oplus m/a) = -(GM_\oplus m/2a)$$

we see that a must *decrease* as TE becomes more negative. Therefore, the orbital radius decreases and the potential energy $PE = -(GM_\oplus m/a)$ becomes more negative. But the satellite *speeds up* as it loses altitude, since the kinetic energy *increases*—$KE = +(GM_\oplus m/2a)$! In an elliptical orbit, the atmospheric drag is greatest at perigee, where the atmosphere is densest; hence, the semi-major axis of the orbit is reduced most at perigee, and the orbit becomes *more circular*.

All deviations from an ideal elliptical orbit are termed *perturbations*. Excluding

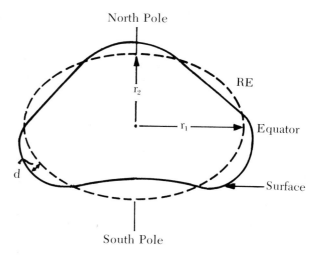

Figure 3–20. *Shape of the Earth.* Perturbations of satellite orbits show the Earth's shape; all distortions are greatly exaggerated here for clarity. The dashed line is the reference oblate spheroid ($r_1 =$ 6378.2 km, $r_2 = 6356.8$ km), while the solid line illustrates the pear-shaped surface with amplitude of oscillation $d \approx 75$ meters.

rocket-thrusting, some sources of orbital perturbation are: atmospheric drag (already discussed); the gravitational attraction from bodies other than the primary; meteoritic impacts; magnetic forces and light pressure; and the *nonspherical shape* of the Earth. Far above the atmosphere the last effect is the most important. The Earth's rotation gives it the shape of an oblate spheroid (Chapter 4), and internal stresses lead to a pear-shaped distortion (an elevation of the southern hemisphere and depression of the northern; see Figure 3–20). These distortions are deduced from their effects upon satellite orbits; they produce a *noncentral* gravitational attraction which changes the direction of perigee and shifts the *line of nodes* (the intersection of the orbital plane and the Earth's equatorial plane).

Problems

3–1. Assume that the greatest elongation of Venus is 48°, and consult Figures 3–3 and 3–6. Draw two sequences of sketches illustrating the phases (i.e., visual appearances of the disk) of Venus that Galileo would have seen as Venus moved from inferior conjunction to superior conjunction, if
(a) the Ptolemaic theory were correct;
(b) the Copernican heliocentric theory were correct.

3–2. Assume that the orbital plane of a superior planet is inclined 10° to the ecliptic, and that the planet crosses the ecliptic moving northward at opposition. Make a diagram similar to Figure 3–4, showing the interesting retrograde path of this superior planet. Can you now generalize our statement about retrograde *loops?*

3–3. A species of intelligent Jovians observes the Earth from Jupiter. What do they observe the Earth's synodic orbital period to be?

3–4. (a) Explicitly carry out the derivation of equation (3–4), showing all the appropriate steps.
(b) On graph paper, plot the polar equation (3–3) for an ellipse, (3–6) for a parabola, and (3–7) for a hyperbola.

3–5. In terms of the gravitational acceleration at the surface of the Earth

(g), find the surface gravitational accelerations of:

(a) the Moon ($M_{\mathrm{)}} = 0.0123 M_{\oplus}$, $R_{\mathrm{)}} = 1738$ km);

(b) the Sun ($M_{\odot} = 2 \times 10^{30}$ kg, $R_{\odot} = 7 \times 10^{10}$ cm); and

(c) Jupiter ($M_{2\!\!\!1} = 318 M_{\oplus}$, $R_{2\!\!\!1} = 5.6 R_{\oplus}$).

3–6.   (a) What are the perihelion and aphelion speeds of Mercury? What are the perihelion and aphelion distances of this planet? Compute the product vr (speed times distance) at each of these two points and interpret your result.

(b) Find the relative position of the barycenter for (i) the Sun-Jupiter system; and (ii) the Earth-Moon system.

3–7.   The Syncom satellite is in circular orbit about the Earth, with a sidereal period of exactly 24 hours. What is the distance from the Earth's surface for such a satellite [use either equation (3–24) or equation (3–37)]? If the satellite appears stationary to an earthbound observer, what is the orientation of the orbital plane?

3–8.   (a) Using the orbital data for Titan given in Table A2–4, find the mass of Saturn.

(b) A stone is released from rest at the Moon's orbit and falls toward the Earth. What is the stone's speed when it is 192,000 km from the center of the Earth?

3–9.   (a) What is the semi-major axis of the *least-energy* elliptical orbit for a space probe from the Earth to Venus?

(b) Relative to the Earth, what is the velocity of such a probe at the Earth's orbit?

(c) When the probe reaches Venus, what is the velocity of the probe relative to that planet?

3–10.   Consult Figures 3–17 and 3–19, and suppose that a projectile has a burn-out speed $\sqrt{2}v$ (where $v_c/\sqrt{2} < v < v_c$) at a distance r from the Earth's center ($r > R_{\oplus}$). If the velocity vector points 45° *above* the local parallel to the Earth's surface, find the semi-major axis (a), the sidereal period (P), and the eccentricity (e) of the resulting elliptical orbit in terms of r, v, and constants. Can you also find θ, the angle from burn-out to the orbit's perigee? (Hint: Section 3–5(c) is very helpful.)

Reading List

Baker, R., and Makemson, M.: *Introduction to Astrodynamics*. New York, Academic Press, 1960.

Birney, D. Scott: *Modern Astronomy*. Boston, Allyn and Bacon, Inc., 1969.

Brouwer, Dirk, and Clemence, Gerald M.: *Methods of Celestial Mechanics*. New York, Academic Press, 1961.

Cajori, Florian: *Newton's Principia*. Berkeley, University of California Press, 1934.

Davidson, Martin: *Elements of Mathematical Astronomy*. New York, The Macmillan Company, 1962.

Drake, Stillman: *Galileo's Dialogue Concerning the Two Chief World Systems*. Berkeley, University of California Press, 1953.

Ebbighausen, E. G.: *Astronomy*. Second Ed. Columbus, Ohio, Charles E. Merrill Publishing Company, 1971.

Halliday, David, and Resnick, Robert: *Physics*. Combined Ed. New York, John Wiley & Sons, 1966.

Hutchins, Robert M. (ed.): *Great Books of the Western World: Ptolemy, Copernicus, Kepler.* Vol. 16. Chicago, Encyclopaedia Britannica Inc., 1952.

McDonald, Robert L., and Hesse, Walter H.: *Space Science.* Columbus, Ohio, Charles E. Merrill Publishing Company, 1970.

Moulton, F. R.: *An Introduction to Celestial Mechanics.* New York, The Macmillan Company, 1959.

Rosen, Edward: *Three Copernican Treatises.* New York, Dover Publications, 1959.

Russell, H. N., Dugan, R. S., and Stewart, J. Q.: *Astronomy.* Vol. 1. Revised Ed. Boston, Ginn and Company, 1945.

Ryabov, Y.: *An Elementary Survey of Celestial Mechanics.* New York, Dover Publications, 1961.

Small, Robert: *An Account of the Astronomical Discoveries of Kepler.* Madison, The University of Wisconsin Press, 1963.

Symon, Keith R.: *Mechanics.* Second Ed. Reading, Massachusetts, Addison-Wesley Publishing Company, 1960.

Van de Kamp, Peter: *Elements of Astromechanics.* San Francisco, W. H. Freeman and Company, 1964.

Chapter 4
Motions of the
Earth

Before we proceed to the Solar System and beyond, let us study the dynamics of our Earth. This chapter and Chapter 6 will serve as the basis for our investigation of the other planets (Chapter 7), just as the Sun and Galaxy will be our stepping-stones to the stars and Universe in later chapters.

4–1 TIME AND THE SEASONS

(a) Terrestrial Time Systems

We are all familiar with the words second, minute, hour, day, week, month, and year. But what exactly do they mean? There are many ways to measure *time*, all of them arbitrary and conventional. Astronomers define the second, minute, hour, and day in terms of the Earth's rotation; the week and month in terms of the Moon's orbital motion; and the year in terms of the Earth's revolution about the Sun.

A *day* is that time interval between two successive upper transits of a given celestial reference object. The vernal equinox (Υ) is the zero-point for *sidereal time*, and the sidereal day is arbitrarily divided into 24 sidereal hours of equal length, each of which consists of 60 sidereal minutes with 60 sidereal seconds per minute. In Chapter 2 we showed that 1^h corresponds to $15°$ of rotation, so the *local* sidereal time is $0^h\ 0^m\ 0^s$ when the vernal equinox lies on our celestial meridian and 2^h when a star with right ascension $\alpha = 2^h$ is in upper transit (and the vernal equinox is 30° west of the celestial meridian). Accordingly, we may *define local sidereal time as the hour angle of the vernal equinox* (see Figure 4–1). In section 4–4(c) we shall see that the vernal equinox precesses westward along the celestial equator by about 46″ per year (0″125 per day); hence, the actual period of the Earth's rotation (measured with respect to the stars) is 0.008 longer than one sidereal day. Since sidereal time is so closely associated with right ascension and hour angle, it is the time used by astronomers.

Timekeeping on the Earth is based in a complicated way upon the position of the Sun. *Apparent solar time* is the hour angle of the Sun plus 12^h, so that the zero-point is midnight at 0^h and local apparent noon always occurs at 12^h. But the length of the apparent solar day is *not* constant during the year, even to a given observer. These variations are caused by the eccentricity of the Earth's orbit and the inclination of the Earth's equatorial plane to the ecliptic (see Figure 4–2). The orbital speed of the Earth is greatest at perihelion (about January 2) and least at aphelion (about July 3), in accordance with Kepler's second law. The Sun reflects this motion by moving eastward along the ecliptic faster at perihelion than at aphelion. To return the Sun to upper transit, the Earth must turn through a greater angle, and the apparent solar day is longer at perihelion than at aphelion. In addition, the Sun moves on the ecliptic, while apparent solar time is measured along the celestial equator. Hence, only that component of the Sun's eastward motion which is parallel to the celestial equator affects apparent solar time. To avoid the inconveniences of this variable solar time, we define *mean solar time* as the time kept by a fictitious point (the *mean Sun*) which moves eastward along the *celestial equator* at the *average* angular rate of the true Sun. The *mean solar day* begins at midnight (with respect to this point) and its length is always 1/365.2564 year. The difference between apparent solar time and mean solar time is called the *equation of time*; since the effects are cumulative, the mean Sun may lead or lag the true Sun by as much as 16 minutes.

Referring to Figure 4–3, we may now compare sidereal time with mean solar time. The Earth's rotation returns the vernal equinox to upper transit as the Earth

Figure 4–1.   *Local Sidereal Time.* An observer's celestial meridian (CM) is shown on the celestial sphere. His local sidereal time is the hour angle (HA) of the vernal equinox (♈), which equals the right ascension (α) of an object in upper transit at that moment. When the vernal equinox is in upper transit, the local sidereal time is 0^h.

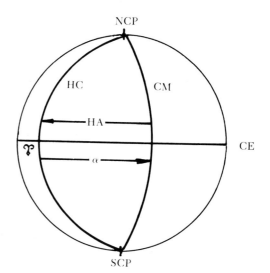

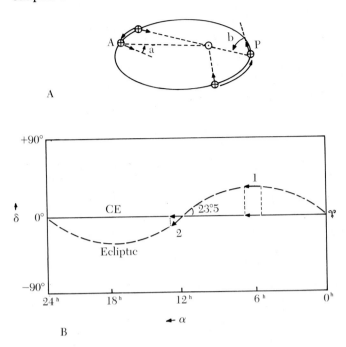

A

B

Figure 4–2. *Variations of Apparent Solar Time.* **A,** The Earth's orbital motion in one day is shown schematically. Near aphelion (A) the Earth must rotate through 360° plus angle *a* to complete one apparent solar day, whereas near perihelion (P) the Earth completes an apparent solar day by turning through 360° plus angle *b*. An apparent solar day is longer at perihelion than at aphelion, since *b > a*. **B,** Our rectangular map of the celestial sphere—coordinates are right ascension (α) and declination (δ)—shows the ecliptic inclined 23°5 to the celestial equator (CE). The Sun's daily eastward motion along the ecliptic is shown schematically at two points, 1 and 2; only the component of this motion parallel to the celestial equator causes apparent solar time to vary.

moves the distance *A* in its orbit—one sidereal day has passed. Since the Earth has moved about (360/365) ≈ 1° around its orbit, it must rotate through this angle before the Sun returns to the local meridian and a mean solar day has passed (*B*). But 1° corresponds to 4^m of sidereal time, so the mean solar day is about four minutes *longer* than the sidereal day. The precise length of the sidereal day is $23^h 56^m 4^s.09$ in mean solar time units. A consequence of this distinction is that the stars appear to rise about four minutes *earlier* each night in terms of the mean solar time which we maintain on Earth. Therefore, a star which is in upper transit at midnight tonight will reach the meridian at 10 P.M. one month from now ($30 \times 4^m = 2^h$).

Mean solar time differs at every longitude on the Earth's surface. The practical difficulties of such a timekeeping system have been avoided by the establishment of 24 *time zones* around the world. Within each longitude zone approximately 15° (or 1^h) wide, all locations have the same *standard time;* the boundaries of each zone are adjusted for maximum convenience (e.g., a city is usually placed wholly within one time zone). The reference zone is centered on Greenwich, England, at 0° longitude. Standard time at Greenwich is referred to as *Greenwich Mean Time* or, equivalently, *Universal Time* (abbreviated U.T.); astronomical events such as total eclipses are usually given in terms of U.T. New York

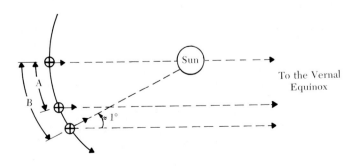

To the Vernal Equinox

Figure 4–3. *Sidereal and Solar Days.* The Earth rotates once with respect to the vernal equinox, and completes one sidereal day, in moving through the distance *A*. To complete a mean solar day, it must rotate 1° farther to bring the Sun back to the meridian; the Earth has then moved a greater distance.

City lies five hours west of Greenwich in the Eastern Standard Time (E.S.T.) zone, so that we subtract 5^h from U.T. to find the local New York time for such events. To take advantage of the extra hours of daylight during the summer [see section 4–1(b)], an hour is *added* to local standard time from mid-spring to mid-fall in many parts of the world. Hence, 11 P.M. *Daylight Saving Time* in San Francisco corresponds to 10 P.M. Pacific Standard Time (P.S.T.), and we *always add* 8^h to P.S.T. to find U.T., since San Francisco is eight time zones west of Greenwich (i.e., U.T. equals Pacific Daylight Saving Time plus 7^h).

In section 4–4 we will see that the Earth's rotation rate is subject to extremely small, but unpredictable, variations. To precisely predict the positions of bodies in the Solar System we require a steady time standard, so Universal Time is replaced by *Ephemeris Time* (E.T.) in celestial mechanics. At the beginning of 1900 A.D., an *ephemeris second* was defined as 1/31,556,925.97474 the length of the tropical year 1900 and both U.T. and E.T. were in agreement, but today these times differ by about 40^s.

The *week* and *month* are of ancient origin and derive from the Moon's synodic orbital period of 29^d53. To fit within the year, months have been given conventional lengths of 28, 30, and 31 days. The week of seven days (each named after a planet) is based upon the quarter phases of the Moon ($29.53/4 = 7.38 \approx 7^d$; see Chapter 6).

The *year* is the time it takes the Earth to orbit the Sun, but different definitions give us three types of years. With respect to the stars the Earth's revolution takes one *sidereal year* of 365.2564 mean solar days ($365^d\ 6^h\ 9^m\ 10^s$), while the *tropical year* of 365.2422 mean solar days ($365^d\ 5^h\ 48^m\ 46^s$) is the period with respect to the vernal equinox which precesses about $50''$ (or $20^m\ 24^s$ of time) westward along the ecliptic each year. Finally, since planetary perturbations cause the Earth's perihelion to precess in the direction of the orbital motion, we call the time between successive perihelion passes the *anomalistic year* of 365.2596 mean solar days ($365^d\ 6^h\ 13^m\ 53^s$).

Today there are two conventional ways to keep track of the passage of time. The *Gregorian calendar*, which attempts to approximate the year of seasons (the tropical year), consists of 365 days per common year and 366 days in years divisible by four (leap years). To achieve an accuracy of one day in 20,000 years, only those century years divisible by 400 are leap years (e.g., 2000 A.D.), while century years divisible by 4000 remain common (e.g., 8000 A.D.). In astronomy, the more convenient, linear, *Julian Day* (J.D.) system is used. Days, and fractions thereof, are counted continuously from noon U.T. on January 1, 4713 B.C. Hence, at 6 P.M. U.T. on January 1, 1971, we have J.D. 2,440,952.25.

(b)   The Seasons of the Earth

The Earth's seasons—spring, summer, autumn, and winter—arise because the Earth's equatorial plane is constantly inclined $23°5$ to the ecliptic plane (see Figure 4–4). The eccentricity of the Earth's orbit is too small ($e = 0.017$) to affect the seasons; note that perihelion occurs during the northern winter (January 2). The number of daylight hours and the noon altitude of the Sun lead to the characteristic temperatures of the seasons, and both depend upon the latitude of the observer and the Sun's position on the ecliptic. Let us consider these two causes independently.

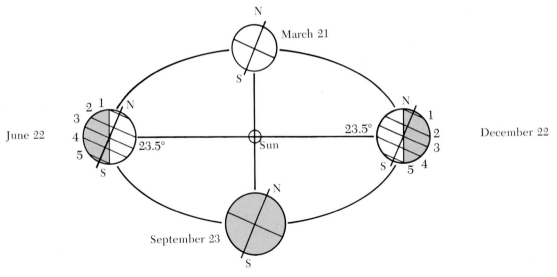

Figure 4–4.   *The Earth's Inclination and Seasons.* The Earth's equator is always inclined 23°.5 to the ecliptic. Day (light) and night (dark) are shown at the vernal equinox (March 21), summer solstice (June 22), autumnal equinox (September 23), and winter solstice (December 22). The latitudes indicated are (1) Arctic Circle (66°.5 N), (2) Tropic of Cancer (23°.5 N), (3) equator (0°), (4) Tropic of Capricorn (23°.5 S), and (5) Antarctic Circle (66°.5 S). The north and south poles are N and S, respectively.

When the Sun is at the *vernal equinox* (about March 21) or the *autumnal equinox* (about September 23), its declination is 0°, and there are 12 hours of day and 12 hours of night at all points on the Earth's surface. The noon altitude of the Sun is 90° (the zenith) at the equator, and diminishes to 0° at the poles. At the *summer solstice* (about June 22), the Sun attains its greatest declination of $+23°.5$, and passes directly overhead at noon for all observers at latitude 23°.5 N—the *Tropic of Cancer.* (About 3000 years ago the Sun was in the constellation Cancer at the summer solstice.) On this date the days are longest in the northern hemisphere, and it is mid-summer there; at the same time it is mid-winter in the southern hemisphere, for the days are shortest there. The situation is reversed at the *winter solstice* (about December 22), when the Sun's declination is $-23°.5$. The Sun then passes directly overhead for all observers at latitude 23°.5 S—the *Tropic of Capricorn* (again, the winter solstice of the ancients was in the constellation Capricornus); it is mid-winter in the northern hemisphere and mid-summer in the southern. Latitude 66°.5 N is called the *Arctic Circle*—as spring becomes summer there, the days lengthen until the summer solstice, when the Sun doesn't set for 24 hours; from fall to winter the days shorten until the 24 hour night at the winter solstice. In the regions north of the Arctic Circle, the Sun does not set for many days (the *midnight Sun*) in the summer, and does not rise for many days during the winter. In fact, at the north pole a six-month "day" begins at the vernal equinox and a six-month "night" begins at the autumnal equinox. South of the *Antarctic Circle* (latitude 66°.5 S), exactly the same events occur six months after they have taken place in the north.

The amount of solar energy received by the Earth is essentially constant from day to day, but the heating effectiveness of this energy—the *solar insolation*— depends upon both latitude and time. The altitude of the Sun determines over what

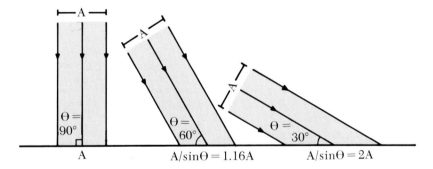

Figure 4–5.   *Solar Insolation.* A unit of solar energy strikes area A when the Sun is at the zenith, but the same energy is spread over the area $(A/\sin\theta)$ when the Sun's altitude is θ.

area a given amount of radiation is spread, as shown in Figure 4–5. Suppose that a unit of energy falls upon the area A when the Sun is at the zenith. When the Sun's altitude is θ, this same amount of energy is spread over the area $(A/\sin\theta)$—the heating efficiency decreases as θ decreases. Hence, it is warmer in the daytime than at night, and the summer is warmer than the winter, since the Sun is higher in the sky for more hours in the summer. The Earth's equatorial regions are always warm because the noon Sun never passes far from the zenith. At the poles we are not surprised to find the *polar icecaps*, since the Sun spends many months below the horizon, and stays at a low altitude when it does rise.

The Earth's surface (especially the oceans) and atmosphere are good thermal insulators, so that they respond slowly to solar heating. As a result, temperature variations are moderated by daily and seasonal *time lags* between the extremes of solar insolation and the extremes of temperature. For example, the early afternoon is usually the warmest part of the day, though solar insolation is greatest at noon. February is the coldest month of northern winter, but December is the month of least insolation.

4–2   PROOFS OF THE EARTH'S ROTATION

That the Earth rotates seems obvious to us, but how can we prove it? The westward circling of the celestial sphere could be a reflection of the daily eastward turning of the Earth, but this is no proof, since the ancients were equally justified in their concept of a rotating celestial sphere centered upon a stationary Earth. Other kinematical arguments, such as the Moon's daily parallax (Chapter 6) and the daily Doppler effect [section 4–3(c)], are also extremely suggestive, but they can be reproduced by a complicated system of epicycles about a nonrotating Earth. We can prove the Earth's rotation only by basing our arguments upon the well-verified *dynamical* laws of Newton.

(a)   The Coriolis Effect

The apparent trajectories of rockets and Earth satellites can be understood only if the Earth rotates. Consider Figure 4–6 where a projectile is launched from

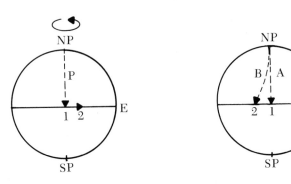

Figure 4–6. *Projectile Trajectories on the Earth.* On the rotating Earth (left), the target moves from 1 to 2, while the projectile moves due south from the north pole (NP) to the equator (E). As seen from the Earth's surface (right), the projectile's trajectory would be *A* if the Earth did not rotate, but is actually *B* for the rotating Earth—the projectile is apparently deflected to its right.

the north pole to impact at the equator. On a nonrotating Earth, the projectile would clearly follow a single meridian of longitude during its entire flight. On a rotating Earth, however, the target on the equator moves eastward at 0.46 km sec^{-1}, and the projectile impacts *west* of the target. Though the projectile's motion is due south, it *appears* to be deflected to the right with respect to the Earth's surface. The fictitious acceleration which produces this effect—the *Coriolis effect*—was deduced by Gaspard Gustave de Coriolis (1792–1843 A.D.) in 1835.

By considering a variety of projectile trajectories, we soon find that moving bodies always appear to be deflected to the right in the northern hemisphere and to the left in the southern hemisphere. If the projectile's velocity is **v**, and the Earth's *vector angular velocity* is **ω** (its direction is toward the north celestial pole, and its magnitude measures the Earth's spin; units = radians sec^{-1}), then these observations are summarized in terms of the *Coriolis acceleration*:

$$\mathbf{a}_{\text{Coriolis}} = 2(\mathbf{v} \times \boldsymbol{\omega}) \tag{4–1}$$

As before, the cross product yields the product of the perpendicular components of **v** and **ω**, and the direction of **a**$_{\text{Coriolis}}$ is "the direction in which your thumb points when you align the fingers of your right hand along **v** and rotate them through the smallest angle to **ω**" (the *right hand rule*). Let us derive equation (4–1), using Figure 4–7. A body moves with a constant radial velocity **v** above a turntable rotating with angular speed ω. At time t the body leaves the origin (A), and it moves a distance dr to point B in an infinitesimal time dt. Meanwhile, point B has revolved through the angle $d\theta = \omega\,dt$ to B'. From the geometry, we have $dr = v\,dt$ and $ds = dr\,d\theta$; therefore, we find $ds = (v\,dt)(\omega\,dt) = v\omega(dt)^2$. But Newton's second law implies that a body moves the distance $ds = a(dt)^2/2$ in time dt when it experiences the *constant* acceleration a, so that

$$a_{\text{Coriolis}} = 2v\omega$$

In addition, the direction of the apparent deflection is to the right, in accordance with equation (4–1).

The Coriolis effect is responsible for the characteristics of large-scale wind patterns in the Earth's atmosphere (as well as for ocean currents). A *cyclone* is a local counterclockwise circulation of air in the northern hemisphere (clockwise in

Figure 4–7. *The Coriolis Effect.* The turn-table rotates at the angular speed ω, while the body moves from A to B (distance dr) at velocity **v** in time dt. At the same time, B rotates through the angle $d\theta$ and the distance ds to B'. The body appears to be deflected to its right relative to the turn-table.

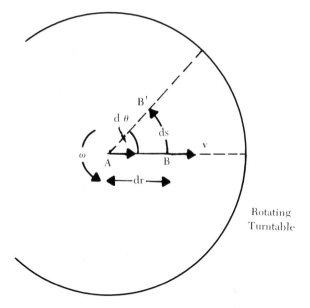

Rotating
Turntable

the southern), produced by the Coriolis deflection to the right of air flowing toward the center of a low pressure region. An *anticyclone* arises when air flowing away from the center of a high pressure region is deflected into a local clockwise circulation in the northern hemisphere (counterclockwise in the southern). As shown in Figure 4–8, solar heating produces large-scale vertical *cells* of wind motion. At the Earth's surface the Coriolis effect causes these winds to flow in the known directions of the easterly trade winds (5–30° N and S), the temperate westerlies (35–50° N and S), and the polar easterlies (60–90° N and S). At low latitudes the bands of relatively calm air are known as the doldrums (0–5° N and S) and the horse latitudes (30–35° N and S).

Figure 4–8. *General Wind Patterns.* Solar heating produces the vertical cells of "rolling" air, while the Coriolis effect deflects the air to give the easterly trade winds (ETW), the temperate westerlies (TW), and the polar easterlies (PE).

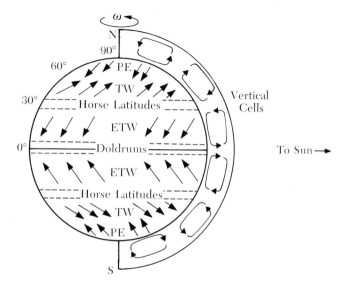

(b) Foucault's Pendulum

In 1851 Jean Bernard Léon Foucault (1819–1868 A.D.) hung a pendulum (a heavy ball at the lower end of 60 meters of wire) from the ceiling of the Pantheon in Paris, and proved the Earth's rotation by noting that the pendulum's plane of oscillation rotated during the day. If the Earth did not rotate, this phenomenon would be impossible, since all forces acting on the ball (the Earth's gravity and the tension in the wire) would lie in the plane of oscillation.

The behavior of *Foucault's pendulum* is explicable in terms of the Coriolis effect; we suggest that the reader try to verify this statement. It is simpler here, however, to merely step off the Earth and observe the pendulum swinging as the Earth turns. At the north pole the pendulum oscillates in a fixed plane, while the Earth rotates below it every 24 sidereal hours; the pendulum appears to rotate *westward* with the period $P = 24^h$. A pendulum swinging in the equatorial plane at the Earth's equator feels no forces perpendicular to the plane of oscillation, so it doesn't rotate at all ($P = \infty$). At an intermediate latitude ϕ (see Figure 4–9), the vertical component of the Earth's angular speed (ω) is $\omega \sin \phi$. But angular speed is inversely proportional to the period of rotation ($\omega = 2\pi/P$), so that the pendulum appears to rotate westward with the period $P = (24^h/\sin \phi)$. Precise measurements of Foucault pendula verify this relation in detail, and permit us to determine the Earth's period of rotation.

(c) The Oblate Earth

The shape of the Earth's surface is that of an *oblate spheroid;* the polar radius ($r_p = 6356.8$ km) is 21.4 km less than the equatorial radius ($r_e = 6378.2$ km), so that the *oblateness* is $(r_e - r_p)/r_e = 21.4/6378.2 = 1/298.3$. Newton knew this, and he took it as a strong *indication* that the Earth rotates. If the Earth were a *fluid* body,

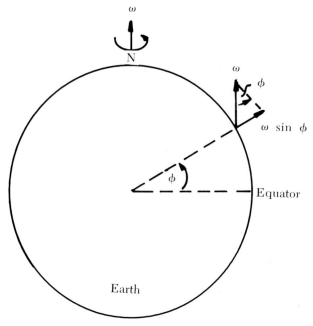

Figure 4–9. *Rotation of Foucault's Pendulum.* At latitude ϕ, the vertical component of the Earth's angular speed is $\omega \sin \phi$, so the Earth turns below the pendulum in a period proportional to $1/\sin \phi$.

Figure 4–10. *Cause of the Oblate Earth.* A fluid mass *m* fixed at latitude ϕ feels the gravitational acceleration *A* and the centripetal acceleration *B*. When the mass is free to move, it experiences the centrifugal acceleration $C = -B$, with the vertical component *a* which lightens the mass' weight and the horizontal component *b* which moves the mass to the Earth's equator (E). The oblate shape is built up as many fluid masses migrate toward the equator.

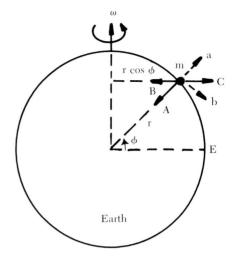

its shape would prove its rotation, since a fluid must adjust its shape to all external forces—a nonrotating fluid body is always spherical. But the Earth is composed of materials which *might* maintain an oblate shape even if no rotation occurs.

Today we know that the Earth's materials have an average strength close to that of steel, but over the eons they are incapable of maintaining any shape other than the *equilibrium* shape of a rotating fluid body. Figure 4–10 shows how this equilibrium shape comes about. A mass *m* at latitude ϕ on a *spherical* rotating Earth experiences two accelerations: (A) the gravitational acceleration $GM_\oplus/r^2$ directed toward the Earth's center, and (B) the centripetal acceleration $\omega^2 r \cos \phi$ maintaining the mass in the circular orbit of radius $r \cos \phi$. If the mass is free to move (i.e., a fluid mass), the centripetal acceleration is not balanced by any force, so the mass feels an apparent fictitious *centrifugal acceleration* of the same magnitude in the opposite direction (C). The vertical component of this centrifugal acceleration, $a = \omega^2 r \cos^2 \phi$, reduces the weight of the mass, while the horizontal component, $b = \omega^2 r \sin \phi \cos \phi$, causes the mass to migrate to the Earth's equator. If we consider many such fluid masses (in fact, the whole Earth), we see that a bulge grows at the Earth's equator until fluid masses can no longer climb this equatorial "hill"—the equilibrium oblate shape is now established.

4–3 PROVING THE EARTH'S REVOLUTION ABOUT THE SUN

As with the Earth's rotation, we must be careful if we want to prove the Earth's revolution about the Sun. The following three proofs were unavailable at the time of Copernicus and Kepler, so their theories were considered suspect and were slow to be accepted. Today these proofs are indisputable evidence for the heliocentric theory of the Solar System.

(a) The Aberration of Starlight

In 1729 the English astronomer James Bradley (1693–1762 A.D.) discovered the *aberration of starlight*, and using the earlier measurement of the finite

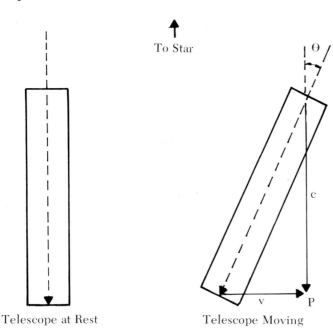

Figure 4–11. *Aberration of Starlight.* The telescope at the left is at rest, and so is pointed vertically upward to see the star. When the telescope moves with speed v (right), it must be tilted through the angle θ where $\tan \theta = (v/c)$ so that the starlight (speed $= c = 3 \times 10^5$ km sec^{-1}) reaches point P when the bottom of the telescope does.

To Star

Telescope at Rest   Telescope Moving

speed of light, $c \approx 3 \times 10^5$ km sec^{-1}, by Ole Römer in 1676, he explained this phenomenon as due to the orbital motion of the Earth.

Suppose that a man walks through a vertically-falling rain with an umbrella over his head. The faster he walks, the farther he must lower the umbrella in front of himself to prevent the rain from striking his face. When starlight enters a telescope, an analogous phenomenon occurs (see Figure 4–11). If the Earth were at rest, we would point our telescope toward the zenith to see a star situated there. But, if the Earth is in motion at speed v, we must tilt the telescope in the direction of motion by an angle

$$\theta \approx \tan \theta = (v/c) \tag{4–2}$$

so that the bottom of the telescope can meet a light ray which has entered the top of the telescope. Bradley observed this very small tilt angle, $\theta = 20\rlap{.}''49$; using equation (4–2), we immediately deduce the Earth's orbital speed as $v = \theta c = (9.934 \times 10^{-5} \text{ radians})(3.0 \times 10^5 \text{ km sec}^{-1}) = 29.80$ km sec^{-1}.

The direction in which we tilt our telescope is constantly changing as the Earth moves around the Sun. Since the Earth's orbit is essentially circular, stars appear to trace out annual *aberration orbits* on the celestial sphere as shown in Figure 4–12. A star at the ecliptic pole is seen to move around a circle of angular radius $20\rlap{.}''49$ once a year. Stars on the ecliptic oscillate to and fro along lines of angular half-length $20\rlap{.}''49$. At an intermediate celestial latitude β (angle from the ecliptic) the aberration orbit is an ellipse, with semi-major axis $20\rlap{.}''49$ and semi-minor axis $(20\rlap{.}''49) \sin \beta$. If the Earth did not revolve around the Sun, this *observed* behavior of the aberration orbits would be totally inexplicable.

Figure 4–12. *Aberration Orbits.* The Earth ($\oplus$) is shown at four positions in its solar ($\odot$) orbit, with a telescope inclined to view a star at the ecliptic pole ($\beta = 90°$). The apparent aberration orbits at three celestial latitudes (β) on the celestial sphere are shown relative to the star's (✷) *true* position.

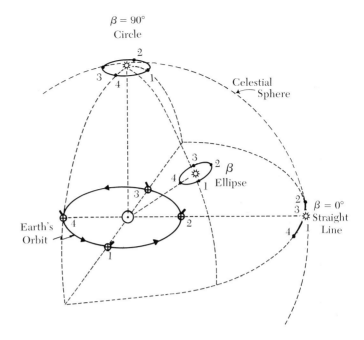

(b)   Stellar Parallax

As we drive along a highway in our car, we notice that nearby objects seem to be moving backward with respect to more distant objects (see Figure 4–13). This perspective effect of our line of sight is termed *parallax*. According to the heliocentric theory of the Solar System, nearby stars should exhibit parallax effects

Figure 4–13. *Parallax.* **A**, A nearby object appears to move backward with respect to the background as we move from 1 to 5. **B**, A stellar parallax angle (π'') is defined by the Earth's solar orbit. Note that $d(\text{AU}) = 1/\tan[\pi(\text{radians})] \approx 206{,}265/\pi''$, when ✷ $\pi \ll 1$ radian.

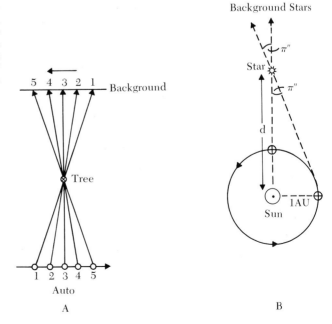

A   B

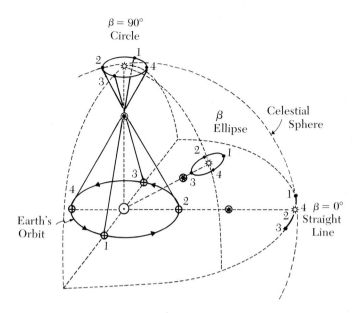

Figure 4–14. *Parallactic Orbits.* This diagram is similar to Figure 4–12, except that here we view the apparent parallactic orbits—relative to the true (✸) celestial position—of three *nearby* stars (⊛) at different celestial latitudes (β). Note how the size of the parallax orbit depends upon the distance to the nearby star, and the phase difference of 90° with respect to the aberration orbits shown in Figure 4–12.

on the celestial sphere, since the Earth is in motion about the Sun (Figure 4–13). If the parallax angle of a star is π'' (in seconds of arc), then, from the figure, that star's distance is

$$d = (206{,}265/\pi'') \text{ AU} \qquad (4\text{–}3)$$

Note that 206,265 is the number of arc-seconds in one radian. (See Chapter 10 for a complete discussion of stellar parallaxes and distances.)

 The heliocentric models of Copernicus and Kepler remained on shaky ground until the first stellar parallax was observed in the nineteenth century. Tycho refused to believe the Earth's motion because he could detect no stellar parallax to an accuracy of 1′. In 1838 Friedrich Wilhelm Bessel (1784–1846 A.D.) published the first observed stellar parallax: 0″.294 for the star 61 Cygni; at about the same time Wilhelm Struve found the parallax of Vega (α Lyrae) and T. Henderson found that of α Centauri. Today we know that the nearest star is Proxima Centauri with a parallax of 0″.764 and a distance of 270,000 AU (about 4 × 10^{13} km); hence, all stellar parallaxes are less than 1″.0.

 As the Earth moves around the Sun in its orbit, each star traces out a yearly *parallactic orbit* on the celestial sphere (Figure 4–14). Stars at the ecliptic poles move in circles with radii dependent upon their distances from the Sun, and stars on the ecliptic oscillate along lines. For the general parallactic ellipse, the ratio of semi-minor to semi-major axis is sin β, where β is the celestial latitude of the star, just as for aberration orbits. Note, however, that for a given direction (β) the aberration orbit is always the same, while the parallactic orbit depends upon a star's distance. In addition, the aberration orbit of a star is a quarter cycle out of phase with the parallactic orbit. In Chapter 14 we shall see that this picture is further complicated by the individual *motions* of the stars through space.

(c) The Doppler Effect

Our final proof of the Earth's revolution is the Doppler effect. In Chapter 8 (see also Chapter 14) we will derive the fact that the wavelength of electromagnetic radiation (e.g., light, radar, etc.) is shifted in proportion to the relative line-of-sight speed of the object observed. If a star emits radiation at the wavelength λ_e, and we observe this radiation at wavelength λ_o, then the Doppler formula is

$$\Delta\lambda/\lambda_e = (\lambda_o - \lambda_e)/\lambda_e = v/c \qquad (4\text{–}4)$$

where c is the speed of light and v is the relative line-of-sight speed (positive for recession; negative for approach) between the observer and the observed.

For stars at the ecliptic pole, there should be no Doppler shift, and indeed none is observed (except that attributable to the space motions of the *stars*; see Chapter 14). At an intermediate celestial latitude β, the shift is sinusoidal with a period of one year, and the amplitude $\Delta\lambda$ varies as $\cos\beta$. The maximum amplitude of this Doppler shift occurs for stars on the ecliptic, where the full magnitude of the Earth's orbital velocity comes into play; a standard method of determining the Earth's speed of revolution is to measure this maximum shift and to use equation (4–4) to deduce $v_\oplus = 29.80$ km sec^{-1}. The spatial motion of a star superimposes a *constant* Doppler shift on the time-varying shift because of the Earth's revolution; the two effects are easily untangled.

The Earth's rotation causes a *diurnal* Doppler shift with an amplitude proportional to $(\cos\phi\cos\delta)$, where ϕ is the terrestrial latitude of the observer and δ is the declination of the source. This effect *must* be taken into account in deducing the precise periods of pulsar emission (see Chapter 16), so that we may consider it a "proof" of the Earth's rotation.

4-4   DIFFERENTIAL GRAVITATIONAL FORCES

We have already seen that two spherical bodies behave gravitationally as point masses. If the bodies are elastic or nonspherical, or if we introduce several more bodies on the scene, *differential gravitational forces* become extremely important. This effect arises because gravitation depends upon distance, and different parts of an extended body (or system) will therefore feel different accelerations. We will illustrate this concept and its consequences throughout the remainder of this chapter.

(a) The Phenomenon of Tides

Anyone who has spent time near a large body of water knows that the level of the water rises and falls twice daily in the *tides*, and also knows that a given tide occurs about an hour later each day. We are assured that the Moon is the principal cause of the tides, by the fact that the Moon returns to upper transit 53 minutes later each day—at about the time of high tide. The solid body of the Earth, which has much greater cohesive strength than water, also responds quickly to the Moon's tidal forces. Albert Abraham Michelson (1852–1931 A.D.) first measured these Earth tides in 1913 by observing water tides in long horizontal pipes. He assumed that the Earth was infinitely rigid and employed Newton's laws to calculate the expected water tides; the observed water tides were only 69% as high as theory said they should be. This difference could be accounted for by assigning the Earth a rigidity somewhat greater than that of *steel*, so that it would respond to the Moon's tidal forces by raising body tides several centimeters in height.

Let us use Newton's laws to explain the tides. We will neglect the Earth tides, and assume that the Earth has its equilibrium rotation shape (an oblate spheroid) which is essentially spherical. We entirely cover this sphere with a uniform depth of water, and ask what effect the Moon's gravitational force has on this water. Figure 4–15A shows the gravitational acceleration due to the Moon at several points of the Earth. By subtracting the vector acceleration at the Earth's center (C) from each of the surface vector accelerations, we obtain the *differential tidal accelerations* displayed in Figure 4–15B. These tidal forces raise water tides about a meter high at points A and B on the line of centers, and the Earth rotates below this configuration once each day. Actual tides arise from forced oscillations

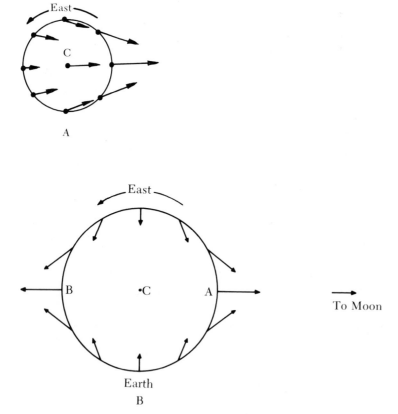

Figure 4–15. *Tidal Gravitational Acceleration.* **A**, The gravitational acceleration due to the Moon is shown at several points of the Earth. **B**, We have subtracted the vector acceleration at the Earth's center (C) from the surface accelerations; now only the *tidal* (differential) accelerations remain, and high tides occur at points *A* and *B*.

in the Earth's ocean basins, so that the height and timing of tides may differ markedly from the theoretical case. In some bays and estuaries, the tidal waters may accumulate to heights greater than ten meters.

Let us now be somewhat more quantitative in our derivation of tidal accelerations. As a first approximation we will neglect the *small* centripetal acceleration at the Earth's surface due to the orbital motion of the Earth and Moon about their barycenter every sidereal month ($27^{d}.32$). Figure 4–16 shows the centers of Earth and the Moon separated by the distance d, and a small particle on the Earth's surface at an angle ϕ to the line of centers. The gravitational acceleration of the Earth's center due to the Moon has the magnitude $A = GM_{\text{☽}}/d^2$, while the particle's acceleration is $B = GM_{\text{☽}}/r^2$. Let us vectorially subtract A from B, just as in Figure 4–15. The component of B perpendicular to the line of centers is unaffected by this subtraction, and has the magnitude

$$a = B \sin \theta = GM_{\text{☽}} R_{\oplus} \sin \phi / r^3 \qquad (4\text{–}5)$$

The component of B parallel to the line of centers is

$$b = B \cos \theta = GM_{\text{☽}}(d - R_{\oplus} \cos \phi)/r^3$$

and the remainder after subtracting A is

$$b' = \left(\frac{GM_{\text{☽}}}{r^3}\right)\left(d - R_{\oplus} \cos \phi - \frac{r^3}{d^2}\right) \qquad (4\text{–}6)$$

Using the law of cosines, and the fact that $(R_{\oplus}/d) \ll 1$, we have

$$r^3 = d^3\big[1 - 2(R_{\oplus}/d) \cos \phi + (R_{\oplus}/d)^2\big]^{3/2} \approx d^3$$

therefore, equation (4–5) finally becomes (to lowest order):

$$a \approx GM_{\text{☽}} R_{\oplus} \sin \phi / d^3 \qquad (4\text{–}7\mathbf{a})$$

Equation (4–6) may be written as

$$b' \approx (GM_{\text{☽}}/d^2)\big[1 - (R_{\oplus}/d) \cos \phi - (r/d)^3\big]$$

and expanding our result for r^3 using the *binomial theorem* (given in the Mathematical Appendix) we obtain

Figure 4–16. *Quantitative Accelerations.* This diagram shows the geometry of the accelerations produced at the Earth by the Moon. See the text, where the symbols are used.

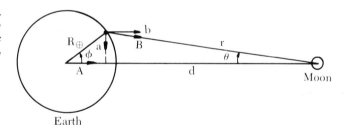

$$b' \approx 2GM_{\mathrm{\text{☽}}}R_{\oplus} \cos \phi/d^3 \qquad\qquad (4\text{–}7\mathbf{b})$$

Equations (4–7) verify the qualitative picture of Figure 4–15B, and tell us that tidal gravitational forces go as (MR/d^3), where R is the size of the body being tidally influenced, M is the mass of the source of tidal force, and d is the separation of the two bodies.

The Sun also produces tidal effects on the Earth. Since the differential accelerations go as (MR/d^3), and $R = R_{\oplus}$ in both cases, the tide-raising force of the Sun relative to that of the Moon is

$$\left(\frac{M_{\odot}}{M_{\mathrm{\text{☽}}}}\right)\left(\frac{r_{\mathrm{\text{☽}}}}{r_{\odot}}\right)^3 = \left(\frac{1.99 \times 10^{30}\ \mathrm{kg}}{7.36 \times 10^{22}\ \mathrm{kg}}\right)\left(\frac{3.84 \times 10^5\ \mathrm{km}}{1.50 \times 10^8\ \mathrm{km}}\right)^3 \approx \frac{5}{11}$$

The tidal effects of the Sun and Moon combine vectorially, so that the resultant tides depend upon the elongation of the Moon. When the Moon is at conjunction or opposition, both forces add to produce the very high *spring tides*, whereas when the Moon is at quadratures, the two effects partially cancel to give us the unusually low *neap tides*.

(b)   Consequences of Tidal Friction

When the Earth and the oceans yield to the tide-raising forces, energy is dissipated (in the form of heat) as a result of friction; most of this energy is lost in shallow seas and at the shorelines where ocean tides abut against the continents. This *tidal friction* reduces the energy of the Earth's rotation, so that the length of the day *increases* at the measurable rate of about $0\overset{s}{.}002$ per century. Tidal friction is the cause of two fascinating phenomena in the Earth-Moon system: the Moon's *synchronous rotation* and *tidal evolution*.

The Earth raises body tides on the Moon which are about $(M_{\oplus}R_{\mathrm{\text{☽}}}/M_{\mathrm{\text{☽}}}R_{\oplus}) \approx$ 20 times higher than the Earth's body tides. The enormous energy dissipation which results slows the Moon's rotation until the Moon always shows but one face to the Earth—just as we see it today. Hence, the Moon was forced into *synchronous rotation*, wherein its sidereal rotation period is exactly the same as its sidereal period of revolution about the Earth.

The average external torques exerted on the Earth-Moon system are negligible, so that the total angular momentum (see Chapter 3) of the system must remain constant. The angular momentum of the Earth decreases as tidal friction slows its rotation; hence, the Moon must increase its angular momentum by *moving away from the Earth*. Kepler's third law then implies that the month must be lengthening. In the distant future the "day" and "month" will become the same, and equal to about 50 present days. Figure 4–17 shows the dynamical reason for this *tidal evolution*. The Earth and its oceans do not respond instantly to the Moon's tidal forces, and as the Earth rotates beneath the Moon, the tidal bulges rise slightly eastward of the Earth-Moon center line. Bulge A is nearer the Moon than bulge B, so A exerts a slightly greater force on the Moon. The resulting *noncentral* force accelerates the Moon, causing it to spiral away from the Earth. At the same time, the Moon's force on the bulges decelerates the Earth's rotation (Newton's third law). If we go backward in time, we see that the Earth must have been spinning

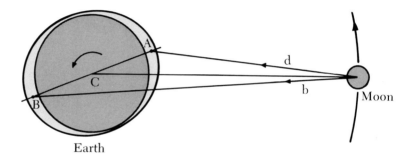

Figure 4–17. *Tidal Evolution.* The tidal bulges *A* and *B* do not respond instantly to the Moon, and so rise slightly eastward of the line of centers. The Moon's force on *A* is greater than on *B*, so that the Earth's rotation slows while the Moon is accelerated forward in its orbit.

faster and the month must have been shorter than today. Indeed, paleontological studies of fossilized corals which lived about 10^8 years ago show that there were 400 "days" in each year, and that ocean tides were more vigorous than today—the Earth *was* rotating faster and the Moon *was* closer in the past. We do not know how far back we can extrapolate this picture, but we do know that large areas of the Earth's surface were melted during some catastrophic event about 3.5×10^9 years ago!

(c) Precession and Nutation

We are all familiar with the behavior of a spinning top or a gyroscope. When the top's spin axis is not aligned with the vertical, we expect the top to fall on its side, but instead the spin axis maintains the same angle to the vertical and simply rotates slowly about the vertical. This *precession* of the top is predicted by Newton's laws of motion.

The Earth's gravitational attraction **F** on the top produces the *horizontal* torque $\mathbf{N} = \mathbf{r} \times \mathbf{F}$ [see section 3–5(a)]. But since torque is just the time rate of change of the top's vector angular momentum **L** ($\mathbf{N} = d\mathbf{L}/dt$), there is no vertical component to topple the top; hence, the top can only rotate or precess about the vertical. Differential gravitational forces acting upon the oblate Earth's (rotational) equatorial bulge produce torques which lead to a similar phenomenon—the precession of the Earth.

The Moon and Sun are the principal causes of the Earth's *luni-solar precession*, with the Moon's effect dominating. The Moon's orbit is inclined about 5° to the ecliptic, but its average force is centered on the ecliptic. Figure 4–18 shows how the Earth's equatorial plane is inclined 23°.5 to the ecliptic, with the prominent equatorial bulges at *A* and *B*. Just as in Figure 4–17, bulge *A* is more strongly attracted to the Moon than bulge *B*, and the *differential* forces are indicated. The torque which results points into the diagram, so that the Earth's angular momentum vector, **L**, precesses toward the *west*. The Sun induces a similar but slightly weaker effect, and the perturbing torques from the other planets produce a *planetary precession* which amounts to about 1/40 the total. As a result of these torques, the celestial poles remain inclined 23°.5 to the ecliptic pole, but trace a circular path around the ecliptic pole once every 26,000 years. At present, the bright star Polaris (α Ursae

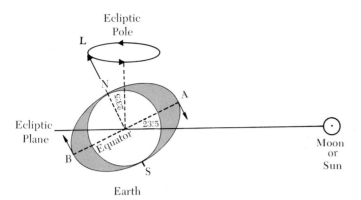

Figure 4–18. *Earth's Precession.* Differential gravitational forces on the Earth's equatorial bulge (*A* and *B*) are shown. They lead to a torque (pointing *into* the diagram) which causes the Earth's angular momentum vector **L** to precess westward at a constant angle of 23°.5 to the ecliptic axis.

Minoris) approximately marks the position of the north celestial pole, but in about 14,000 A.D. the pole star will be Vega (α Lyrae).

As the celestial poles precess, the intersection of the celestial equator and the ecliptic (vernal and autumnal equinoxes) progresses westward at the rate of $360°/26,000$ yr $\approx 50''$ per year *along the ecliptic* (about $50'' \cos 23°.5 \approx 46''$ per year along the celestial equator; see Figure 4–19). This phenomenon of the *precession of the equinoxes* has several important effects: it affects terrestrial time-keeping systems through the definition of the day and the year; it causes the constellations to be in positions quite different from their ancient places; it changes those stars which a given observer would consider circumpolar; and it significantly affects the right ascension and declination of all celestial objects. Since the precession effect is readily observable, the positions (α, δ) of objects in the celestial equatorial coordinate system must be constantly updated to the current epoch by the use of precession equations and tables.

Since the Moon and Sun move above and below the Earth's equatorial plane, there are periodic variations in the torques acting on the Earth's equatorial bulge. These variations lead to a *nutation* or wobbling of the Earth's rotation axis. Small contributions to the nutation have monthly and yearly periods, but the principal contribution (discovered by James Bradley of aberration fame) of amplitude $9''$ and period 18.6 years, is due to the regression of the nodes of the Moon's orbit (see Chapter 6).

A final phenomenon which we should mention is the *variation in latitude.* The axis of the Earth's shape (or figure) is not aligned with the rotation axis, but orbits it erratically at a distance of about 15 meters at the poles. The primary

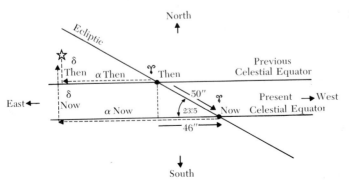

Figure 4–19. *Precession of the Equinoxes.* The Earth's luni-solar precession slides the celestial equator westward along the ecliptic by $50''$ per year. Therefore, the vernal equinox moves, and the right ascension (α) and declination (δ) of stars constantly change with time.

part of this "precession" is known as the *Chandler wobble* with a period of 13 to 14 months; it is probably due to major earthquakes exciting the elastic Earth out of alignment with its rotation axis. Annual perturbations associated with seasonal redistributions of snow, air, and water are also detectable. This motion of the figure axis (a) causes changes in the Earth's rotation rate by about 0.0025 seconds in a year, and (b) implies that the latitude of points on the Earth's surface may change by as much as about $0\rlap{.}''5$.

(d)   The Roche and Instability Limits

In Chapter 3 we discussed the relative orbits of two gravitating (rigid) spherical masses, and obtained Kepler's laws. Here let us consider the effects of differential gravitational forces upon nonrigid bodies and upon systems of more than two bodies. We shall find in general that a satellite cannot approach its primary planet too closely (the *Roche limit*) or stray too far from the planet (the *instability limit*) without dire consequences.

Figure 4–20A shows a spherical satellite of mass m and radius r orbiting its very massive $(M \gg m)$ primary of mass M and radius R at a distance d. If the satellite is large enough $(r \gtrsim 500$ km), its self-gravitation dominates all cohesive forces, and determines its shape and strength. In 1850 Edouard Roche (1820–1883 A.D.) demonstrated that such a satellite would be torn asunder by tidal forces if it approached the primary closer than

$$d = 2.4554(\rho_M/\rho_m)^{1/3} R \qquad (4\text{–}8)$$

where ρ_M is the average *density* (units = kg m^{-3}) of the primary and ρ_m is the average density of the satellite—this is the *Roche limit*. For example, if our Moon were to come closer than $d \approx 2.9 R_\oplus \approx 18{,}500$ km from the center of the Earth, it would disrupt into small fragments.

Equation (4–8) was derived for a fluid satellite which assumed the shape of a *prolate* spheroid (i.e., a "football") in response to the primary's tidal forces. Let us use Figure 4–20A for a *rigid* spherical satellite to *approximately* derive this result. Since the satellite's centripetal acceleration $\omega^2 d$ is produced by the gravitational attraction of primary and satellite GM/d^2, the angular speed of the satellite about the massive primary is (Kepler's third law):

Figure 4–20. *Roche and Instability Limits.* **A**, When the differential gravitation A and differential centripetal acceleration B between points 1 and 2 exceed the satellite's self-gravitation, the tidal forces from the primary disrupt the satellite—the Roche limit. **B**, The satellite m escapes its primary M_1, as a result of differential perturbations $a = (A - B)$ from M_2, when d exceeds the instability limit.

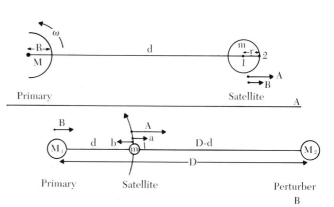

$$\omega = (GM/d^3)^{1/2}$$

The differential gravitational acceleration between the center of the satellite (point 1) and the outer edge (point 2) due to the primary [see section 4–4(a)], is

$$A = (GM/d^2) - GM/(d + r)^2 \approx 2GMr/d^3$$

while the differential centripetal acceleration between these two points is

$$B = \omega^2(d + r) - \omega^2 d = \omega^2 r = GMr/d^3$$

The combination $A + B = 3GMr/d^3$ must be balanced by the satellite's self-gravitational acceleration, Gm/r^2, if the satellite is not to be pulled apart, so that disruption occurs at

$$d = r(3M/m)^{1/3} \tag{4-9}$$

Since average density is defined as total mass divided by volume, we have $\rho_M = 3M/4\pi R^3$ and $\rho_m = 3m/4\pi r^3$, and substituting into equation (4–9) yields the desired result:

$$d = R(3\rho_M/\rho_m)^{1/3} \approx 1.44(\rho_M/\rho_m)^{1/3}R \tag{4-10}$$

The slight numerical difference between equations (4–8) and (4–10) is due entirely to our assumption of a rigid (nonfluid) satellite.

All natural satellites in the Solar System orbit beyond the Roche limit of their primaries, though some (such as Saturn's Janus with $d = 2.60R_h$) are close to it. Saturn's beautiful system of rings lies about 80,000 to 136,000 km from the center of the planet, and so is *entirely within* the Roche limit of about 150,000 km; no single satellite can exist or could have been formed there without disrupting. The fact that these rings are composed of multitudes of solid particles implies that the *cohesive strength* of these particles is greater than the disrupting tidal forces. Artificial satellites certainly do orbit within the Earth's Roche limit, but they are held together by the tensile strength of their materials. For example, a solid sphere of steel one meter in diameter can approach within 100 meters of a *point* of mass $M_\oplus$ before the tidal forces overwhelm its internal tensile strength!

As a body orbits farther and farther from its primary, *differential perturbations* from other bodies become more important (see Figure 4–20B). Beyond the *instability limit*, the body escapes from its primary. In the figure we see that the perturber M_2 produces the differential acceleration

$$a = A - B = [GM_2/(D - d)^2] - GM_2/D^2 \approx 2GM_2d/D^3$$

between the orbiting body and its primary (when $d \ll D$). When this acceleration equals the gravitational acceleration from the primary $b = GM_1/d^2$, the orbiting body is at the instability limit:

$$\boxed{d = (M_1/2M_2)^{1/3}D} \tag{4-11}$$

Note that equation (4–11) is valid only when $M_1 \ll M_2$; when $M_1 \gtrsim M_2$, we must use the more exact relation

$$d^3(2D - d) = (M_1/M_2)D^2(D - d)^2 \tag{4–12}$$

to solve for d in terms of D. From equation (4–11), the instability limit for our Moon—with the Sun as perturber—is 1,700,000 km; this is four times the Moon's present distance from the Earth, so our Moon is stable against escape. In Chapter 5 we shall see that comets may escape from the Solar System as a result of perturbations from other stars $(M_1 \approx M_2)$, when their aphelia lie at distances greater than about 10^5 AU [since equation (4–12) implies $d \approx D/2$ and $D \approx 2 \times 10^5$ AU].

Problems

4–1. (a) How much does a sidereal clock gain (or lose) on a mean solar clock in five mean solar hours?
(b) What is the approximate sidereal time when it is noon apparent solar time on the following days: (i) the first day of spring, (ii) the first day of summer, (iii) April 21, and (iv) January 2.

4–2. The right ascension of the star Sirius is $\alpha = 6^h43^m$, and its declination is $\delta = -16°39'$.
(a) What is the local sidereal time when Sirius is at an hour angle of 2^h17^m (west)?
(b) Explain how to determine the sidereal time in Greenwich, England, at this instant, and give a numerical example.

4–3. In terms of the horizon system (azimuth and altitude), describe the Sun's daily path across the sky during every season of the year at:
(a) the equator,
(b) latitude 38° N, and
(c) the north pole.
Use such descriptive terms as noon altitude, sunrise azimuth, sunset azimuth, and angle at which the Sun meets the horizon.

4–4. Cape Kennedy is at longitude 80°23' W and latitude 28°30' N. A rocket is launched from there due south, and it impacts on the equator ten minutes later. What is the longitude of impact?

4–5. In discussing the Coriolis effect in section 4–2(a), we mentioned that a body which suffers the constant acceleration a will travel a distance $s = at^2/2$ in a time t. Show that the speed of the body is proportional to time, and that the body's acceleration is indeed a.

4–6. The Earth's *rotation* also produces an aberration of starlight.
(a) What is the maximum value of this daily aberration?
(b) Where on the Earth is this effect maximum?
(c) For what stars (location on celestial sphere) is the effect maximum?

4–7. (a) If the apparent centrifugal acceleration due to the Earth's rotation is $\omega^2 R_\oplus$ at the equator, and $\omega = 2\pi/P$, where P is one day, then by what per cent is a man's weight reduced as he walks from the north pole to the equator? Ignore the Earth's oblateness.
(b) Let us now ignore the Earth's rotation, and ask by what per cent a man's weight increases as he walks from equator to pole on the oblate Earth?

(c) Compare your results from (a) and (b) above, and combine them to deduce how the *effective g* varies from the Earth's poles to the equator.

4–8.   The Earth's orbital speed is approximately 30 km sec^{-1}. A star emits a spectral line at wavelength $\lambda_e = 5173$ Ångstroms (1 Å $= 10^{-10}$ meter). Over what amplitude does this wavelength oscillate as the Earth orbits the Sun when the star is located at celestial latitude $\beta = 0°$ (the ecliptic)?

4–9.   In section 4–4(a) we computed the Moon's differential tidal forces at the Earth's surface, ignoring the motion of the Earth-Moon system about its center of mass every sidereal month. Working in a coordinate system centered upon this barycenter and rotating eastward with the angular speed ω (due to this sidereal monthly motion), include the apparent centrifugal acceleration at the Earth's surface to deduce the *correct* dependence of the total tidal acceleration at the Earth's surface. Ignore the daily rotation of the Earth. (Hint: the barycenter is located *within* the Earth.)

4–10.   On a large piece of graph paper, plot *to scale* the distances of
(a) the Roche limit,
(b) the instability limit, and
(c) the orbits of the planetary satellites and the rings of Saturn for the following planets: (i) the Earth, (ii) Mars, (iii) Jupiter, (iv) Saturn (the rings go here), and (v) Uranus.

Assume that $\rho_M = \rho_m$ in every case. Write a brief summarizing statement about your results.

Reading List

(See also the Reading Lists of Chapters 2 and 3.)

Allen, C. W.: *Astrophysical Quantities*. Second Ed. London, The Athlone Press, 1963.

The American Ephemeris and Nautical Almanac. Superintendent of Documents, U.S. Government Printing Office, Washington, D.C. 20402. Published yearly.

Blanco, V. M., and McCuskey, S. W.: *Basic Physics of the Solar System*. Reading, Massachusetts, Addison-Wesley Publishing Company, 1961.

Caputo, Michele: *The Gravity Field of the Earth*. New York, Academic Press, 1967.

Darwin, Sir George Howard: *The Tides*. Cambridge, The University Press, 1898.

————: *Scientific Papers*. Vols. I–III. Cambridge, The University Press, 1907–1910.

Fedorov, Eugenii P.: *Nutation and Forced Motion of the Earth's Pole*. New York, The Macmillan Company, 1963.

Halliday, David, and Resnick, Robert: *Physics*. Combined Ed. New York, John Wiley & Sons, 1966.

Jeans, Sir James H.: *Astronomy and Cosmogony*. New York, Dover Publications, 1961.

Kopal, Zdeněk (ed.): *Physics and Astronomy of the Moon*. New York, Academic Press, 1962.

Kuiper, Gerard P. (ed.): *The Earth As A Planet*. Chicago, The University of Chicago Press, 1954. (Especially Chapters 1 and 2.)

Olcott, William Tyler, and Putnam, Edmund W.: *Field Book of the Skies*. Third Ed. Rev. New York, G. P. Putnam's Sons, 1936.

Page, Thornton, and Page, Lou Williams (eds.): *Wanderers in the Sky*. New York, The Macmillan Company, 1965.

Stumpff, Karl: *Planet Earth*. Ann Arbor, The University of Michigan Press, 1959.

Wyatt, Stanley P.: *Principles of Astronomy*. Second Ed. Boston, Allyn and Bacon, Inc., 1971.

Chapter 5
The Solar System
in Perspective

Our Earth is but one member of the family of the Sun—the *Solar System*. In attendance to our star are nine planets, their thirty-two natural satellites or moons, multitudes of asteroids, an unknown number of comets, and an abundance of meteoroids. Each class of objects has distinguishing characteristics which we will elucidate in this chapter, systematic behavior which requires explanation in terms of the origin of the Solar System, and individual idiosyncrasies which will concern us in Chapters 6 and 7. Here we will describe and correlate what has been observed in the Solar System; the theories advanced to account for the origin of these features are set forth at the end of the chapter.

5–1 CONTENTS OF THE SOLAR SYSTEM

(a) Planets

Today we place the Sun at the center of the heliocentric system of nine *planets* (in order from the Sun): Mercury, Venus, Earth, Mars, Jupiter, Saturn, Uranus, Neptune, and Pluto. These most massive children of the Sun comprise a total mass of but $0.0014 M_\odot$, and they shine only by reflected sunlight. Figure 5–1 displays the relative sizes of the Sun, the planets, and the largest moons. We will see that our Solar System is essentially the Sun, in the midst of a cold vast emptiness wherein swarm interesting pebbles, e.g., the planets.

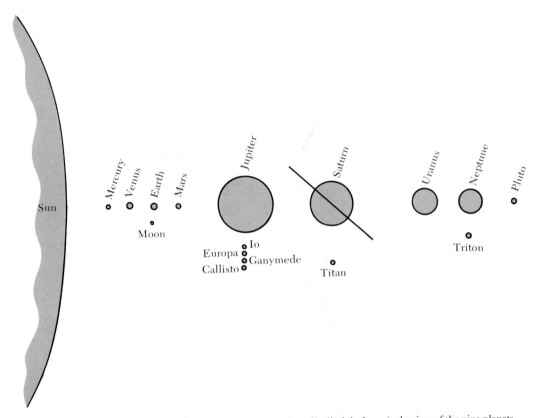

Figure 5–1.   *Planetary Sizes.* Relative to the Sun (only a portion of its limb is shown), the sizes of the nine planets, Saturn's rings, and the seven largest moons are shown. All other members of the Solar System are of insignificant size and mass.

(i)   *Motions*

In Table A2–2 we show the principal *orbital* data for the nine planets. The planets obey the laws of Kepler and Newton as they move in elliptical orbits about the Sun, but the orbital semi-major axes are otherwise arbitrary. In 1766 (before Uranus, Neptune, and Pluto were discovered) Titius of Wittenberg found an approximate empirical rule relating mean Sun-planet distances; Johann Bode publicized this relation in 1772, and it is now called *Bode's Law* or the *Titius-Bode Rule*. The prescription is to write down the series of numbers

$$4, \quad 4+(3 \times 2^0), \quad 4+(3 \times 2^1), \quad 4+(3 \times 2^2), \quad 4+(3 \times 2^3), \dots,$$

divide each result by 10, and arrive at the sequence

$$0.4, 0.7, 1.0, 1.6, 2.8, 5.2, 10.0, 19.6, 38.8, 77.2, \dots$$

With the Sun-Earth distance (1 AU) as the unit of length, Table A2–2 gives the mean Sun-planet distances as:

Mercury	Venus	Earth	Mars	—	Jupiter	Saturn	Uranus	Neptune	Pluto
0.39	0.72	1.00	1.52		5.20	9.54	19.2	30.1	39.5

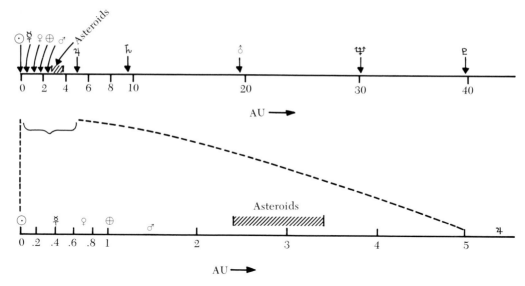

Figure 5–2.   *Relative Orbital Distances.* Here distances from the Sun (☉) are given in AU (1 AU = 1.5 × 10⁸ km) for Mercury (☿), Venus (♀), Earth (⊕), Mars (♂), the asteroid belt, Jupiter (♃), Saturn (♄), Uranus (♅), Neptune (♆), and Pluto (♇). The bottom scale expands the inner planet region eight times, and on this scale the Sun is as large as the dot in the symbol (☉).

Except for the "gap" at 2.8 AU, Bode's law works surprisingly well for the inner seven planets—though numerological, it is a regularity to be remembered. Figure 5–2 illustrates the relative distances of the planets from the Sun [and includes the asteroid belt; see section 5–1(c)]. Note the close spacing from the Sun to Mars, and the more open and regular distribution from Jupiter outward; even on the expanded lower scale our Sun looks no larger than the period at the end of this sentence.

The orbital data in Table A2–2 reveal three striking features of planetary motion. First, all planets orbit the Sun counterclockwise as seen from the north ecliptic pole—we say that the orbits are *direct* or *prograde*. Second, the orbital planes lie very close to the ecliptic plane, so that all planets (except Pluto with an inclination of 17°) are always found in the 16°-wide band of the *zodiac*. And third, the orbital eccentricities are less than 1/10, except for the closest planet to the Sun, Mercury, and the most distant, Pluto. As a consequence of its large eccentricity ($e = 0.249$), the orbit of Pluto ranges from 29.68 AU at perihelion to 49.36 AU at aphelion; Figure 5–3 shows that Pluto passes closer to the Sun than Neptune on occasion. The conjecture [see section 5–3(b)] that *Pluto is an escaped satellite of Neptune* is made plausible by this observation.

The known rotational data for the planets and our Moon (see Chapter 6) are also presented in Table A2–2. The sidereal rotation period refers to a planet's rotation with respect to the stars; these periods are determined in a variety of ways. Mercury and Venus were assumed to be in synchronous rotation about the Sun (i.e., their sidereal rotation periods were believed to be *equal* to their sidereal orbital periods) until the mid-1960s, when Doppler radar measurements (see Chapter 7) unambiguously fixed their present periods. For the Earth, the Moon, and Mars, there are clear and abundant surface markings which may be followed to deduce the rotation rates. The giant planets Jupiter, Saturn, Uranus, and Neptune show

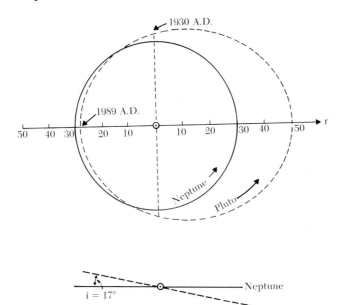

Figure 5–3. *Pluto's Orbit.* At the top, the distance scale is marked in AU. Neptune's orbit is essentially circular at 30 AU, while Pluto moves from 29.7 AU at perihelion (*inside Neptune's orbit*) to 49.4 AU at aphelion. The lower diagram shows Neptune's orbital plane within 2° of the ecliptic plane, while Pluto moves far above and below the plane at an inclination of 17°. The dashed line in the top diagram indicates where Pluto's orbital plane intersects the ecliptic plane in the line of nodes.

us only their thick gaseous atmospheres, so that their rotation is found by following atmospheric mottlings or spots or by observing the Doppler-shifted radiation from their limbs. The atmospheres of these big planets rotate slightly faster at their equators than at their poles; hence, no unambiguous rotation rate exists and only approximate averages can be given. Finally, Pluto appears as a point of light even in the largest telescopes, but since that planet's surface is not perfectly spherical and uniform, the amount of sunlight it reflects varies as it rotates. By observing these fluctuations, we infer Pluto's rotation period.

A few generalizations concerning the rotation periods are in order. In Chapter 4 we said that tidal friction had locked our Moon into synchronous rotation; we will indicate in this chapter how *solar* tidal friction might have produced the rotation rates of Mercury and Venus via spin-orbit coupling. The similarity of the periods of the Earth and Mars is only a coincidence, considering the great difference in their masses and the tidal evolution of the Earth-Moon system. Note that the large gaseous planets (Jupiter, Saturn, Uranus, and Neptune) spin much faster than the small solid planets (Mercury, Venus, Earth, Mars, and Pluto). It has been speculated that at the beginning of the Solar System all planets had a *primordial* rotation period of about 10 hours, and that the planets closest to the Sun slowed their rotation on account of the solar tidal friction and the loss of their huge proto-atmospheres. The slow rotation of Pluto is easier to understand if that planet escaped from a satellite orbit about Neptune [see section 5–3(b)]; this hypothesis can also (in theory) account for Neptune's relatively slow rotation.

In Chapter 4 the oblate spheroidal shape of the Earth's surface was taken as an indication of its rotation; there we defined a planet's *oblateness* ϵ by the following equation:

$$\epsilon = (r_e - r_p)/r_e$$

where r_e is the equatorial radius and r_p the polar radius. Note that a perfect sphere has $\epsilon = 0$. If all planets adjust to an equilibrium fluid shape, we expect that ϵ will increase as the rotation rate increases; the trend in Table A2–2 corroborates this idea. Though no observations are available for Mercury, Venus, and Pluto, their slow rotations imply oblatenesses near zero. We have already seen that accurate measurements of the Earth yield an oblateness $\epsilon_\oplus = 1/298.3$. For the other planets, oblateness is determined by measuring the visible disk of the planet or by analyzing oblateness perturbations on the orbits of the planet's moons. The large gaseous planets exhibit enormous oblatenesses, so large that Jupiter and Saturn appear distinctly oval in large telescopes.

The equators of the planets are inclined to their orbital planes by varying amounts. The rotation axes of Mercury, the Moon, and Jupiter are nearly aligned with their revolution axes, while those of the Earth, Mars, Saturn, and Neptune are tilted through about 25°. The *equatorial inclination* is less than 90° for these bodies, so that they all rotate from west to east or *direct* (in the same sense that all the planets orbit the Sun). As shown in Figure 5–4A, Venus and Uranus rotate from east to west or *retrograde*. The rotation axis of Uranus lies essentially in its orbital plane, so that if Uranus presents its pole toward us now, in 21 years we will lie in its equatorial plane (see Figure 5–4B). This strange phenomenon is still unexplained.

The main features of planetary rotation are intricately tied in with the origin of our Solar System (see section 5–3, Cosmogony). But let us be naive and ask how *solar tidal forces* might affect rotation rates through the eons. In general, the Sun's tidal force goes as $M_\odot R_p/a^3$, where $M_\odot$ is the mass of the Sun, R_p is the radius of the planet, and a is the planet's orbital semi-major axis. If we take $R_\oplus = 1$ and $a_\oplus = 1$ (AU), we find the following relative spin-down torques:

Mercury	Venus	Earth	Mars	Jupiter	Saturn	Uranus	Neptune
2.4	2.4	1.0	0.08	0.9	0.1	0.002	0.0004

These very crude estimates indicate that solar tidal friction is most important for Mercury and Venus; if Jupiter's rotation rate is primordial, then this effect is clearly *not* the reason for slow rotation.

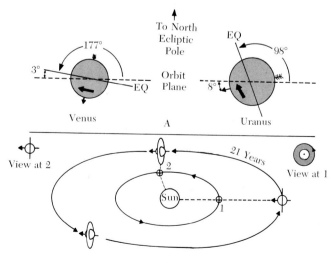

Figure 5–4. *Equatorial Inclinations.*
A, Venus and Uranus rotate retrograde since their equator-orbit inclinations are greater than 90°. Each equator is indicated by EQ.
B, Uranus' rotation axis and orbital plane are nearly aligned. Hence, at 1 we see the planet pole-on, and 21 years later, at 2, we see its equatorial plane (and satellite orbits) edge-on.

We will return to the question of planetary spins in section 5–3(b), but let us now consider Peter Goldreich's recent controversial theory of *spin-orbit coupling and resonance* as it applies to Mercury and Venus. Just as we found the relation between the sidereal and synodic orbital periods of planets in Chapter 3, so also can we obtain the synodic *rotation* period of a planet. The synodic rotation period of Venus (its rotation period with respect to the Earth) is 146 days, which is exactly one-quarter of Venus' synodic orbital period. From Table A2–2 we see that the sidereal rotation period of Mercury is exactly two-thirds its sidereal orbital period. Hence, Venus appears to be in a 4:1 synodic resonance with the Earth, while Mercury seems to exhibit a 3:2 sidereal lock on the Sun. In analogy with the slight nonequilibrium *ellipticity* of our Moon (Chapter 6), Goldreich assumes that Mercury and Venus maintain small elliptical deformations of their mass. Such non-equilibrium shapes might arise during the formation of these planets, or might be supported by internal stresses. Solar tidal forces interact with these deformations, slowing each planet's direct rotation until a resonance is achieved. Mercury's highly eccentric orbit resulted in the Sun-Mercury resonance illustrated in Figure 5–5A. Mercury points alternate ends of its long axis toward the Sun at each perihelion, rotating three times in two sidereal orbital periods. Apparently, tidal dissipation was not great enough to carry Mercury through the 3:2 resonance to synchronous rotation. Venus seems to have slightly overshot the point of synchronous rotation, so that it began to rotate retrograde until caught in the 4:1 synodic resonance with the Earth (see Figure 5–5B). This planet always points the *same* face toward the Earth at inferior conjunction, rotating four times during each orbit with respect to the Earth. Remember that while Goldreich paints an attractive picture of the rotations of the inferior planets, his theory is not universally accepted and may even be wrong.

(ii)   *Interiors*

Sundry *physical* data for the planets are tabulated in Table A2–3. On the basis of these characteristics, the planets naturally divide into two groups: (1) the small solid *terrestrial* planets (Mercury, Venus, Earth, Moon, Mars, and Pluto) with masses no greater than $M_\oplus$, and (2) the large gaseous *Jovian* planets (Jupiter, Saturn, Uranus, and Neptune) which range in mass from 15 to $318M_\oplus$. Figure 5–1 emphasizes this dichotomy. It is sometimes advantageous to consider the larger planetary satellites in the class of terrestrial planets.

Planetary *masses* are determined (a) by applying Kepler's third law to satellite orbits, and, when there are no natural satellites, (b) by observing a planet's gravitational perturbations on the orbits of other planets, asteroids, comets, and artificial space probes. Planetary *radii* are deduced (a) by measuring the apparent optical size of the planetary disk, (b) by accurately timing the *occultations* (when the planet passes in front of an object) of stars, the planet's moons, and space probes, and, for the closer planets, (c) by precisely timing radar pulses reflected from various points on the planet's surface. By dividing a planet's mass (M) by its total volume ($4\pi R^3/3$), we may define its *average density* ($\bar{\rho}$; units = gm cm^{-3})

$$\bar{\rho} = M/(4\pi R^3/3)$$

Figure 5–5. *Spin-Orbit Resonance.* **A**, Mercury rotates direct in its eccentric orbit. The planet's long axis points toward the Sun at perihelion (P) and at a right angle to the Sun at aphelion (A). One-and-one-half sidereal rotations occur per sidereal revolution. **B**, In its circular synodic orbit (note that the Earth is motionless in the diagram), Venus rotates four times, so that the same face of the planet points toward the Earth ($\oplus$) at inferior conjunctions.

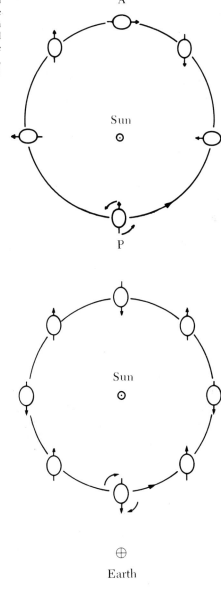

The unit gm cm^{-3} is convenient, since the density of pure water is 1 gm cm^{-3}. The high densities of the terrestrial planets, in the range 3.4 to 5.5, are consistent with their compositions of heavy, nonvolatile elements such as iron, silicon, and magnesium. The very low densities of the Jovian planets (note that Saturn could *float* on water) imply a composition similar to that of our Sun, with hydrogen and helium predominating.

The internal structure of a planet depends upon the distribution of its chemical composition, density, temperature, and pressure. In every case, the central pressures must support the weight of the overlying layers of matter. Density and pressure increase as we approach the center of a planet, and the temperature also rises as a result of greater pressures and the heat released by decaying radioactive elements. Different materials are stable under different conditions, so that the chemical composition is *stratified*. The smaller terrestrial planets (Mercury, Moon, Mars, and Pluto) are probably not massive enough to cause their interiors to melt, so their compositions should be fairly homogeneous throughout. The Earth is known to have a molten core above which float the lighter silicates (e.g., basalt; see Chapter 6), and we expect Venus to exhibit a similar interior. The Jovian planets have thick atmospheres of hydrogen and helium, with a small admixture of ammonia (NH_3) and methane (CH_4); nearer their centers is probably a liquid slush, which might become a *solid* hydrogen core at the centers of Jupiter and Saturn.

Using the observed average density, composition, and oblateness of a planet, we may construct physically consistent *models* of its interior (just as we do for stars in Chapter 15). Usually a range of models accommodates our data, so that only improved theory and observation can lead us to a unique picture. Since we cannot probe planetary interiors directly, we must proceed by inference and induction.

(iii)   *Surfaces*

The individual peculiarities of each planet's surface are described in Chapters 6 and 7; here we enumerate the general properties. While the Jovian planets and Venus show us only their cloudy atmospheres, the terrestrial planets and larger moons reveal fixed surface markings. These markings are best seen visually at a telescope or on telescopic motion pictures, so that the brief periods when the Earth's atmosphere is stable—the times of *good seeing*—may be used to advantage. The most important general data on planetary surfaces include color, albedo, phase effect, polarization, and temperature.

A planet's *color* tells us something about its surface composition. The oceans and continents give the Earth a blue color mottled with green, brown, and orange; large areas of cloud or snow cover appear white. The basaltic surface of the Moon looks dark gray with some tan, while the deserts of Mars give its characteristic brown-orange color. By spreading the light from a planet into a spectrum (see Chapter 8), we may deduce the presence of atoms and molecules similar to water on that planet.

The *albedo* of an object is the fraction of the incident sunlight reflected by the object. For planets with little or no atmosphere (Mercury, Moon, Mars, and Pluto), the albedo is usually very low, since rocks are inefficient reflectors. The high reflectivity of clouds leads to the high albedos of the Jovian planets and Venus. The Earth's albedo is variable, since it depends upon the season and upon snow and cloud cover.

The *phase effect* refers to the way in which a planet's brightness varies with its geocentric phase (and elongation; see Chapters 2 and 3). Conventionally, $0°$ phase occurs when the hemisphere facing the Earth is fully sunlit (at opposition for superior planets, and at superior conjunction for all planets), while $180°$ phase is when the facing hemisphere is completely dark (only at inferior conjunction). Inferior planets and our Moon exhibit all possible phases, while superior planets show only a small phase range (e.g., Mars ranges from full at $0°$ phase to gibbous at $47°$ phase). Mercury exhibits essentially the same phase effect shown for the

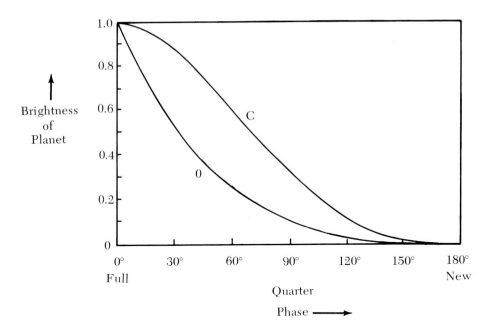

Figure 5–6. *Lunar Phase Effect.* As the Moon moves from new phase (180°) to full phase (0°), its observed brightness (O) increases less rapidly than that calculated (C) for a smooth sphere. Shadows due to surface granularity cause this effect, and lead to the "almost-spike" at 0° phase when no shadows are cast. Mercury shows a very similar phase effect.

Moon in Figure 5–6; here O is the observed effect, while C is the theoretical curve for a *smooth* surface. The granularity of the Moon's surface leads to shadows which disappear at the "almost-spike" at 0° phase. The thin atmosphere of Mars, and the dense scattering atmosphere (of molecules or small particles) of Venus, are noticeable in the phase effects of these planets.

In Chapter 8 we discuss electromagnetic waves and *polarization* in detail. Here we need only remark that the unpolarized light from the Sun becomes partially polarized when it is reflected, scattered, or absorbed by a medium. Since polarization depends upon the medium, and upon the angles of the incident sunlight and the observed light, we may deduce some of the properties of planetary surfaces by observing the polarization of planetary light as a function of geocentric phase. We detect polarization by measuring the *intensity* (I) of light transmitted through a Polaroid filter; such a filter passes a maximum intensity $(I_{\parallel})$ in one orientation, and a minimum intensity $(I_{\perp})$ when rotated through 90°. We define the fractional *polarization* of light by

$$FP = (I_{\parallel} - I_{\perp})/(I_{\parallel} + I_{\perp})$$

For completely unpolarized light we have $(I_{\parallel} = I_{\perp}$ and $FP = 0$, while for *linearly* polarized light $I_{\perp} = 0$ and $FP = 1$. Again, the Moon and Mercury give similar polarization-phase data, indicating that both have rough powdery surfaces (like sand or ash). Polarization measurements of Venus and Jupiter reveal thin atmospheres mixed with or overlying their dense cloud layers.

An important characteristic of planetary surfaces is *surface temperature.*

Let us now examine how surface temperatures are observed and computed. In Chapter 9 we will see that our Sun radiates energy at an effective surface temperature $T_\odot = 5800$ K (Kelvin or absolute temperature; see Appendix 5), supplying light and heat energy to the rest of the Solar System. The Sun and the planets behave much like *blackbody radiators* (see Chapter 8), which absorb essentially all radiation incident upon them and reradiate a like amount. But the planets (and their satellites) intercept only a small fraction of the energy radiated by the Sun, so their surface temperatures are very low compared to $T_\odot$. A blackbody radiates most of its energy at wavelengths near

$$\lambda_{\max} = (0.0029 \text{ meters})/T = (2898 \ \mu)/T = (2.9 \times 10^7 \ \text{Å})/T \qquad (5\text{–}1)$$

where T is the absolute temperature of the blackbody; one micron (μ) equals 10^{-6} m and one Ångstrom (Å) equals 10^{-10} m. Equation (5–1) is *Wien's law*, which tells us that solar radiation peaks at 5000 Å (*green* light); at normal room temperature, $T \approx 290$ K ($\approx 70°$ Fahrenheit), the peak is at the longer wavelength $\lambda_{\max} = 100,000$ Å $= 10 \ \mu$. Figure 5–7 shows the wavelength distribution of radiation from the Sun and from a planet like the Earth (see the Planck distribution, Chapter 8). Hence, we may determine a planet's surface temperature by observing the planet's radiation (*not* reflected sunlight) in the micron infrared and at centimeter wavelengths.

Radiation of different wavelengths comes from different parts of a planet's surface. For example, centimeter waves originate from the solid surface of Venus, millimeter waves can escape only from its lower atmosphere, and micron waves come from its upper atmosphere; the thick atmosphere of Venus is completely opaque to infrared radiation ($\lesssim 1 \ \mu$). Therefore, we can probe the temperature and composition at varying levels of a planet's surface and atmosphere. Since the Moon has no atmosphere, infrared rays come directly from its visible surface, while longer wavelength radiation originates several centimeters beneath the surface. Using this phenomenon, we may determine the *thermal conductivity* of the Moon's surface materials. High thermal conductivity implies rapid response to heating, so that the maximum surface temperature occurs at 0° phase; low thermal conductivity leads to slow heating and cooling, and introduces a *phase lag* at longer wave-

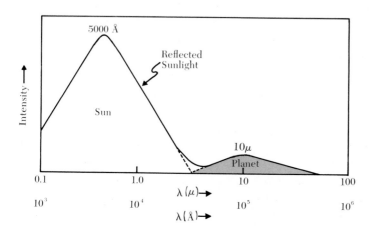

Figure 5–7. *Blackbody Radiation: Sun and Planet.* At the Sun's surface temperature of 5800 K, its radiation distribution peaks at 5000 Å and spreads from 0.1 to 3.0 μ. The shaded portion represents a planet at 290 K, with the peak at 10 μ, and the spread from about 2 to 20 μ.

Figure 5–8. *Lunar Heating and Phase Lag.* Visible and geocentric lunar phases are shown, with the temperature variations plotted at three wavelengths. The response of the lunar surface is fast at short wavelengths, slow with a phase lag at 1 cm, and practically nil at long wavelengths ($\gtrsim 10$ cm).

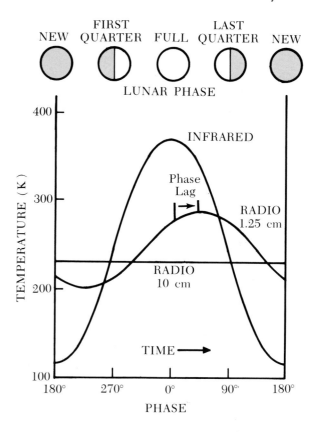

lengths. Figure 5–8 illustrates these points for our Moon. During lunar eclipses, the surface infrared temperatures drop precipitously, while the subsurface radiowave temperatures vary only slightly.

If a planet radiates more energy per second than it receives from the Sun, that planet must be generating heat internally. Delicate measurements at the Earth's surface reveal heat flowing from the hot interior. The only other planets *known* to produce excess heat are Jupiter and Saturn. In theory, a planetary contraction of about *one centimeter per year* converts enough gravitational potential energy into heat to account for these observations.

Let us now show how the theoretical blackbody temperatures tabulated in Table A2–3 are found. The basic idea is to find the temperature at which a small blackbody (e.g., a planet) must radiate to balance the energy input from the Sun. We will need only one new concept, *Stefan's law* of blackbody radiation from Chapter 8. This law relates the *energy flux E* (energy radiated per unit area per unit time or ergs cm^{-2} sec^{-1}) to the temperature T of a blackbody:

$$E = \sigma T^4 \qquad (5-2)$$

where the proportionality constant is $\sigma = 5.67 \times 10^{-5}$ ergs cm^{-2} sec^{-1} (K)$^{-4}$. Since the area of a spherical surface is $4\pi R^2$, our Sun radiates energy at the rate ($T_\odot = 5800$ K):

$$4\pi R_\odot{}^2 E_\odot = 4\pi R_\odot{}^2 \sigma T_\odot{}^4 = 3.8 \times 10^{33} \text{ ergs sec}^{-1}$$

This same energy flows through a sphere of area $4\pi r_p{}^2$ at the Sun-planet distance r_p, so that the energy flux there is

$$E_p = 4\pi R_\odot{}^2 E_\odot / 4\pi r_p{}^2 = (R_\odot/r_p)^2 E_\odot$$

If a square meter of blackbody material intercepts this energy head-on, it will clearly be raised to the *subsolar temperature*

$$T_{ss} = (R_\odot/r_p)^{1/2} T_\odot \approx 395(r_p)^{-1/2} \text{ K} \tag{5-3}$$

since $E_p = \sigma T_{ss}{}^4$; in the last equality we have inserted the appropriate solar values and expressed r_p in AU. The subsolar temperatures in the last column of Table A2–3 follow from equation (5–3).

While subsolar temperatures apply to the subsolar point (where the Sun is at the zenith) on the surfaces of very slowly rotating planets (Mercury, Moon, and Pluto), they are not appropriate for planets with atmospheres or for planets in rapid rotation. For these latter planets, let us assume that the effective absorbing area is $\pi R_p{}^2$ and that the effective radiating area is $4\pi R_p{}^2$, where R_p is the radius of the planet. The albedo (A) is the fraction of incident solar radiation which is reflected, so that only the fraction $(1 - A)$ is absorbed. Therefore, the energy absorbed per second is $(1 - A)\pi R_p{}^2 E_p = (1 - A)\pi R_p{}^2 (R_\odot/r_p)^2 \sigma T_\odot{}^4$, while the planet radiates the power $4\pi R_p{}^2 \sigma T_p{}^4$; the result is:

$$T_p = (1 - A)^{1/4}(R_\odot/2r_p)^{1/2} T_\odot \approx 277(1 - A)^{1/4}(r_p)^{-1/2} \text{ K} \tag{5-4}$$

In equation (5–4), r_p is expressed in AU, and the equilibrium blackbody temperatures of Table A2–3 follow when we let $A = 0$. By including the albedo effect we may more closely approximate the observed planetary temperatures, but we must remember that our crude estimates neglect such important complications as thermal conductivity, the opacity, circulation, and convection of planetary atmospheres, and the heat retention (*greenhouse effect*) of certain types of atmospheres. Nature is much more intricate than our simple models.

(iv)   Atmospheres

Of the terrestrial planets, Mercury, the Moon, and Pluto seem to have no atmospheres; Venus and Mars possess carbon dioxide (CO_2) atmospheres; and the Earth's atmosphere is primarily nitrogen (N_2) and oxygen (O_2). The principal constituents of the enormous atmospheres of the Jovian planets are hydrogen (H) and helium (He). In general, planetary atmospheres are densest near the planet's surface, and thin rapidly with increasing altitude. The composition of an atmosphere may be stratified, with the heaviest gases residing lowest, but turbulent mixing and winds can lead to regions of homogeneous composition. Far from a planet's surface, incoming solar ultraviolet and x-radiation usually ionize atmospheric atoms or dissociate molecules to form the layered *ionosphere* (see Chapter 6).

To gain some understanding of planetary atmospheres, let us once again be naive and consider the simplest and crudest theory—the theory of *the retention of atmospheres*. In Chapter 8 we will study the *thermal equilibrium of a perfect gas*. To a first approximation, an atmosphere behaves like a perfect gas, that is, as a tenuous collection of particles which interact only through elastic collisions. While the speeds of these particles (atoms or molecules) are distributed over a large range, and change violently in each collision, the *average* kinetic energy per particle is

$$\overline{KE} = (m/2)\overline{v^2} = 3kT/2 \tag{5-5}$$

Here m is the particle's mass, T is the *kinetic* absolute temperature of the gas, and the Boltzmann constant is $k = 1.380 \times 10^{-16}$ ergs $(K)^{-1}$. From equation (5-5) we obtain the *root mean square speed* v_{rms}, as

$$v_{rms} = \sqrt{\overline{v^2}} = (3kT/m)^{1/2} \tag{5-6}$$

hence, the "mean" speed of the particles increases with temperature and decreases with mass. In the very thin upper regions of atmospheres, a particle which moves outward with the *escape speed* v_e (see Chapter 3) has an excellent opportunity to leave the atmosphere and the planet:

$$v_e = (2GM/R)^{1/2} \tag{5-7}$$

here M is the planet's mass and R is its radius. If $v_{rms} = v_e$ for a given particle species, that gas will leave the atmosphere in only a few days. To retain an atmosphere for several billion years (approximately the age of the Solar System), a planet must be such that $v_e \gtrsim 10v_{rms}$. Therefore, a given type of molecule is retained indefinitely when [equations (5-6) and (5-7)]

$$\boxed{T \lesssim (GMm/100kR)} \tag{5-8}$$

In Figure 5-9 we plot *points* corresponding to the equilibrium blackbody temperature (Table A2-3) and v_e for the planets and some moons; the *dashed lines* represent $10v_{rms}$ for various molecular species. In terms of this crude theory, a planet retains all gases with lines passing below its point, while the other gases escape. In spite of its simplicity and neglect of all complicating factors, this theory is reasonably consistent with the observations that (a) the Jovian planets have retained all gases, (b) the Earth, Venus, and Mars have lost their hydrogen and helium, but retained nitrogen and carbon dioxide, (c) Mercury and the Moon have no atmospheres, and (d) the largest moons have *some* atmosphere (Titan is known to have methane!).

(b) Moons

There are 32 natural planetary satellites or *moons* in the Solar System; together they comprise a total mass of but $(1/10)M_\oplus$. Our Moon was known in

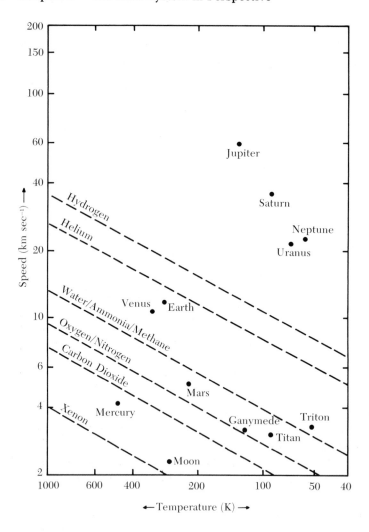

Figure 5–9. *Retention of Gases.* Molecular speeds and escape speeds are plotted as a function of temperature. See the text for interpretation of this diagram.

prehistory, and the four Galilean moons of Jupiter were found by Galileo in 1610 A.D.; the rest were discovered between 1671 and 1966. Here we describe the general features of these bodies, reserving a detailed description for Chapters 6 and 7. Only our Moon has a known sidereal *rotation* period of $27^d.32$; the other moons are too small and too distant for their rotations to be found. In Table A2–4 we list known data for satellites.

Seven satellites are approximately the size of our Moon, while the other 25 are much smaller and resemble large asteroids. Our Moon, the two satellites of Mars, the five inner satellites of Jupiter, the eight innermost of Saturn, and all five of Uranus' moons have nearly circular orbits lying essentially in their planet's equatorial plane; it is reasonably safe to conjecture that these 21 moons exhibit synchronous rotation due to tidal friction. All moons orbit between the Roche and instability limits of their planet: exceptional cases are Phobos, Jupiter V, Janus, and Miranda which are near the Roche limit; and our Moon and Jupiter IX which orbit nearest the instability limit. Note the *retrograde* orbits of Jupiter VII, IX, XI, and XII, Saturn's Phoebe, and Neptune's Triton.

From Table A2–4 we see that only three moons circle the terrestrial planets, while the Jovian planets possess 29. The larger the *mass* of a planet, the greater is its sphere of gravitational influence (i.e., the instability limit). To see if satellite number is correlated with planetary mass, let us arrange the planets in order of decreasing mass (placing the number of satellites in parentheses): Jupiter(12), Saturn(10), Neptune(2), Uranus(5), Earth(1), Venus(0), Mars(2), Mercury(0). Apparently mass *is* an important factor, but Figure 5–10 shows the even better correlation of satellite number with a planet's *distance from the asteroid belt*. It appears that the Jovian satellites with small masses and highly eccentric and inclined orbits (and retrograde motion) are probably *captured asteroids*.

(c) Asteroids

Bode's law "predicted" the existence of a planet at 2.8 AU, between Mars and Jupiter, but it was not until 1801 that Giuseppe Piazzi discovered the *minor planet* Ceres in this region. Though Ceres fit the Titius-Bode relation, its radius was exceedingly small (389 km) and its orbital inclination anomalously large (10°37′). By 1890 more than 300 small planetoids or *asteroids* had been found with semi-major orbital axes between 2.3 and 3.3 AU—this is the *asteroid belt*. Today the orbital elements of more than 2000 asteroids are known, and each such body is numbered in order of orbit determination and given a name, such as asteroid 1000 Piazzia. An estimated 45,000 asteroids should be visible with a 254-cm (100-inch) telescope.

Asteroids are too small to retain atmospheres, for their observed radii range from the largest, Ceres (389 km), Pallas (244 km), and Vesta (200 km), to the more abundant smaller ones (about 1 km); certainly a multitude of small "rocks" (meteoroids?) also swarm in the asteroid belt. The total *mass* in asteroids is probably less than $1/500 M_\oplus$—only enough for a small moon. The largest asteroids tend to reside farthest from the Sun, and the smallest closest; they are all in direct orbit about the Sun. Asteroids exhibit orbital eccentricities up to 0.83 with the average being 0.15, and orbital inclinations as large as 52° but an average of 10°. The belt shows distinctly *depleted* regions, called the *Kirkwood gaps*, at semi-major axes where the orbital period would be a simple fraction (e.g., 1/2, 1/3, 1/4, 2/5, 3/7) of Jupiter's orbital period. It seems clear that periodic gravitational perturbations from Jupiter

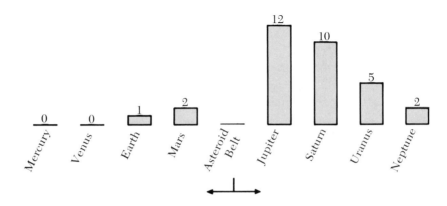

Figure 5–10. *Satellite Numbers.* This diagram shows the suggestive correlation between the number of planetary satellites and the planet's distance from the asteroid belt. Note that the mass of the planet is also an important factor.

have removed all asteroids from these gaps. Where the ratio of periods is 2/3 and 1/1, there are *accumulations* of asteroids, called *groups* or *families;* the most interesting are the 15-member Trojan groups situated at Jupiter's orbit on the vertices of equilateral triangles with the Sun and Jupiter at the other two vertices. The Trojans are bound at these *Lagrangian points,* for they are stable to small perturbations from their equilibrium positions. In addition, the work of K. Hirayama and Dirk Brouwer has led to the identification of 29 *asteroid streams* with members (about one-third of all known asteroids) which seem to be physically related.

Some asteroids, such as Eros, Icarus, and Geographos, have perihelia within the Earth's orbit and pass near the Earth at times. Accurate observations of their orbits are used to determine the Astronomical Unit. Icarus ($a = 1.1$ AU, $e = 0.83$) has been detected by radar, and observations of the precession of its perihelion have confirmed Einstein's general theory of relativity to within 5%.

Just as with Pluto, the rotation periods of asteroids are determined from fluctuations in their reflected sunlight. The observed *light curves* exhibit two maxima and two minima per cycle, corresponding to *oblong* bodies which tumble every three to 20 hours (the average rotation period for asteroids is seven hours). The rotation axes are randomly oriented and no wobbling (nutation) has been detected. Thus, we have the picture of irregular fragments of rock, several kilometers in size, spinning in a few hours and colliding with one another no more frequently than once every million years.

(d)   Comets

We have been discussing the *major* constituents of our Solar System; comets are small and infrequent interlopers. Though faint telescopic comets far outnumber the bright visual comets, the latter are spectacular phenomena which can stretch 30° across the celestial sphere. In the superstitious past, comets were heralded as dire omens in the Earth's atmosphere, until Tycho Brahe showed that they move far beyond the Moon's orbit.

A comet is named after its discoverer (Comet Oterma) or codiscoverers (Comet Ikeya-Seki), and also by year in the order of its discovery (1971a is the first discovered in 1971, 1971b the next, etc.). After an orbit has been computed, the numbering is in the order of perihelion passage (Comet 1971 I was the first to pass perihelion in 1971, Comet 1971 II the second, etc.). Although all comets move in elliptical orbits about the Sun, and are members of the Solar System, there are two distinct classes of comets: (1) the *long-period* comets which are in the great majority and which have orbital eccentricities very close to 1 (almost parabolic), and (2) the much smaller group of *short-period* comets which are periodic in their returns. Most comets reach perihelion around one or two AU from the Sun, and the inclinations of their orbits range through *all* values; hence, comets move in a spherical volume centered on the Sun while the planets move near the ecliptic plane. The aphelia of long-period comets lie in the range from 10^4 to 10^5 AU, so that their orbital periods are one to 30 million years; these comets are only seen during one perihelion passage. In some cases, planetary perturbations make comet orbits hyperbolic ($e > 1$) so that they escape the Solar System, while in other cases the orbits are changed into small ellipses. In this way Jupiter has captured a family of about 45 comets which now orbit the Sun direct in about six years with aphelia nearJupiter's orbit. Short-period comets move around within the planetary system: Comet Encke has the shortest period of 3.3 years ($a \approx 2.2$ AU), while Comet Rigollet has the longest of 151 years ($a \approx 28$ AU). The most famous comet of short period is *Halley's comet* which returns every 76 years (the next appearance is around 1986).

Current *models* of comets are based upon their appearance and their spectra (see Chapter 8). Far from the Sun (at great elongation) we see only a star-like *nucleus* which shines by reflected sunlight. Since these nuclei do not noticeably perturb the orbits of planets or satellites, their masses must be less than 10^{17} kg $\approx 10^{-8} M_{\oplus}$; since they are not seen when passing across the face of the Sun, their radii are smaller than 50 km. Hence, the density of the nucleus is approximately 1 gm cm^{-3}. Such a nucleus might be a tight swarm of ice and dust particles (a "flying gravel bank"), but Fred Whipple's picture of a *dirty iceberg* (dust and stones bound in a matrix of frozen water, carbon dioxide, methane, and ammonia) is the more commonly held opinion. A few AU from the Sun, rising temperatures [see blackbody temperatures, section 5–1(a),iii)] begin to vaporize the ices at the nucleus' surface, producing a tenuous gas halo or *coma* about 10^5 km across. The nucleus-coma combination is called the *head* of the comet. Closer to the Sun, when

Figure 5–11. *Cometary Tails and Structure.* **A**, As the comet passes around the Sun ($\odot$), its tail always points away from the Sun. **B**, The various parts of a comet are indicated; see the text.

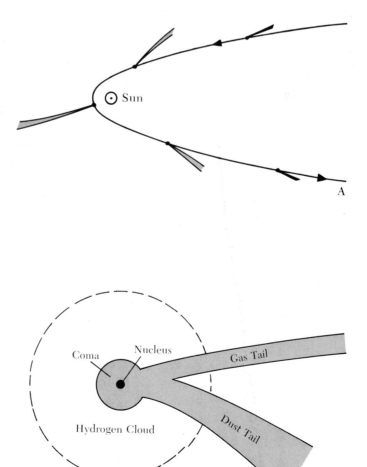

A

Figure 5–12.   *Comet Mrkös.* Dust tail is fainter and more diffuse than gas tail. Note structure of gas tail. **A,** August 22, 1957. **B**, August 26, 1957. (48-inch Schmidt, Hale Observatories)

the comet can be seen only in the twilight sky of sunrise or sunset, several *tails* as long as 1 AU form. These tails always point *away from* the Sun (see Figure 5–11A), and they are of two types: (1) a curved *dust* tail formed of solid particles escaping the nucleus into their own solar orbits, and (2) a straight gas tail pushed back from the comet's head by solar radiation and the solar wind (see Chapter 9). Occasionally, explosive vaporization of the nuclear ices produces secondary tails. Figure 5–11B

B

Figure 5-12B.

illustrates our general picture of a comet's structure, and Figure 5-12 shows the beautiful example of the gas and dust tails of Comet Mrkös.

The thin dust tail of a comet shines by reflected sunlight, and it may be seen as a *counter-tail* pointed toward the Sun when we look along the comet's orbital plane; Figure 5-13 displays this phenomenon for Comet Arend-Roland. In addition to the sunlight reflected from its nucleus and dust, a comet glows by the *fluorescence* of its gases. From the coma we observe the molecular spectral bands emitted by

Figure 5–13.   *Comet Arend-Roland.* Note forward spike. (Lick Observatory)

CN, C_2, CH, NH, NH_2, and OH *radicals* which result from the dissociation of cyanogen (C_2N_2), methane (CH_4), ammonia (NH_3), and water (H_2O) by solar ultraviolet radiation. At very close perihelion passes, the atomic emission lines of Na, Fe, O, Cr, and Ni (characteristic of metallic rocks or dust) are occasionally seen. When a comet is within 1.5 AU of the Sun, the gas tail emits radiation from the *ionized* molecules CO^+, N_2^+, OH^+, CO_2^+, and CH^+. Satellite observations have shown that some comets are surrounded by enormous hydrogen clouds with diameters exceeding that of the Sun. These clouds emit the ultraviolet spectrum line of hydrogen, Lyman alpha (1216 Å), and therefore are not detectable from the Earth's surface.

The observed *brightness B* of a comet depends upon its distance from the Sun *R* (which determines its fluorescence and the amount of sunlight reflected) and its distance from the Earth *r* (which determines the *flux* of radiation which we receive):

$$B \propto R^{-n} r^{-2} \tag{5–9}$$

Far from the Sun, no fluorescence occurs, so $n = 2$; but near the Sun $n \approx 4$, and it may range from 2 to 6 depending upon the particular comet.

With each perihelion passage, a comet loses gas, dust, and rocks as a result of the intense solar heating and tidal forces; eventually the comet vanishes. Some short-period comets have been observed to split into several pieces or even disintegrate. First discovered in 1772, Comet Biela (with a period of 6.6 years) was seen to break in two in 1846, then was observed as a double comet in 1852, and disappeared thereafter. In 1872 a brilliant *meteor shower* [see section 5–1(e)] was seen as the Earth crossed Biela's original orbit; the cometary debris strewn along the orbit had entered the Earth's atmosphere as meteors. The Bielid shower now occurs on November 14 as the Andromedids. In 1965 Comet Ikeya-Seki emerged from the solar corona as two pieces; this bright comet was visible even during the daytime. Since the comet's period is 880 years, it will be centuries before it is known if a meteor stream was formed.

Since comets quickly waste away on a geological time scale, the supply must be constantly replenished if we are to see them today. It is presumed that cometary nuclei were formed with the Solar System about five billion years ago. To account for their continued existence, Jan Oort has hypothesized a spherical *comet cloud* between 3×10^4 and 10^5 AU from the Sun wherein reside about 10^{11} nuclei. At its outer edge (the instability limit), this reservoir loses nuclei on account of perturbations from passing stars, while at the inner edge, planetary (and stellar) perturbations deflect nuclei toward the Sun, so that about one nucleus in 10^5 is close-in at any time. Further perturbations cause the nucleus to either escape from the Sun altogether or enter a periodic orbit where it is slowly disintegrated. Sad to say, few theories of cometary origin exist, and fewer still are even partially convincing.

(e)   Meteoroids

Roaming throughout the Solar System with orbits of all inclinations are the rocks called *meteoroids*. These range in size from small asteroids (10 km) down to micrometeoroids (<1 mm) and interplanetary dust ($\approx 1\ \mu$). The small-size tail of the asteroid distribution resulting from *asteroid collisions* is probably the source of the larger rocks, while the smaller rocks and dust come from *disintegrated comets*. Although we think that we know the source of meteoroids, the origin of their *composition* is still a complete mystery.

When a meteoroid enters the Earth's atmosphere, friction heats it to incandescence and visibility at an altitude of about 120 km—we call this a *meteor* or *shooting star*. By 60 km, most meteoroids are completely consumed; from their brightness we deduce that their average density is 0.2 to 1 gm cm^{-3} (similar to a comet's nuclear material!). A meteoroid loses mass by *ablation* (e.g., vaporization, melting, and fragmentation) and ionizes the air around it, so that we observe the spectral lines characterizing ionized air molecules and such common elements as H, N, O, Mg, Ca, and Fe. Some meteors are bright enough to cast shadows—these are called *fireballs*.

The trajectories and speeds of meteors are found by employing wide-angle cameras such as the Baker Super-Schmidt (Figure 5–14). By using two widely-separated cameras, the altitude of the meteor may be found by triangulation, and by interrupting the meteor trail with a rotating shutter, its speed may be deduced. Such observations, in combination with visual sightings, show that there are two types of meteors: (1) the *sporadics* which occur randomly and which come from all directions, and (2) the *shower meteors* which come from one direction and which arise from a stream of meteoroids. A swarm of meteoroids produces meteor trails which diverge from a single point on the celestial sphere—this is the *radiant* of a meteor shower. Meteor showers are named after the constellation in which their radiant lies (e.g., the Perseids from Perseus). Some meteoroid streams move along *comet orbits:* in addition to the Andromedids from Biela's comet, we have the Orionids (October 20) and the Eta Aquarids (May 4) due to Halley's comet. Meteor showers are spectacular but irregular when cometary debris is bunched in the orbit; planetary perturbations tend to spread the rocks out along the orbit (and even produce sporadic meteoroids), leading to more consistent shower displays.

Meteoroids enter the atmosphere with speeds ranging from about 12 to 72 km sec^{-1}, and more meteors are seen after midnight than before. Figure 5–15 illustrates the cause of these phenomena. Meteors belong to the Solar System, so that their speed at the Earth's orbit cannot exceed the solar escape speed there of 42 km sec^{-1}. Before midnight, only those meteoroids moving faster than the Earth (30 km sec^{-1}) can catch up with it from behind; the relative speed of the fastest such meteor is 12 km sec^{-1}. After midnight, all meteoroids, except those moving faster than the Earth along its orbit, will be seen; now the velocities add to give a maximum relative speed of 72 km sec^{-1}.

Meteoroids which are not totally consumed in the atmosphere strike the ground as *meteorites*. Large meteorites produce craters, such as those preserved on the Moon and the kilometer-wide Arizona Meteor Crater (Figure 5–16), which will eventually be obliterated by erosion. Smaller meteorites do not annihilate themselves upon reaching the ground, and the *micrometeorites* (0.5–200 μ) simply drift through the atmosphere to deposit a total mass of about 10^6 kg per day on the Earth's surface. In the past, meteorites were considered evil condensations of the atmosphere raining down upon the heads of sinners, but today we carefully examine these rocks to learn more about the formation and early history of our Solar System. Meteorite *compositions* are of three types (see Figure 5–17): (1) low-density *stones* which are the most abundant, (2) *stony-irons* with stones in a metal matrix, and (3) the dense metallic *irons*. Some of these objects are found to be as old as 4.5 × 10^9 years, and all have undergone a complex thermal and chemical history; nevertheless, we still do not understand why meteoroids are chemically differentiated. Our current ideas suggest that neither asteroids nor comets are sufficiently massive to

Figure 5–14. **A**, *Baker Super-Schmidt Meteor Camera.* This camera has a field of 55°. **B**, *Photograph of Three Meteors Taken by Super-Schmidt Camera.* A rotating shutter produces breaks in the meteor trails. The shaft of the shutter obstructs the center of the field. Note the image of the Andromeda galaxy at the right edge of the field! (Smithsonian Astrophysical Observatory)

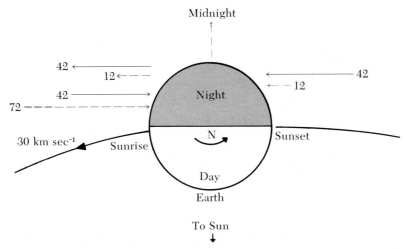

Figure 5–15. *Meteor Speeds and Frequency.* Solar escape speed at the Earth's orbit is 42 km sec^{-1}; meteoroids cannot move faster than this. In the early evening, meteoroids must catch the Earth, while after midnight, the Earth (30 km sec^{-1}) catches up to all but the fastest meteoroids moving *along* its orbit.

Figure 5–16. *Arizona Meteor Crater.* (Official U.S. Air Force Photo)

Figure 5-17. *Meteorites.* **A,** The exterior of the Lost City, Oklahoma, stony meteorite, a chondrite that fell in January 1970. Note the flow lines resulting from the fall through the Earth's atmosphere. **B,** A cross section of the same meteorite, circular structures are chondrules. **C,** An iron meteorite, again showing fusion crust and aerodynamic shaping. **D,** Polished and etched cross section of meteorite shown in C. The structure, characteristic of iron meteorites, is named Widmanstatten pattern. (Courtesy of the Smithsonian Institution)

lead to compositional stratification, such as that found within the Earth—how then did the irons arise?

Let us finally consider the *interplanetary dust* (size $\approx 1-100\ \mu$), which probably comes from the dust tails of comets. Mariner space probes reveal dust clouds around the Earth and Mars. All planets have probably captured dust clouds by gravitational attraction. The faint band of *zodiacal light* results from the sunlight reflected by a marked concentration of dust in the ecliptic plane; the intensity of this light diminishes with distance from the Sun. At opposition in the night sky we can detect a faint fuzzy brightness called the *gegenschein* or *counterglow*, which is probably due to a dust concentration at this Lagrangian point of stability. Although dust is constantly being fed into the Solar System by disintegrating comets (and perhaps asteroid collisions), there are two physical processes which efficiently remove the dust: radiation pressure and the Poynting-Robertson effect.

   Radiation pressure arises because electromagnetic radiation carries momentum at the speed of light. In Chapter 8 we will show that the radiant energy flux E (erg cm^{-2} sec^{-1}) corresponds to a *momentum flux $P = E/c$* (erg cm^{-3}), where c is the speed of light. When solar radiation interacts with a dust particle of effective area or cross-section A, the rate of change of the particle's momentum is the *radiation force*

$$F_R = PA = AE/c$$

When we discussed planetary temperatures above, we found $E = (R_\odot/d)^2 \sigma T_\odot^4$, where d is the distance from the Sun. If we assume that $A = \pi r^2$, where r is the particle's radius, then the radiation force pushing the particle *away* from the Sun is

$$F_R = (\pi \sigma r^2 R_\odot^2 T_\odot^4/c)/d^2 \tag{5–10}$$

But the Sun's gravitational attractive force on the particle is

$$F_G = GM_\odot(4\pi r^3 \rho/3)/d^2 \tag{5–11}$$

where ρ is the density of the particle. When we form the *ratio* of equations (5–10) and (5–11), and substitute the appropriate numbers we find:

$$\frac{F_R}{F_G} = \frac{3\sigma R_\odot^2 T_\odot^4}{4cGM_\odot(\rho r)} = \frac{5.78 \times 10^{-5}}{(\rho r)} \tag{5–12}$$

where the ratio is dimensionless if we express r in cm and ρ in gm cm^{-3}. Note that the particle's distance d from the Sun has disappeared. We see that $F_R = F_G$ for reasonable densities ($\rho \approx 1$–6) when $r \approx 0.1$–1 μ; hence, dust particles smaller than one micron are *blown out of the Solar System*. The solar wind (Chapter 9) aids radiation in this cleansing operation.

   For larger particles $F_G \gg F_R$, but now the *Poynting-Robertson effect* becomes important. Just as the Earth's orbital motion led to the aberration of starlight, so also will the particle's Keplerian orbit cause solar radiation to appear to be coming from slightly in front of the particle. If v is the speed of the particle in circular solar orbit, the angle between the incoming radiation and the radius vector to the Sun is clearly $\theta \approx v/c$ (as before), so that the component $(v/c)F_R$ of the radiation force is *impeding* the particle's motion. Therefore, the particle will spiral into the Sun, just as atmospheric friction lowers the orbits of Earth satellites. A straightforward calculation shows that a particle originally orbiting at distance d (AU) will fall into the Sun in a time

$$t = (7 \times 10^6)(\rho r)d^2 \text{ years}$$

For example, a particle of size $r = 1\ \mu$ and density $\rho = 4.3\ \mathrm{gm\ cm}^{-3}$ takes only 3×10^3 years to spiral into the Sun from a distance of 1 AU, and about 5×10^6 years from 40 AU. In this fashion the Solar System is purged of its dust. Remember that the complications of solar wind and planetary perturbations have been neglected in this simple presentation.

5–2   UNANSWERED QUESTIONS

Our brief survey has revealed the structure and content of the Solar System, but the story is not closed. We can successfully interpret many of the observed features, such as the Moon's synchronous rotation and the sources of meteoroids. However, a plethora of unanswered questions remains concerning the *origin* of these features. Let us ask some of these unresolved questions to put our meager knowledge into perspective; tentative suggestions as to the answers will be outlined when we discuss Cosmogony in section 5–3.

The fundamental question is how and why the Solar System originated. Why are there five tiny terrestrial planets and four huge Jovian planets—a total of nine? What produced the distribution of orbits, and is Bode's law fundamental? Why do the planets move in direct orbits near the ecliptic plane? The *angular* momentum distribution of the Solar System is also a paradox. The rotational angular momentum of the Sun is

$$(2M_\odot R_\odot{}^2 \omega_\odot / 5) \approx 10^{42}\ \mathrm{kg\ m^2\ sec^{-1}}$$

while each planet has an orbital angular momentum $(2\pi m a^2 / P)$, where m is the planet's mass, a is its orbital semi-major axis, and P is its sidereal orbital period; the planetary total is about $33 \times 10^{42}\ \mathrm{kg\ m^2\ sec^{-1}}$. Hence, the planets contribute 97% of the angular momentum of the system—why doesn't the much more massive Sun provide the dominant contribution?

Was there a primordial rotation rate, and if so, why do the terrestrial planets rotate much more slowly than the Jovian planets? What caused the rotation axis of Uranus to lie in its orbital plane?

Are any planetary atmospheres primordial? Why do Venus and the Jovian planets have such dense atmospheres? Did the surface features of the planets originate at the beginning of the Solar System?

How did planetary satellites originate, with their observed number, masses, and orbital distributions? Why Saturn's rings and no others, and what caused the asteroid belt? How were comets formed, and how many are out there? What produced the meteoritic compositions which we observe?

These are the most basic questions which must be answered by any theory pretending to account for our Solar System. Until such a theory is forthcoming, we can only accept what we see and attempt to decipher the riddle before us.

5–3   COSMOGONY

The science of the origin of our Solar System is *cosmogony*; in a broader sense this word refers to the origin of the structure of the entire Universe. Let us first present the history of cosmogonic theories, and then end this chapter with a composite sketch of the possible development of our Solar System.

(a)   Theories of the Origin of the Solar System

In ancient times the Earth was thought to have been born of a goddess (the fertility symbol) or out of primordial chaos (the Biblical picture). The Greeks derived the world from their four basic elements: fire, air, water, and earth. The heavy and base earth settled lowest, followed by water, with the airy atmosphere above; fire rose upward to form the heavenly lights at the highest level. The first truly scientific theories of cosmogony did not appear until the European Renaissance. In 1644 René Descartes assumed that all of space was filled with matter and ether moving in circular vortices; the matter accumulated at the edges of the vortices, where it formed the Sun and planets. The cosmogonic theories which have been advanced in the 300 years since this courageous but inauspicious beginning may be divided into three general categories:

(1) theories with a normal Sun, from which the planetary material later emerges or is extracted by a passing star;

(2) theories with a normal Sun, which captures interstellar material to form the planets; and

(3) theories in which the Sun and planets form essentially simultaneously from the interstellar medium.

The theories of category (1) began in 1745 with G. L. L. Buffon's suggestion that a comet collided with the Sun to eject the material which condensed into the planets. Collision theories were also put forward by A. W. Bickerton (1878) and T. J. J. See (1910). In 1916 and 1929 J. H. Jeans proposed that the tidal forces from a passing star drew gas streamers from the Sun, which became the planets; T. C. Chamberlin (1901) and F. R. Moulton (1905) had a similar idea, but said that the gas first formed asteroid-sized *planetesimals* which later combined or *accreted* into cold planets. Many variations on this theme were played until (a) R. Gunn (1932) concluded that the probability of a close stellar passage is negligible, (b) F. Nölke (1932) showed that the detached material is unstable to solar tidal forces, and (c) L. Spitzer (1939) showed that the hot gases of the streamers would expand and dissipate rather than condense.

Theories of category (2) are even less acceptable than those just discussed. Though a star may capture an interstellar nebula, it is difficult to see why this nebula would be in rapid direct rotation in the ecliptic plane, as it must be in order to reproduce the Solar System. In addition, there is no obvious way for such a capture to trigger the formation of planets, moons, and comets. The Sun's slow rotation (Chapter 9) also remains unexplained. Most theories of category (2) assume the existence of an appropriate nebular disk, and then proceed from there.

Only theories of category (3) remain viable. Immanuel Kant (1755) and P. S. de Laplace (1796) proposed the first *nebular* theories of the origin of the Solar System: a large cloud of gas contracted gravitationally into a disk as a result of its slight initial rotation, and the Sun and planets condensed from this gaseous disk. Unfortunately, these simple first efforts (a) could not account for the small angular momentum of the Sun, and (b) inadequately described the condensation of the planets. More recent attempts, within the context of the nebular hypothesis, are associated with the names (see the Reading List for particulars): H. P. Berlage (1932–1953), H. Alfvén (1942–1954), C. F. von Weizsäcker (1944), D. Ter Haar (1950), G. P. Kuiper (1949–1951), and R. A. Lyttleton (1961). F. L. Whipple (1946–1948) assumed gas-streaming within the nebula to obtain the planets, but found that the most massive planets would be the closest to the Sun. The only reasonably successful and plausible nebular cosmologies today are those of F. Hoyle (1955–1960) and W. H. McCrea (1957–1960). Hoyle postulates a slowly-rotating gas cloud which collapses gravitationally until a protosun at the verge of rotational instability results. Rings of material come off as the Sun contracts further, and *magnetic coupling* between the ionized rings and the Sun transfers angular momentum from the Sun outward to the rings. Dust particles form in the rings and accrete to larger masses, but only the nonvolatile materials can remain close to the Sun, so that the terrestrial and Jovian planets naturally result. McCrea assumes that the original cloud immediately condensed into 10^5 small cloudlets or *floccules*, which moved about at random. Collisions between floccules led to larger floccules, the largest of which grew by gravitational attraction into the slowly-spinning Sun. A few hundred floccules remained behind to form the ecliptic disk, and thence the planets. The distant floccules formed the Jovian planets, but those closer to the Sun were within the Sun's Roche limit, so that only their solid cores were stable (the terrestrial planets).

(b)   A Composite Picture

To reproduce the observed features of the Solar System, let us combine the most promising parts of various nebular theories into a composite picture of the origin. The overall features of this picture are probably reasonably correct, while the fine details are still vague and uncertain.

In the beginning we have an inhomogeneous cloud of interstellar gas and dust, with a mass slightly greater than $1 M_\odot$ and a radius of about 2×10^5 AU (≈ 1 parsec). The gas consists of about 75% hydrogen by mass, 25% helium, and a small admixture of the heavier elements. Because of turbulence, the cloud has a small net rotation and angular momentum. Under the influence of its self-gravitation, the cloud begins to contract. Perhaps at this early stage some gas and dust accrete to form the cometary nuclei at the outer boundary of the cloud.

As the collapse proceeds, the cloud assumes an oblate spheroidal shape (see Figure 5–18) as a result of its rotation, and the rotation rate increases to conserve angular momentum. The density at the center of the cloud increases most rapidly, leading to a massive bulge—the protosun. The collapse converts gravitational potential energy into heat, which is radiated from the dense cloud into interstellar space. As the nebular density increases, the dust particles accrete into larger particles which bind water, carbon dioxide, ammonia, and methane as ices into their matrix. Those particles closer to the rapidly-heating solar bulge lose their volatile materials, while those farther out are shielded from the central heat by the intervening cloud. Soon asteroid-sized planetesimals have accreted by the billions. The larger planetesimals grow rapidly in mass owing to their stronger gravitational attraction; they begin to sweep up the nebular materials in the regions of their orbits and grow into the *protoplanets*. Smaller planetesimals enter satellite orbits

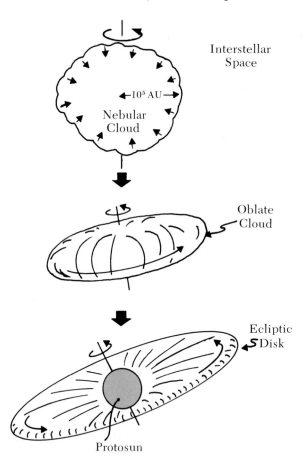

Figure 5–18. *Collapse of Nebular Cloud.* An interstellar gas and dust cloud about 2×10^5 AU in radius collapses as a result of self-gravitation. Its rotation leads first to an oblate spheroidal shape, then to a flat disk about the central protosun.

about the larger protoplanets, and gravitational perturbations might deflect vast numbers of other planetesimals into Oort's cometary reservoir of nuclei (an alternative hypothesis).

All protoplanets have attracted enormous atmospheres by this stage. Perhaps the tidal forces from the protosun upon the atmospheres of the terrestrial proto-planets slow their rotations significantly. Gravitational perturbations from the Jovian protoplanets might prevent the formation of a planet at the region of the asteroid belt, but it is more likely that only a few planetesimals remained there to collide and form the belt. The protosun has not yet ignited its thermonuclear reactions, but its energy output has surged to 10^4 times that of the present Sun. This radiation blows away the proto-atmospheres of the terrestrial planets, ionizing the remaining nebular gases which couple magnetically to the Sun to transfer angular momentum away from the Sun (this momentum transfer might occur at an earlier stage).

As the Solar System is swept clean, the protosun contracts to thermonuclear ignition and becomes a stable star. The planets consolidate their interiors, with Venus, the Earth, and Mars producing secondary atmospheres. Collisions slowly increase the number of asteroid fragments, and comets begin to be deflected into the inner parts of the Solar System. Planetary orbits stabilize, and Mercury and Venus move toward their equilibrium rotation states. Two planets, the Earth and

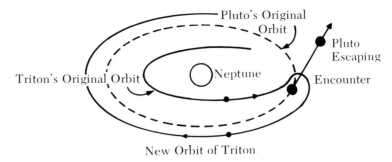

Figure 5–19.   *Pluto Escapes from Neptune.* During a chance encounter in their satellite orbits about Neptune, Pluto is ejected into its own solar orbit escaping Neptune while Triton is deflected into a retrograde orbit. Neptune's rotation rate might also decrease slightly to conserve angular momentum at this stage.

the Moon, might be orbiting independently near the Earth's present orbit, with the Moon being captured by the Earth after many orbits; tidal evolution now begins.

As the Solar System settled into its present configuration, it is possible that Pluto was ejected from a satellite orbit around Neptune. Figure 5–19 shows how this might have come about. Initially, both Pluto and Triton orbited Neptune direct, with Pluto closer to the planet. During a chance near-encounter of the two satellites, Triton's orbital motion was reversed, while Pluto was launched into its own solar orbit at a high inclination (see also Figure 5–3).

This composite picture is not perfect, but it is extremely suggestive. The reader may form his own theory (an adequate theory is sorely needed), but he should remember the many observational constraints which must be satisfied. The discipline of cosmogony remains a fertile field for research and thought.

Problems

5–1.   (a) Describe the apparent path of the Sun across the sky of Mercury during one "day," as seen by an observer at that planet's equator.
(b) Describe the seasons of Uranus, giving numbers where appropriate.

5–2.   Consider the planets Uranus, Neptune, and Pluto.
(a) On a graph, compare their orbital semi-major axes with the distances predicted by Bode's law.
(b) Show that the ratios of their orbital periods $[$e.g., $(P_\Psi /P_\delta)]$ are approximately *commensurable*, that is, nearly fractions like 3/2.

5–3.   Show that the satellites of Mars, Phobos and Deimos, obey Kepler's third law; deduce the mass of Mars from the orbits of these two moons.

5–4.   (a) How much does a spaceship of 10^3 kg mass weigh on the Earth's surface?
(b) Remembering Jupiter's rapid rotation, how much will this craft weigh at the equator of Jupiter's surface? At the poles?

5–5.   Derive the formula which gives the synodic rotation period of Venus with respect to the Earth, and verify the number quoted in the text.

5–6.   Ole Römer computed the speed of light by observing that the satellite Jupiter I was occulted by the planet approximately $16\frac{2}{3}$ minutes *earlier* when Jupiter was in opposition than when Jupiter was near superior conjunction. Repeat his calculation.

5–7.   How far from the star would we have to be (in AU) to find a substellar temperature comparable to that of the Earth for
(a) Rigel (surface temperature $= T = 12,000$ K, radius $= R = 35 R_\odot$)?
(b) Barnard's star ($T = 3000$ K, $R = 0.5 R_\odot$)?

5–8.   Formaldehyde (H_2CO) has recently been discovered in interstellar space.
(a) Calculate its "mean" molecular speed for $T = 280$ K. Would our Moon retain this gas for billions of years?
(b) Would Saturn's satellite Mimas retain formaldehyde? (For Mimas, radius ≈ 300 km, mass $\approx 4 \times 10^{19}$ kg.)

5–9.   Consider a comet with an aphelion distance of 5×10^4 AU and an orbital eccentricity of 0.995.
(a) What are the perihelion distance and orbital period?
(b) What is the comet's speed at perihelion and at aphelion?
(c) What is the speed of escape from the Solar System at the comet's aphelion, and what do you conclude from this result?

5–10.   Draw a diagram to explain why some meteor showers are consistent from year to year, while others are spectacular on occasion and feeble at other times.

5–11.   What effects do radiation pressure and the Poynting-Robertson effect have upon an artificial space probe of mean density $\rho \approx 1$ gm cm^{-3} and radius $r \approx 1$ meter? Justify your answer quantitatively and state your assumptions.

5–12.   At the origin of the Solar System, the Sun's tidal forces and Roche limit might have played an important role. To avoid tidal disruption, what is the minimum density ρ (in gm cm^{-3}) of a protoplanet at the distance d (in AU) from the Sun? Comment upon your result.

Reading List

Alfvén, Hannes, and Arrhenius, Gustaf: "Structure and Evolutionary History of the Solar System. I," *Astrophysics and Space Science*, Vol. 8, pp. 338–421 (1970).

Allen, C. W.: *Astrophysical Quantities*. Second Ed. London, The Athlone Press, 1963.

Anders, Edward: "Meteorites and the Early Solar System," *Annual Review of Astronomy and Astrophysics*, Vol. 9, pp. 1–34 (1971).

Blanco, V. M., and McCuskey, S. W.: *Basic Physics of the Solar System*. Reading, Massachusetts, Addison-Wesley Publishing Company, 1961.

Brandt, John C., and Hodge, Paul W.: *Solar System Astrophysics*. New York, McGraw-Hill Book Company, 1964.

Dollfus, A. (ed.): *Moon and Planets*. Vols. I & II. Amsterdam, North-Holland Publishing Company, 1967–8.

Hartmann, William K.: *Moons and Planets: An Introduction to Planetary Science*. Tarrytown-on-Hudson, N.Y., Bogden & Quigley, Inc., 1972.

Hawkins, G. S.: *Meteors, Comets, and Meteorites*. New York, McGraw-Hill Book Company, 1964.

Hynek, J. A. (ed.): *Astrophysics (A Topical Seminar)*. First Ed. New York, McGraw-Hill

Book Company, 1951. See "On The Origin of the Solar System," by Gerard P. Kuiper, Chapter 8, pp. 357–424.

Jastrow, Robert, and Cameron, A. G. W. (eds.): *Origin of the Solar System.* New York, Academic Press, 1963.

Kaula, William M.: *An Introduction to Planetary Physics (The Terrestrial Planets).* New York, John Wiley & Sons, 1968.

Kuiper, Gerard P. (ed.): *The Atmospheres of the Earth and Planets.* Revised Ed. Chicago, The University of Chicago Press, 1952.

Kuiper, Gerard P., and Middlehurst, Barbara M. (eds.): *Planets and Satellites.* Chicago, The University of Chicago Press, 1961.

Lovell, A. C. B.: *Meteor Astronomy.* Oxford, The Clarendon Press, 1954.

Middlehurst, Barbara M., and Kuiper, Gerard P. (eds.): *The Moon, Meteorites, and Comets.* Chicago, The University of Chicago Press, 1963.

Page, Thornton, and Page, Lou Williams (eds.): *The Origin of the Solar System.* New York, The Macmillan Company, 1966.

Porter, J. G.: *Comets and Meteor Streams.* New York, John Wiley & Sons, 1952.

Roth, Günter D.: *The System of Minor Planets.* London, Faber and Faber, Ltd., 1962.

Ter Haar, D.: "On the Origin of the Solar System." *Annual Review of Astronomy and Astrophysics,* Vol. 5, pp. 267–278 (1967).

Watson, Fletcher G.: *Between the Planets:* Revised Ed. Cambridge, Massachusetts, Harvard University Press, 1956.

Whipple, Fred L.: *Earth, Moon, and Planets.* Third Ed. Cambridge, Massachusetts, Harvard University Press, 1968.

Williams, I. P., and Cremin, A. W.: "A Survey of Theories Relating to the Origin of the Solar System." *Quarterly Journal of the Royal Astronomical Society,* Vol. 9, pp. 40–62 (1968).

Wood, John A.: *Meteorites and the Origin of Planets.* New York, McGraw-Hill Book Company, 1968.

Chapter 6
The Earth-Moon
System

Our detailed description of the Solar System begins with our home planet, the Earth, and its nearest neighbor, the Moon. Centuries of probing the Earth (*geophysics*) have lead to a relatively complete understanding of this planet, while the study of the Moon (*selenophysics*) has been revolutionized by the space age since 1957. Let us outline what we know about the terrestrial neighborhood.

6-1 DIMENSIONS

The size of the Earth was first determined by the Egyptian astronomer Eratosthenes (276–195 B.C.), who noted that the noon altitude of the Sun at the

Figure 6–1. *Earthrise as Photographed from Orbiter I.* The crescent Earth is seen rising over the limb of the Moon. (NASA)

summer solstice differed by $7°12'$ when viewed from Syene and Alexandria. Assuming that the Earth was spherical, and noting that Alexandria was 5000 stadia (1 stadium $\approx$ 0.16 km) due north of Syene, he found the Earth's circumference to be $(360°/7°2)5000 = 250,000$ stadia ($\approx$ 40,000 km or 25,000 miles). This value gives a radius within one per cent of the Earth's equatorial radius, $R_\oplus = 6378.2$ km (3963.4 mi). The same technique has been used to find the Earth's oblateness (see Chapter 4) of 1/298.3.

The Earth is not alone in space, but is a partner in the unique *Earth-Moon system*. We say unique because the Moon is by far the largest and most massive satellite in the Solar System, *relative to its primary planet* (see Figure 6–2A). The Moon's radius is 1738 km $(0.272R_\oplus)$ and its mass is 7.35×10^{22} kg $(0.0123M_\oplus)$; the closest runners-up are Neptune's Triton $(0.090R_\psi, 0.0013M_\psi)$ and Saturn's Titan $(0.040R_\hbar, 0.00025M_\hbar)$. The mean distance between the centers of the Earth and the Moon is 384,405 km, or about $60.3R_\oplus$, as shown to scale in Figure 6–2B. The *barycenter* (center of mass) of the system is located $M_\mathrm{D}a_\mathrm{D}/(M_\oplus + M_\mathrm{D}) = (0.0123)(384,405)/(1.0123) = 4671$ km from the Earth's center; as shown in Figure 6–2C, the Earth and Moon orbit this point buried 1707 km *below* the Earth's surface once each month.

The *mass* of the Earth was found in Chapter 3. If we assume that the Moon's mass is small compared to that of the Earth, we may use Kepler's third law: $M_\oplus = 4\pi^2 a_\mathrm{D}^3/GP_\mathrm{D}^2$; this law may also be applied to artificial satellite orbits. The result is always

$$M_\oplus = 5.977 \times 10^{24} \text{ kg}$$

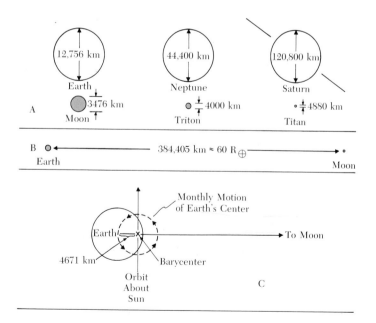

Figure 6–2. **A**, *Satellite Sizes.* The three largest moons, relative to their planets, are shown to scale. **B**, *Earth-Moon System.* The dimensions of the Earth-Moon system are drawn to scale. **C**, *Barycenter.* The location of the Earth-Moon barycenter (X) is shown, and the Earth's monthly motion about this point is depicted.

The Moon's mass may be determined by observing the Earth's motion about the barycenter; by knowing the Earth's mass and the location of the barycenter we may compute

$$M_{\mathbb{)}} = (d_{\oplus}/d_{\mathbb{)}}) M_{\oplus} = 7.35 \times 10^{22} \text{ kg} = (1/81.3) M_{\oplus}$$

where the distance from the barycenter to the Earth's (Moon's) center is $d_{\oplus}(d_{\mathbb{)}})$. Today the orbits of lunar spacecraft are used to accurately deduce $M_{\mathbb{)}}$, as well as the internal mass distribution of the Moon. We find that (a) the oblateness of the Moon is 0.0006, (b) the Moon is slightly *oblong*, with the long axis (tidal bulge!) pointed towards the Earth, and (c) there are high-density concentrations of matter beneath the giant, ringed maria—the *mascons*.

Using Kepler's third law to determine the distance to the Moon, we may find the *size* of the Moon by (a) timing occultations of stars and planets, and (b) measuring the apparent angular size of the Moon in the sky. Since its apparent diameter is 31' of arc, or about 1/2°, the Moon's diameter is 3476 km and its radius 1738 km. Today, the Moon's radius is determined by reflecting radar pulses from the surface (accuracy of several meters) or laser light pulses from mirrors set up by the Apollo astronauts (accuracy of about 1 cm).

6–2   DYNAMICS

(a)   Motions

Let us summarize the principal motions of the Earth, which were discussed in Chapter 4. The barycenter of the Earth-Moon system orbits the Sun at about 1 AU in one sidereal year of 365.2564 mean solar days (3.156×10^7 seconds). The Earth rotates once every 24 sidereal hours, and its center orbits the barycenter in one sidereal month of 27.322 days. (Note that the orbits of the centers of the Earth and Moon are *always concave* toward the Sun.) The Earth's rotation axis precesses about the ecliptic pole in 26,000 years, while the body axis wobbles slightly about the rotation axis. Finally, tidal friction slows the Earth's rotation by a few seconds per millennium.

The Moon's motions are both interesting and informative. With respect to the stars, the Moon orbits the barycenter in one *sidereal month* (27^d322); with respect to the Sun, this orbit takes one *synodic month* (the month of phases, 29^d531); with respect to its line of nodes (see below), the orbital period is one *nodical* or *draconic month* (27^d212); and with respect to its perigee, the period is one *anomalistic month* (27^d555). The Moon's *synchronous* rotation period is one sidereal month; hence, we see only one hemisphere of the Moon—the face of the "man in the Moon." The *farside* or hidden face was not seen until first photographed by the Russian space-craft Luna 3 on October 7, 1959. Actually we can see about 59% of the total lunar surface, on account of the Moon's *librations* (first known and interpreted by Galileo). Lunar librations are geometrical effects caused by the inclination of the Moon's orbit and equator (see Figure 6–3). The Moon rotates at a fairly steady rate, but it moves at different speeds in its eccentric orbit; this leads to an apparent east-west

Figure 6–3. *Lunar Librations.* **A**, The 6°17′ libration in longitude results from the Moon's steady rotation (indicated by shading) and eccentric orbit about the Earth. **B**, The 6°41′ libration in latitude results from the 5°9′ inclination of the Moon's orbit to the ecliptic and the inclination of its equator to its orbit. **C**, The 57′ (1°54′ total) diurnal libration is a parallactic effect caused by the displacement of an earthbound observer due to the Earth's rotation.

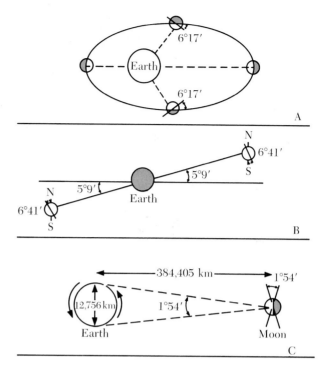

"rocking" of the Moon by about 6°17′—the *libration in longitude*. While the Moon's equator lies essentially parallel to the ecliptic plane, the lunar orbit is inclined 5°9′ to the ecliptic; hence, we see an apparent north-south "nodding" motion of about 6°41′—the *libration in latitude*. Finally, an observer at the Earth's surface sees the Moon from different angles during the day, since a point at the Earth's equator is displaced 12,756 km in 12 hours by the Earth's rotation. This results in lunar parallax and the *diurnal libration* of about 57′. The Moon's slightly nonspherical shape also produces a small *physical libration*, due to irregularities in the lunar rotation rate.

Differential solar forces on the Earth-Moon system produce three significant effects: (a) they tend to elongate the Moon's orbit at inferior conjunction and opposition, (b) they cause the perigee of the Moon's orbit to precess eastward (direct) with a period of 8.85 years, and (c) they produce a torque on the inclined orbit which causes the *line of nodes* (intersection of the Moon's orbital plane and the ecliptic plane) to regress westward along the ecliptic with a period of 18.6 years. Because of its eccentricity ($e = 0.055$), the Moon's orbital distance ranges from 363,263 km at perigee to 405,547 km at apogee, while the angular diameter of the Moon varies from 32′.9 to 29′.5.

(b) Phases

The Sun always illuminates one hemisphere of the Moon, but from the Earth we see varying fractions of the sunlit hemisphere depending upon the Moon's elongation. Hence, as shown in Figure 6–4, the cycle of *geocentric phases* of the Moon lasts one *synodic* month; these phases occur in the sequence: new (inferior conjunction), waxing crescent, first quarter (quadrature), waxing gibbous, full (opposition),

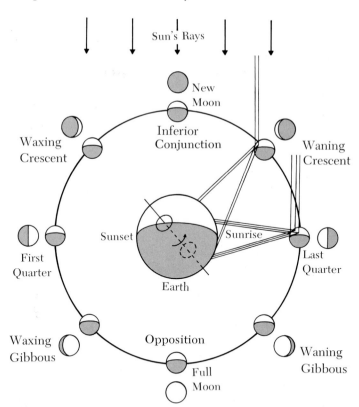

Figure 6–4. *Lunar Phases.* The Moon is shown at eight points on its synodic orbit, and the corresponding phases are depicted on the outer circle of eight disks.

waning gibbous, third quarter (quadrature), waning cresent, and back to new. The Moon's position in the sky is correlated with its phase, as the following example shows. The first quarter Moon rises in the east at noon, transits the local meridian at sunset, and sets in the west at midnight. At other phases, the simple geometry is similarly interpreted; the reader is encouraged to work out several examples. As seen from the Moon, the Earth exhibits similar phases.

(c) Eclipses

Let us briefly discuss solar and lunar eclipses, while referring the reader to the more detailed and complete presentations in the reading list. In general, an *eclipse* occurs when one celestial body passes between us and another body; in Chapter 7 we will mention the eclipses of Jupiter's moons by the planet, and in Chapter 16 we will see that stars in a binary system may eclipse one another. Eclipses in the Earth-Moon system depend upon the orientation of the *line of nodes* of the Moon's inclined orbit, since the Moon must be near the ecliptic plane (a) to pass directly between the Sun and Earth in a solar eclipse, and (b) to pass through the Earth's shadow in a lunar eclipse (see Figure 6–5).

Solar Eclipses. The Sun's apparent angular diameter is 32′, almost the same as that of the Moon! The line of nodes of the Moon's orbit points toward the Sun twice each year, so that new Moon can occur close enough to the ecliptic for

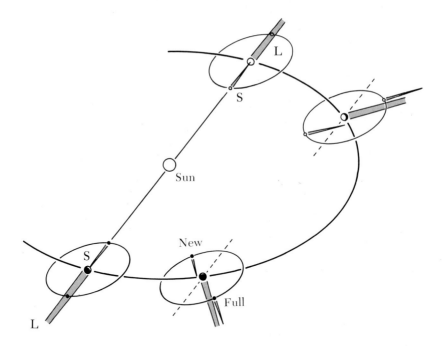

Figure 6–5. *The Line of Nodes and Eclipses.* The Sun, Earth, and Moon are shown during the year, with the shadows of the Earth and Moon (shaded). The lines of nodes (dashed) must point toward the Sun if an eclipse is to occur. Lunar (L) and solar (S) eclipses are indicated. (From G. Abell)

the Moon to cover the Sun—this is a *solar eclipse*. When the Moon passes to one side of the center of the Sun, we have a *partial* solar eclipse. If the Moon is not near perigee in its orbit, its angular size is slightly smaller than that of the Sun, so an *annular* solar eclipse can take place with a ring of the Sun's disk visible. Near perigee, when the Moon's center crosses the Sun's center we have a spectacular *total* solar eclipse (see Figures 9–14 and 9–15). The duration of *totality* at a given point on the Earth's surface cannot be longer than about 7.5 minutes.

Lunar Eclipses. At the Moon's orbit, the Earth's shadow (where the Sun is totally hidden) is easily found to be 9212 km across, or an angular diameter of $1°22.'4$. When the line of nodes points down this shadow, a *lunar eclipse* can take place at full Moon. Both *partial* and *total* lunar eclipses are possible, with the Moon remaining dark for up to 1^h40^m in a *central* total lunar eclipse.

The Greek astronomer Meton (c. 400 B.C.) noted that the Moon exhibits the same phases on the same day of the month (e.g., full Moon on April 13) at intervals of 18.6 years—this is known as the *Metonic cycle*. A similar phenomenon, also due to the *regression of the nodes* of the Moon's orbit, is the *Saros cycle*. Similar solar and lunar eclipses take place at intervals of 223 synodic months (18 years, 10 days); since the precise Saros interval is 6585.32 days, we must wait three Saros cycles to see an eclipse repeat *at the same place on the Earth*. To see how the Saros cycle comes about, the reader is urged to check and think about the near equality of 223 synodic months, 242 nodical months, and 239 anomalistic months.

6–3 INTERIORS

(a) The Earth

The *average density* of the Earth is

$$\bar{\rho} = (3M_\oplus/4\pi R_\oplus^3) = 5.52 \text{ gm cm}^{-3}$$

(remember that the density of pure water is 1 gm cm^{-3}). Since the density of rocks at the Earth's surface is about 3 gm cm^{-3}, the interior of the Earth must be *very* dense. Figure 6–6A shows the stratified structure of the Earth's interior: a 35-km-thick *surface* layer of density 3.3, then the *solid mantle* of silicates like olivine $[(Mg, Fe)_2, SiO_4]$ with $\rho \approx 3$–6 to a depth of 2900 km, next the 2200-km-thick *liquid outer core* with densities from about 9–11, and finally the (*solid?*) *inner core* of radius 1300 km with densities around 17 gm cm^{-3}.

This picture of the Earth's interior has been deduced from the study of the propagation of earthquake waves (*seismology*). Earthquakes generate both *longitudinal compression* (*P*) waves and *transverse distortion* (*S*) waves, which travel on paths determined by the density and composition of the interior materials (see Figure 6–6B). Both types of waves are refracted (bent) by changes in density, and reflected

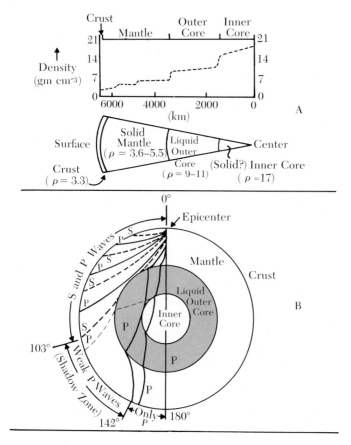

Figure 6–6. A, *The Earth's Interior*. Above the pie-slice of the Earth, the interior density is plotted versus the distance from the Earth's center. B, *Seismic Waves*. Both *S* (heavy dashed) and *P* (solid) waves are shown, with the *S* waves absorbed by the liquid outer core (shaded) and weak *P* waves (light dashed) reflected into the shadow zone by the inner core.

by density discontinuities; only the P waves, however, can propagate through a liquid region. By measuring the type of wave, and its time of travel from the focus or *epicenter* of an earthquake, we may infer the wave's path through the Earth. Up to 103° from the epicenter, both S and P waves are observed, while between 103° and 180° no S waves are seen (they are absorbed by the liquid outer core). The abrupt refraction of P waves at the mantle-core interface results in a *shadow zone* from 103° to 142° wherein only weak P waves reflected by the *solid* core are observed. From 142° to 180°, strong P waves are again in evidence.

While part of the Earth's core is known to be liquid,[*] the *composition* of the core is not known. Some geophysicists attribute the density jump to a change from silicates in the mantle to nickel-iron in the core, while others believe that *phase changes* in a uniform composition (probably olivine) are the cause. A phase change occurs when the crystalline (?) structure of a material abruptly alters its *stable* form as pressure and temperature are increased. Sulfur has a low melting point and may play a critical role in the discontinuous properties of core and mantle. Only continued research can answer the question of the Earth's interior composition.

(b) The Moon

Our knowledge of the lunar interior is very sparse. The Moon's average density is 3.37 gm cm^{-3}, surprisingly similar to the density of the Earth's crust. The Apollo missions have returned lunar surface samples with densities near 3 gm cm^{-3}, and the composition of basaltic silicates, so the density cannot increase appreciably toward the Moon's center. The seismic stations set up by the Apollo astronauts have revealed the following fascinating features: (a) the Moon is seismically "quiet," except for meteorite impacts, occasional crater slumping, a few feeble vibrations originating from a focus about 800 km below the surface and triggered by the Earth's tidal forces at perigee, and the occurrence of moonquake "swarms," a series of quakes lasting several days, of unknown origin but apparently unrelated to tidal effects; and (b) vibrations caused by the deliberate impacts of Apollo components on the lunar surface damp very slowly (the Moon "rings" for almost an hour!). One current model of the Moon's interior based on this lunar seismology includes a solid, inactive core, a mantle in which partial melting has occurred, and a crust at least 60 to 70 km deep in some places. Thus, the Moon, like the Earth, is stratified and differentiated.

6–4 SURFACE FEATURES

The surfaces of the Earth and Moon are radically different. The Earth's surface undergoes rapid and dynamic evolution as a result of the active interior (e.g., mountain building, continental drift, etc.) and the thick erosive atmosphere; since the Moon's interior is "dead" and there is little or no atmosphere, the Moon's surface has preserved a relatively clear picture of its origin and primordial development. Let us briefly outline the characteristics of the Earth's surface (more detailed

---

[*] Possibly the source of the Earth's magnetic field; see section 6–6(b).

expositions may be found in the reading list), and then thoroughly discuss the lunar surface.

(a)   The Earth's Surface

The surface of the Earth is the interface between the low-density *crust* and the *atmosphere*. Figure 6–7 shows how the crust is composed of (a) the solid *lithosphere* of average density 3.3 gm cm^{-3}, with light granitic continental blocks reaching to a depth of 35 km and the denser basaltic ocean floors extending down only 5 km, and (b) the liquid water *hydrosphere*, which covers approximately 70 per cent of the Earth's surface area to an average depth of 3.5 km. The crust "floats" on the mantle, in isostatic (i.e., buoyant) equilibrium, and is separated from the convective low-viscosity mantle by the Mohorovičić discontinuity. Slow motions of the upper mantle are responsible for continental drift, mountain building, and earthquakes. The surface of the lithosphere, with an observed temperature range of 200 to 340 K,* is the home of mankind.

To compare the Earth's surface with that of another planet, we must ask how the Earth appears from space. The Mercury and Vostok astronauts were the first men to see this view; Figure 6–8 shows a photograph of the Earth taken from an Apollo spacecraft. Cloud cover and the polar icecaps appear bright (high albedo), while the continents show light green and tan. The oceans are very dark blue, except for the bright spot of *specular reflection* where the Sun's image has been mirrored by the water. Our home is a beautiful but complex and fragile ship in space.

---

* K stands for Kelvin, the unit of *absolute* temperature which is closely related to degrees Celsius (Centigrade); see Appendix 5.

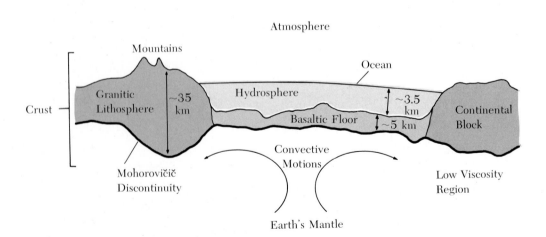

Figure 6–7.   *The Earth's Surface.* The upper mantle, crust (lithosphere and hydrosphere), and atmosphere of the Earth are shown schematically.

Figure 6–8. Apollo photo of Earth. (NASA)

(b) The Lunar Surface

While the lunar interior is still *terra incognita*, the visible surface and first few meters of the soil have been carefully studied. The surface is made up of gray-tan materials of basaltic composition with a very low average albedo (0.07). As the Moon's *terminator* (line separating dark and sunlit hemispheres) moves along the lunar equator at 15.4 km hr^{-1}, the surface temperature drops rapidly from its maximum of 390 K to a minimum of 110 K.

(i) *The Visible Structure*

Between 1966 and 1968 five American lunar Orbiters photographed 99.5 per cent of the entire lunar surface with high resolution. Since prehistoric times lunar surface markings have been noted as the "face of the man in the Moon." In 1609 A.D. Galileo first discerned the dark lunar *maria* or "seas" (which were given Latin names by Riccioli in 1651), the light lunar *highlands* with their numerous *craters*, and lunar *mountains*—all through his primitive telescope (see Figure 6–9 for Earth-based photographs of the Moon). With the photographs from the Orbiters, and the soft-landings of the American Surveyor and Apollo missions and the Russian Luna series, we now know a great deal about the lunar surface. Let us describe what has been observed on the Moon. The lunar maps inside the front and back covers of the book may be used to locate the features referred to in the following paragraphs.

Craters. The entire lunar surface is pockmarked with *craters:* concave depressions with rims elevated above the exterior level and the crater *floor*, sometimes with a *central peak*. Craters range in size from microscopic pits to the largest on the nearside, Bailly, with a diameter of 295 km. Figure 6–10 depicts the cross-section of the large crater Copernicus, with the Grand Canyon shown for comparison. The

Figure 6–9A. *Western Quarter of Lunar Near-Side.* See inside cover for identification of lunar features. (Lick Observatory)

juxtaposition of different features permits us to determine relative *ages*, since young craters will clearly overlap older craters. We will see that the lunar maria have far fewer craters than the lunar highlands, so the maria must be younger than the highlands. Many craters, especially Tycho and Copernicus (see Figure 6–11), exhibit radial patterns of *rays* which have been resolved into numerous *secondary craters* on Ranger photographs.

Craters evolve and disappear slowly as material slides down their walls, and the walls themselves *slump*; meteoritic bombardment produces new craters which fill, obliterate, and degrade the older craters (see Figure 6–12). The lifetime against such "erosion" has been estimated as several million years for 1 cm diameter craters, and longer than the age of the Moon for large craters (tens of km in diameter). Although large craters do not disappear, their walls do slump and their rims and rays are eroded by meteoritic impact, temperature changes, and moonquakes. In addition, isostatic adjustments cause the crater floor to rise and the rim to lower.

Figure 6–9B. *Eastern Quarter of Lunar Near-Side.* (Lick Observatory)

The accepted theory of crater origin is *meteorite impact*, with the largest craters arising from small asteroids which hit the Moon. Many lunar features, including some craters like Ptolemaeus and Alphonsus (Figure 6–13), betray evidence of lava flows or volcanic origin (*vulcanism*), but it appears that few craters are not impact craters. The long historic controversy of impact versus volcanic origin has virtually ended by vindicating both points of view, though heavily favoring the impact theorists. The latter, however, were surprised by the extensive evidence of vulcanism, which is discussed below. The largest impact "crater" is

Figure 6–10. *Crater Cross-Section.* The dashed line indicates the Moon's surface level. Copernicus is seen to be eight times broader than the Grand Canyon and more than twice as deep. Note the elevated rim, depressed floor, and central peak.

90 km — Raised Rim

11 km

3840 m

1600 m

Floor Central Peak

Copernicus

Grand Canyon

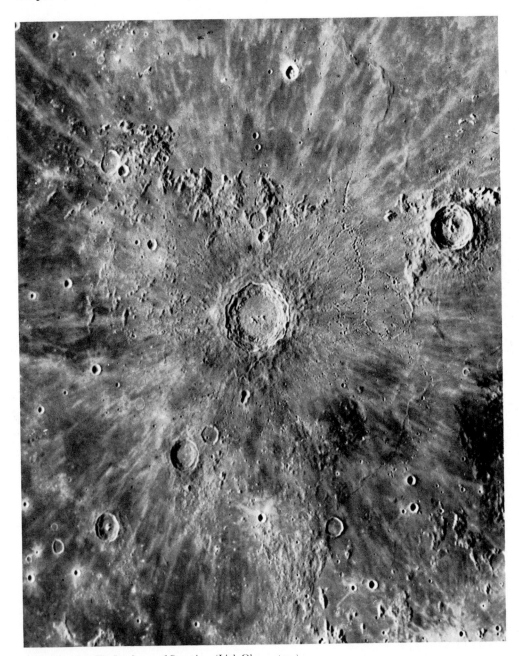

Figure 6–11. *The Ray System of Copernicus.* (Lick Observatory)

Mare Orientale with a diameter of 1000 km, at the western limb of the Moon; Figure 6–14 shows this magnificent triple-ringed "bull's-eye" structure.

An impact crater is produced when an incoming meteoroid hits the surface and converts much of its kinetic energy into explosive thermal energy; the remainder of the kinetic energy is used to fracture surface materials and create seismic waves. Lunar material is scattered in analogy to the explosion of a shallow underground nuclear blast (see Figure 6–15). A wave of heavy material flaps over to evacuate a hole, pile up an elevated rim, and fling an *ejecta blanket* of debris beyond the rim.

Figure 6–12. *Slumping and Ejecta Blanket.* This vertical view of the north wall of Copernicus shows several ledges produced by slumping and downward flow of material. Note the ejecta scattered beyond the crater rim. (NASA)

Sprays of high-velocity particles land farther on to produce rays of secondary craters, and the rebound of the initial shock frequently raises a central peak. An asteroid 1 km in radius contains about 2×10^{16} gm of rock, and when it hits the Moon at a speed of 30 km sec^{-1}, the kinetic energy of $(mv^2/2) = 10^{29}$ ergs vaporizes the rock to create a crater about the size of Copernicus. Large craters such as Tycho, Clavius, and Mare Orientale require energies in the range 10^{28} to 10^{33} ergs for their excavation. Figure 6–16 may be taken as a strong justification of the impact theory, for lunar craters follow the same correlation of depth to size as do ordinary terrestrial impact pits; much more data on *small* lunar craters is available today, and the trend is even better verified.

 Highlands. Most of the lunar surface consists of the light-colored, heavily-cratered *lunar highlands* (or continents, uplands, terrae, etc.), especially the backside. The highlands exhibit much bare rock and a chaotic terrain, which has been age-dated at 4.6×10^9 years; apparently this represents the primordial lunar surface, for it is much older than the maria and about the age of the Solar System. In general, the highlands have a greater elevation than the maria, with a possible analogy to

Figure 6–13. *The Craters Ptolemaeus and Alphonsus.* Ptolemaeus is the hexagonal crater at the top of the picture. Note how one wall lines up with the wrinkle that crosses Alphonsus. Several sunken and ghost craters are seen on the floor of Mare Nubium. The feature at the center of the lower edge of the photograph is the Straight Wall. (Hale Observatories)

the terrestrial continental blocks. Figure 6–17 is an Orbiter photograph of the lunar farside, which shows typical highland topography and cratering.

Maria. The large dark areas of the Moon's surface are called maria or "seas" (the singular is mare), though they are really flat plains of dark basaltic lava. Of the 30 maria known, only four are found on the backside of the Moon. Several are roughly circular in form and range in diameter from about 300 to 1000 km. These circular *mare basins* are believed to be impact features that subsequently filled with molten lava. The presence of old *ghost craters* and *flooded craters* implies that some cratering occurred between the time of formation of the mare basin and the extrusion of the lava. The impact theory of mare basins is supported by the curved shape of some of the mountain chains (e.g., the Apennines) that form their borders and rise above the maria, and by the observation of ejecta sculpting of the nearby highlands (see Figure 6–18). The Apennine Mountains, therefore, are presumed to result from uplifting of the crust at the time of the impact that produced Mare Imbrium.

The maria have relatively few craters compared to the lunar highlands; this feature is consistent with the ages of maria, which are about 3.5×10^9 years. Fresh "young" craters, such as Copernicus (see Figures 6–11 and 6–20), are seen superimposed on the maria. In addition, the maria exhibit wrinkle ridges, scarps, domes, rills, and valleys, many of which are evidence of some type of volcanic activity. They are described as follows:

Wrinkle ridges—Rope-like raised structures, hundreds of meters high and many kilometers long, are seen running across the maria (see Figure 6–21). These may be the tops of mountain ranges covered by the mare lavas or lines of lava extruded by cracks in the mare surface.

Figure 6–14. *Mare Orientale.* This remarkable three-ringed basin lies just beyond the western limb of the nearside. In this Orbiter 4 photograph, Mare Procellarum covers the top right portion of the Moon, and the flooded crater Grimaldi lies on the right, almost half way from the top. Beyond the centermost of the concentric circular scarps, one can clearly see the ejecta blanket deposited by the base surge. (NASA)

Scarps—Abrupt discontinuities in the level of a mare, such as the famous Straight Wall, are believed to be the results of faults. Frequently, *straight rills* with lines of *subsidence craters* are seen to trace out fault or fracture systems.

Domes—Convex blister-like structures, such as those near the crater Marius, rise above the maria, and probably represent upwellings of molten lava.

Rills—Sharply defined narrow trenches that occur in several forms, such as the straight rills already referred to, and *sinuous rills*. The latter resemble snake-like

Figure 6–15. *Nuclear Cratering.* **A**, *The Explosion.* The base surge is seen as a low-level cloud of dust; rocks and other ejecta escape on various trajectories, thus causing secondary craters. **B**, *The Resulting Crater.* A raised rim, an extended ejecta blanket, and several secondary craters resulted from the nuclear explosion, but we also see characteristics of lunar impact craters. (NASA)

Figure 6–16. *Diameter-Depth Relation for Impact Craters.* This log–log graph shows the correlation between diameter and depth for *impact* craters ranging in size from 1-meter pits to 100-km lunar craters. The lunar craters fit the impact curve closely, and so are probably of meteorite-impact origin. (After Baldwin, 1949.) (Key: × = missile pit, ○ = bomb and explosion pit, ● = lunar crater, ■ = terrestrial meteorite crater.)

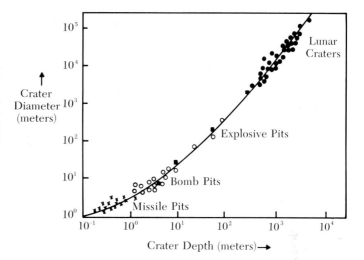

rivers (see Figure 6–22), meandering over the mare surface, but predominantly bordering the highland boundary. Although some appear to run "downhill' from crater to mare, others fade to nothing at both ends. Some sinuous rills wind down the centers of *valleys* the size of the Grand Canyon (e.g., the Alpine and Schröter's Valleys). These phenomena are still a mystery. Are they a result of water, or lava, or particles carried by hot gases? Some selenologists consider them to be analogous to collapsed lava tubes as found on Earth—tunnels through which lava flowed long ago.

Gravitational perturbations of the lunar Orbiters imply that the mare basins on the nearside are filled with *dense* material—these gravitational anomalies are called *mascons* (mass concentrations). While it is possible that a large dense meteorite may be buried beneath each basin, the more popular view is that the lava filling the basins is of higher density and different composition than the rest of the lunar surface.

The origin of the lava is still controversial. That the mare material is lava-like is well documented, particularly by composition studies which show it to be basaltic. One line of thought would have the lava formed by the energy of impact. Flooding of craters within the mare basins is difficult to explain according to this theory, for the lava flow should have occurred at the time of impact. Others argue that the delayed flooding requires an internal source for the lava. Moreover, the layered appearance of Hadley Rille (see Figure 6–22C) points to a series of lava flows rather than a single event. Molten lava apparently existed in the interior of the Moon, either in a hot center or in the mantle. Some scientists maintain that the lunar interior is still hot.

(ii) *Composition of Surface Materials*

American Surveyor and Russian Luna soft-landers have probed and analyzed the lunar surface, while the Apollo missions have returned hundreds of kilograms of surface soil and rocks. We now know that the bedrock of the lunar surface is covered by a thin, well-mixed layer of loose soil and rocks—the *regolith*. The regolith was produced by meteorite impacts which fragmented and scattered the bedrock,

(*Text continued on page 136.*)

Figure 6–17.   *The "Hidden Face" of the Moon.* Part of the southeastern nearside is visible in this photograph. Mare Smythii located on the eastern limb of the nearside is seen here near the top right, while Humboldt is in the center, about a third of the way from the top. Crater Tsiolkovsky, first seen by Lunik 3, is the prominent dark-floored crater near the eastern limb on this picture. Note also the double-walled crater with the radial clefts or ridges in the lower part of the photograph. (NASA)

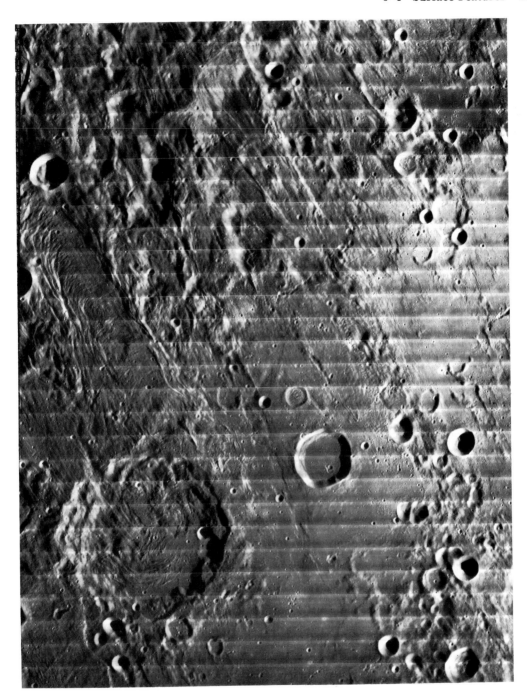

Figure 6-18. *Ejecta Sculpting of Lunar Highlands.* This region lies southeast of the center scarp of Mare Orientale. Sculpting by the base surge is most clearly seen in the upper left of the picture, which also includes some of the ejecta blanket. The lower right portion of the large crater, Inghirami, is hardly affected, in contrast to the opposite wall. (NASA)

Figure 6–19.   *Silver Spur* in the Mount Hadley region in the Lunar Apennines, photographed on the Apollo 15 mission. Note the sloping strata or layers. (NASA)

Figure 6–20.   *Copernicus: A Young Post-Mare Crater.* This oblique view shows the southern rim and ejecta blanket in the foreground, and the floor, central mountains, and slumped northern wall toward the top of the photograph. (NASA)

Figure 6–21. *Structural Details of the Mare Surface.* Volcanic-type domes and cones and related wrinkle ridges are clearly seen in the Marius Hills area in Oceanus Procellarum. (NASA)

and pulverized the soil over the eons. This soil ranges in depth from 2 to 10 meters in the maria and is estimated to be 10 cm deep in the highlands; large chunks of bedrock are frequently seen as boulders at the bottoms of and near deep impact craters. The lunar soil has the texture and strength of dusty cohesive sand-gravel, and its color is dark gray with a shade of tan. Churning due to meteorite impacts mixes the regolith thoroughly in about 50 million years.

Materials brought back from the Moon include: (a) samples of the cohesive fine-grained soil containing particles of average size 10 μ, with an admixture of small (50 μ) glass or obsidian spheres (see Figure 6–23); (b) small *igneous* rocks which show evidence of melting, recrystallization, and shock fracture; small glass-lined pits are also seen on the surfaces of these rocks; and (c) small chunks of coarse-grained composite *breccias*, which resemble a compressed conglomerate of many materials. The igneous rocks appear to be fragments of the lunar bedrock brought to the surface by cratering, while the breccias were produced when the regolith was compressed into agglomerates by the meteorite impact. Most samples returned from the maria have been age-dated at 3 to 4 billion years, while some fragments which apparently originated in the highlands have ages as high as 4.6 billion years. Typical age-dating procedures are the krypton-argon, rubidium-strontium, and uranium-lead radioactive decay methods.

One of the most important and surprising conclusions derived from these age-datings is that the major part of the evolution of the Moon's surface must have occurred within a very short time span—within the first billion years after the origin of the Moon itself. We have seen that the mare basins must have been formed well after the highlands, and that the flooding by lava came still later. Yet dating of maria material gives it an age exceeding 3 billion years.

The chemical composition of the lunar surface is basically *basaltic*, with the following interesting features: (a) the maria are lavas of igneous (molten) origin, (b) the content of free iron is low, with the maria containing more iron, cobalt, and nickel than the highlands, (c) most of the volatile elements are underabundant, and some chemical differentiation has apparently taken place, (d) the lunar lavas have a higher content of iron and titanium *bound* into minerals than do terrestrial basalts, (e) the lunar materials are *anhydrous;* that is, they were not formed in the presence of water or water vapor, and (f) no signs of fossilized or viable life-forms have been detected. The areas sampled indicate that the maria are *chemically* homogeneous, and differ somewhat from the highlands in that they have a considerably lower abundance of aluminum.

6–5 ATMOSPHERES

(a) The Moon

No lunar atmosphere has ever been detected. Indeed, the Moon appears to be too small to undergo extensive internal differentiation, though rare instances of outgassing (like volcanic vapors) have been observed. Any atmosphere generated by the lava flows of the maria has long since escaped as a result of the Moon's low surface gravity and high surface temperatures. In addition, the solar wind is very efficient in sweeping away any trace gases which might emerge at the surface. Hence, there is practically a perfect vacuum at the Moon's surface.

(b) The Earth's Atmosphere

We believe that the Earth lost its original proto-atmosphere at the origin of the Solar System, but produced a *secondary* atmosphere as a result of internal vulcanism and outgassing. The large mass and gravitational attraction of the Earth permit it to retain this atmosphere, which we breathe today. Table 6–1 lists the chemical composition by volume of the Earth's atmosphere near ground level.

Table 6–1. Chemical Composition of the Earth's Atmosphere Near Ground Level (by Volume)

Gas	Percentage
Nitrogen (N_2)	78.08
Oxygen (O_2)	20.95
Argon (Ar)	0.934
Carbon dioxide (CO_2)	0.033
Neon (Ne)	0.0018
Helium (He)	0.00052
Methane (CH_4)	0.00015
Krypton (Kr)	0.00011
Hydrogen (H_2), carbon monoxide (CO), xenon (Xe), ozone (O_3), radon (Rn), etc.	< 0.0001
Water vapor (H_2O)	variable (0 to 4%)

Figure 6–22. *A Sinuous Rille—Hadley Rille as Photographed by Apollo 15.* **A,** Overview from command module. (NASA)

(i)   *Atmospheric Structure*

As shown in Figure 6–24, the Earth's atmosphere consists of characteristic layers. Nearest the surface is the dense, well-mixed *troposphere* where most weather takes place. As evidenced by snow-capped mountains, the temperature decreases steadily with height until we reach the *tropopause* at 15 km. Then the temperature rises slightly in the thin stable *stratosphere*, which extends to 40 km, where we reach the *mesosphere*. A second temperature minimum occurs near 90 km (at about $-80°C \approx 190$ K), and then the temperature increases steadily through the *thermosphere* (90 to 250 km) until it levels off near 1500 to 2000 K at the base of the *exosphere*. The exosphere is the lower portion of the tenuous *protonosphere*, which consists mainly of hydrogen from the dissociation of H_2O and CH_4 near 80 km; the protonosphere extends to about 10 $R_\oplus$ where it meets the solar wind.

Figure 6–22. **B**, Close-up view of rille, showing sharp bend near middle of photograph A, just above crater (St. George) and mountain (Hadley Delta). **C**, Far wall of Hadley Rille, showing layers of exposed rocks. Note, for example, the well-defined thin layer just below and at a slight angle to the second row of crosses. (NASA)

Figure 6–23. *Glassy Spherules of Lunar Soil from Apollo 11.* The spherules were placed on an aluminum plate to be photographed. Some are opaque, others are almost transparent. While some are colorless, others are gray, green, brown, or deep red. The largest is 0.5 mm in diameter. (NASA)

While atmospheric density decreases steadily with height, the temperature profile exhibits two minima and three maxima. Solar ultraviolet and X-radiation interact with the compositionally stratified atmosphere to dissociate oxygen and nitrogen molecules, and ionize some atoms to produce the *stratified density of free electrons in the ionosphere* (about 50 to 320 km). The regions of maximum electron density are called the D-, E-, and F-layers. The ionization balance of the ionosphere is self-regulating, since free electrons *recombine* faster with ions as more electrons and ions become available; hence, a steady-state equilibrium is set up. Concentrations of atomic oxygen (O), atomic nitrogen (N), ozone (O_3), and nitric oxide (NO, NO^+) are found at different levels of the ionosphere, and the entire system acts as a shield which absorbs most of the solar radiation dangerous to life.

(ii)   *Observed Effects of the Atmosphere*

Electromagnetic radiation is absorbed, scattered, and refracted by the Earth's atmosphere. The characteristics of these phenomena depend upon the wavelength λ of the incoming radiation. The atmosphere is opaque to most wavelengths, transmitting only (a) radio waves with λ in the range 1 cm to 20 m (which are detected by radio telescopes), and (b) visible light from 2900 Å (near ultraviolet) to about 10,000 Å (near infrared)—this is the "window" through which we *see* the Universe with our eyes and telescopes. We have already mentioned how short-wavelength radiation (less than 2900 Å) dissociates and ionizes the atmosphere,

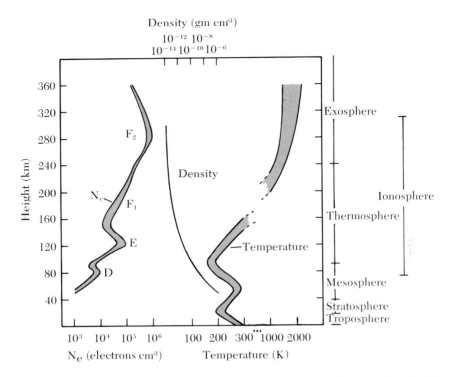

Figure 6–24. *The Earth's Atmosphere.* The various levels of the atmosphere are indicated, with their density, temperature, and ionization (electron density) profiles as a function of height above the surface.

while being absorbed in the process. For example, the ozone layer (10 to 40 km) absorbs radiation from 3000 to 2100 Å. In the infrared region (about 1 μ–1 cm) radiation is absorbed when it excites molecules like N_2, O_2, CO_2, and especially H_2O. At wavelengths longer than about 20 meters, the ionosphere acts like a conducting shield which absorbs and reflects the incoming radiation.

Let us now discuss how the atmosphere affects the *visible light* which ranges from violet (4100 Å) to red (6500 Å). The most important effects are scattering, extinction, refraction (and seeing), and dispersion.

Light is *scattered* when it interacts with a "particle," and the characteristics of the scattering depend upon the wavelength λ of the light and the size L of the scatterer. When light is scattered by atmospheric molecules ($L \ll \lambda$), the *intensity* of the scattered radiation obeys the *Rayleigh scattering law*:

$$I_{\text{scat}} \propto 1/\lambda^4$$

Hence, more blue light is scattered out of incident sunlight than red, and we see a *blue sky*. This causes the Sun (or a star) to appear *reddened* when viewed through a considerable thickness of atmosphere, such as at sunset. Note that some radiation is scattered at every wavelength, even the red, so that a star's brightness is dimmed by the atmosphere—this we term *extinction* of the light. Corrections to astronomical observations must be made to account for reddening and extinction, although they can be reduced by viewing near the zenith or from a mountaintop. When $L \approx \lambda$, such as for 1 μ dust particles scattering red light, the scattering law becomes:

$$I_{\text{scat}} \propto 1/\lambda$$

Finally, when $L \gg \lambda$, as for water droplets in clouds, the scattering is essentially *independent of wavelength*—clouds appear white since they scatter sunlight this way.

When light passes from one medium to another, it is *refracted* or bent toward the "denser" medium. We characterize a medium by its *index of refraction*, and say that the index of refraction of air increases with its density. Hence, starlight coming from a star at a true altitude θ is refracted downward toward the Earth, so that it appears to reach the surface at the altitude angle $\theta' > \theta$ (see Figure 6–25). This effect, which vanishes at the zenith, becomes increasingly important as we approach the horizon. At the horizon, a celestial object appears to be elevated about 35′ above its true position, so that the setting Sun is actually *below* the horizon when we "see" its bottom resting on the horizon. Density inhomogeneities in the atmosphere cause randomly-fluctuating amounts of refraction for incoming light, so that stars appear to "dance about" (scintillate or *twinkle*) on the celestial sphere. This effect limits the *angular resolution* of telescopes (even large telescopes) to about 1″ (0.″25 in exceptional cases), and we term this *astronomical seeing*. We say that the seeing is good when the atmosphere is stable and the scintillation is small. Planets and satellites with angular diameters exceeding the seeing limit shine steadily, since only their edges can be observed to scintillate.

Dispersion results when the index of refraction depends upon the wavelength of the radiation refracted. The most familiar example of dispersion is the *rainbow*, caused by light of different colors refracting through raindrops at different angles. In the Earth's atmosphere, blue light is refracted more strongly than red light, so that stars near the horizon appear as tiny *spectra* (in a telescope) with the red end nearest the horizon.

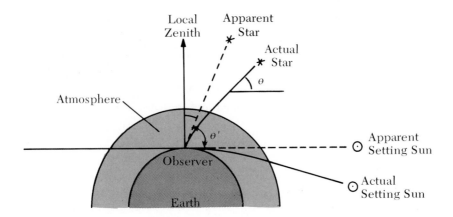

Figure 6–25. *Atmospheric Refraction.* An observer "sees" a star at altitude $\theta' > \theta$, when the star is actually at altitude θ. The refraction effect is zero at the zenith ($\theta' = \theta = 90°$), and reaches a maximum of 35′ at the horizon.

With all of these effects to be taken into account, and especially with the opaqueness of the Earth's atmosphere to most wavelengths of radiation, it is not surprising that observers locate their instruments on mountaintops or strive to view the Universe from orbiting satellites. The ideal observatory is clearly one located above the Earth's obscuring and confusing atmosphere. It is a tribute to the persistence and skill of astronomers that so much has already been learned from the bottom of our murky sea of air.

6–6 MAGNETIC FIELDS

(a) The Lunar Magnetism

While the Earth's magnetic field, with a strength of about 1/2 Gauss at the surface, has been known for several hundred years, the lunar magnetism could only be sought with space probes during the last decade. Magnetometers placed on the Moon's surface by the Apollo astronauts reveal that the intrinsic lunar magnetic field is *less than* 10^{-5} Gauss (essentially zero), but that *localized* sources of magnetism (meteorites?) might exist with strengths around 10^{-4} Gauss. In any case, the Moon's magnetism is negligibly small.

(b) The Earth's Magnetosphere

The Earth's *dipole* magnetic field has been extensively investigated. The axis of the field is inclined 12° to the Earth's rotation axis, so that the field lines emerge (north magnetic pole) and re-enter (south magnetic pole) the Earth's surface about 1336 km from the rotation poles. The close alignment of the magnetic and spin axes implies that the Earth's rotation is an important factor in producing the magnetism. It is believed that fluid motions in the metallic (conducting) outer core are responsible for the magnetic field, through some type of *dynamo* action. Paleomagnetic investigations have shown that the field *reverses its direction* in a random way with an average period of about 10^4 to 10^5 years, and today we can observe that the strength of the field is slowly decreasing.

Since the strength of the magnetic field drops as $(1/r^3)$ with distance r from the Earth, while the density of the atmosphere falls exponentially, the magnetic field is important far beyond the atmosphere. The domain of the Earth's magnetic field is the *magnetosphere*, which is delimited by the *magnetopause* boundary where it meets and interacts with the solar wind (see Chapter 9) in a shock wave. Figure 6–26 shows the comet-like structure of the magnetosphere, caused by the solar wind "pushing" the Earth's magnetic field away from the Sun. Within 8 to $10R_\oplus$ of the Earth the dipole field is evident, while beyond there is the abrupt shock-and-transition region on the sunward side and the magnetic tail longer than $1000R_\oplus$ pointing away from the Sun.

Our knowledge of the complex structure of the magnetosphere has been achieved only in the last decade, when magnetometers were first carried beyond the magnetopause by artificial space probes on highly eccentric orbits. These probes have shown that the Moon perturbs the particle flow of the solar wind, as well as the Earth's magnetotail. Spacecraft passing Venus and Mars have found

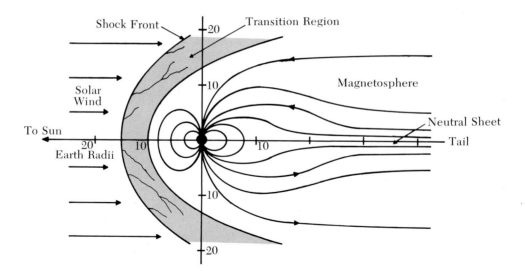

Figure 6–26. *The Earth's Magnetosphere.* The diagram shows the Earth's dipole magnetic field running into the solar wind which creates a shock wave and pushes out a long magnetotail. The dimensions of the magnetosphere are indicated in Earth radii ($R_\oplus$).

similar effects, but nothing that could be attributed to intrinsic magnetism of either planet.

(c)   The Van Allen Radiation Belts and Aurorae

The Van Allen Belts.   The magnetopause deflects the solar wind away from the Earth, but many particles (mainly protons and electrons) leak into the magnetosphere. There they are trapped by the Earth's dipole magnetic field in toroidal radiation belts, concentric to the magnetic axis (see Figure 6–27). Although F. C. M. Störmer (1874–1957) had predicted the existence of such a phenomenon around 1905 on the basis of auroral observations (see The Aurorae), the discovery of the belts (*Van Allen radiation belts*) by the group of James A. Van Allen from satellite observations in 1958 came as a complete surprise.

Satellites carrying an assortment of *particle detectors* have now mapped the radiation belts in detail. Though both protons and electrons are found throughout the magnetosphere, there are two belts of particular concentration: (1) the small *inner belt* between 1 and $2R_\oplus$ where protons of energy 50 MeV* and electrons with energies greater than 30 MeV tend to reside, then a distinct gap or *slot*, and (2) the large *outer belt* from 3 to $4R_\oplus$ where less energetic protons and electrons are concentrated. The inner belt is relatively stable, while the outer belt shows variability in its number of particles by as much as a factor of 100. The origin of the particles trapped in the belts is still controversial, but it is suspected that most

---

* The *electron Volt* (eV) is a convenient energy unit in atomic and particle physics. 1 eV $= 1.6021 \times 10^{-12}$ erg; 1 keV $= 10^3$ eV; and 1 MeV $= 10^6$ eV.

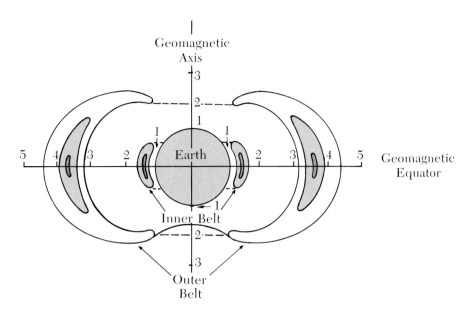

Figure 6–27. *The Van Allen Radiation Belts.* The radiation belts are shown to scale with the Earth. All dimensions are given in Earth radii ($R_\oplus$).

come from the solar wind and from cosmic ray interactions in the Earth's upper atmosphere.

The charged particles trapped in the radiation belts are not stationary, but spiral along lines of magnetic force while bouncing (with periods from 0.1 to 3 seconds) between the northern and southern *mirror points* (see Figure 6–28). Particles in the inner belt may interact with the tenuous upper atmosphere at these mirror points to produce the aurorae (see The Aurorae); such particles are lost from the belts. In addition to spiralling and oscillating north-south, the particles *drift in longitude* owing to the decreasing strengths of both the magnetic and gravitational fields with increasing distance from the Earth. High-energy *protons drift westward* around the Earth in about 0.1 second, while low-energy *electrons drift eastward* in about 1 to 10 hours. This drift leads to longitudinal uniformity of the radiation belts.

Let us investigate these charged particle motions quantitatively. The *Lorentz force law*

$$\mathbf{F} = \frac{q}{c}\,(\mathbf{v} \times \mathbf{B}) \tag{6–1}$$

tells us the force $\mathbf{F}$ (in dynes) experienced by a particle of *charge q* (in esu*) moving with velocity $\mathbf{v}$ (in cm sec^{-1}) through a magnetic field of strength $\mathbf{B}$ (in units of Gauss). Here the speed of light is $c \cong 3 \times 10^{10}$ cm sec^{-1}. Note that the magnetic field is represented by the *vector* $\mathbf{B}$, and that $\mathbf{F}$ is perpendicular to both $\mathbf{v}$ and $\mathbf{B}$, in accordance with the "right-hand rule" (see Figure 6–29A). While the mass of

---

* The fundamental unit of *charge* in the centimeter-gram-second (cgs) system is the electrostatic unit or esu; see Appendix 5 at the end of the book.

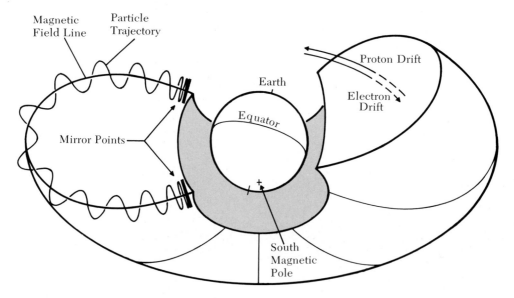

Figure 6–28. *Particle Motion in the Radiation Belts.* Charged particles spiral along the magnetic field lines, while bouncing between the northern and southern mirror points. Since both the magnetic and gravitational fields decrease with distance from the Earth, the protons drift westward and the electrons drift eastward at the same time.

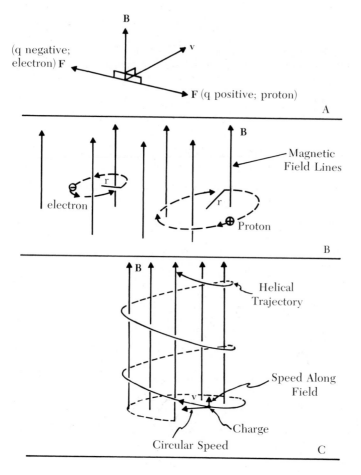

Figure 6–29. *Charged Particles in Magnetic Fields.* **A,** Vector diagram illustrating the Lorentz force law. **B,** Circular orbits of proton (+) and electron (−) in uniform magnetic field (**B**). **C,** Helical orbit of a charge with some speed along the uniform magnetic field lines (**B**).

the proton $(1.673 \times 10^{-24}$ gm) is much larger than the mass of the electron $(9.109 \times 10^{-28}$ gm), the charge of the proton $(e = 4.8029 \times 10^{-10}$ esu) is equal to the charge of the electron $(e = -4.8029 \times 10^{-10}$ esu) but of *opposite sign*. No *work* can be done on a charged particle by the magnetic field, since $\mathbf{F} \cdot d\mathbf{r} = (\mathbf{F} \cdot \mathbf{v})\, dt = 0$ ($\mathbf{F}$ is perpendicular to $\mathbf{v}$!); hence, *the kinetic energy of a particle is not changed by a magnetic field* (i.e., the total speed v remains constant). Figure 6–29B shows how such a particle moves in a circular orbit in a *uniform* magnetic field. The Lorentz force [equation (6–1)] provides the centripetal force for the orbit of radius r, so that

Figure 6–30. *Aurora Borealis.* (National Research Council of Canada)

$$mv^2/r = qvB/c \Rightarrow \boxed{r = mvc/qB}$$ (6–2)

For example, a proton with speed $v = 10^8$ m sec^{-1} circles at a radius $r = 10$ kilometers when $B = 1$ Gauss.

If the particle has some speed *along* a magnetic field line, it will follow a spiral or *helical* trajectory (Figure 6–29C). When the particle moves into a region of higher field strength (B increases), the circular orbit shrinks while the circular speed increases. Since the particle's total kinetic energy cannot change, its motion *along* the field line must slow down; eventually the forward motion will reverse at the *mirror point*. We can now understand some of the motions of protons and electrons in the radiation belts.

The Aurorae.   The inner radiation belt interacts with the Earth's upper atmosphere to produce the colorful *aurorae*: aurora borealis or "northern lights" in the high northern latitudes, and aurora australis in the high southern latitudes. Diffuse luminous forms over large areas of the sky have been observed since ancient times in northern regions between 15° and 30° from the magnetic pole. Auroral displays may be flickering streaks or "curtains" of light, or steady lacework "draperies," which change intensity and color (pale pink, blue, and green) in the course of hours (see Figure 6–30). Triangulation measurements of the heights of aurorae show that most occur between 80 and 160 km, with a few as high as 1000 km.

Auroral emissions result when low-energy electrons precipitate out of the inner radiation zone, and collisionally excite and ionize (see Chapter 8) atmospheric gases. As these excited gases, principally oxygen and nitrogen, return to their stable forms, visible light is emitted with characteristic colors. The solar wind plays an important role in aurorae, by either supplying the necessary electrons or by perturbing the magnetosphere so that the particles trapped in the radiation belts are "dumped" into the atmosphere near the mirror points. We know that solar activity (especially solar flares) is important, since it is so strongly correlated with the appearance of auroral displays.

6–7   ORIGIN AND EVOLUTION OF THE EARTH-MOON SYSTEM

Now that we understand the nature of the Earth-Moon system, let us discuss its origin and early evolution. Just as the birth of the Solar System is not well understood, so also is the beginning of the Earth-Moon system shrouded in mystery and controversy. We will attempt to present an interesting and consistent picture here, but the reader is forewarned that no single view is accepted by all scientists.

One hypothesis of the origin, which has passed out of favor, is the *fission theory*. Its proponents argued that since the Moon's average density is about the same as that of the Earth's crust, the Moon must have formed when the crust overlying the Pacific Ocean (!) broke away from the hot fluid proto-Earth. This idea is easily disproven: (a) the Earth must rotate in *2 hours* to shed material from its equator owing to rotational instability; it is *dynamically impossible* for a single body (the Moon) to have been cast off in this way; (b) if the Moon had emerged from the Earth, it would have been within the Earth's Roche limit, and so would have disrupted completely into tiny fragments; (c) the Pacific Ocean is straight-forwardly explained in terms of the theory of continental drift (see end of this section), so this "loophole" no longer exists.

A more likely picture of the origin is the following: after planetesimals formed and accreted in the solar nebula, there were *two* protoplanets—the Earth and Moon—orbiting at essentially the same distance from the Sun. Solar heating and solar tidal forces quickly swept away the proto-atmospheres of both bodies before the Sun became a stable star (as we see it today). As the Earth consolidated, melted, and differentiated its interior, the crust was formed and a thick *secondary* atmosphere of carbon dioxide, nitrogen, ammonia, methane, and water vapor (similar to Venus' atmosphere) was outgassed at the surface. The Moon was too small to differentiate extensively, and any secondary atmosphere it formed quickly escaped. Numerous planetesimals were swept up by the Earth and the Moon, but the resulting scars (craters) were quickly eroded away on the Earth's surface while being preserved as the multitudinous highland craters on the lunar surface. Age dating of meteorites and lunar rocks shows that all of these events took place about 4.6 billion years ago.

About 1 billion years after the formation of the Solar System—or about 3.5 billion years ago—a catastrophic event seems to have taken place. Geological evidence and the lunar mare samples show that the Earth's surface melted and the lunar maria were formed (*lava flows*) at this time. Apparently the Moon passed close enough to the Earth to be *captured* (see Figure 6–31). Enormous tidal bulges were raised, and fantastic amounts of energy were dissipated in tidal friction, as the Moon passed close to the Earth's Roche limit. Early in this period, large bodies collided with the Moon to form the mare basins; these bodies may have been planetesimals left circling the Earth after its formation or perhaps parts of the Moon detached during the capture (or fission fragments?). Somewhat later, but within a half billion years, immense lava flows occurred on both bodies, leading to the flooding of the maria. Further tidal evolution circularized the Moon's orbit, and slowed its spin to synchronous rotation. As the Earth's surface rapidly cooled, rains fell to fill the ocean basins. Much carbon dioxide was absorbed into solution in this water, and life appeared. Photosynthesis by primitive plants consumed CO_2

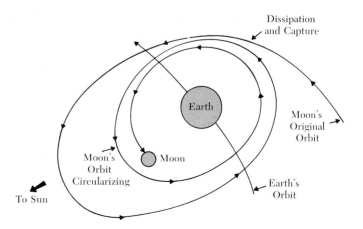

Figure 6–31. *The Capture of the Moon.* The Moon might have been captured from an independent solar orbit by the Earth, due to tidal dissipation, about 3.5×10^9 years ago. See the text.

to produce O_2, and as this molecular oxygen accumulated in the atmosphere, ultraviolet radiation from the Sun produced ozone (O_3). Land animals (oxygen-consumers) evolved when the oxygen content became high enough, and when the ozone layer blocked most of the lethal solar UV from the Earth's surface. Some of the most primitive animals to appear in the oceans were diatoms and corals, which could efficiently bind carbon dioxide into the mineral calcium carbonate ($CaCO_3$). Eventually an equilibrium was achieved, with about 80% N_2, 20% O_2, and traces of CO_2 and H_2O in the atmosphere.

After the Moon's capture (?), the granitic continental blocks of Earth bore little resemblance to today's continents. Convective motions in the upper mantle conveyed huge *plates* of the crust about on the Earth's surface at speeds of about 2 cm per year, rending into pieces the primeval "continents" along *rift* zones and causing continental collisions. This *continental drift* drastically altered biological evolution, as species of flora and fauna were dispersed and then brought together in different configurations. Today we speak with assurance of this process of *plate tectonics*, for the basaltic ocean floors have recorded the numerous reversals of the Earth's magnetic field during the last 250 million years—in which time the Atlantic Ocean opened up and the Himalaya Mountains were raised as India ran into Asia from the south.

As the active Earth evolves and the far more quiescent Moon recedes, we arrive at the present. All that we have said is based upon inference and gross extrapolation into the past, so it would not be surprising if many of the details are incorrect. The general picture, however, seems to be correct; we leave it to the reader to mull over this and other possibilities, and to begin to appreciate the beauty of the awesome pageant which has preceded us.

Problems

6–1.   (a) An Apollo spaceship is headed for the Moon. At what point between the Earth and Moon will the ship experience no net gravitational acceleration?
(b) How long will it take the ship to circumnavigate the Moon in a circular orbit 50 km above the lunar surface?
(c) At what speed must the spacecraft leave the lunar surface in order to escape from the Moon?

6–2.   Using the length of the sidereal month ($27^{d}.322$), and the periods of the Earth's revolution ($365^{d}.26$), the regression of the nodes (18.6 years), and the precession of the Moon's perigee (8.85 years), compute the lengths of
(a) the synodic month.
(b) the nodical month.
(c) the anomalistic month.

6–3.   Assume that the Moon's orbit is circular and that it lies in the ecliptic plane. Find the difference in solar attraction on the Moon's *orbit* at opposition and inferior conjunction, and compare this with the gravitational attraction from the Earth. Can you now understand why the lunar orbit is not a simple ellipse?

6–4. Lunar and solar eclipses take place *near* the line of nodes of the Moon's orbit. How close (angular distance) to the line of nodes must the Moon's center be

(a) for any type of solar eclipse to take place?

(b) for a total lunar eclipse to take place?

Your results are related to the *eclipse limits* (see Abell).

(Hint: The orbital inclination of the Moon is 5°9′.)

6–5. (a) If the Moon had a core of radius $R_{\leftmoon}/10$, and the remainder of the Moon's interior had a uniform density $\rho = 3$ gm cm^{-3}, what must be the uniform density of this core?

(b) Compare the Earth's tidal forces on the Moon at the perigee and apogee of the Moon's orbit; comment on your results.

6–6. When we discussed the retention of atmospheres in Chapter 5, we said that the mean kinetic energy per molecule in a gas at temperature T is $(mv^2/2) = (3kT/2)$, where $k = 1.380 \times 10^{-16}$ erg (K)$^{-1}$. In Chapter 3 we saw that a particle with a vertical speed v at the Earth's surface will rise to a height $h = v^2/2g$ before it falls back to the Earth.

(a) Show that the *characteristic height* for a molecule of mass m at temperature T is given by $h = (3kT/2mg)$.

(b) At $T = 250$ K, compute the characteristic heights for nitrogen (N_2), oxygen (O_2), carbon dioxide (CO_2), and hydrogen (H_2); what does this tell you about the *composition* structure of the Earth's atmosphere?

6–7. (a) If a star emits the same intensity of radiation at all visible wavelengths, what will be the apparent color of the star at the Earth's surface?

(b) Explain why the Sun appears *flattened* (like an ellipse) at sunset.

6–8. Show that the circular period P (seconds) for a charge in a uniform magnetic field B does not depend upon the radius of the orbit. Evaluate this period for a proton moving at speed $v = 10^9$ cm sec^{-1} in a magnetic field of 1 Gauss.

6–9. Draw a diagram and use the Lorentz force law to explain why electrons in the radiation belts *should* drift *westward* owing to the decrease in magnetic field strength with distance from the Earth. Try to explain why the electrons actually drift *eastward* (observationally).

6–10. Do some research in the library, and write a 1000-word essay on the characteristics, merits, and deficiencies of at least two *theories of the origin of the Earth-Moon system*. Compare and contrast the theories where possible.

Reading List

Abell, George: *Exploration of the Universe*. Second Ed. New York, Holt, Rinehart and Winston, 1969.

Baldwin, Ralph B.: *The Measure of the Moon*. Chicago, The University of Chicago Press, 1963.

Baldwin, Ralph B.: *A Fundamental Survey of the Moon*. New York, McGraw-Hill Book Company, 1965.

Calder, Nigel: *The Restless Earth*. New York, Viking Press, 1962.

Carovillano, Robert L., McClay, John F., and Radoski, Henry R. (eds.): *Physics of the Magnetosphere*. New York, Springer-Verlag, 1968.

Fielder, Gilbert: *Structure of the Moon's Surface*. New York, Pergamon Press, 1961.

Hess, Wilmot N., Menzel, Donald H., and O'Keefe, John A. (eds.): *The Nature of the Lunar Surface*. Baltimore, The Johns Hopkins Press, 1966.

Kopal, Zdeněk: *The Moon*. Second Ed. New York, Academic Press, 1964.

———, (ed.): *Physics and Astronomy of the Moon*. New York, Academic Press, 1962.

Krauskopf, Konrad, and Beiser, Arthur: *Fundamentals of Physical Science*. Fifth Ed. New York, McGraw-Hill Book Company, 1966.

Kuiper, Gerard P. (ed.): *The Earth As A Planet*. Chicago, The University of Chicago Press, 1954.

Levinson, A. A.: *Proceedings of the Apollo 11 Lunar Science Conference*. New York, Pergamon Press, 1970. See also *Science, 167*, No. 3918, January 30, 1970.

Marsden, B. G., and Cameron, A. G. W. (eds.): *The Earth-Moon System*. New York, Plenum Press, 1966.

Mason, Brian, and Melson, William C.: *The Lunar Rocks*. New York, Wiley Interscience, 1970.

Middlehurst, Barbara M., and Kuiper, Gerard P. (eds.): *The Moon, Meteorites, and Comets*. Chicago, The University of Chicago Press, 1963.

Mutch, Thomas A.: *Geology of the Moon*. Princeton, Princeton University Press, 1970.

Namowitz, Samuel N., and Stone, Donald B.: *Earth Science*. Third Ed. Princeton, D. Van Nostrand Company, 1965.

Oppolzer, Theodor von: *Canon of Eclipses*. New York, Dover Publications, 1962.

Stacey, Frank D.: *Physics of the Earth*. New York, Wiley & Sons, 1969.

Whipple, Fred L.: *Earth, Moon, and Planets*. Third Ed. Cambridge, Massachusetts, Harvard University Press, 1968.

Wyllie, Peter J.: *The Dynamic Earth*. New York, Wiley & Sons, 1971.

Chapter 7
Inferior and
Superior Planets

In Chapter 5 we discussed the general properties of the Solar System, and in Chapter 6 the Earth-Moon system was described in detail. Let us now consider the individual characteristics of the other eight planets. The relevant data are given in Tables A2–2 through A2–4 in the Appendix.

7–1 MODERN PLANETOLOGY

The Solar System, from the Sun out to Saturn, has been observed visually since prehistoric times. Since the human eye alone cannot discern details smaller than 1′, it was not until the invention of the *telescope* in about 1600 A.D. that the investigation of the planets—*planetology*—began in earnest. In 1610 Galileo published his *Sidereus Nuncius*, wherein he described telescopic observations of: (a) the four large (Galilean) moons of Jupiter, (b) the phases of Venus, (c) the cratered surface of the Moon, (d) the "dark" spots on the Sun, and (e) the strange nonspherical shape of Saturn (its rings). At visible wavelengths, both visual and photographic observations at the telescope have continued to the present, telling us much about the behavior of the planets and their satellites. But the Earth's atmosphere (*seeing*) limits the visible resolution of telescopes to about 1″, so that the surface markings and atmospheres of only those objects of much greater angular size can be studied in any detail at all. Note that the maximum angular diameter of Mercury is only 12″, while that of Uranus is about 4″; in such cases, optical discrimination is marginal at best.

New technologies and techniques characterize *modern planetology*. The "optical window" has been expanded to the near ultraviolet (photography) and near infrared (solid-state detectors). Radio telescopes and radar now peer through the "radio window" from 1 cm to 20 m in wavelength. High-altitude balloons, rockets, artificial satellites, and space probes circumvent the obscuration of the Earth's atmosphere to give us a broader and clearer view of the Solar System. Spacecraft have been sent to the Moon, Venus, and Mars to dramatically improve our picture of these bodies; soon such craft will visit Mercury, Jupiter, and the planets beyond; eventually probes will rendezvous with comets and some will escape to interstellar space.

With these new tools, our understanding of the other planets has improved more in the last few decades than in all of previous history. In this chapter, we will describe what we know today. Many uncertainties still remain, but the excitement of the new planetology is contagious and the future optimistic.

7–2 MERCURY

Mercury (☿) is the closest planet to the Sun, and probably the smallest of the terrestrial planets. Each year it appears about three times as a bright evening star near the sunset horizon, and as a predawn morning star. Because of the swiftness of its celestial motion it is named after the mythological god of flight. At times Mercury rivals Venus in brightness, but it is usually obscured by the brilliance of the Sun.

(a) Motions

Mercury revolves about the Sun in a highly eccentric ($e = 0.2056$) and inclined ($7°00$ to the ecliptic plane) orbit of semi-major axis $a = 0.3871$ AU, with a sidereal orbital period of 87^d96. The greatest elongation of this planet ranges from 18° (perihelion) to 28° (aphelion). From Mercury all other planets exhibit retrograde loops in their apparent orbits, and the angular diameter of the blazing Sun ranges from $1°1$ to $1°7$.

Planetary perturbations cause the perihelion of Mercury's orbit to precess direct, but, as early as 1859, U. J. Leverrier found an excess *perihelion advance* over that predicted by Newtonian theory. To account for this anomaly, the hypothetical planet Vulcan was postulated to orbit closer to the Sun than Mercury, and observers occasionally claimed to have spotted this new planet transitting the Sun's face. The need for Vulcan disappeared when Albert Einstein (1879–1955) explained Mercury's anomalous motions in 1915 in terms of his *theory of general relativity*. The observational work of Simon Newcomb (1898), and later R. L. Duncombe (1956), on the perihelion advances of the terrestrial planets is compared with the predictions of Einstein's theory in Table 7–1. Note the close agreement.

The sidereal *rotation* period of Mercury was guessed to be 24^h in 1813 by F. W. Bessel, and by 1889 the observations of G. V. Schiaparelli seemed to fix it firmly at the *synchronous* value of 88^d. However, in the early 1960s, *radar* pulses reflected from

Table 7–1. Excess Perihelion Advances of Terrestrial Planets*

	Mercury	Venus	Earth	Mars
Observed	$43\overset{''}{.}11 \pm 0\overset{''}{.}45$	$8\overset{''}{.}4 \pm 4\overset{''}{.}8$	$5\overset{''}{.}0 \pm 1\overset{''}{.}2$	—
Einstein's Theory	$43\overset{''}{.}03$	$8\overset{''}{.}6$	$3\overset{''}{.}8$	$1\overset{''}{.}35$

* The excess angular advance (direct) is given in arc-seconds per century.

the surface of Mercury were first detected, and by 1965 G. H. Pettengill and R. B. Dyce had unambiguously shown the rotation period to be $59^d.3 \pm 2^d$ direct, using Doppler radar techniques (see following paragraph). A re-analysis of the old data, and new visual data obtained from the Pic-du-Midi Observatory in France, verified the radar period, and in 1967 A. Dollfus announced that Mercury rotates in $58^d.65 \pm 0^d.01$ with its equator essentially parallel to its orbital plane. Recall that in section 5–1(a, i) Peter Goldreich accounted for this period by saying that Mercury is in a spin-orbit resonance with the Sun (see Figure 5–5); its sidereal rotation period is two-thirds (or $58^d.64$) its sidereal orbital period. The earlier periods were probably in error because surface features on Mercury can only be observed visually near greatest elongation, which occurs at the *synodic* orbital period of $115^d.88$—essentially twice the sidereal rotation period.

Radar astronomy permits us to deduce a planet's distance, size, and rotation by simply analyzing radar pulses (wavelength $\approx$ 1 to 10 cm) reflected from the planetary surface. To date, the Moon (1946), Venus (1961), Mercury (1963), and Mars (1965) have been detected by radar. While lunar surface features, such as craters and mountains, are easily resolved, a radar pulse to the terrestrial planets is usually larger than the planet itself. Hence, we must resolve the return pulse into its *time-delayed* and *Doppler-shifted* components (see Figure 7–1). The shortest time-delay occurs at the sub-Earth point on the planet, and tells us the planet's distance from the Earth. Successively longer time-delays originate from circular annuli concentric about the sub-Earth point, with the longest delay defining the planet's *limb* (edge) and therefore its radius. As the planet rotates, the radar waves are

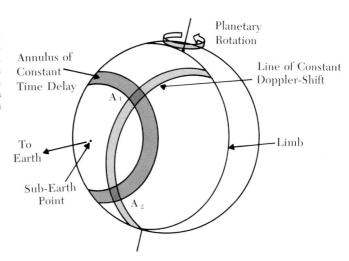

Figure 7–1. *Doppler Astronomy.* A radar pulse illuminates and reflects from the hemisphere of the planet facing the Earth. Concentric annuli of constant time delay intersect lines of constant Doppler-shift in two places, A_1 and A_2, leading to a two-fold ambiguity in the position of feature A.

Doppler-shifted in frequency (see Chapters 4 and 8), so that higher frequencies than the original return from the approaching limb and lower frequencies from the receding limb; these Doppler-shifts tell us the rotation speed of the planet. Radar mapping of the planetary surface is limited only by uncertainties in signal strength, time-delay, and frequency shift; in addition, a two-fold positional ambiguity shown in Figure 7–1 also exists, but this may be resolved as the Earth-planet direction changes.

(b) Characteristics

Telescopic observations of Mercury are best made at greatest elongation, when the planet is the greatest angular distance from the Sun. No sign of an atmosphere has been detected, though some observers have reported seeing a vague "haze"; the *upper limit* to the atmospheric pressure on Mercury is 10^{-4} Earth atmosphere or about 0.1 millibar (the sea-level pressure of the Earth's atmosphere is 1 *atmosphere* = 1013 millibars = 14.7 pounds per square inch). The absence of an atmosphere is consistent with Mercury's small mass and high surface temperatures.

Very subdued surface markings, reminiscent of the lunar maria and high-lands, have been seen and mapped by M. Camichel and A. Dollfus. The surface albedo is very low (0.06 visible; 0.06 to 0.07 at the radio wavelength 70 cm), indicative of dark rock; this albedo, the phase effect, and the polarization characteristics of Mercury are very similar to the properties of the Moon's surface. Surface temperatures vary from about 700 K at the perihelion subsolar point to about 110 K on the dark side. These tremendous temperature changes, as well as bombardment by meteorites and the solar wind, are probably responsible for the low relief of the surface—no well-defined shadows have ever been observed on Mercury.

Using radar, G. H. Pettengill and his associates in 1967 found Mercury's radius to be 2434 ± 2 km. The planet's mass of $0.055 M_{\oplus}$ is deduced from its gravitational perturbations upon other planets and asteroids—Mercury has no natural satellites. Hence, the average density is 5.45 gm cm^{-3}, typical of a terrestrial planet but high for Mercury's size. Solar heating has probably left Mercury with only the denser nonvolatile materials, such as iron and magnesium.

7–3   VENUS

Venus (♀) is the second (terrestrial) planet from the Sun, and, as a morning and evening star which rises to a greatest elongation of 48°, its maximum brilliance is exceeded only by the Sun and Moon. Named for the goddess of Love, Venus closely approximates the Earth in size, but the surface of this "mystery planet" is completely shrouded by clouds.

(a) Motions

The nearly-circular $(e = 0.0068)$ orbit of Venus is inclined $3°39$ to the ecliptic, with a semi-major axis of 0.7233 AU and a sidereal orbital period of

224^d.70. At the orbit of Venus, Mercury and the Earth are spectacularly brilliant planets, and the Sun's angular diameter is about 44′.

Radar contact with Venus was achieved in 1961, and in 1962 R. L. Carpenter and R. M. Goldstein discerned that Venus rotates *retrograde* in approximately 250 days. Recent Doppler radar studies at Goldstone, California, and Arecibo, Puerto Rico, carried out by R. M. Goldstein and G. H. Pettengill, show the planet to rotate retrograde with a sidereal period of 243^d.0 ± 0^d.2 and with its equator inclined but 3° to its orbital plane (conventionally this inclination is written as +177° or −87° to indicate the retrograde behavior). Again, P. Goldreich's spin-orbit coupling theory [section 5–1(a, i)] explains this rotation as a 4:1 synodic resonance with the *Earth* (see Figure 5–5): if Venus has a retrograde sidereal rotation period of 243^d.16, it will rotate synodically exactly four times during each synodic orbit with respect to the Earth.

(b)   Characteristics

The brightness of Venus is a result of the high albedo (0.76) of its clouds and thick atmosphere. Some structure and motion in the clouds may be seen in ultraviolet photographs of the planet taken near greatest elongation (see Figure 7–2); the planet's thick atmosphere causes its twilight zone to reach beyond the terminator into the dark side. The observed phase effect and polarization data for Venus indicate a gaseous atmosphere, as well as suspended particles with sizes near 1 μ. Water vapor and ice crystals (H_2O) have been invoked to explain the clouds, but only negligible traces of H_2O have ever been detected on Venus (see following paragraph). G. P. Kuiper conjectured that the clouds might be formed of carbon suboxide (C_3O_2). In 1967 P. and J. Connes used high-resolution infrared spectroscopy (see Chapter 8) to detect carbon dioxide (CO_2) and traces of carbon monoxide (CO), hydrogen chloride (HCl), and hydrogen fluoride (HF) in the Venusian atmosphere. Radio-frequency measurements show the planet's temperature to vary from 650 ± 70 K at the subsolar point to 500 ± 100 K at the poles, while the atmosphere of Venus is cooler.

It is exceedingly difficult to detect water on Venus (or on other planets) from the Earth. Water vapor in the Earth's atmosphere strongly absorbs infrared radiation to produce the *telluric bands* of H_2O. Only by using stratospheric balloons to get above the obscuring water, and by observing the Doppler-shifted H_2O features from the other planet, may we escape the telluric obscuration. This is how traces of water vapor have been discovered on Venus and Mars.

Our present picture of the Venusian environment is mainly the result of American Mariner and Russian Venera space probes to that planet (the journey takes about 4 months): Mariner 2 (December 1962), Venera 4 (October 1967), Mariner 5 (October 1967), Venera 5 and 6 (May 1969), Venera 7 (December 1970), and Venera 8 (July 1972). The Mariners (a) detected no Venusian magnetic field (< 0.005 Gauss at the surface or < 500 gammas, where 1 *gamma* = 10^{-5} Gauss) or radiation belts, but did detect the shock wave of Venus in the solar wind; (b) observed radiation from the atomic hydrogen (H) in Venus' protonosphere; (c) found a two-peaked Venusian ionosphere similar to but much less pronounced than the E and F layers of the Earth's ionosphere; (d) sent radio waves through the atmosphere of Venus to the Earth during occultation to deduce the pressure-

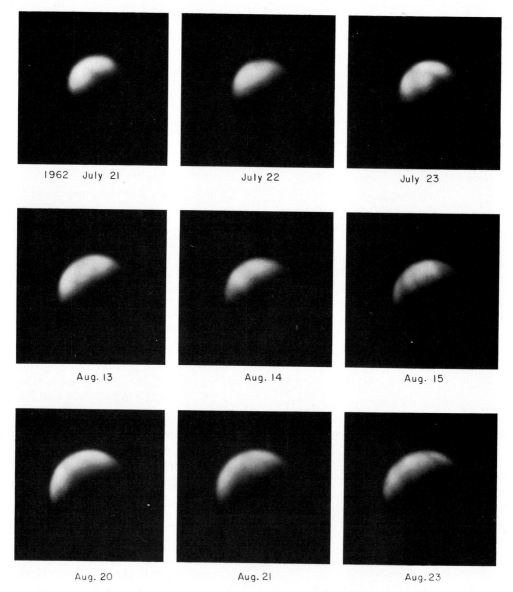

Figure 7–2. *Photograph of Venus.* Venus photographed in ultraviolet light on different dates shows the slightly changing nature of the clouds. (Lick Observatory)

temperature profile of the Venusian atmosphere; and (e) measured the surface temperature of Venus to be in the range 650 to 800 K. The Venera probes entered the atmosphere of Venus, and reported pressures, temperatures, and composition during the descent and after landing on the Venusian surface. The atmospheric composition is 90 to 95 per cent CO_2, less than five per cent N_2, one per cent O_2, and less than one per cent H_2O. As indicated in Figure 7–3, the *extrapolated* surface conditions are a pressure around *100 atmospheres* and a temperature near *700 K*—very extreme and inhospitable indeed.

How are we to understand the Venusian atmosphere? If the CO_2 were taken away, the atmosphere would closely resemble that of the Earth both in composition

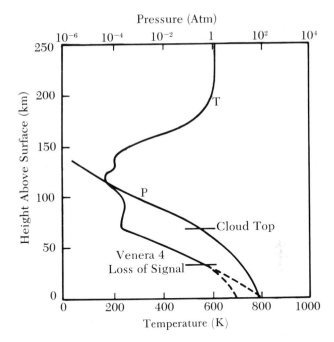

Figure 7–3. *Atmospheric Profile of Venus.* The observed (Mariner and Venera) temperature and pressure profiles of the Venusian atmosphere are shown, with extrapolated temperatures below 30 km where the Venera probes ceased operating.

and volume. But the higher surface temperature on Venus precludes the condensation of surface water, so that the CO_2 cannot precipitate out and be bound as, e.g., calcium carbonate ($CaCO_3$) as on Earth. The temperature of about 700 K is fairly uniform over the entire surface of the planet, even on the dark side. There is certainly some truth to the *Greenhouse Model*, which postulates that visible solar radiation reaches the surface of Venus, is re-radiated in the infrared ($\approx$ heat), and this longer-wavelength radiation is trapped near the surface by the thick, opaque CO_2 atmosphere—this would explain the high temperatures. The dense atmosphere would not have to move fast (convection and winds) to distribute this temperature over the entire surface; hence, Ernest J. Öpik's *Aeolosphere Model* with its deep turbulent dust layer heating the Venusian surface, while not ruled out, is now less attractive than when it was first put forward. Life, as we know it, is practically impossible on the surface of Venus, but there is the intriguing possibility that it could originate in the cooler (H_2O?) cloud layers above the planet.

Radar penetrates through the clouds to the surface of Venus, and crude *radar maps* of about one-third of the surface have been constructed (see Figure 7–4). Regions rough on a scale of about 10 cm have a different radar albedo than do smoother areas, and the resulting contrast indicates large-scale ($\approx$ 500 × 500 km) surface features which are suggestive of Venusian highlands and "circular maria."

While optical observations of (a) occultations of Regulus and (b) solar transits by Venus give the planet's radius as 6150 $\pm$ 25 km, the radar measurements of G. H. Pettengill in 1967 imply the definitive value of 6056 $\pm$ 1 km. Gravitational perturbations on the trajectories of the Mariner spacecraft yield the well-defined mass for Venus of $0.815 M_\oplus$; hence, the mean density is 5.25 gm cm^{-3}—very reasonable for the size of this terrestrial planet. We know nothing about the interior of Venus, but can conjecture that the structure is similar to that of our Earth, with much differentiation and a liquid core. The fact that Venus has an exceedingly small magnetic field (if any) is probably coupled to the planet's slow rotation.

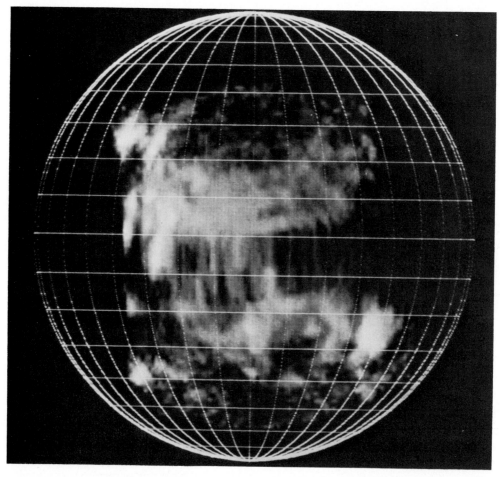

Figure 7–4.   *Radar Map of Surface of Venus, April, 1969.* This map shows the features according to their reflectivity at 3.8 cm. Reflectivity is a function of surface roughness. The area mapped is limited to the central portion of the disk. (From A. E. E. Rogers and R. P. I. Ingalls, Lincoln Laboratory: *Science,* Vol. 165, p. 798, August 22, 1969. Copyright 1969 by the American Association for the Advancement of Science)

7–4   MARS

As we pass beyond the Earth, we come to the first of the superior planets, Mars ($\male$). The ruddy red-orange color of this terrestrial planet led to its name, after the god of war. This fourth planet from the Sun runs the Earth a close race in its orbit, so that it exhibits the most pronounced retrograde motion in its apparent celestial path near opposition. Opposition occurs at the synodic period of $779^{d}.9$, or about 26 months, and at this time the $\approx 18''$ disk of Mars may be viewed at *full phase* throughout the entire night. In the Martian sky, Jupiter, Saturn, and the Earth are very bright planets, and the Sun's angular diameter is about $21'$.

(a)   Motions

Mars' orbit of semi-major axis 1.5237 AU is inclined only $1°.85$ to the ecliptic, and the modest orbital eccentricity of $e = 0.0934$ implies that *at opposition* Mars may range from about 10^8 km (an angular diameter of 14″) to 5.5×10^7 km (25″) from the Earth. Martian surface markings are best seen telescopically during the close *favorable oppositions* which take place around August and September when Mars is near its perihelion; the most recent favorable oppositions occurred in 1956 and 1971.

The tenuous atmosphere of Mars permits well-defined surface markings to be followed unambiguously; the Martian sidereal rotation period is $24^h37^m22^s\!.6$ and the rotation axis is inclined $25°12'$ to the orbital plane. This near-coincidence with the Earth's rotational properties ($23^h56^m4^s\!.09$ and $23°27'$) implies that an earthling would feel very much at home on Mars, except that the Martian seasons last about twice as long as the Earth's.

(b)   Satellites

Mars has two exceedingly small (radius < 10 km) natural satellites, Phobos (meaning Fear) and Deimos (Panic), which were discovered in 1877 by Asaph Hall. It is interesting to note that Jonathan Swift "predicted" these two Martian moons in his 1727 tale, *Gulliver's Travels*. Note that Phobos and Deimos are the *only* moons possessed by a terrestrial planet, if we consider the Earth and Moon as a binary planetary system. Both satellites move in almost-circular orbits near the equatorial plane of Mars, Phobos at a distance of 9380 km with a sidereal period of 7^h39^m and Deimos at 23,500 km with the orbital period of $1^d\!.26$. It is conjectured that both moons are in synchronous rotation. From the surface of Mars, Phobos appears to rise in the *west* and set in the *east* due to its rapid orbital motion; Phobos orbits just beyond the Martian Roche limit [see section 4–4(d)].

(c)   Surface

Interior.   Precise values for the Martian radius (3380 km) and mass ($0.107 M_\oplus$) have been determined from radar ranging, space probe trajectories, and the two moons of Mars. Mars is intermediate in size between the Earth and the Moon, and its average density of 3.97 gm cm^{-3} is similar to that of our Moon. The planet's oblateness (0.0052) reflects its rotation. We expect the interior of Mars to be somewhat more differentiated than that of the Moon, to have *no* liquid core, and to be seismically 'quiet." The magnetic field at the surface of Mars must be less than 100 gammas, since the Mariner spacecraft detected no radiation belts or magnetic field around Mars; only the Martian shock wave in the solar wind was seen.

Surface Markings.   Figure 7–6 is a telescopic photograph showing some of the abundant surface markings of Mars. Such photographs are best taken when the planet is near a favorable opposition, for then its angular diameter approaches 25″. The most prominent surface features are (a) the bright (high albedo) reddish-orange *deserts*, such as Hellas, (b) the darker grayish-brown *maria*, such as Syrtis Major, and (c) the brilliant white Martian *polar caps*. Recent radar studies show that surface elevations on Mars range over 5 to 10 km, with the deserts occupying *flat* regions while the "maria" tend to occur on *steep slopes*. At times of exceptional seeing, some observers have noted *thin dark lines*—called *canals*—on the Martian surface;

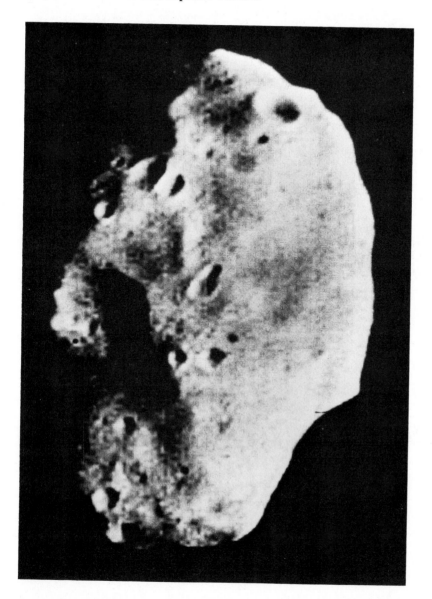

Figure 7–5. *Phobos.* Computer-enhanced photograph of the Martian satellite taken by Mariner 9 shows a heavily cratered surface. (NASA)

these were first observed by G. V. Schiaparelli in 1877 (he called them *canali*, meaning "channels"), and later popularized by Percival Lowell. The nature of these linear features is still unknown, and their existence is questionable. Many conjectures have been advanced concerning the *composition* of the Martian surface: the "maria" look like dark basalts (lavas?); A. Dollfus hypothesizes that the deserts are made of the iron ore Limonite, $Fe_2O_3(nH_2O)$. Measurements of the thermal variations of the deserts indicate a dry, dusty surface. Enormous yellowish "dust storms" are seen to originate in the deserts (especially just after Mars has passed perihelion) and to spread over the "maria." The dark regions return to their former hue once the yellow haze has passed. A major dust storm obscured the

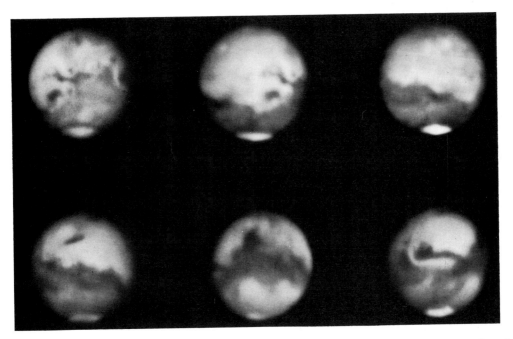

Figure 7–6. *High-Resolution Red Photographs of Mars at 1971 Opposition.* The photographs are composites of several images to increase resolution of surface markings. (Courtesy of R. Minton and S. Larson, Lunar and Planetary Laboratory, University of Arizona)

entire surface of Mars just when the Mariner 9 spacecraft arrived. Infrared spectroscopy during the storm provisionally identified SiO_2 as a major constituent of the dust.

Seasons. Figure 7–7 shows the seasonal variations on the surface of Mars. Consider the northern Martian hemisphere: during the Martian spring, increasing solar insolation causes the north polar cap to shrink rapidly, while a "wave of darkening" spreads from its periphery toward the equator; the "maria" become darker and browner. By mid-summer, the north polar cap may have completely disappeared. The dark regions begin to lighten progressively from the high latitudes to the equator, and in the fall the polar cap begins to grow beneath a hazy cloud cover. By mid-winter the polar cap reaches its maximum extent, and the "maria"

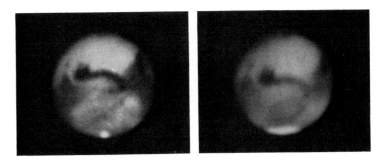

Figure 7–7. *Seasonal Changes on Mars.* Two photographs comparing the appearance of Mars in spring and summer, respectively. In the summer photograph, the polar cap is much smaller and the dark gray regions have darkened. (Lowell Observatory)

have returned to their original light-gray tint. The southern hemisphere undergoes similar changes, displaced in time by half a Martian "year," but the south polar cap rarely disappears in the summer.

The polar caps are layers of carbon dioxide and perhaps water ice (see below) several centimeters deep, similar to hoarfrost. This material vaporizes rapidly in the thin Martian atmosphere as the surface temperature rises slightly with the coming of summer. The "wave of darkening" is probably due to *mineralogical effects*, such as traces of the polar water vapor combining with basaltic rocks; another possibility is that primitive plants (lichens?) grow as water vapor moves toward the equator, but the extremely low atmospheric pressure and content of O_2 and H_2O argue against this. The prospects for finding life (in any form) on Mars are remote.

Mariner Results. Four American space probes have reached Mars after a journey of about seven and one-half months. Mariner 4 (July 1965), Mariner 6 (July 1969), and Mariner 7 (August 1969) all flew by the planet, while Mariner 9 (November 1971) was placed in orbit around it, and is still returning data at this writing. Two Soviet space probes reached Mars shortly after Mariner 9 and are also orbiting the planet. Both ejected soft landers, but most data come from the orbiters' photometers and radiometers. New and exciting deductions concerning Mars based on Mariner 9 and the Soviet Mars 2 and 3 are only now beginning to be published; hence, our discussion must be limited.

The Mariner space probes transmitted *television pictures* of the Martian surface. Far-encounter pictures from Mariners 6 and 7 show the entire disk of the planet (see Figure 7–8). Although Mariner 4 first revealed the lunar-like Martian surface covered with shallow *craters*, the pictures obtained by later Mariners are far more detailed. The tenuous Martian atmosphere and the proximity of Mars to the asteroid belt had led R. B. Baldwin and E. J. Öpik to predict a cratered surface, but its discovery came as something of a shock.

Martian craters resemble lunar highland craters, but are less numerous and are very shallow (no slopes $> 10°$ have been found). Some craters have central peaks, and slumping and terracing of crater walls are evident (see Figure 7–10). Numerous craters in the dark "maria" lend a mottled appearance to these regions. The absence of sharp relief indicates that Martian craters are subject to more *erosion* than their lunar counterparts; in fact, the desert region Hellas shows no trace of craters; apparently some process, such as erosive dust storms, has obliterated all craters there. The number and size distributions of Martian craters imply that some of the larger craters are *several billion years old*. If this is true, then Mars never had a dense atmosphere and water erosion never played a significant role.

Craters on the periphery and interior of the south polar cap have dark floors and bright rims, which are interpreted as the effect of differential melting of a thin layer of frost according to the amount of incident sunlight (see Figure 7–11). The central regions of the polar caps consist of a laminated terrain—a series of plates superimposed on each other and partially eroded away. The equatorial surface temperature of Mars ranges from 300 K (day) to 200 K (night), but at the poles the temperature is about 150 K—near the freezing point of CO_2 under Martian atmospheric conditions. Both ground-based and spacecraft observations indicate traces of H_2O in the atmosphere of Mars, with especially high concentrations over the south polar cap, so we can assume that the polar caps are a frozen mixture of CO_2 and H_2O.

(*Text continued on page 168.*)

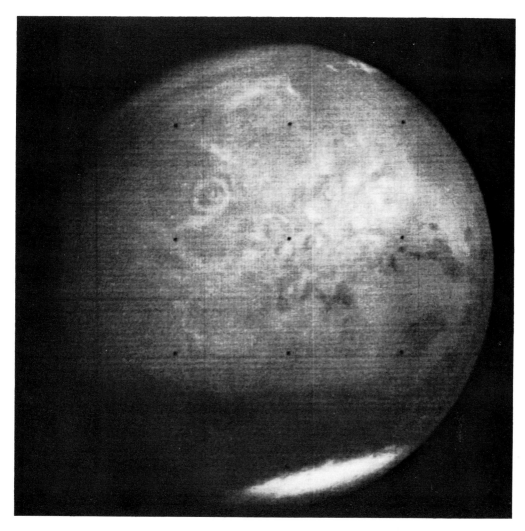

Figure 7–8. *Mariner 7 Far-Encounter Photograph of Mars.* The light circular region near the center of the planet is Nix Olympica. (NASA)

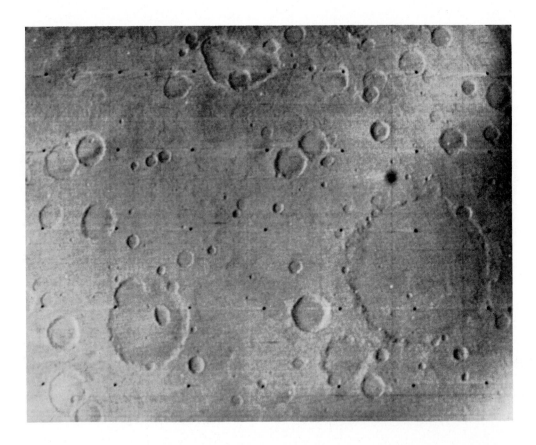

Figure 7–9. *Craters on Mars as Photographed by Mariner 6.* Note the low relief (i.e., shallowness) of the Martian craters, as compared with lunar craters. (NASA)

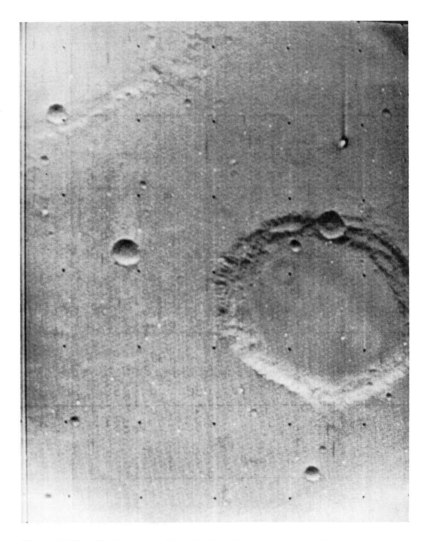

Figure 7–10. *Martian Crater.* This Mariner 6 near-encounter photograph shows a crater, about 100 km in diameter, with well-defined slumped walls. The small crater in the left corner is about the same size as the Arizona Meteor Crater. (NASA)

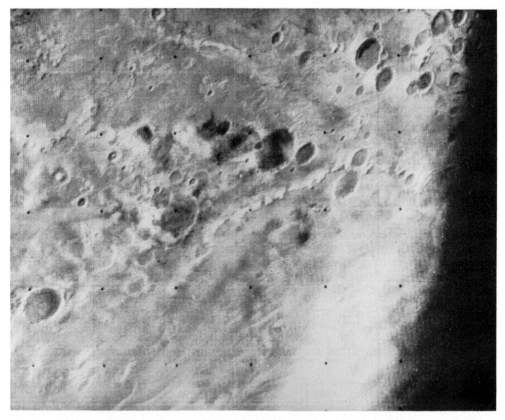

Figure 7–11. *Photograph of Martian "Snow Field."* A Mariner 7 picture of the south polar cap region showing the unevenness of the "snow" layer and irregular formations including craters. Sunlight comes from the top left, as seen from shadows in some craters. The terminator lies on the right side of the picture. (NASA).

Impact craters are not the only type of crater to be found on Mars; Mariner 9 revealed several large impressive craters as they protruded through the settling dust cloud. Nix Olympica is a prime example of such an elevated crater system (see Figure 7–12). The complex of overlapping rimless craters resembles terrestrial volcanic collapse calderas (e.g., some of the Hawaiian craters) far more closely than lunar impact craters with their sharp, ridged rims. Similarly, the 100 km diameter "South Spot," with its concentric faults and smooth floor, is more suggestive of vulcanism than of an impact phenomenon (see Figure 7–13).

Among the most fascinating sites shown by Mariner 9 is a long (5000 km) chasm with subsidiary valleys, appropriately nicknamed Feather River (see Figure 7–14). Here indeed is evidence of tectonic or internal processes producing faults and rifts, perhaps with accompanying flow of material. Even stronger indications of flow are sinuous rilles similar to those on the Moon; Martian rilles, however, differ in that they have branches. Although some water has been detected in the atmosphere, this is insufficient to cause such rivers. Hence, these rilles are presumed to be associated with lava flows. Mars may not be as geologically active as Earth, but it may well be more so than the Moon.

(*Text continued on page 172.*)

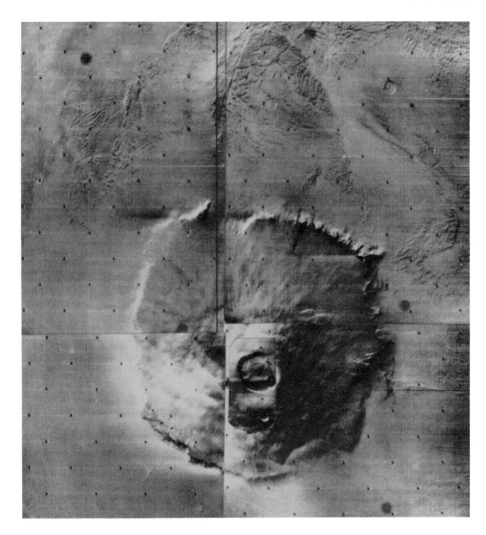

Figure 7-12. *Nix Olympica.* A volcanic mountain on Mars, with a total diameter of 500 km. The multiple crater is 65 km across, larger than any volcanic mountain on Earth. (Mariner 9 composite photo, NASA)

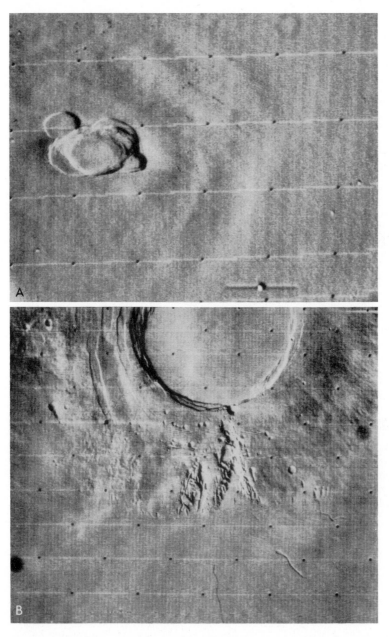

Figure 7–13.   *Two Large Volcanic Craters on Mars.* **A**, This is one of the early Mariner photographs, taken as the 1971 dust storm was subsiding and just revealing this complex crater. **B**, South Spot Crater with concentric faults and smooth floor. (Mariner 9 photos, NASA)

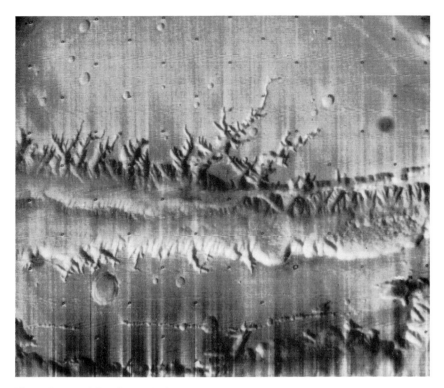

Figure 7–14. *A Vast Chasm with Branching Canyons.* (Mariner 9 photo, NASA)

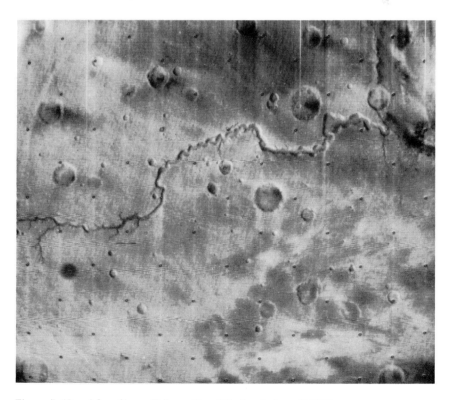

Figure 7–15. *A Long Sinuous Valley on Mars.* (Mariner 9 photo, NASA)

(d)   Atmosphere

The albedo of Mars is 0.16, intermediate in reflectivity between the Moon and the Earth. In addition to the yellowish haze described above, white and bluish clouds and haze are seen in the Martian atmosphere, especially over the polar regions and near the terminator. Moreover, recurrent afternoon clouds form in the vicinity of the volcanic mountains such as Nix Olympica. The nature of these clouds is not understood; they may be composed of CO_2 or H_2O ice crystals or water vapor. Although planet-wide dust storms such as that of 1971 are rare, localized dust storms occur quite frequently. Hellas particularly appears to be a "dust bowl" in the real sense, for it is a low-lying dusty area. Strong winds are responsible for lifting particles of 10 to 20 μ and greater into the atmosphere.

Observations from the Earth have shown that the Martian atmosphere contains CO_2 as a major constituent (G. P. Kuiper, and P. and J. Connes), with traces of CO (P. and J. Connes) and H_2O (G. P. Kuiper and A. Dollfus). Analyses of the observed spectral bands of CO_2 led to estimates of the *surface atmospheric pressure* on Mars, which ranged from 85 millibars to H. Spinrad's 1966 value of 8 ± 5 mb (recall the surface pressure of 1013 mb on the Earth). Direct probing of the Martian atmosphere during the Mariner flybys has greatly clarified the atmospheric structure and composition. Spectrometers found the radiation of atomic hydrogen in the Martian protonosphere. A double-peaked ionosphere at a temperature of 200 K was observed 100 to 150 km above the planet, with a peak density of 9×10^4 electrons per cubic centimeter. More than 90 per cent of the atmospheric abundance is CO_2, with some O, CO, and H_2O detected. Neither N_2 nor Ar were seen, but inert argon may be the second most abundant constituent of the atmosphere. Mariner 9 also detected ozone in the Martian atmosphere, in amounts small compared to terrestrial abundances. The amounts of both ozone and water show seasonal variations, but in a complementary sense. Ozone is most abundant when water is scarce, namely, during the winter when water has settled out of the atmosphere onto the surface of the polar cap.

All the Mariners passed behind Mars as seen from the Earth, so that their radio signals were occulted by the planet's atmosphere and disk. The observed radio waves showed changes due to refraction and dispersion in the Martian atmosphere, implying a surface atmospheric pressure of three to seven mb. Different surface elevations at the points of occultation account for the difference between the Mariner 6 result of 6.5 mb and the Mariner 7 value of 3.5 mb. Note that these pressures are about 5×10^{-3} Earth atmosphere, similar to the conditions in the Earth's upper *stratosphere*.

7–5   JUPITER

Beyond Mars we pass the *asteroid belt* near 3 AU, and finally come to the lord of the Jovian planets, Jupiter (♃), named after the king of the Olympian gods. Because of its enormous size and high albedo (0.51), Jupiter is a very bright planet in the Earth's night sky; it has been known since prehistoric times. From Jupiter, the other Jovian planets (especially Saturn) are bright in the sky, but the inferior terrestrial planets are never farther than 19° from the small disk (6′) of the Sun.

(a) Motions

Jupiter's orbit about the Sun has a small eccentricity (0.0484) and is inclined only 1°.31 to the plane of the ecliptic; at a semi-major axis of 5.2028 AU, the planet completes one sidereal orbit in 11.862 years. The synodic period of 398^d.88 implies that Jupiter returns to opposition (at full phase) about one month later each year.

Although the mass of Jupiter is only about one-thousandth that of the Sun, the planet's *orbital angular momentum* (see section 5–2) of about 2×10^{43} kg m^2 sec^{-1} comprises 60 per cent of the total angular momentum of the Solar System. Therefore, during the formation of the Solar System, most of the original angular momentum may have been transferred to Jupiter (and the other Jovian planets).

Since we can see only the dense atmosphere of Jupiter (see Figure 7–16), the planet's rotation period is determined by (a) following atmospheric features

Figure 7–16. *Photograph of Jupiter.* Note the banded structure of the atmosphere, and the prominent Great Red Spot at the upper left. (Hale Observatories)

such as the Great Red Spot, and (b) noting the Doppler-shifts of radiation coming from the approaching and receding limbs. We find that Jupiter's rotation axis is inclined $3°7'$ to its orbital axis, but the sidereal rotation period varies from 9^h50^m near the equator to 9^h55^m at higher latitudes. Hence, the gaseous Jovian atmosphere exhibits *differential rotation*; an equatorial band about 20,000 km wide moves about 0.1 km sec^{-1} faster than the rest of the atmosphere. In Chapter 9 we will see that the Sun also rotates differentially. The rotation structure of Jupiter's atmosphere is reminiscent of the Hadley cells and characteristic "tradewinds" in the Earth's atmosphere [see section 4–2(a)]. Jupiter's extremely rapid rotation is responsible for the large oblateness (0.062) of the planet's shape.

(b)   Structure

The equatorial radius $(11.19R_{\oplus})$ and mass $(318M_{\oplus})$ of Jupiter have been accurately determined by (a) observing the orbits and occultations of its twelve moons, (b) noting its gravitational perturbations upon the orbits of comets and asteroids, and (c) measuring its visible disk (47″ at opposition). This giant prototype of the Jovian planets has a mean density of only 1.33 gm cm^{-3}. Apparently the composition of Jupiter is similar to the *solar* mass abundances of about 75 per cent hydrogen, 25 per cent helium, and less than one per cent of all heavier elements.

The disk of Jupiter shows bands of clouds colored white, blue, red, and yellow (see Figure 7–16). These bands change their structure with time, but the relatively stable Great Red Spot has been observed since 1831. This enormous atmospheric feature measures about 20,000 km by 50,000 km, and it changes shape, position, and intensity. The Spot's intensity seems to be correlated with solar flare events (see Chapter 9). The most promising hypothesis says that the Spot is the top of a *Taylor column*: a vertical atmospheric vortex rising above some disturbance on the planet's surface. Jupiter's cloudy atmosphere is responsible for the planet's high albedo.

Infrared measurements of the clouds reveal an atmospheric temperature of about 130 K; this temperature is slightly higher than the computed equilibrium blackbody temperature, so that internal radioactivity or gravitational contraction must be supplying heat. Stellar occultations by the Jovian atmosphere imply a mean molecular weight of 3.3, intermediate between hydrogen and helium. H. Spinrad concludes that the atmospheric composition (by number) is about 60 per cent hydrogen, 36 per cent helium, perhaps three per cent neon, with traces of methane (< one per cent) and ammonia (< 0.05 per cent). Spectroscopic observations have revealed CH_4, NH_3, and H_2 (molecular hydrogen). The coloration of the Jovian atmosphere may be due to organic compounds (hydrocarbons)—possibly even the building blocks of life!

Radar contact with Jupiter has not yet been attained, so we must construct *models* to represent the structure of Jupiter. The dense atmosphere of hydrogen and helium is probably about 700 km deep, with the clouds (methane and ammonia?) occurring about 250 km above the "surface." As atmospheric density, pressure, and temperature increase toward the center, we encounter the liquid (hydrogen?) or "slushy" surface of the planet, which probably becomes solid about 250 km down. The oblateness of Jupiter indicates a highly-condensed core, possibly of *liquid*

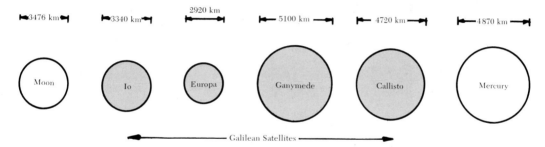

Figure 7–17. *The Galilean Moons of Jupiter.* Here we compare the size of our Moon to the sizes of Io, Europa, Ganymede, and Callisto. The planet Mercury is also shown.

hydrogen with some admixture of the heavier elements. Conditions in the lower atmosphere of Jupiter might be right for life to appear there.

(c) Satellites

Jupiter's family of twelve satellites is the largest in the Solar System. These moons occur in three groups: (1) the innermost five with circular orbits of very small eccentricity; these five satellites probably rotate synchronously; (2) Jupiter VI, VII, and X, which orbit *direct* at a mean distance of about 12×10^6 km; and (3) the outermost four, Jupiter VIII, IX, XI, and XII, which revolve *retrograde* in highly eccentric and highly inclined (to Jupiter's equator) orbits at about 23×10^6 km; these are probably captured asteroids or cometary nuclei.

The inner family or group (1) consists of Jupiter V, which was discovered by E. E. Barnard in 1892 and which orbits closest to Jupiter in about 12^h, and the four large *Galilean* satellites (within 2×10^6 km of Jupiter) discovered by Galileo in 1610 and used by him as a counterexample to the geocentric theory of Ptolemy (see Figure 7–17). The shadows of the Galilean satellites may be seen to cross the face of Jupiter; recall that Ole Römer first deduced the speed of light by observing the eclipses of these bodies by Jupiter's shadow.

The Galilean satellites appear as disks in telescopes, so that their dimensions may be measured (a) from their apparent angular diameters (note that Ganymede subtends $1\rlap{.}''7$) and the known distance to Jupiter, and (b) by observing their occultations by Jupiter. The sizes of the other moons are deduced from their brightnesses by assuming reasonable albedos. No atmospheres have ever been seen on the Galilean satellites, and photometric studies indicate (a) that their surfaces are rough like the lunar surface, (b) that some appear to be covered with "ice" or hoarfrost, and (c) that Callisto has an anomalously low density.

(d) Magnetic Field

Jupiter exhibits *radio emissions* which have been linked to a *Jovian magnetic field* of about 10 Gauss at the surface. This strong magnetic field is conjectured to arise from a dynamo mechanism in a rapidly-rotating liquid core of metallic hydrogen within the planet. At wavelengths from 3 cm to 75 cm, the planet is observed to radiate *nonthermally*; this *decimeter or DIM* radiation (1 decimeter = 10^{-1}

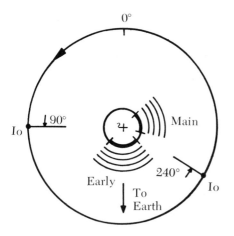

Figure 7–18. *DAM Radio Bursts from Jupiter.* Jupiter and the orbit of Io are shown, with the direction to superior conjunction indicated. The sporadic decameter (DAM) radiation is greatly enhanced when Io is either at 90° or 240°, and either the "early" or "main" longitude zone on Jupiter points toward the Earth. In this diagram, the "early" source (90° wide) points in the direction of the Earth. (After J. W. Warwick)

meter) is apparently the *synchrotron radiation* (see Chapter 16) from relativistic electrons spiralling at speeds very near the speed of light in Jovian *radiation belts* trapped by Jupiter's magnetic field. Radio interferometer measurements reveal belts similar to the Earth's Van Allen belts (see Figure 6–27), extending to three Jovian radii at the magnetic equator; the magnetic axis is inclined about 10° to Jupiter's rotation axis.

In 1955, scientists at the Carnegie Institute observed *sporadic radio bursts* at decameter or DAM wavelengths (1 decameter = 10 m); these bursts were found to be correlated with the *transit of Jupiter*. In 1964, it was discovered that this DAM radiation is associated with the position of the Jovian satellite Io relative to Jupiter's magnetic axis. As shown in Figure 7–18, the probability of radio bursts is greatly enhanced when Io is in one of two positions relative to Jupiter and the Earth: either 90° or 240° from the direction to superior conjunction. The DAM emissions arise when one of two longitude zones on Jupiter faces the Earth: the "early" source (90° wide) or the "main" source (60° wide). Though no adequate theory of this phenomenon has appeared, it is probable that Io interacts with the Jovian magnetic field and radiation belts, perturbing them and dumping high-energy particles into the upper atmosphere of Jupiter. Perhaps the radiation arises from the strong Jovian "aurorae" thus produced.

7–6 SATURN

Beyond Jupiter orbits the last of the "seven planets" known to the ancients, Saturn (♄), which is named after the father of Jupiter. Saturn is a gaseous Jovian planet, the second largest in the Solar System, and it is girdled by a magnificent system of *rings*. In our night sky Saturn is a bright planet because of its large size and the high albedo (0.50) of its atmosphere. The splendor of the Saturnian sky is characterized by the bright bands of its rings, its ten moons, and the impressively bright sister Jovian planets. The terrestrial planets are always crowded within 11° of the tiny (3.'4) Sun.

(a) Motions

At the orbital semi-major axis of 9.539 AU, Saturn's sidereal revolution period is 29.458 years, in a moderately eccentric (0.0557) orbit inclined 2°.49 to the ecliptic. From the Earth, Saturn's angular diameter at opposition is about 20″.

Like Jupiter, Saturn has a thick cloud-filled atmosphere which rotates differentially (see Figure 7–19). By observing the Doppler-shifts across the planet, and accurately timing atmospheric markings and the strange *white spots*, the sidereal period is found to be 10^h14^m near the equator and 10^h38^m at high latitudes. Saturn's equator is inclined 26°45′ to its orbital plane, so that alternate poles of the planet are tilted toward the Earth at intervals of about 15 years; the seasons on Saturn are similar to those on Earth, but last thirty times longer. The rapid rotation causes the large oblateness (0.096) of Saturn; the polar and equatorial radii of the planet are in the ratio of about 9 : 10.

(b) Rings

At the telescope, Saturn is a breathtaking sight (see Figure 7–19), for it is the only planet with *rings*. The entire ring system lies within the planet's Roche limit (near 148,000 km) in the range 71,000 to 138,000 km from the center of Saturn. The rings exhibit varying degrees of brightness, with dark gaps, as illustrated in Figure 7–20. From the moderately bright *outer ring* (ring A) we cross the dark *Cassini division* inward to the very bright *middle ring* (B), then across a smaller gap to the dim *crepe ring* (C), and finally to the newly-discovered, nearly-invisible

Figure 7–19. *Photograph of Saturn.* (Hale Observatories)

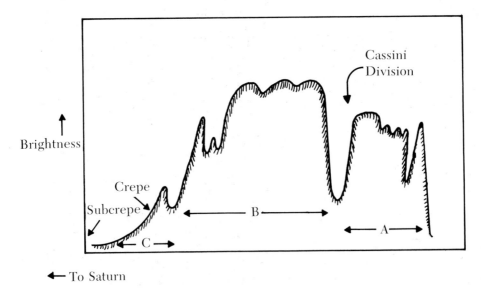

Figure 7–20. *Brightness Profile of Saturn's Rings.* A photometric (brightness) profile of Saturn's Ring System shows the outer (A), bright (B), crepe (C), and subcrepe rings, with the Cassini division. (After M. Camichel and A. Dollfus)

subcrepe ring. From the Earth, the appearance of the rings changes as Saturn moves in its orbit; when the rings are viewed edge-on they are practically invisible, while about 7-1/2 years later they are inclined about 27° to the line of sight and are spectacular. The rings were edge-on in 1966, and will be at maximum tilt in 1974.

Galileo first observed Saturn's rings in 1610, but C. Huygens was the first to explain their true form in 1655. The physical nature of the ring system follows from these facts: (a) stars may be seen *through* the rings, so they are not a solid body; (b) the rings shine simply by reflected sunlight with no absorption features characteristic of gases, so they are composed of *solid particles*; (c) spectra of the rings (see Figure 7–21) show motional Doppler-shifts consistent with these particles being in Keplerian orbits about Saturn; and (d) the thickness of the rings is much less than 10 km. In 1857 J. C. Maxwell proved that rings composed of myriad solid particles could orbit Saturn stably within the Roche limit. The size of these particles is probably between 1 mm and several meters, and they seem to be composed of ammonia or water *ice*. The numerous "gaps" in the rings are analogous to the Kirkwood gaps in the asteroid belt [see section 5–1(c)], arising from gravitational perturbations due to the moons of Saturn. A particle orbiting in Cassini's division has a period of $11^h 17^m.5$—half the orbital period of Mimas—and so is periodically perturbed and eventually removed from such an orbit. A. Dollfus used this analogy, and one of the observed gaps in the rings, to find the moon Janus (which orbits closest to Saturn) in December 1966 when the rings appeared edge-on.

(c) Satellites

Saturn's ten moons occur in three groups: (1) the inner six move in nearly-circular direct orbits of low inclination within 550,000 km of the center of Saturn (they probably rotate synchronously); (2) the next three orbit (direct) more

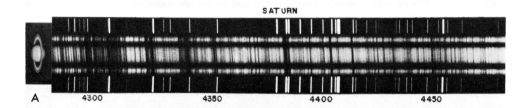

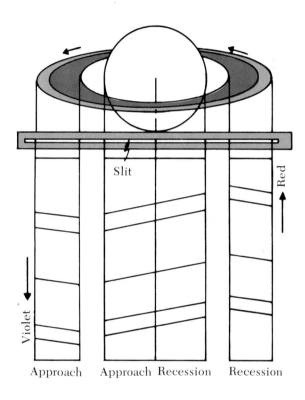

B

Figure 7–21. *The Rotation of Saturn and its Rings.* **A**, A spectrum across the equatorial plane of the planet shows the Doppler-shifted radiation from both the planet and its rings. (Lowell Observatory) **B**, The diagram shows the orientation of the slit and the resulting Doppler shifts. The rings clearly do not rotate as a solid body.

eccentrically in the range 1 to 4 × 10⁶ km; and (3) the outermost—Phoebe—moves in a highly eccentric and inclined *retrograde* orbit of semi-major axis 13 × 10⁶ km, and is probably a captured asteroid. The moons are of small to modest size, except Titan (radius = 2440 km) which is as large as Mercury and which has an *atmosphere* containing methane (detected by G. P. Kuiper in 1944). Note that Titan is the *only* planetary satellite for which an atmosphere has been observed.

(d)   Structure

Saturn's equatorial radius ($9.47R_\oplus$) and mass ($95M_\oplus$) imply an average density of 0.68 gm cm^{-3}—*less dense than water*. The structure of Saturn is probably similar to that of Jupiter, with hydrogen and helium predominating. Radio observations indicate Saturnian *radiation belts* and a strong *magnetic field*; hence, we can conjecture that Saturn has a rotating fluid core. The observed surface temperature of the planet is slightly higher than the computed equilibrium blackbody temperature, so some internal heat source must be active (just as for Jupiter).

The bands of atmospheric clouds on Saturn are colored white, yellow, gray-green, and orange (see Figure 7–20). Though hydrogen and helium are the major atmospheric constituents, only methane (CH_4) and molecular hydrogen (H_2) have been observed spectroscopically. The apparent absence of ammonia (NH_3) is consistent with the low temperatures of Saturn's atmosphere; ammonia has probably condensed as "frost" or "snow." Below Saturn's atmosphere and clouds there is probably a liquid region and then the solid body of the planet, in direct analogy to Jupiter's structure.

7–7   URANUS

Uranus (♅) is named after the progenitor of the Titans; it is the seventh planet from the Sun and the third Jovian planet. Though this body was plotted on star maps as early as 1690, it was first discovered to be a planet by William Herschel in 1781; at first Herschel thought that it was a comet, but his observations implied a low-eccentricity elliptical—hence, planetary—orbit about the Sun. Uranus is just at the limit of naked-eye visibility from the Earth, with an angular diameter at opposition of only 3''.6. The Jovian planets figure prominently in the Uranian sky, but the terrestrial planets are within 5° of the small (100''), dim Sun.

(a)   Motions

The solar orbit of Uranus has a semi-major axis of 19.182 AU, an eccentricity of 0.0472, and an inclination of only 0°.77 to the ecliptic; the sidereal orbital period is 84.013 years.

In Chapter 5 we discussed the bizarre rotational behavior of Uranus; with its equatorial plane inclined 98° to its orbital plane, Uranus rotates *retrograde* in 10^h49^m. Since the rotation axis lies essentially in the ecliptic plane, we will observe the following phenomena (see Figure 5–4): if we see one pole of the planet now, the orbital plane of Uranus' satellites will appear face-on. In 21 years the planet's equatorial plane will be seen edge-on, and in 42 years the opposite pole will point toward the Earth. The large oblateness (0.06) of Uranus results from its high rotation rate and low density.

(b)   Satellites

The five moons of Uranus move in small orbits (128,000 to 586,000 km from the center of Uranus) of very low eccentricity and inclination to the equatorial

Figure 7–22. *Photograph of Uranus with Satellites.* (Lick Observatory)

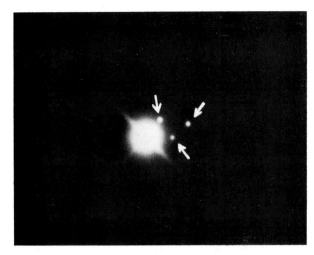

plane; apparently all are in synchronous rotation. The outer four range in diameter from about 400 to 1000 km, while the innermost, Miranda, which was discovered by G. P. Kuiper in 1948, is smaller. All five satellites orbit in the same sense that Uranus rotates—*retrograde*.

(c) Structure

Uranus is a Jovian planet, with a large equatorial radius ($3.73R_\oplus$), a large mass ($14.5M_\oplus$), and a low mean density of 1.56 gm cm^{-3}. Though hydrogen plays a less predominant role than it did in Jupiter and Saturn, we expect the structure of Uranus to be analogous to that of the larger planets. Our knowledge of the Uranian magnetic and temperature environment is sparse and uncertain.

The high albedo (0.66) atmosphere of Uranus is cloudy, and appears telescopically with a greenish color attributable to methane. Both methane (CH_4) and molecular hydrogen (H_2; discovered in the infrared in 1952) have been observed spectroscopically. The atmospheric temperature seems to be around 100 K, so that ammonia (NH_3) has probably solidified out of the atmosphere. Owing to Uranus' rotational inclination, we expect interesting *periodic* variations in its atmospheric circulation pattern.

7–8 NEPTUNE

The last of the Jovian planets, and the eighth planet from the Sun, is Neptune (Ψ). This near-twin to Uranus is named for the god of the sea. Between 1790 and 1840 the orbit of Uranus exhibited perturbations from an unknown source, and the existence of a more distant planet was suspected. J. C. Adams (1843) and U. J. Leverrier (1846) independently used Newtonian celestial mechanics to deduce the mass and orbit of this "eighth planet" from the observed perturbations on Uranus. In 1846 J. G. Galle at the Berlin Observatory found Neptune within 1° of the predicted position!

Neptune's angular diameter is only 2″ at opposition, and it can only be seen telescopically from the Earth. In the Neptunian sky, all the bright planets are near the Sun, and appear as morning and evening stars; the Sun has shrunk to an angular diameter of 64″, and the terrestrial planets stay within 3° of it.

(a) Motions

Neptune moves around its low eccentricity (0.0086), low inclination (1°.77) orbit at a semi-major axis of 30.058 AU in the sidereal period of 164.79 years. Since its discovery, Neptune has traversed only three-quarters of its orbit.

The rotation of Neptune is determined by the Doppler effect, and it is conventional: the planet rotates direct in 15^h40^m with its equator inclined 29° to its orbital plane. The planet's oblateness is moderately large at 0.02.

(b) Satellites

Neptune has only two satellites: the large moon Triton (radius = 2000 km) which orbits *retrograde* only 353,000 km from the center of Neptune, and the very small moon Nereid (radius $\approx$ 100 km) in direct orbit at a semi-major axis of 5.6×10^6 km. Both moons have highly inclined orbits ($\approx$ 25° to the equator of Neptune), but the eccentricity of Triton's orbit is negligible (implying synchronous rotation?) while that of Nereid is the largest in the planetary system (0.75). Recall the *hypothesis* [section 5–1(a, i)] that these satellite orbits resulted when Pluto escaped from an orbit about Neptune; can there be another reasonable explanation of Triton's orbit?

(c) Structure

Neptune's moons tell us the planet's radius ($3.50R_\oplus$) and mass ($17.2M_\oplus$); the average density is 2.25 gm cm^{-3}. While the structure of Neptune is probably similar to that of Uranus, the heavy elements are becoming more important. Undoubtedly, hydrogen and helium are still the major constituents. Ammonia appears to be absent in Neptune's cold, nondescript, blue-green atmosphere, and only methane and hydrogen have been detected spectroscopically. The thick atmosphere is probably filled with methane clouds, leading to the observed high albedo of 0.62.

7–9 PLUTO

Pluto ($♇$), the ninth planet from the Sun, is named after the god of the underworld (Hades). This small terrestrial planet is suspected to be an escaped satellite of Neptune. From the Earth, Pluto presents only a faint stellar image at the telescope; from Pluto, the rest of the Solar System is distant and close to the weak Sun, which appears only as a bright star in the sky.

After the discovery of Neptune, there still seemed to be small unexplained perturbations in the orbit of Uranus. Between 1900 and 1930 A. Gaillot, P. Lowell, and W. H. Pickering computed the characteristics of a ninth planet which would cause these perturbations. Since the predicted position was very uncertain, the initial attempts to find this planet were unsuccessful, and it was not until 1930 that Clyde Tombaugh found Pluto near the position predicted by Lowell (see Figure 7–23). Today we know that this discovery was a fluke, since Pluto could not have caused the apparent perturbations of Uranus. If Pluto had not been near the

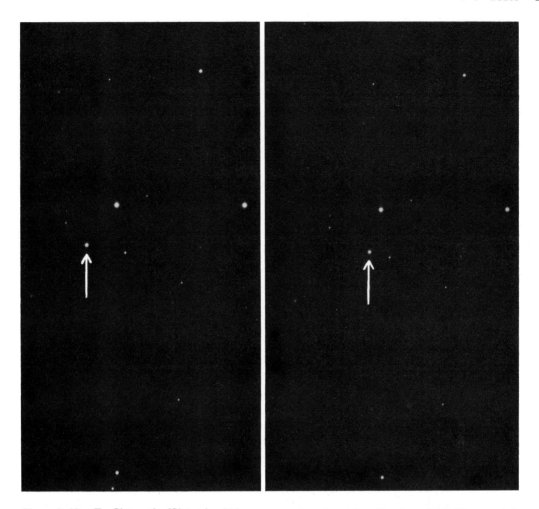

Figure 7–23. *Two Photographs of Pluto*, taken 24 hours apart to show the motion of the planet. (Hale Observatories)

ecliptic in 1930, it would never have been seen by Tombaugh. Pluto's discovery was an accidental coincidence!

(a) Motions

Pluto's orbit is highly eccentric (0.249) and highly inclined (17°.15) to the ecliptic plane. At a semi-major axis of 39.518 AU, the planet takes 248.43 years to complete one sidereal orbital period. Though Pluto is *nearer* the Sun at perihelion than Neptune is [see section 5–1(a, i) and Figure 5–3], gravitational perturbations never permit Pluto to approach closer than 18 AU to Neptune. Hence, Pluto can never return to collide with or orbit about Neptune.

Pluto's long sidereal rotation period of 6.3867 $\pm$ 0.0003 days was determined by M. Walker and R. Hardie, who photometrically observed periodic brightness fluctuations of the planet. The reason for this slow rotation is unknown.

(b) Characteristics

The physical nature of Pluto is very uncertain, since the planet is exceedingly difficult to observe. Pluto has no known satellites, so its mass must be determined solely from its gravitational perturbations on the other planets. The most recent results indicate a mass less than $0.18 M_\oplus$. From G. P. Kuiper's measurements at the 508-cm (200-inch) Palomar telescope, the near occultation of a star by Pluto in 1965, and the fact that Pluto's albedo must be less than 1.0, we find that the planet's radius lies between $0.16 R_\oplus$ and $0.46 R_\oplus$. Indeed, Pluto *may* resemble our Moon in its dimensions. We can only obtain an *upper limit* to the average density of Pluto; if, for example, we use a diameter of 5500 km and a mass of $0.18 M_\oplus$, the mean density of Pluto would be 7.7 gm cm^{-3}—an absurdly high value. Clearly the mass of the planet must be closer to $0.01 M_\oplus$.

We suspect that Pluto is too small to retain any atmosphere. Spectral analysis of the planet's light reveals no gases; at the expected surface temperature of about 60 K most common gases would be frozen out.

We do not know what planets might orbit beyond Pluto. Since the discovery of Pluto was a chance affair, there may be a tenth planet. Clyde Tombaugh has just ended a long systematic search for any trans-Plutonian planets near the ecliptic. His survey indicates that no such planet as large as Neptune orbits within 270 AU of the Sun. But recall that the Solar System does not end at Pluto, for numerous cometary nuclei reside beyond at aphelion distances near 10^5 AU.

Problems

7–1. Describe the appearance of Mercury and Venus as morning and evening stars, giving the time intervals between their first appearance (e.g., superior conjunction) and greatest elongation, and between greatest elongation and their disappearance (e.g., inferior conjunction). Support your statements with appropriate diagrams.

7–2. Why do Mercury and Venus only occasionally transit the face of the Sun at inferior conjunction? Indicating your assumptions, attempt to calculate the recurrence period for such transits as seen from the Earth.

7–3. A radar pulse is sent from the Earth to Venus at inferior conjunction. If the radar wavelength is 3 cm,
   (a) what is the *maximum* time-delay observed between the sub-Earth point and another point on Venus?
   (b) what is the total Doppler-shift (spread) of the return pulse?

7–4. Construct a table showing all the species of atoms and molecules *observed* on other planets. Indicate which species belong to which planets.

7–5. Draw a scale diagram of Mars and its satellites showing
   (a) the size of Mars.
   (b) the orbits of Phobos and Deimos.
   (c) the Martian Roche limit (assume $\rho_M = \rho_m$).
   (d) the "synchronous orbit," where a satellite would remain fixed over one longitude on the Martian surface.
   Comment upon your results.

7–6. A native of Jupiter would consider the Sun-Jupiter distance to be one Jovian Astronomical Unit (JAU). Express the orbital semi-major axes of all the planets in the Solar System in JAUs.

7–7. Plot the orbits of Jupiter's 12 satellites to scale, including the effect of the orbital eccentricities. Are you convinced that there are three groups of Jovian satellites? (Indicate prograde and retrograde motion.)

7–8. Find the orbital periods of particles at the inner and outer edges of Saturn's rings. At what distance from the center of Saturn will a particle orbit the planet in 10^h14^m? Show that the inner particles of the rings rise in the *west* and set in the *east* of Saturn's sky, while the outer particles rise in the *east* and set in the *west*. Is this result paradoxical? Explain.

7–9. Show how the orbits of Uranus' moons appear from the Earth over a period of 100 years.

7–10. Show that the moons of Neptune obey Kepler's third (harmonic) law, and thence deduce the mass of Neptune. (Hint: Use appropriate units or ratios.)

7–11. If Pluto's radius is $0.46R_{\oplus}$, what must be the *mass* of Pluto to give the planet the same density as our Moon? As the Earth?

Reading List

Alexander, A. F. O'D.: *The Planet Saturn*. New York, The Macmillan Company, 1962.
———: *The Planet Uranus*. London, Faber and Faber, 1965.

Brandt, John C., and Hodge, Paul W.: *Solar System Astrophysics*. New York, McGraw-Hill Book Company, 1964.

Brandt, John C., and McElroy, Michael B. (eds.): *The Atmospheres of Venus and Mars*. New York, Gordon and Breach Science Publishers, 1968.

Carr, Thomas D., and Gulkis, Samuel: "The Magnetosphere of Jupiter." *Annual Review of Astronomy and Astrophysics*, Vol. 7, pp. 577–618 (1969).

Dollfus, A. (ed.): *Moon and Planets*. Vols. I and II. Amsterdam, North-Holland Publishing Company, 1967–1968.

Evans, John V., and Hagfors, Tor (eds.): *Radar Astronomy*. New York, McGraw-Hill Book Company, 1968.

Firsoff, V. Axel: *The Interior Planets*. London, Oliver & Boyd, 1968.

Grosser, Morton: *The Discovery of Neptune*. Cambridge, Massachusetts, Harvard University Press, 1962.

Hartmann, William K.: *Moons and Planets: An Introduction to Planetary Science*. Tarrytown-on-Hudson, N.Y., Bogden & Quigley, Inc., 1972.

Ingersoll, Andrew P., and Leovy, Conway B.: "The Atmospheres of Mars and Venus." *Annual Review of Astronomy and Astrophysics*, Vol. 9, pp. 147–182 (1971).

Kuiper, Gerard P., and Middlehurst, Barbara M. (eds.): *Planets and Satellites*. Chicago, The University of Chicago Press, 1961.

Page, Thornton, and Page, Lou Williams (eds.): *Wanderers in the Sky*. New York, The Macmillan Company, 1965.

Peek, Bertrand M.: *The Planet Jupiter*. London, Faber and Faber, 1958.

Pickering, William H.: "The Grand Tour." *American Scientist*, Vol. 58, pp. 148–155 (1970).

Runcorn, S. K. (ed.): *Mantles of the Earth and Terrestrial Planets*. New York, Interscience Publishers, 1967.

Sagan, C., Owen, T. C., and Smith, H. J. (eds.): *Planetary Atmospheres*. (IAU Symposium 40). Dordrecht, Holland, D. Reidel, 1971.

Sky and Telescope. Cambridge, Massachusetts, Sky Publishing Corporation. Published monthly.

Slipher, Earl C.: *A Photographic Study of the Brighter Planets*. Flagstaff, Lowell Observatory, 1964.

Whipple, Fred L.: *Earth, Moon, and Planets*. Third Ed. Cambridge, Massachusetts, Harvard University Press, 1968.

Part 3

Basic Stellar Characteristics

Chapter 8
Electromagnetic Radiation and Matter

Having surveyed the local menagerie which we call our Solar System, let us cross the threshold to the ubiquitous stars. The nearest star is our own Sun, which is described in considerable detail in Chapter 9. But the other stars are merely points of light, scattered like dust across the night sky—how can we study such objects? Stars betray their presence by the radiations which they emit, and only by detecting and deciphering these radiations may we infer the properties of the emitting object.

In this chapter we will study the characteristics of *electromagnetic radiation* (most familiar to us in the form of visible *light*), the atomic structure of matter, and the all-important interactions between matter and radiation. Sir Isaac Newton showed how light could be dispersed into a rainbow-hued *spectrum*, and Gustav Kirchhoff (1824–1887) in 1859 stated the following three empirical rules or "laws" relating light spectra to their material source:

(1) a solid, a liquid, or a gas under high pressure, when heated to incandescence, will produce a *continuous spectrum*;

(2) a gas under low pressure, but at a sufficiently high temperature, will give a spectrum of *bright emission lines*; and

(3) a gas at low pressure (and low temperature), lying between a hot continuum source and the observer, causes an *absorption line spectrum*, i.e., a number of dark lines superimposed on the continuous spectrum.

Although Kirchhoff's laws simply describe spectral phenomena, and are incomplete as written in the preceding paragraph,* the recognition that *exactly* the same spectral lines appeared when a given gas (e.g., oxygen) was used in experiments (2) and (3) provided the Rosetta Stone of stellar astronomy. Each element has its own characteristic set of spectral lines. By comparing laboratory and stellar spectra, men like Joseph Fraunhofer (1787–1826), Gustav Kirchhoff, and Sir William Huggins (1824–1910) could deduce the presence of certain elements in stars. The way was now open for an astrophysical interpretation of the compositions and properties of the distant stars.

8–1   ELECTROMAGNETIC RADIATION

Keeping in mind our ultimate goal of truly understanding stellar spectra, let us begin by investigating the nature of *light*.

(a)   The Undulatory Nature of Light: Waves

Since the beginning of recorded history, mankind has been aware of both electrical and magnetic phenomena. In 1864, following the groundwork laid by C. A. de Coulomb (1736–1806), A. M. Ampère (1775–1836), K. F. Gauss (1777–1855), and Michael Faraday (1791–1867), the brilliant James Clerk Maxwell (1831–1879) created the modern theory of *electromagnetism*. This theory predicts that *electromagnetic waves* will propagate through a vacuum at a speed of $c = 299,793$ km sec^{-1}—the *speed of light*.

What is a *wave*? We are all familiar with water waves, which are undulatory disturbances travelling along the surface of the liquid. If we place corks along the line of the wave's motion, we may characterize such a wave by the height h of the corks above the average surface level (see Figure 8–1) in the mathematical form:

$$h = h_0 \sin \left[\frac{2\pi}{\lambda} (x - vt) \right] \tag{8–1}$$

Equation (8–1) represents a sinusoidal wave of *amplitude* h_0 progressing with time t along the positive x-axis at the speed v; the distance between successive wave crests is called the *wavelength* λ. Therefore, at a given time (say, $t = 0$), the corks betray the oscillatory pattern

$$h = h_0 \sin (2\pi x/\lambda) \tag{8–2}$$

whereas, if we observe the cork at $x = 0$, it rises and falls periodically. From equation (8–1) we see that a given cork completes one oscillation in the time (λ/v) seconds; hence, in one second (v/λ) wave crests will have passed a given point, so that the *frequency* (ν) of oscillation is just $\nu = (v/\lambda)$ sec^{-1}. Therefore, we may completely characterize such a wave by the fundamental relation

---

* For example, in (3) the interposed gas need not be cooler than the continuum source.

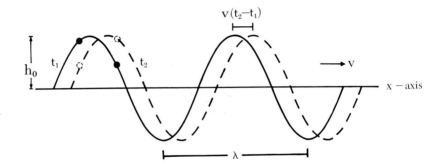

Figure 8–1. *A Travelling Wave.* A water wave of wavelength λ moves with speed v to the right, and is shown at two instants of time, t_1 and t_2. Floating corks mark the oscillating surface $h = h_0 \sin\left[(2\pi/\lambda)(x - vt)\right]$.

$$\lambda v = v \tag{8–3}$$

For example, water waves of wavelength $\lambda = 5$ cm, moving at a speed $v = 10$ cm $\sec^{-1}$, will pass a given point at a frequency $v = 2 \sec^{-1}$ (i.e., twice per second).

Most of the waves with which we are familiar require a material medium in which to be transmitted: water waves travel along the surface of water, sound waves move through air, and earthquake waves, both compressional and transverse [see section 6–3(a)], propagate through the solid Earth. Electromagnetic waves, however, may propagate through a pure vacuum at speed c.* What is it, then, that is propagating? The space around an electric charge may be characterized by an *electric field vector*, **E**, which manifests itself as a force on a test charge placed nearby. If an electromagnetic wave encounters such a test charge, that charge will perform oscillatory motion, so that we may ascribe the sinusoidal electric field

$$\mathbf{E} = \mathbf{E}_0 \sin\left[\frac{2\pi}{\lambda}(x - ct)\right] \tag{8–4}$$

to an electromagnetic wave travelling along the x-axis. But Maxwell's equations tell us that a time-varying electric field produces a perpendicular time-varying *magnetic field*, **B**, so that an electromagnetic wave is a self-propagating disturbance of **E** and **B** fields in a vacuum (see Figure 8–2). If **E** always oscillates in a single plane, we say that the wave is *linearly polarized*; otherwise, the wave is *elliptically polarized* (a special case is *circular* polarization). From equation (8–3) we see that

$$\lambda v = c \tag{8–5}$$

is the fundamental relation between the wavelength and frequency of an electromagnetic wave (in vacuum).

---

* It is a proven experimental fact, and a fundamental postulate of the special theory of relativity, that this speed is invariant; that is, every observer will locally measure exactly the same value for c whatever his velocity with respect to the source of light.

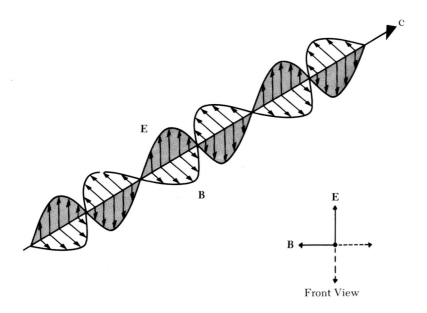

Figure 8–2. *An Electromagnetic Wave.* Perpendicular oscillatory electric **E** and magnetic **B** fields propagate (in phase) at speed *c* through a vacuum. The front view illustrates that this is a linearly polarized wave.

(i)   *The Electromagnetic Spectrum*

What is *light*? Light is a class of electromagnetic radiations, occupying the wavelength* range $\lambda \approx 3900$ Å (violet) to $\lambda \approx 7200$ Å (far red)—we call this the *visible spectrum*. Electromagnetic waves of different wavelengths are detected in different ways (see Table 8–1), so that we give characteristic names to various parts of the entire electromagnetic spectrum: gamma rays ($\lambda \lesssim 10^{-2}$ Å), X-rays (10^{-2} Å–10 Å), ultraviolet (10 Å–3000 Å); visible light (4000 Å–8000 Å), infrared (1μ–$10^3 \mu$), radar (1 mm–10 cm), and radio waves ($\lambda \gtrsim 1$ cm). While twenty-one decades of wavelength are shown in Table 8–1, visible light occupies less than one decade. Yet light is of paramount importance to astronomy, since (a) it easily penetrates the Earth's atmosphere and (b) it is readily perceived by the human eye.

(ii)   *Reflection and Refraction*

Let us consider some of the properties of light which are well-described by the wave picture of electromagnetic radiation. If we shine a ray (or beam or pencil) of light upon a mirror, we know that the ray is *reflected* in accordance with the rule (see Figure 8–3): "the angle of reflection (*r*) equals the angle of incidence (*i*)." Note that these angles are defined with respect to the *normal* (or perpendicular) to the reflecting surface. Reflecting telescopes utilize the law of reflection to collect, direct, and focus light.

---

* A wavelength (λ) is a *length*, with units of cm or m (or km or miles); two convenient units of length in astronomy are the micron μ (or micrometer), $1 \mu = 10^{-6}$ m, and the Ångstrom (Å), 1 Å $= 10^{-10}$ m $= 10^{-4} \mu$. The basic frequency unit is the Hertz (Hz) where 1 Hz $= 1$ sec^{-1}. See Appendix 5.

Figure 8–3. *Reflection of Light.* The angle of reflection (*r*) equals the angle of incidence (*i*).

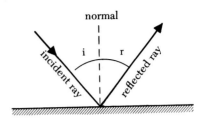

Table 8–1. The Electromagnetic Spectrum

λ	*v* or *hv*	Type of E-M Radiation	Detectors
10^{-5} Å	1,240 MeV	Gamma Rays	Geiger Counters
10^{-4} Å			Scintillators
10^{-3} Å	12.4 MeV		Nuclear Emulsions
10^{-2} Å			Proportional Counters
10^{-1} Å		X-Rays	Spark Chambers
1 Å = 10^{-8} cm	12.4 keV		Cerenkov-light detectors
10 Å			Photographic and Photoelectric Detectors
100 Å	124 eV	Ultraviolet	Image Forming Telescopes and Spectrographs and Spectrometers
1000 Å			
10,000 Å = 1 μ	1.24 eV	Visible	
10 μ			Photoconductive Detectors
100 μ	0.012 eV	Infrared	Thermal Detection Radiometers
1000 μ = 1 mm			
10 mm = 1 cm	30,000 MHz	Radar	
10 cm		UHF	Radio Telescopes
100 cm = 1 m	300 MHz	FM	
10 m		Short Wave	RADIO
100 m	3 MHz		Radio Receivers
1000 m = 1 km	300 kHz	Broadcast	
10 km		Long Wave	
100 km	3 kHz		
1000 km			

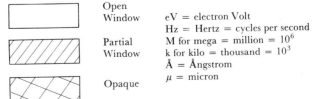

Open Window

Partial Window

Opaque

eV = electron Volt
Hz = Hertz = cycles per second
M for mega = million = 10^6
k for kilo = thousand = 10^3
Å = Ångstrom
μ = micron

When light is reflected, both the incident and reflected waves travel in the same medium at the same speed. But what happens when light passes from one medium into another? The speed of light (v) in a medium is generally different from the speed of light (c) in a vacuum. Hence, we may characterize a given medium by its *index of refraction*: $n = c/v$. The index of refraction of air is $n = 1.0003$, which is practically the same as that of a vacuum $(n \equiv 1.0)$. Crown glass, however, has $n = 1.5$ and $v = 2c/3 \approx 2 \times 10^5$ km sec^{-1}. Figure 8–4 shows how light passing from one medium (with index of refraction n_1), into another (n_2) is bent or *refracted* in accordance with *Snell's law*:

$$n_1 \sin i = n_2 \sin r \qquad (8-6)$$

where i is the angle of incidence in medium 1 and r is the angle of refraction in medium 2 (both with respect to the normal to the interface between the media). A refracting telescope is a practical embodiment of this phenomenon.

In general, the index of refraction of a medium depends upon wavelength, $n = n(\lambda)$. That is, light of different wavelengths (or *colors*) is refracted through different angles, $r(\lambda)$, when the incident beams all have the same angle of incidence i. Hence, an incident mixed (or white) beam is dispersed into separate beams of pure colors, just as Newton demonstrated. The phenomenon of *dispersion* enables us to decompose light (e.g., starlight) into its component colors—in the form of a spectrum—and is the basis of spectrographs and the discipline of spectroscopy.

(iii)   *Diffraction and Interference*

When water waves encounter an island, a distinct and sharp shadow is *not* formed on the lee side of the island. The waves *diffract* (or curve) around the sides of the island to converge, and when these converging waves meet one another, they *interfere*. In the case of sound waves, we are also familiar with these phenomena of *diffraction* (sound "bends" around sharp corners) and *interference* (recall the "dead" or silent spots in a large auditorium). The early optical experiments of Christiaan Huygens (1629–1695), A. J. Fresnel (1788–1827), and Thomas Young (1773–1829) demonstrated that light exhibits these same phenomena, which are understandable only in terms of a wave theory.

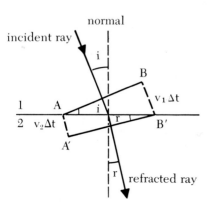

Figure 8–4.   *Refraction or Bending of Light.* $\overline{AB}$ represents a plane wavefront incident on the interface between medium 1 and medium 2 with angle of incidence i. Point A of the wavefront will travel to A' in time Δt, a distance of $v_2 \Delta t$. Point B, still in medium 1, travels a distance $v_1 \Delta t$. The ray of light is bent so that $\sin i/\sin r = v_1/v_2$. The common hypotenuse of the two triangles indicated is $\overline{AB'}$.

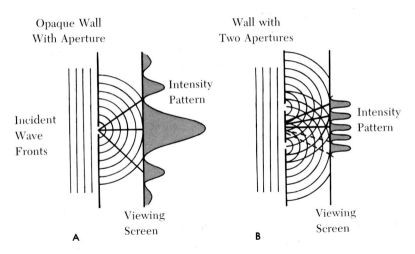

Figure 8–5. *Diffraction and Interference*. **A**, Diffraction with single aperture. **B**, Interference with two apertures.

Consider the situations illustrated in Figure 8–5, where light waves meet an opaque wall having one or two apertures, with a viewing screen placed beyond. The sharp images of the apertures, which we expect to see on the viewing screen, are distorted (a) by the diffraction or spreading-out of the light waves and (b) by the constructive and destructive interference of different light waves. We may understand both processes by considering how the intensity patterns at the viewing screen are formed. The intensity (I) of light is proportional to the square of the electric field (**E**), or

$$I \propto |\mathbf{E}|^2 \tag{8–7}$$

At an arbitrary point on the viewing screen of Figure 8–5B, the electric field of the wave is given by [see equation (8–1)]

$$E = E_0\big[\sin{(a)} + \sin{(a + b)}\big] \tag{8–8}$$

where a refers to the wave from one aperture and $a + b$ denotes the wave from the other aperture. The *phase shift b* corresponds to the difference in path lengths from the apertures to the point on the viewing screen. From equation (8–8), the intensity is

$$I \propto E_0^2\big[\sin^2{(a)} + \sin^2(a + b) + 2\sin{(a)}\sin{(a + b)}\big] \tag{8–9}$$

The first two terms on the right are just the intensities attributable to the separate apertures; the third term is responsible for the oscillatory *interference* pattern we see on the screen (as b changes with position).

Every point within the single aperture of Figure 8–5A contributes to the *diffraction* pattern at that viewing screen. A relatively straightforward application (which we leave to textbooks on optics) of equations (8–7) and (8–8) shows that the *angular width* (θ) of the principal diffraction image is given by

$$\theta \approx \lambda/d \tag{8–10}$$

where λ is the wavelength of the light and d is the size of the aperture. No optical image can be smaller than the *diffraction limit* of equation (8–10), so we say that this is the optimum *angular resolution* of the system. For example, the resolution of a telescope of aperture $d = 1$ meter, viewing light of wavelength $\lambda = 5000$ Å, is

$$\theta \approx \frac{5 \times 10^3 \text{ Å}}{10^{10} \text{ Å}} = 5 \times 10^{-7} \text{ radian}^* = 0.1 \text{ arc-second}^*$$

(iv) The Doppler Effect

Thus far we have considered waves in situations where both the source and the observer are relatively at rest. When either is in motion along the line connecting them, both the wavelength and frequency of the wave are altered by the famous *Doppler effect*—named for the Austrian physicist, C. J. Doppler (1803–1853). Again, we are familiar with the Doppler phenomenon in water and air waves; for example, as a siren approaches us, the pitch (frequency) of its sound is high, but the frequency drops noticeably to a lower tone as the siren passes and then recedes from us. Most familiar wave motions are associated with a material medium, which may complicate the picture by itself being in motion. Let us, therefore, consider electromagnetic waves propagating in a vacuum—Armand Fizeau correctly explained the classical Doppler effect of light in 1848, and Albert Einstein gave the relativistic explanation in 1905.

Figure 8–6 depicts a light source E receding at speed v from an observer O, while emitting radiation of wavelength λ_e and frequency v_e. In the time $t = 1/v_e$, one wavelength (λ_e) emerges from the source, but, as seen by the observer, that wave has the length

$$\lambda_o = (c + v)t = c\left(1 + \frac{v}{c}\right)\bigg/v_e = \lambda_e\left(1 + \frac{v}{c}\right) \tag{8–11}$$

since the source has travelled the distance vt to the right. Note that the fundamental equation (8–5) has been used in the last equality. The observed frequency is

$$v_o = c/\lambda_o = v_e\bigg/\left(1 + \frac{v}{c}\right) \tag{8–12}$$

---

* See Appendix 5.

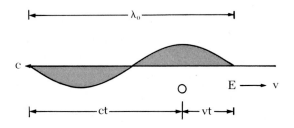

Figure 8–6. *The Doppler Effect of Light.* The source E recedes from the observer O at speed v, while emitting electromagnetic waves of wavelength λ_e and frequency v_e. The observer sees the wavelength λ_o and the frequency v_o.

so that $\lambda_o > \lambda_e$ and $\nu_o < \nu_e$, and we say that the light is *redshifted* (i.e., shifted to longer wavelengths, or toward the red). When the source *approaches* the observer, v changes sign to give $\lambda_o < \lambda_e$ and $\nu_o > \nu_e$, or we say that the radiation is *blueshifted*.

From equation (8–11) we may extract the useful *shift in wavelength* (see Chapters 11 and 14)

$$\Delta\lambda/\lambda_e \equiv (\lambda_o - \lambda_e)/\lambda_e = v/c \qquad (8\text{–}13)$$

Remember that the sign of v is positive $(+)$ for recession and negative $(-)$ for approach.

When v becomes comparable to c, Einstein's theory of special relativity must be invoked. The two basic postulates of this theory are: (1) the speed of light (c) is independent of the motion of either source or observer and (2) only relative motions are observable. Consequently, v is the *relative* speed of the source and the observer, no relative speed greater than the speed of light is possible (i.e., $v \leq c$), and equations (8–11) and (8–12) necessarily take the symmetrical forms

$$\lambda_o = \lambda_e \left[\left(1 + \frac{v}{c}\right) \middle/ \left(1 - \frac{v}{c}\right) \right]^{1/2} \qquad (8\text{–}14a)$$

$$\nu_o = \nu_e \left[\left(1 - \frac{v}{c}\right) \middle/ \left(1 + \frac{v}{c}\right) \right]^{1/2} \qquad (8\text{–}14b)$$

As an exercise (see the Problems), the reader may show that the shift in wavelength [equation (8–13)] follows from equation (8–14a) in the limit when $v \ll c$.

(b)   The Quantum Nature of Light: Photons

As early as the beginning of the eighteenth century, Newton proposed a particle or *corpuscular* theory of light. It was not until the end of the nineteenth century that the particle-like manifestations of light were clearly discerned—light is neither a particle nor a wave, but it can manifest itself as either or both! This apparently paradoxical behavior is necessary to explain such phenomena as the photoelectric effect, Compton scattering, and blackbody radiation [see section 8–6(a)]. In addition, the interaction of light with atoms and molecules is understandable only if electromagnetic energy propagates in the form of discrete bundles, which we call *photons* or *quanta*. The energy of a light quantum (ϵ) is proportional to the frequency (ν) characterizing the light wave:

$$\epsilon = h\nu \qquad (8\text{–}15)$$

where $h = 6.625 \times 10^{-27}$ erg sec is known as Planck's constant. Hence, we may *crudely* picture a classical light wave of wavelength λ and frequency ν as being composed of multitudes of quanta, each with the energy given by equation (8–15). Our discussion of photons will continue throughout the remainder of this chapter.

(c) Intensity Versus Flux

When we attempt to detect the energy or to count the photons coming from a distant light source, we must be careful to distinguish between intensity and flux.

Intensity depends upon direction, in the sense that the intensity (I) of a source is the amount of energy emitted per unit time (Δt), per unit area (ΔA) of the source, per unit frequency interval ($\Delta \nu$), per unit solid angle ($\Delta \Omega$) in a given direction. The concept of *solid angle* is illustrated in Figure 8–7A, where the solid angle of the beam ($\Delta \Omega$) is related to the area (Δa) intercepted by the beam at the spherical surface of radius r via

$$\Delta \Omega = \Delta a / r^2 \qquad (8\text{–}16)$$

The unit of solid angle is the *steradian** (ster), with the entire spherical surface subtending 4π steradians—since $\Delta a = 4\pi r^2$ for the surface area of a sphere. Therefore, the common units of intensity are $[\text{erg cm}^{-2} \text{ sec}^{-1} \text{ Hz}^{-1} \text{ ster}^{-1}]$. For example, the total energy flow from a spherical star of surface area A is just

$$\left[4\pi A \int_0^\infty I(\nu) \, d\nu \right] \text{erg sec}^{-1}$$

where $I(\nu)$ is called the monochromatic intensity.

Flux (F) relates directly to what we measure. The flux of energy through a surface (or into a detector) is the amount of energy per unit time (Δt) passing through a unit area (ΔA) of the surface, per unit frequency interval ($\Delta \nu$), or $F(\nu) = \text{energy}/\Delta A \cdot \Delta t \cdot \Delta \nu$ (see Figure 8–7B). Hence, the unit of monochromatic flux is $[\text{erg cm}^{-2} \text{ sec}^{-1} \text{ Hz}^{-1}]$. For example, if a star emits energy at the *rate* $\mathscr{E}$ erg sec^{-1}, then the flux of energy through a concentric spherical surface of radius R is

$$F = (\mathscr{E}/4\pi R^2) \text{ erg cm}^{-2} \text{ sec}^{-1} \qquad (8\text{–}17)$$

---

* A steradian is essentially one square radian; since 1 radian = 57°.3, 1 ster = 3283 square degrees of arc.

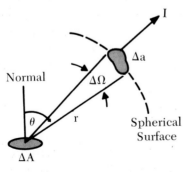

Figure 8–7A. *Intensity versus Flux.* Intensity depends upon direction (θ) and solid angle ($\Delta \Omega$).

Normal

$\Delta \Omega$

Δa

I

r

Spherical Surface

θ

ΔA

A

Figure 8–7B. Flux depends only upon the energy (or numbers of particles) passing through an area (ΔA) per unit of time.

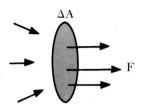

B

Equation (8–17) is a representation of the well-known inverse-square law (i.e., R^{-2}) for the diminution of radiant flux with distance.

Notice that we may speak equally well of the intensity and flux of numbers of moving particles (e.g., photons). Also note that a one-dimensional beam has flux, but no intensity (since $\Delta\Omega = 0$). We may relate intensity and flux by noting that the energy passing through a certain area is composed of beams entering at different angles (θ) to the normal to the surface; to a beam entering at the angle θ, the surface area appears reduced by the factor $\cos\theta$ so that

$$F(v) = \int I(v, \theta) \cos\theta \, d\Omega$$

8–2 ATOMIC STRUCTURE

(a) Atomic Building Blocks

If we divide matter into smaller and smaller pieces, we find that, at scales near 1 Å (10^{-8} cm), all forms of matter are composed of *atoms* (and molecules, which are aggregates of atoms). An elemental substance, such as gold, consists solely of atoms of the *element* gold; there are but 92 natural elements (hydrogen to uranium; see Appendix 3), but this list has been extended to 105 with the addition of the man-made trans-uranic elements.

What is an atom, and how may we characterize the elements? An atom is composed of a small *nucleus* which contains most of the atom's mass in a size of the order of 10^{-12} cm, surrounded by a diffuse "cloud" of *electrons* extending out to about 10^{-8} cm. The atomic nucleus consists of *protons* and *neutrons* bound together by the strong interaction. These three elementary particles—the proton, the neutron, and the electron—are the fundamental building blocks of an atom. Table 8–2 lists the important parameters of each particle:

Table 8–2. Parameters of Atomic Particles

Name	Mass (GM)	ELECTRON MASSES	Electric Charge
Proton	1.6725×10^{-24}	1836	$+e$
Neutron	1.6748×10^{-24}	1838	0
Electron	9.1091×10^{-28}	1	$-e$

Both the proton and neutron are about 2000 times more massive than the electron. While the neutron is uncharged, the proton (positive) and the electron (negative) have electric charges of opposite sign and equal magnitude, $e = 4.803 \times 10^{-10}$ electrostatic units (esu) or 1.602×10^{-19} Coulomb. A normal *neutral* atom consists of an equal number of protons and electrons, with approximately the same number of neutrons as protons; hence, we may characterize an element by the number of protons, Z, in the atomic nucleus. For example, hydrogen ($Z = 1$) is one electron "circling" one proton; uranium ($Z = 92$) has 92 protons and 92 electrons, with about 145 neutrons in its nucleus.

A given element may exist in several different forms, which we call *isotopes*. All isotopes of an element have the same number (Z) of protons in the nucleus, but differing numbers of neutrons (N). We term the different isotopic nuclei, *nuclides*. To characterize a nuclide, the notation $_Z X^{Z+N}$ or $_Z X^A$ is used, where X is the symbol of the element with Z protons, and $A = Z + N$ is the *atomic number* (number of protons plus neutrons). For example, three isotopes of hydrogen are known: ordinary hydrogen, $_1 H^1$; deuterium, $_1 H^2$; and tritium, $_1 H^3$. Note that X and the subscript Z are redundant, conveying the same information. The mass of an atom is conveniently given in terms of *atomic mass units* (amu); since 1961, the standard has been $_6 C^{12}$ with a mass of exactly 12 amu. Since 1 amu is essentially the mass of a proton, we see that an atom's mass is A amu. When atomic masses are tabulated (see Appendix 3), they frequently depart from integer numbers because it is the average mass of the naturally-occurring isotopes which is listed.

(b)   The Bohr Atom

What is the dynamical configuration of the electrons about the nucleus which leads to a stable atom? Since the attractive electric force between a proton and an electron (Coulomb's law) is analogous to Newton's law of gravitation, scientists envisaged electrons orbiting the nucleus just as the planets orbit our Sun. But electrons are *charged* particles, and charged particles radiate energy when they are accelerated (e.g., in circular orbit). Hence, the very property which permits a bound atom seems to cause it to collapse almost at once! In 1913 Niels Bohr (1885–1962) advanced a simple theory which offered a way out of this dilemma, and which led to the modern theory of *quantum mechanics* in the 1920's. Let us discuss Bohr's atomic theory, and apply it to the hydrogen atom (which it accounts for most satisfactorily).

(i)   *Quantized Orbits*

In 1911 Sir Ernest Rutherford (1871–1937) had proposed the "nuclear" or Solar System model of the atom. Bohr spent a year with Rutherford, and then stated two astounding postulates which breathed life into atomic theory. Of the infinite number of possible electron orbits in Rutherford's model, Bohr first postulated that *only a discrete number of orbits is allowed to the electron, in which orbits the electron cannot radiate.* The permitted orbits are those in which the orbital angular momentum of the electron is an *integer* multiple of $(h/2\pi)$, where h is Planck's constant.

Let us apply Bohr's postulate of quantized orbits to an electron in circular orbit (of radius r) about a nucleus of charge Ze. The centripetal force maintaining the orbit (where m is the electron's mass)

$$mv^2/r$$

is provided by the Coulomb attraction between the electron and the nucleus

$$(Ze)e/r^2$$

But Bohr's postulate implies the additional constraint

$$mvr = n(h/2\pi) \quad \text{(with } n = 1, 2, 3, \ldots) \tag{8–18}$$

Combining these equations, we find

$$r = Ze^2/mv^2 = nh/2\pi mv$$

or

$$\boxed{r = n^2(h^2/4\pi^2 me^2 Z)} \tag{8–19}$$

Therefore, the permitted discrete orbits occur at geometrically increasing (n^2) distances, with the smallest Bohr orbit occurring when the *principal quantum number n* equals one.

We may now find the *total energy* (E) of these orbits; if E is negative, the system is bound and we have an atom. From Chapter 3 we know that E = kinetic energy + potential energy, and, in direct analogy to the case of universal gravitation, we may write

$$E = \frac{mv^2}{2} - \frac{Ze^2}{r} \tag{8–20}$$

By evaluating equation (8–20), using equations (8–18) and (8–19), we easily find:

$$\boxed{E(n) = -\frac{2\pi^2 me^4 Z^2}{n^2 h^2}} \tag{8–21}$$

The smallest Bohr orbit $(n = 1)$ is the most strongly bound, as we might well expect, and all orbits are bound until $n \to \infty$ where $E \to 0$. For $E > 0$, a *continuum* of (unbound) hyperbolic orbits is available to the electron.

(ii)   *Quantized Radiation*

Bohr's second postulate concerns the absorption and emission of radiation by an atom—the fundamental interaction between matter and radiation. According to the first postulate, the electron cannot radiate while in one of the allowed discrete orbits, so the second postulate states: (a) *radiation in the form of a single discrete quantum is emitted or absorbed as the electron jumps from one orbit to another, and (b) the energy of this radiation equals the energy difference between the orbits.*

When an electron makes a *transition* (jumps) from a higher orbit (n_a) to a lower orbit (n_b), a photon is emitted, and the energetics of this process may be symbolized by:

$$E(n_a) = E(n_b) + h\nu \quad (emission) \tag{8-22a}$$

where $n_a > n_b$. For the electron to make a transition from the lower orbit to the upper orbit, the atom must absorb a photon of exactly the correct energy:

$$E(n_b) + h\nu = E(n_a) \quad (absorption) \tag{8-22b}$$

In both cases [equations (8–22)], the frequency of the photon involved is [see equations (8–15) and (8–21)]

$$\boxed{\nu_{ab} = \frac{E(n_a) - E(n_b)}{h} = \left(\frac{2\pi^2 m e^4}{h^3}\right) Z^2 \left(\frac{1}{n_b^{\,2}} - \frac{1}{n_a^{\,2}}\right)} \tag{8-23}$$

Note that only one quantum is emitted or absorbed, even if $n_a > n_b + 1$; the electron may jump over several intermediate orbits. On the other hand, an electron may *cascade* to the lowest orbit, emitting several photons of different energy as it jumps through a series of adjacent orbits. The lowest orbit, called the *ground state*, corresponds to $n = 1$.

(c)   The Bohr Model of the Hydrogen Atom

Let us apply Bohr's picture to the simplest atom—hydrogen $(Z = 1)$. The electron's permitted orbital energies are, from equation (8–21),

$$E(n) = -\left(\frac{2\pi^2 m e^4}{h^2}\right)\left(\frac{1}{n^2}\right) \equiv -R'\left(\frac{1}{n^2}\right) \tag{8-24}$$

where $R' = 2.18 \times 10^{-11}$ erg incorporates all the other constants. Since we customarily observe the wavelengths of the radiations involved in electron transitions, it is convenient to express equation (8–23) in terms of the *wave number* (reciprocal wavelength)

$$\boxed{\frac{1}{\lambda_{ab}} = \frac{\nu_{ab}}{c} = \frac{R'}{ch}\left(\frac{1}{n_b^{\,2}} - \frac{1}{n_a^{\,2}}\right) = R\left(\frac{1}{n_b^{\,2}} - \frac{1}{n_a^{\,2}}\right)} \tag{8-25}$$

The *Rydberg constant*, R, has the value $109{,}677.6$ cm^{-1}.

Equation (8–25) tells us that there is a *series* of wavelengths for *each* value of n_b, when we consider a sequence of increasing values of n_a beginning with

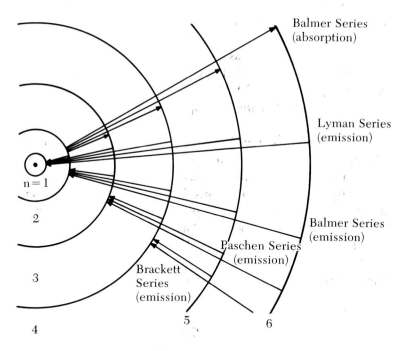

Figure 8–8. *Bohr Model of the Hydrogen Atom.*

$n_a = n_b + 1$ (see Figure 8–8). Transitions of the *Lyman* series all have the ground state ($n_b = 1$) as their lowest orbit, with $n_a \geq 2$. The visible spectral lines of the *Balmer* series have $n_b = 2$ and $n_a \geq 3$. Other important series are the Paschen, Brackett, and Pfund series, with $n_b = 3$, 4, and 5, respectively. All of these series are named for the physicists who first observed the spectral lines corresponding to the indicated transitions.

The first hydrogen lines discovered were the Balmer series, and they are designated H_α for $n_a = 3$, H_β for $n_a = 4$, H_γ for $n_a = 5$, etc. Let us use equation (8–25) to compute the wavelength of the H_α line, which corresponds to $n_b = 2$ and $n_a = 3$:

$$\frac{1}{\lambda_{H_\alpha}} = 109{,}677.6(\tfrac{1}{4} - \tfrac{1}{9}) \text{ cm}^{-1} = 15{,}233.0 \text{ cm}^{-1}$$

or

$$\lambda_{H_\alpha} = 0.00006563 \text{ cm} = 6563 \text{ Å}$$

Similarly, the Lyman α line ($n_b = 1$, $n_a = 2$) has the wavelength $\lambda_{Ly_\alpha} = 1216$ Å.

(d) The Energy Level Diagram

We must warn the reader that Bohr's simple theory is an approximation to the actual dynamics of atomic phenomena. The theory encounters insurmountable difficulties when extended to atoms more complicated than hydrogen, and the full machinery of *quantum mechanics* is necessary to understand the atomic domain in detail.

Atomic particles (e.g., the electron) also exhibit a wave nature like photons; hence, there is an inherent uncertainty in both the position and velocity of an electron (the famous *Heisenberg uncertainty principle*). The electron in the hydrogen atom may be portrayed as a *cloud* surrounding the proton, with the most probable position of the electron being one of the Bohr orbits. A multi-electron atom has several such clouds, with the electrons tending to occupy fuzzy *shells* about the nucleus. The simplest shells are spherical, but, in general, more complicated shapes occur.

We may sidestep the sophistications of quantum mechanics in this book by abandoning all reference to spatial models of atoms. Instead, we will represent atoms abstractly by means of an *energy level diagram* (see Figure 8–9). Such a diagram is directly related to atomic transitions, so it can be constructed even for complicated atoms. As an example, let us explain the energy level diagram for hydrogen.

The permitted energies of a bound electron in the hydrogen atom [equation (8–24)] are negative. Since we observe the positive-energy photons corresponding to electronic transitions, let us normalize to a positive energy scale by subtracting the ground state energy $E(1)$ from all energies $E(n)$, to obtain

$$E(n) = R'\left(1 - \frac{1}{n^2}\right)$$

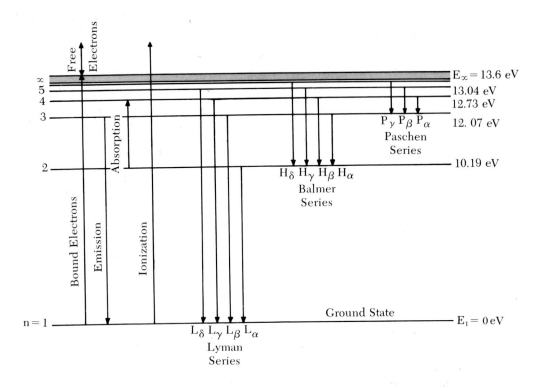

Figure 8–9. *Hydrogen Energy Level Diagram.* Several transitions of the first three hydrogen series are shown in emission. The left-hand side of the diagram illustrates the several types of atomic transitions that occur.

Note that $E(1)$ now equals zero. Finally, let us change our energy units from ergs to *electron Volts* (eV), where

$$1 \text{ eV} = 1.602 \times 10^{-12} \text{ erg}$$

The electron Volt is the energy acquired by an electron (or any particle of charge e) when accelerated through a voltage difference of one Volt; this unit is clearly of importance and convenience in atomic and particle physics. We now note that

$$E(\infty) = R' = 2.18 \times 10^{-11} \text{ erg} = 13.6 \text{ eV}$$

The energy levels for several values of n are shown in Figure 8-9 $[E(1) = 0,$ $E(2) = 10.2 \text{ eV}, E(3) = 12.1 \text{ eV}, \text{ etc.}]$. When the atom is in any level above the ground state, we say that it is in an *excited state*, and the energy of such a level is called its *excitation potential*. To reach a higher level, the atom must be *excited*—an excessive energy of excitation leads to *ionization*—and in dropping back down toward the ground state the atom is *de-excited*.

(i) *Excitation*

An atom may be excited to a higher energy level in two ways: radiatively or collisionally. *Radiative excitation* occurs when a photon is absorbed by the atom; the photon's energy must correspond exactly to the energy difference between two energy levels of the atom. This process produces absorption lines in astrophysical spectra.

In section 8-2(d, ii) we will see that an atom generally remains in an excited state for only an extremely short time (about 10^{-8} sec) before re-emitting a photon. How then can an absorption line be produced? Let us recall that the electron may cascade through several energy levels on its way to the ground state; hence, several lower-energy photons may be emitted for each photon absorbed. The input wavelength is converted to longer wavelengths, depleting the spectrum at the input wavelength. Furthermore (see Figure 8-10), the absorbed photons come predominantly from one direction—the direction of their source—while the emitted photons can travel in any direction. Hence, fewer photons at the absorption wavelength reach an observer than photons at other wavelengths. An absorption line is dark in comparison to the unabsorbed continuum, but not completely black, since some photons of the critical wavelength still reach the observer. Thus is Kirchhoff's law of *absorption spectral lines* explained.

Collisional excitation takes place when a free particle (an electron or another atom) collides with our atom, giving part of its kinetic energy to the atom. Such an inelastic collision does not involve any photons. A particle approaching the atom with speed v_i, and leaving with speed v_f, has deposited the energy $E = m(v_i^2 - v_f^2)/2$ with the atom; if E corresponds to the energy of an electronic transition, the atom is collisionally excited to a higher state. Such an excited atom returns to its ground state by emitting photons producing an *emission line spectrum* in the process.

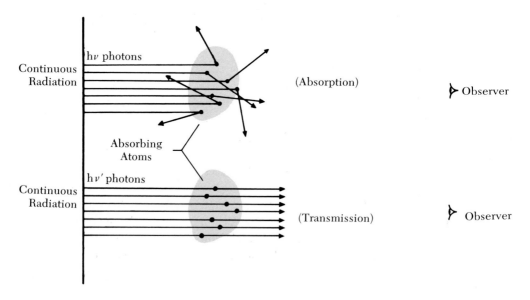

Figure 8–10. *Absorption of Radiation. E* is the energy of excitation of a given atom, and therefore photons with the frequency *v*, such that $E = hv$, can be absorbed. Only a fraction of the re-emitted photons travel in the direction to the observer. Photons with the energy $E' = hv'$ cannot be absorbed, and therefore reach the observer without diminution.

(ii)   *De-excitation*

Atoms are always interacting with the electromagnetic field. This interaction (which leads us again to quantum mechanics) causes an excited atom to jump *spontaneously* to a lower energy level—or de-excite—in a characteristic time of order 10^{-8} second. A photon is emitted, so we call the process *radiative de-excitation*. A more "silent" form of de-excitation, where the phenomenon is not announced by a photon, is *collisional de-excitation*; this is the exact inverse of collisional excitation, for the colliding particle *gains* kinetic energy in the exchange (i.e., a super-elastic collision). In actual astrophysical situations, both modes of de-excitation compete with one another.

(iii)   *Ionization*

If we provide sufficient energy (either radiatively or collisionally) to liberate an electron from a neutral atom, the atom is *ionized*. Schematically, this reaction is just

$$X + \text{energy} \rightarrow X^+ + e^-$$

where X represents the atom. To denote neutral hydrogen (with *no* electrons removed) we may write H or H I, where the Roman numeral I signifies the first *ionization state*; similarly, we have neutral helium, He or He I. The singly-ionized atom (with *one* electron removed) is in its *second* ionization state, such as $H^+ = $ H II or $He^+ = $ He II. Hydrogen possesses but one electron, so that it may not be ionized beyond H II, but we can have O^{++} or O III (doubly-ionized oxygen) and O^{+++}

or O IV (oxygen with three electrons removed). For high states of ionization (e.g., Fe XIV), the Roman numeral system is more convenient and common than the multiple-sign convention.

The energy required to ionize an atom depends upon (a) the ionization stage of the atom, (b) the particular electron to be liberated, and (c) the excitation level of that electron. For simplicity, let us consider the hydrogen atom with its sole electron. An electron in the ground state ($n = 1$; see Figure 8–9) is removed from the atom when we supply it the energy $E \geq 13.6$ eV (the *ionization potential*); note that a *continuum* of energy states is available to the free electron above $E(n = \infty)$. If the electron is in the first excited state ($n = 2$), it will be freed when given energies $E \geq E(\infty) - E(2) = 13.6 - 10.2 = 3.4$ eV. In general (for hydrogen), the ionization potential for an electron in excitation level n is simply:

$$IP(n) = E(\infty) - E(n) = (13.6/n^2) \text{ eV} \qquad (8\text{–}26)$$

The kinetic energy available to the departing electron is the difference between the energy provided (E) and the ionization potential (IP), or ($E - IP$).

Radiative ionization (via photons) leads to spectral absorption continua. For example, hydrogen atoms in the ground state absorb discrete-wavelength photons to produce the Lyman absorption series, but this series ends at the *series limit* $\lambda = 912$ Å. For wavelengths shorter than the series limit, we observe the Lyman *absorption continuum*, corresponding to photons which can ionize hydrogen from its ground state. Similarly, we have the Balmer, Paschen, Brackett, and Pfund series limits, and their associated absorption continua for ionizations from levels $n = 2, 3, 4$, etc.

Free electrons can *recombine* with ions by emitting a photon of the appropriate energy. Since this process is just the reverse of ionization, the various hydrogen emission series may end in an *emission continuum* if conditions are right.

8–3 THE SPECTRA OF ATOMS, IONS, AND MOLECULES

(a) Atomic Spectra

In multi-electron atoms, quantum mechanics and the *Pauli exclusion principle* dictate that only two electrons may occupy the innermost shell, eight the next, 18 in the third, etc. When a shell contains the maximum allowed number of electrons, it is filled or closed; in this case, the atom is extremely stable and hard to excite (e.g., helium, neon, argon, . . .). Innermost shells tend to be the first to be filled with electrons, and any excess electrons (*valence electrons*) are available for chemical interactions. For example, calcium (with 20 electrons) acts as if there are only two electrons in the outermost shell, since the two inner shells (2 + 8) are closed and eight of the electrons in the third shell form two stable *subshells* (2 + 6).

Closed shells are tightly bound and shield the nucleus from the outer electrons, so that these excess electrons are only lightly bound (i.e., easily excited and ionized). The spectra of atoms with one outer electron, such as lithium and sodium, are similar to that of hydrogen, but do show the effects of the inner closed shells.

Atoms with more than one outer electron have increasingly complex spectra. To illustrate the various possibilities in the *periodic table*, Table 8–3 lists, for several representative atoms, (a) the excitation potential of the first excited state, (b) the ionization potential from the ground state, and (c) the wavelengths corresponding to these transitions.

(b)   The Spectra of Ions

Ions possessing at least one bound electron behave spectrally just like neutral atoms—they may be excited, de-excited, and ionized further. The spectrum of an ion closely resembles that of a neutral atom with the same number of outer electrons, except for overall wavelength modifications imposed by the greater charge of the nucleus (e.g., He II and H I).

Consider ions with but one electron remaining, such as He II, Li III, O VIII, and even Fe XXVI. The Bohr wave-number relation for these cases is clearly

$$\frac{1}{\lambda_{ab}} = RZ^2 \left(\frac{1}{n_b^{\,2}} - \frac{1}{n_a^{\,2}} \right) \tag{8–27}$$

By analogy, each such ion exhibits Lyman, Balmer, and the other series, but the wavelengths differ from those of the hydrogen spectral lines by the factor Z^{-2}. Thus, He II Lyman α lies at 304 Å instead of 1216 Å (since $Z = 2$).

(c)   Molecular Spectra

Molecules are formed when atoms bind together (the *chemical bond*). Quantum mechanics applies to such a union, and three types of discrete energy levels are

Table 8–3.   Excitation and Ionization Potentials for Selected Atoms

	Excitation* Potential	$\lambda(\text{Å})$	Ionization† Potential	$\lambda_{series}(\text{Å})$ limit
Hydrogen (one electron)	10.2	1216	13.6	912
Helium (one closed shell)	20.9	584	24.5	488
Lithium (1 filled shell, 1 outer electron)	1.8	6708	5.4	2250
Neon (2 filled shells)	16.6	735	21.5	576
Sodium (2 filled shells, 1 outer electron)	2.1	5890	5.1	2430
Magnesium (2 filled shells, 2 outer electrons)	2.7	4571	7.6	1630
Calcium (2 filled shells, 2 filled subshells) (2 outer electrons)	1.9	6573	6.1	2030

* From ground state to first excited state.

† From ground state of neutral atom.

exhibited by molecules. (a) There are electronic states in the combined electron cloud surrounding the nuclei. *Electronic transitions* similar to those in an atom can take place between these states, leading to excitation, de-excitation, and ionization of the molecule (e.g., $H_2 \rightarrow H_2{}^+ + e^-$). (b) The internuclear distances are quantized in discrete *vibrational states*, with the consequent vibrational transitions. When the separation becomes so great that the atoms are no longer bound together, we say that the molecule has *dissociated*. (c) A molecule may rotate about various axes in space, resulting in discrete *rotational states*.

The three classes of molecular transitions lead to numerous spectral lines superimposed upon one another. Since the vibrational and rotational transitions involve *small* differences in energy (usually much less than 1 eV), their spectral lines are closely spaced in wavelength and appear as *bands*. Therefore, molecular spectra are much more complex than atomic spectra, but the former are easily recognized by their banded structure.

8–4   SPECTRAL LINE INTENSITIES

We have spoken of emission and absorption spectral lines. Let us now be more quantitative, and explain the *strengths* of these lines (see Figure 8–11). The intensity

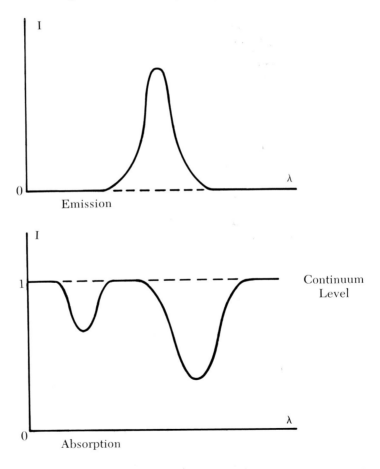

Figure 8–11.   *Spectral Line Intensity Profiles.* Typical intensity (I) profiles of emission and absorption lines.

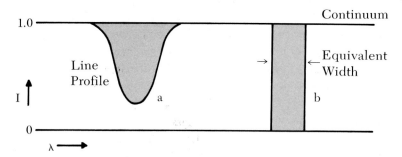

Figure 8–12. *Equivalent Width.* The area of *b* is identical to that contained in profile *a*. The intensity of the continuum is arbitrarily set equal to unity.

of an emission line is proportional to the number of photons emitted in that particular transition. Similarly, the strength of an absorption line—relative to the adjacent continuum—depends upon the number of photons absorbed. Since the majority of astronomical spectra are absorption spectra, we will consider the case for absorption; very similar arguments follow for emission spectra.

An absorption line is never infinitely sharp (see section 8–5). It exhibits a *profile*, the *intensity* (the residual radiation in terms of the continuum as unity) of which varies with wavelength (see Figure 8–11). The total strength of a line is proportional to its area, which may be represented by the line's *equivalent width* (see Figure 8–12). We replace the line profile by a rectangle of *equal* area, with one dimension of the rectangle being the height of the continuum and the other dimension being the equivalent width (in Ångstroms or milliÅngstroms, mÅ). Note that the equivalent width increases with the strength of the line. When the center of the line profile reaches zero intensity, we say that the line has become *saturated*; any further increase in the strength comes only from the wings of the line. The *curve of growth* summarizes these considerations. (For further information, consult a textbook on astrophysics; see reading list for suggestions.)

(a) Excitation Equilibrium: Boltzmann's Equation

The strength of a spectral line depends directly upon *the number of atoms in the energy state from which the transition occurs*. Therefore, we would like to know the fraction of all the atoms of a given element which are excited to that energy state. Recall that excitation and de-excitation may occur collisionally and/or radiatively (a radiative de-excitation is also termed a *spontaneous transition*). Both processes depend upon temperature T, since (a) the mean kinetic energy of a gas particle goes as (see Chapter 5)

$$\overline{(mv^2/2)} = 3kT/2 \qquad (8\text{–}28)$$

(m is the particle's mass, v its speed, $k = 1.380 \times 10^{-16}$ erg K^{-1} is *Boltzmann's constant*, and T is the gas temperature), and (b) the number of photons of a given energy increases rapidly with increasing temperature [see section 8–6(c)]. Consequently, absorption lines originating from excited levels tend to be stronger in hot gases than in cool gases.

For simplicity, let us investigate the situation where *thermal equilibrium* prevails and the average number of atoms in a given state remains unchanged in time (*steady state*). This means that each excitation is, on the average, balanced by a de-excitation. In this case, statistical mechanics shows us that the *number density* (number per unit volume) of atoms in state B is related to the number density in state A ($B > A$) by *Boltzmann's equation*:*

$$(N_B/N_A) = (g_B/g_A) \exp\left[(E_A - E_B)/kT\right] \qquad (8-29)$$

In equation (8-29), N is the number density in the level, g is the multiplicity of the level (an intrinsic characteristic), and E is the energy of the level. The term *exp* [ ] denotes exponentiation; that is, raise $e = 2.71828\ldots$ (the base of the natural logarithms) to the power [ ] (see the Mathematical Appendix). Since $E_B > E_A$, the bracketed quantity is always negative, so that the ratio (N_B/N_A) increases as the temperature increases ($N_B/N_A \to 1$ as $T \to +\infty$). For a given temperature, the *excitation ratio* (N_B/N_A) increases as we decrease the excitation potential $(E_B - E_A)$ between the two energy levels.

Recalling that $\exp[-\infty] = 0$, $\exp[-1] = 0.368$, and $\exp[0] = 1$, we see that significant population of the upper level occurs when $T \approx (E_B - E_A)/k$; an excitation potential of 1 electron Volt corresponds to a temperature $T \approx 11,600$ K. As an example, consider a volume of gas which contains the *same* number of hydrogen, helium, and sodium atoms at a temperature such that the number of hydrogen atoms in the first excited state (N_2) equals one-tenth the number in the ground state (N_1), or $(N_2/N_1) = 0.1$. By referring to Table 8-3, we see that $(N_2/N_1) \approx 1$ for the sodium atoms at this temperature (in fact, much of the sodium will be ionized, but let us ignore that complication for now). On the other hand, the ratio (N_2/N_1) for helium will be very low. So, at a given temperature, the fraction of atoms in the second level differs widely from element to element, depending on the excitation potential. In this case, the absorption lines arising from $n = 2$ to $n = 3$ transitions will be strongest for sodium, less strong for hydrogen, and very weak for helium. It is easy to see that the strength of the line is a function of both the *abundance* of the particular element and the *temperature*. In this example we have unrealistically ignored the ionization of the sodium and hydrogen atoms; for a complete picture, we must include both the ionization and excitation to other levels.

(b) Ionization Equilibrium: Saha's Equation

As the temperature of a gas is increased, more and more energy becomes available (either radiative or collisional) to ionize the atoms. In general, the hot gas consists of neutral atoms, ions, and free electrons. The greater the *electron*

---

* Named after the Austrian physicist, Ludwig Boltzmann (1844-1906), who discovered the relation.

density (N_e = number of electrons per unit volume), the greater is the probability that an ion will capture an electron to recombine to a neutral atom. These two competing processes—ionization ($\rightarrow$) and recombination ($\leftarrow$)—are expressible as a chemical equation

$$X \rightleftarrows X^+ + e^-$$

A steady-state condition of *ionization equilibrium* is achieved in the gas when the rate of ionization equals the rate of recombination. A quantitative expression of this ionization equilibrium is given by *Saha's equation*:*

$$\boxed{\left(\frac{N_+}{N_0}\right) = A\,\frac{(kT)^{3/2}}{N_e}\,\exp\left[-\chi_0/kT\right]} \qquad (8\text{–}30)$$

where N_+ is the number density of ions, N_0 is the number density of neutral atoms in the ground state, the constant A includes several atomic constants, N_e is the electron density, T is the absolute temperature, and χ_0 is the *ionization potential* (in eV) from the ground state of the neutral atom. Equation (8–30) is very similar to Boltzmann's equation (8–29), except for the dependence upon N_e, and the additional factor of $T^{3/2}$ which arises because (a) a continuum of energies above χ_0 will ionize the atom, and (b) the liberated electron is more likely to escape the ion as we increase the electron's kinetic energy.

Let us note that the Boltzmann excitation equation (8–29) applies to *any two levels of excitation*, those of an ion as well as those of a neutral atom. Similarly, the Saha ionization equation (8–30) can be generalized to give the ratio (N_{i+1}/N_i) for any *stage of ionization* ($i + 1$) and the next lower stage (i). The appropriate form of Saha's equation is

$$\left(\frac{N_{i+1}}{N_i}\right) = A\,\frac{(kT)^{3/2}}{N_e}\,\exp\left[-\chi_i/kT\right] \qquad (8\text{–}31)$$

where χ_i is the ionization potential of the lower stage; i.e., χ_i is the energy needed to ionize i from its ground state to ($i + 1$). For example, equation (8–31) applies to the ionization balance between Ca III ($i + 1 = 3$) and Ca II ($i = 2$). We see that the relative population of the upper ionization stage increases rapidly with increasing temperature T or smaller values of χ_i.

(c)   The Boltzmann and Saha Equations Combined

The *Boltzmann equation* gives the number of atoms in an excited state relative to the number in the ground state; this applies both to neutral and ionized atoms. The *Saha equation* tells us the relative populations of two adjacent stages of ionization.

---

*Named after the Indian physicist, Meghnad N. Saha (1893–1956).

We must combine these two equations if we wish to calculate the number of atoms available to make a certain transition and thus produce a given spectral line.

Consider the Balmer absorption lines of neutral hydrogen. Their strength is proportional to the number of atoms in the first excited state (N_2) of the neutral atom, relative to the total number of hydrogen atoms in *all* stages of ionization (N). But hydrogen has only two stages of ionization: neutral (N_0) and singly-ionized (N_+); hence, we know that $N = N_0 + N_+$. The proportion which we seek is, therefore,

$$\frac{N_2}{N} = \frac{N_2}{N_0 + N_+} \cong \frac{(N_2/N_1)}{1 + (N_+/N_0)} \tag{8–32}$$

where we have used the reasonable approximation $N_0 \cong N_1$ in the last equality. The Boltzmann equation yields (N_2/N_1), the ratio of neutral atoms in the first excited state to those in the ground state; the Saha equation gives the ratio of ionized to neutral atoms, (N_+/N_0). The approximate form of equation (8–32) is accurate enough for our present purposes, but if the reader desires greater accuracy, he may use the Boltzmann equation to evaluate, for example, the ratio:

$$\frac{N_2}{N_0} = \frac{N_2}{N_1 + N_2 + N_3 + \cdots} = \frac{(N_2/N_1)}{1 + (N_2/N_1) + (N_3/N_1) + \cdots}$$

A plot of (N_+/N_0) as a function of temperature T (see Figure 8–13) shows that most of the hydrogen is neutral at temperatures below 7000 K, but at higher temperatures ionization increases to the point where the number of neutral atoms becomes negligible. The exponential increase of (N_2/N_1) with increasing temperature is, therefore, countered by the paucity of neutral atoms at high temperatures. As a result, the (N_2/N) curve (see Figure 8–13) has a *maximum* around 10,000 K. The strength of the Balmer absorption lines of hydrogen is greatest near $T \approx 10,000$ K, decreasing at both higher and lower temperatures. At $T = 6000$ K—the approximate surface temperature of the Sun—the ratio (N_2/N_1) is about 10^{-8}, but reasonably strong Balmer absorption lines are seen as a result of the great abundance of hydrogen in the Sun. At $T = 20,000$ K—the temperature of very hot stars—the ratio is about 10^{-2}, but the Balmer lines are similar in strength to those of the Sun, since most of the hydrogen atoms are now ionized!

As we shall see in the chapters which follow, the Boltzmann and Saha equations have wide application in astrophysics. Through them we may interpret stellar absorption (and emission) spectra, in order to deduce the surface temperatures and pressures of stars. For example, at temperatures in the range 5000 to 7000 K, calcium should be predominantly in the form Ca II (singly-ionized). Stars (like our Sun) with strong Ca II lines, but weak Ca I lines, must, then, have temperatures of this order. On the other hand, a star of much *lower density* than the Sun (e.g., a red giant) but which produces equally strong Ca II lines, must actually have a *lower temperature* to compensate for the smaller electron density [N_e; see equation (8–31)], assuming that the chemical composition is the same.

We end this section by generalizing our previous formulae to obtain the

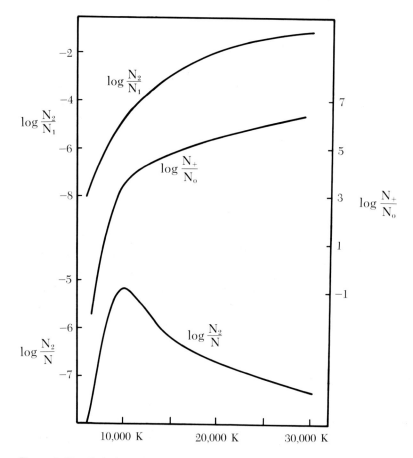

Figure 8–13.   *Excitation and Ionization Equilibrium Curves for Hydrogen Balmer Lines.* The relative populations of atomic energy levels (N_2/N_1) are calculated from the Boltzmann equation, and of ionization stages (N_+/N_0) from the Saha equation. The lowest curve results from the combined effects of the upper two, and gives the number of second-level hydrogen atoms relative to all hydrogen, N_2/N, where $N = N_0 + N_+$.

relative number of atoms in *any* state of excitation s of a stage of ionization i: $N_{i,s}$. The ratio of interest is $(N_{i,s}/N)$, where N is summed over *all stages of ionization,*

$$N = N_0 + N_1 + N_2 + \cdots + N_n = \sum_{i=0}^{n} N_i$$

In general, n is the number of electrons in the neutral atom, but, in practice, only two or three stages of ionization need be considered, for the number of ions in other stages will be negligible. To a reasonable first approximation we may write:

$$\frac{N_{i,s}}{N} \cong \frac{N_{i,s}}{N_{i-1} + N_i + N_{i+1}} = \frac{(N_{i,s}/N_i)}{(N_{i-1}/N_i) + 1 + (N_{i+1}/N_i)} \tag{8–33}$$

The numerator of the last expression is given by the Boltzmann equation

$$(N_{i,s}/N_i) \propto \exp\left(-\chi_s/kT\right)$$

and the denominator can be found from the Saha equation:

$$(N_{j+1}/N_j) \propto (kT)^{3/2} N_e^{-1} \exp(-\chi_j/kT)$$

While equation (8–33) is useful for the strongest spectral lines of a gas, in the general case extensive numerical computations are necessary to accurately reproduce spectral line strengths. Our approximation holds only when i is the *predominant* stage of ionization for the prevailing temperature.

8–5   SPECTRAL LINE BROADENING

Spectral lines are never infinitely sharp; their profiles always have a finite width. The fundamental tenets of quantum mechanics account for the *minimum width* (natural broadening) of a spectral feature, while various physical processes—which we will now discuss—further broaden the line profile. By interpreting the observed profile of a spectral line in terms of these broadening mechanisms, we may deduce some characteristics of the radiation source (e.g., a star).

(a)   Natural Broadening

Quantum mechanics ascribes a wave nature to all atomic particles; an electron in an atomic energy level is such a particle. As a consequence, the *Heisenberg uncertainty principle** implies that the energy of a given state may not be specified more accurately than

$$\Delta E = \frac{1}{2\pi} \frac{h}{\Delta t} \tag{8–34}$$

where h is Planck's constant and Δt is the *lifetime* of the state. Therefore, an assemblage of atoms will produce an absorption or emission line with a minimum spread in photon frequencies—the *natural width*—of order $\Delta v = \Delta E/h \approx 1/\Delta t$. Typical excited states live about 10^{-8} second before decaying (note: a ground state may last forever), so that a normal natural width is near 0.05 mÅ for visible light. Much smaller natural widths occur for the so-called *metastable states*, some of which last more than a second ($\Delta t \gtrsim 1$ sec).

(b)   Thermal Doppler Broadening

While natural broadening depends only upon the intrinsic lifetime of an energy level, *thermal Doppler broadening* depends upon the temperature and composition of a gas. When we say that a gas is at a certain temperature T, we mean that the

---

* Due to one of the founders of quantum mechanics, the German physicist Werner K. Heisenberg (1901–   ).

gas particles (each of mass m) move about with random velocities, characterized by the mean kinetic energy of equation (8–28):

$$\overline{(mv^2/2)} = 3kT/2$$

Atomic motions along our line of sight imply *Doppler shifts* in the radiation absorbed or emitted in atomic transitions. At a given temperature, the spectral lines of heavy elements are narrower than those of light elements since, on the average, the heavy particles move slower than the light particles. For example, neutral hydrogen at $T = 6000$ K moves at the mean speed $v \approx 12$ km sec^{-1}, corresponding to a fractional Doppler broadening of $\Delta\lambda/\lambda \approx v/c \approx 4 \times 10^{-5}$; hence, the thermal Doppler width of the Balmer α line (6563 Å) is approximately 1/4 Å.

(c)   Collisional Broadening

The energy levels of an atom are perturbed (i.e., shifted) by neighboring particles, especially charged particles like ions and electrons (the *Stark effect*). In a gas, these perturbations are random, and they result in a broadening of spectral lines. Since the perturbations are larger the nearer the perturbing particle, there is a direct dependence of this *collisional* (or *pressure*) *broadening* upon the density of particles. The greater the density (and, hence, pressure) of the gas, the greater is the width of the spectral lines.

(d)   The Zeeman Effect

When an atom is placed in a *magnetic field*, the atomic energy levels each separate into three or more sublevels—this is the *Zeeman effect*.* Where before we had a single atomic transition and a single spectral feature, we now have three or more closely-spaced lines (the spacing is proportional to the magnetic field strength). If the Zeeman components are not resolved, we see only a broadened spectral line. In those cases in which the magnetic field is very strong, and is highly ordered and uniform (e.g., sunspots and magnetic stars), we may resolve the Zeeman splitting and deduce the magnetic field strength and orientation at the source.

(e)   Other Broadening Mechanisms

Finally, let us mention three macroscopic broadening mechanisms based upon the Doppler effect. Consider a typical star whose image cannot be resolved by our instruments. Large-scale random motions at the surface of such a star imply Doppler shifts which appear as *turbulence broadening* of spectral lines. If the atmosphere of the star is expanding, we simultaneously see gas moving in all directions; the integrated effect of all the Doppler shifts is *expansion broadening* of the observed spectral lines. A rapidly rotating star (not seen pole-on) will have *rotationally broadened* lines, since one limb of the star is approaching us while the other limb

---

* After the Dutchman, Pieter Zeeman (1865–1943).

recedes. Since all spectral features exhibit the *same* rotational broadening, we may identify the stellar rotation and determine its rate (or period).

8–6 BLACKBODY RADIATION

Thus far we have spoken of individual atomic transitions and the spectral lines which they produce. Where then does a *continuous spectrum* come from? [Recall that spectral absorption lines result when photons are selectively deleted (absorbed) from such a continuum.] We noted that various emission and absorption continua can originate from individual atoms, and that spectral features become more broadened as atoms interact more strongly with one another. When an aggregate of atoms interacts so strongly (e.g., in a solid, a liquid, or a very dense gas) that all characteristic spectral features are washed out, a unique *thermal continuum* becomes possible.

We are thus led to consider the continuum spectrum of a *blackbody*, a spectrum which depends only upon the absolute temperature T. A blackbody is so named because it absorbs *all* electromagnetic energy incident upon it—it is completely *opaque* or *black*. But to be in perfect thermal equilibrium, such a body must radiate energy at exactly the same rate that it absorbs energy; otherwise, the body will heat up or cool down and its temperature will change. Ideally, a blackbody is a perfectly-insulated enclosure, within which radiation has come into thermal equilibrium with the walls of the enclosure. Practically, this *blackbody radiation* may be sampled by observing the enclosure through a tiny pinhole in one of the walls. The gases in the interior of a star are opaque (i.e., highly absorbent) to radiation (otherwise, we would see the stellar interior at some wavelength!); hence, the radiation there is blackbody in character. We sample this radiation as it slowly leaks from the surface of the star—to a good approximation, the continuum radiation from stars is blackbody.

Two important characteristics of blackbody radiation are demonstrated by the following example. If we heat a bar of iron, it is merely warm at first (infrared radiation or heat) and appears black to our eyes. Soon it is brighter, glowing a cherry-red. As the bar grows hotter—higher temperature T—its brightness increases rapidly, and its color changes to orange, yellow, white, and finally blue-white. Let us now turn to a quantitative discussion of these blackbody properties.

(a) Planck's Radiation Law

After Maxwell's theory of electromagnetism appeared in 1864, many attempts were made to understand blackbody radiation theoretically. None succeeded until, in 1900, Max K. E. L. Planck (1858–1947) postulated that electromagnetic energy can only propagate in discrete quanta or *photons*, each of energy $\epsilon = h\nu$. This brilliant German physicist then derived the *spectral intensity relation* or *Planck blackbody radiation law*:

$$I_\nu \, \Delta\nu = \frac{2h\nu^3}{c^2} \frac{1}{e^{(h\nu/kT)} - 1} \, \Delta\nu \qquad \text{(8–35a)}$$

Here $I_v \, \Delta v$ is the *intensity* (i.e., erg cm^{-2} sec^{-1} ster^{-1}) of radiation from a blackbody at temperature T in the frequency range between v and $v + \Delta v$; h is Planck's constant, k is Boltzmann's constant, and c is the speed of light. Note the now familiar exponential in the denominator.

Since the frequency (v) and wavelength (λ) of electromagnetic radiation are related by $\lambda v = c$, we may also express Planck's formula, equation (8–35a), in terms of the intensity emitted per unit wavelength interval:

$$I_\lambda \, \Delta\lambda = \frac{2hc^2}{\lambda^5} \frac{1}{e^{(hc/\lambda kT)} - 1} \, \Delta\lambda \qquad \text{(8–35b)}$$

Equation (8–35b) follows because the intensity $I_v \, \Delta v$ *equals* the intensity $I_\lambda \, \Delta\lambda$ in the corresponding wavelength interval, where

$$v = \frac{c}{\lambda} \Rightarrow |\Delta v| = \left| \frac{c \, \Delta\lambda}{\lambda^2} \right|^*$$

Equation (8–35b) is illustrated in Figure 8–14, for several values of T. Note that both I_λ and I_v increase as we increase the blackbody temperature—the blackbody becomes *brighter*. This effect is easily interpreted in terms of equation (8–35a), when we note that $I_v \, \Delta v$ is directly proportional to the *number of photons* emitted per second near the energy hv.

To better understand the Planck blackbody formulae, equation (8–35), let

---

* This expression follows by differentiation, or by noting that

$$\lambda v = (\lambda + \Delta\lambda)(v + \Delta v) = c$$

where $\Delta\lambda \cdot \Delta v$ is negligible.

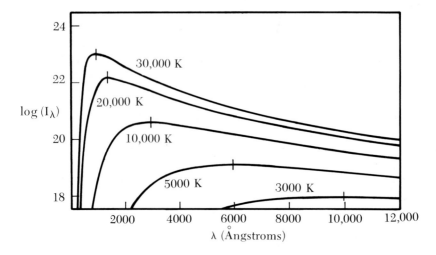

Figure 8–14.   *Planck Radiation Curves for Several Temperatures.*

us outline a derivation by Albert Einstein. At a given frequency v, we want radiation to be in thermal equilibrium with the atoms which compose the walls of our blackbody cavity. Because of the strong atomic interactions in the walls, all possible energy states are available there. Consider the upper (U) and lower (L) energy states corresponding to the quantum of energy $(E_U - E_L) = hv$. The Boltzmann equation relates the number of atoms in each state via:

$$N_U = (g_U/g_L) N_L e^{-(hv/kT)} \tag{8-36}$$

If the probability per atom for a downward transition $(U \to L)$ is a_{UL}, and for an upward transition $(L \to U)$ is a_{LU}, then a steady-state equilibrium is attained when

$$N_L a_{LU} = N_U a_{UL} \tag{8-37}$$

But a_{LU}—"radiative excitation"—depends upon the energy density (ρ_v) of radiation at the frequency v, and upon the atomic parameter B_{LU} (the *Einstein absorption coefficient*): $a_{LU} = B_{LU}\rho_v$. "De-excitation" (a_{UL}) can occur in two ways: spontaneously (A_{UL}) or through the stimulating influence of the radiation field $(B_{UL}\rho_v)$; hence, we have $a_{UL} = A_{UL} + B_{UL}\rho_v$. Combining these various results in equations (8-36) and (8-37), and solving for ρ_v, we soon find:

$$\rho_v = \left[\frac{(A_{UL}/B_{UL})}{\left(\dfrac{B_{LU}g_L}{B_{UL}g_U}\right) e^{(hv/kt)} - 1} \right]$$

the result is independent of the number of atoms involved. Einstein then substituted suitable expressions for the atomic coefficients A_{UL}, B_{UL}, and B_{LU} to obtain the *Planck blackbody radiation-density law*:

$$\rho_v = \frac{8\pi h v^3}{c^3} \frac{1}{e^{(hv/kT)} - 1} \tag{8-38}$$

In equation (8-38), $\rho_v \, \Delta v$ is an *energy density* (erg cm^{-3}); in converting to intensity radiated from the blackbody (I_v), a factor of $(c/4\pi)$ is introduced [equation (8-35a)].

(b) The Wien Law

We have already said that the color of a blackbody appears to become bluer as its temperature increases. From Figure 8-14, we see that this phenomenon corresponds to the bulk of the radiation being emitted at shorter wavelengths as T becomes larger. Wilhelm Wien (1864–1928) expressed the wavelength (λ_{max}) at which the maximum intensity of blackbody radiation is emitted—the *peak** of

---

* That wavelength for which $(dI_\lambda/d\lambda) = 0$.

the Planck curve—by the *Wien displacement law*:

$$\lambda_{max} = \frac{0.2898}{T}$$

$$(8\text{--}39)$$

where λ_{max} is given in centimeters (cm) when T is given in Kelvins (K). For example, the continuum spectrum from our Sun is approximately blackbody, peaking at $\lambda_{max} \approx 5000$ Å; therefore, the surface temperature must be near $T_\odot \approx 5800$ K.

(c)   The Law of Stefan and Boltzmann

The *area* under the Planck curve represents the total *energy flux* F (erg cm^{-2} sec^{-1}) emitted by a blackbody, when we sum over all wavelengths and solid angles:

$$F = \sigma T^4$$

$$(8\text{--}40)$$

where $\sigma = 5.669 \times 10^{-5}$ erg cm^{-2} sec^{-1} K^{-4}. The strong temperature dependence of equation (8–40) was first deduced from thermodynamics in 1879 by Josef Stefan (1835–1893), and was derived from statistical mechanics in 1884 by Boltzmann; hence, we call the expression the *Stefan-Boltzmann law*. The brightness of a blackbody is seen to increase as the fourth power of its temperature. If we may approximate a star by a blackbody, the energy output (erg sec^{-1}) of the star is just $4\pi R^2 \sigma T^4$, since the surface area of a sphere of radius R is $4\pi R^2$.

We must end this chapter with a word of caution concerning *temperature*. If we are dealing with a true blackbody, we may establish its temperature using (a) the shape of the Planck curve or at least two points on the curve, (b) Wien's law, or (c) the Stefan-Boltzmann law. Since no astrophysical object is a perfect blackbody, we will obtain slightly different temperatures when we use each of these three methods. In addition, temperatures based on such continuum spectra may differ from the temperatures derived from the relative intensities of spectral lines by using the Boltzmann and Saha equations. It is, therefore, a good practice to stipulate *what kind of temperature* one is using (see the examples in Table 8–4).

Table 8–4.   The Varieties of Temperatures

Temperature Type	Basic Law or Equation	Necessary Observations
Color	Planck curve	Brightness at two or more wavelengths
Effective or radiation	Stefan-Boltzmann law	Bolometric* magnitude and radius
Excitation	Boltzmann equation	Relative intensities of spectral lines of the same element
Ionization	Saha equation	Relative intensities of spectral lines of adjacent stages of ionization
Kinetic	Thermal Doppler broadening	Widths and profiles of spectral lines

* See Chapter 10 for definition of bolometric magnitude.

Problems

8–1.   (a) Show that a beam of light obliquely incident upon and passing through a plane-parallel piece of glass is simply displaced, without changing direction, when it emerges from the glass.

(b) If the glass has thickness d and index of refraction n, what is the linear displacement of the beam?

8–2.   What "telescope" aperture is required to give 1 arc-second $(1'')$ *resolution* at wavelengths of

(a) 5000 Å,

(b) 21 cm (radio)?

Can you make a generalization from your results?

8–3.   (a) At what wavelengths will the following spectral lines be observed:

(i) A line emitted at 5000 Å by a star moving toward us at 100 km sec^{-1};

(ii) The Ca II line (*undisplaced* wavelength of 3970 Å) emitted by a galaxy receding at 60,000 km sec^{-1}?

(b) A cloud of neutral hydrogen (H I) emits the 21-cm radio line (at a rest-frequency of 1420.4 MHz) while moving away at 200 km sec^{-1}. At what *frequency* will we observe this line?

8–4.   A simple form of the *binomial theorem* states that

$$(1 + x)^n = 1 + nx + \frac{n(n-1)x^2}{2} + \frac{n(n-1)(n-2)x^3}{6} + \cdots$$

when $x^2 < 1$. Use this theorem to derive equation (8–13) for the Doppler shift in wavelength, from the relativistic expression, equation (8–14a), in the case in which $v \ll c$.

8–5.   (a) What is the energy of one photon of wavelength $\lambda = 3000$ Å? Express your answer both in ergs and in electron Volts.

(b) An atom in the *second* excited state $(n = 3)$ of hydrogen is just barely ionized when a photon strikes the atom. What is the wavelength of the photon, if *all* of its energy is transferred to the atom?

8–6.   The emission line of He II at 4686 Å corresponds to what electronic transition?

8–7.   By applying the Boltzmann equation to the *neutral* hydrogen atom (neglect ionization), derive an expression for the population of the nth energy level relative to that of the ground state—at temperature T. Now, assuming that the multiplicity of each level is unity $(g_n \equiv 1)$, construct an appropriate graph showing your results for a temperature of $T = 6000$ K.

8–8.   To understand the relative importance of the different parameters in the Saha equation, perform the following experiment. First assume that $T = 5000$ K, $N_e = 10^{15}$ cm^{-3}, and the ionization potential $= 12$ eV. By what factor does the *ionization ratio* (N_+/N_0) change when we separately:

(a) double the temperature?

(b) double the electron density?

(c) double the ionization potential?

Which was more important during the temperature change, the exponential term or the $T^{3/2}$ term?

8–9.   Let N_2 be the number of second-level (first excited state) hydrogen atoms, and N_1 be the number in the ground state. Using Figure 8–13, find the excitation ratio (N_2/N_1) and the excited fraction (N_2/N) for each of the following stars:

(a) Sirius, $T = 10,000$ K;

(b) Rigel, $T = 15,000$ K;

(c) the Sun, $T = 5700$ K.

Which star will exhibit the strongest Balmer absorption lines? Explain your reasoning in arriving at this answer.

8–10.   (a) What is the *speed* of an electron with just sufficient energy to ionize by collision a sodium atom in the ground state?

(b) What would be the speed of a *proton* ionizing this atom?

(c) What is the corresponding gas temperature?

(d) At this temperature, what is the fractional *thermal Doppler broadening* of a sodium spectral line (i.e., $\Delta\lambda/\lambda$)?

8–11.   (a) How much more energy is emitted by a star at 20,000 K than one at 5000 K?

(b) What is the predominant *color* of each star in part (a)? Use the Wien displacement law, and express your answers as wavelengths.

8–12.   Deduce an *approximate* expression for the Planck radiation law, equation (8–35a):

(a) at high frequencies (i.e., $(h\nu/kT) \gg 1$),

(b) at low frequencies (i.e., $(h\nu/kT) \ll 1$).

You may use the approximation, $\exp(h\nu/kT) \approx 1 + (h\nu/kT)$, when $(h\nu/kT) \ll 1$. To what *wavelength* does $(h\nu/kT) = 1$ correspond?

Reading List

Abell, George: *Exploration of the Universe.* Second Ed. New York, Holt, Rinehart and Winston, 1969.

Aller, Lawrence H.: *Astrophysics: The Atmospheres of the Sun and Stars.* New York, The Ronald Press Company, 1953.

Brandt, John C.: *The Physics and Astronomy of the Sun and Stars.* New York, McGraw-Hill Book Company, 1966.

Dufay, Jean (Gingerich, O., translator): *Introduction to Astrophysics: The Stars.* New York, Dover Publications, 1964.

Feynman, Richard P., Leighton, Robert B., and Sands, Matthew: *The Feynman Lectures on Physics.* Vols. I and II. Reading, Massachusetts, Addison-Wesley Publishing Company, 1963, 1964.

Herzberg, Gerhard (Spinks, J. W. T., translator): *Atomic Spectra and Atomic Structure,* Second Ed. New York, Dover Publications, 1944.

Hynek, J. A. (ed.): *Astrophysics (A Topical Symposium).* First Ed. New York, McGraw-Hill Book Company, 1951.

Jenkins, Francis A., and White, Harvey E.: *Fundamentals of Optics.* Third Ed. New York, McGraw-Hill Book Company, 1957.

McCrea, W. H.: *Physics of the Sun and Stars.* London, Hutchinson, 1950.

Motz, Lloyd: *Astrophysics and Stellar Structure.* Waltham, Massachusetts, Ginn and Company, 1970.

Richards, James A., et al.: *Modern University Physics.* Reading, Massachusetts, Addison-Wesley Publishing Company, 1960.

Rosseland, Svein: *Theoretical Astrophysics.* Oxford, The Clarendon Press, 1936.

Strand, Kaj Aa (ed.): *Basic Astronomical Data.* Chicago, University of Chicago Press, 1965.

Swihart, Thomas L.: *Astrophysics and Stellar Astronomy.* New York, John Wiley and Sons, 1968.

Unsöld, Albrecht (McCrea, William H., translator): *The New Cosmos.* New York, Springer-Verlag, 1969.

Weidner, R. T., and Sells, R. L.: *Elementary Modern Physics.* Boston, Allyn and Bacon, 1960.

Chapter 9
The Sun

Our Sun (*Sol*) is the nearest star, and it dominates the Solar System. The fascinating properties and phenomena of the solar surface layers are easily observed and have been studied in great detail. Since the Sun is a fairly typical star, and since it is the only star which does not appear to us as a mere point of light, the discussion presented in this chapter may serve as the basis for our later investigation of the other stars.

9–1 THE STRUCTURE OF OUR SUN

At one astronomical unit from the Earth resides that blazing pivot point of the Solar System, our Sun, which provides the energy necessary to sustain life and which is our primary celestial timepiece. This gigantic sphere of *gas*, with a radius of $R_\odot = 6.96 \times 10^5$ km ($\approx 109 R_\oplus$) and a mass of $M_\odot = 1.99 \times 10^{33}$ gm ($3.33 \times 10^5 M_\oplus$), has a *luminosity* or rate of energy radiation of $L_\odot = 3.90 \times 10^{33}$ erg sec^{-1}. The average density of the Sun is only 1.41 gm cm^{-3}, a value consistent with the fact that this star consists mainly of gaseous hydrogen.

Knowing these gross characteristics, we may now inquire into the details of solar structure. Let us begin with a brief look at the unobservable interior of the Sun. Our knowledge of stellar interiors is completely inferential,* being based upon known physical laws and theoretical models (see Chapter 15). Newton's law of universal gravitation tells us that the pressure, and therefore also the temperature, in the solar interior must assume extremely large values just to support the mass of the Sun. No solids or liquids can exist under these conditions, so that the body of the

---

Sun must be gaseous; in fact, the gases (primarily hydrogen and helium) must be almost completely ionized. Temperature, pressure, and density increase from the Sun's surface inward to the center where energy is liberated via thermonuclear reactions. As hydrogen is transmuted into helium in the Sun's *core*, vast quantities of energy are released in the form of photons (see Figure 9–1). These photons diffuse outward through the large *radiative zone* until they reach the thin outer *convective zone*, where most of the transport of energy takes place by means of "boiling" motions of the gas. The visible surface of the Sun (the photosphere) occurs at the top of the convective layer, where the complex and extended solar atmosphere begins.

The base of the solar atmosphere is the *photosphere* (see Figure 9–2), an extremely thin layer of gas which represents the greatest depth to which we may observe and from which emanates the bulk of the radiation constituting the solar luminosity ($L_\odot$). Sunspots and faculae are phenomena associated with the photosphere. The next layer outward has a characteristic pinkish hue, so we term it the *chromosphere*; here we find plages, and from the top of the chromosphere emerge the enigmatic spicules and the awesome prominences. Beyond this region lies the tenuous, extended *corona* with its ghostly light merging into the outward-flowing *solar wind* and the interplanetary medium. Of the entire Sun, only the solar atmosphere is directly accessible to observation, but what a rich arena of activity it is. The remainder of this chapter is devoted to a detailed examination of the phenomena observed in these outer atmospheric layers of our Sun.

9–2   THE PHOTOSPHERE

(a)   Granulation

We cannot peer deeper into the Sun than the base of the photosphere. Here (Figure 9–3) we see a patchwork pattern of small (average diameter about 700 km), transient (average lifetime from five to tens of minutes) *granules*: bright irregular formations surrounded by darker lanes.

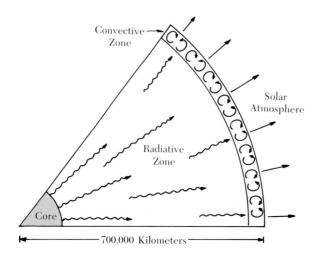

Figure 9–1.   *The Solar Interior.* This pie-slice of the interior of our Sun shows the thermonuclear *core*, the broad *radiative zone*, the thin "boiling" *convective zone*, and the tenuous solar atmosphere.

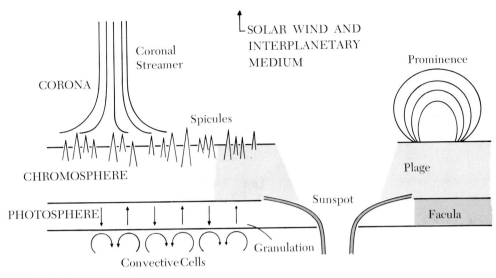

Figure 9–2. *The Sun's Surface and Atmosphere.* The various layers and their associated phenomena are shown in highly schematic form.

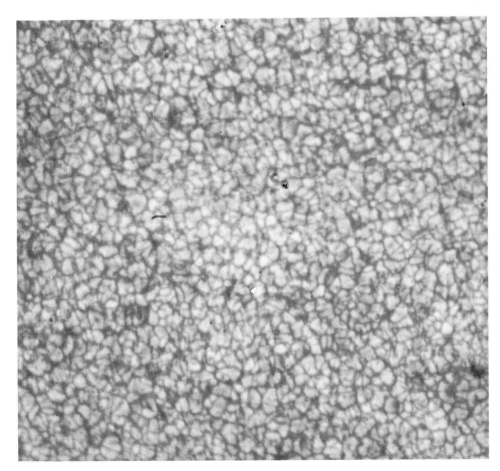

Figure 9–3. *The Solar Granulation.* The patchwork of granules (top of the convective zone) at the base of the photosphere, as photographed from a balloon. (Project Stratoscope of Princeton University)

This solar granulation is the surface manifestation of the Sun's *convective zone*, a thin gaseous layer (some tens of thousands of kilometers thick) located just below the base of the photosphere. In this zone, heat energy is transported by *convection*; hot volumes of gas (*convection cells*) rise, appear as bright granules and dissipate their energy at the photosphere; the cooler gases (dark lanes) sink back into the zone. The resulting transfer of energy imparts motions of the order of tenths of a km sec^{-1} to the lower layers of the photosphere.

(b) Photospheric Temperatures

When we study the continuum spectrum of the entire solar disk, we find a Stefan-Boltzmann effective temperature (as defined in Table 8–4) of 5700 K for the photosphere. Our interest here, however, is in the detailed *temperature profile* of the photosphere. How may we determine this profile? A clue to the answer is evident on any white-light photograph of the Sun (see, for example, Figure 9–20), for we see that the brightness of the solar disk decreases from the center to the limb—this is termed *limb darkening*.

Putting aside all complications for the moment, we can say that limb darkening arises because we see deeper, hotter gas layers when looking directly at the center of the disk, and higher, cooler layers when we look near the limb. Figure 9–4A illustrates this geometrical effect, assuming that we can see only a fixed distance d through the solar atmosphere. Since the limb appears *darkened*, we say that the temperature *decreases* as we move from the lower to the upper

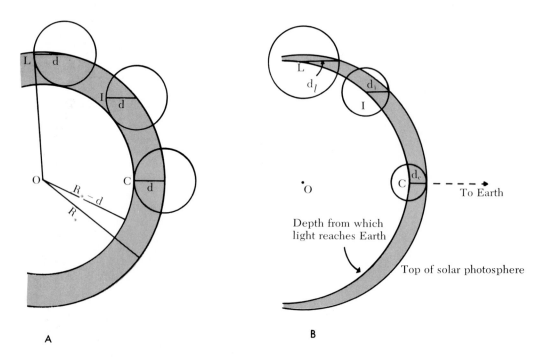

A   B

Figure 9–4. *Limb Darkening.* The center of the disk is denoted by C, the limb by L, and an intermediate position by I. **A,** All circles have the same radius corresponding to geometrical distance d, but at the limb the line of sight terminates higher in the atmosphere than at the center: $OL > OI > OC$. **B,** Here we take into account the fact that the high-density gas low in the atmosphere is more opaque than the rarified gas in the upper atmosphere, and each circle corresponds to the same *optical depth*. The geometrical distances are such that $d_c < d_i < d_\ell$. Still, $OL > OI > OC$.

photosphere—from the Stefan-Boltzmann law (Chapter 8), cool gas radiates less energy than does hot gas.

The temperature profile of the photosphere is extracted by comparing continuum spectra, taken at different points on the solar disk, with the Planck curves for different temperatures. The results are displayed in Figure 9–5, where the top of the photosphere (bottom of the chromosphere) is defined as height = 0 km; the temperature minimum defines the photosphere-chromosphere interface. As we move outward through the photosphere, the temperature drops rapidly (more than 2000 K in the last 300 km), then turns around and *rises* in the chromosphere. The profile can be determined this accurately only because we can resolve the solar disk, an advantage generally lacking for other stars.*

At this point, the reader may have discerned an apparent paradox in Figure 9–5: how can the solar limb appear darkened when the temperature rises rapidly through the chromosphere? A quantitative answer to this question requires the concepts of *opacity* and *optical depth*, both of which are discussed in the following paragraphs. We can, however, simply say that the chromosphere is almost optically transparent compared to the photosphere; the gas density outside the photosphere is so low that comparatively little radiation can be either absorbed or emitted there. Hence, the Sun appears to terminate sharply at its photospheric "surface"—within the outer 300 km of its 700,000 km radius ($R_\odot$).

Our line of sight penetrates the solar atmosphere only to the depth from which radiation can escape unhindered. Interior to this point, solar radiation is constantly

---

* The limb darkening of some eclipsing binary stars can be inferred from their light curves (see Chapter 11), and the limb darkening of several nearby giant stars has recently been measured with a novel stellar interferometer.

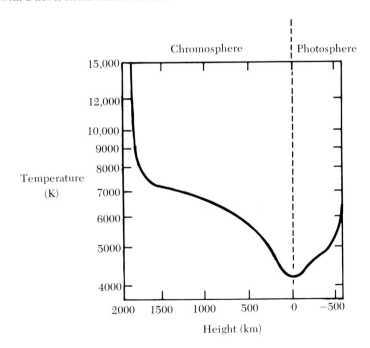

Figure 9–5. *Temperature Structure of the Photosphere and Chromosphere.* The solar interior is to the right. The temperature reaches a minimum at the photosphere-chromosphere boundary (dashed line), then rises rapidly through the chromosphere into the corona. (After J. Linsky and E. Avrett)

absorbed and re-emitted (or scattered) by the gas atoms and ions. To characterize this absorption by the gas at a given wavelength (λ), we speak of the *opacity* κ_λ (units $= $ cm^2 gm^{-1}) of the gas. (This is also referred to as the mass absorption coefficient.) The simplest way to understand opacity is to consider what happens when radiation of flux F_λ strikes a slab of gas of thickness dx; let the mass density ρ (units $=$ gm cm^{-3}) describe this gas. Part of this flux is absorbed by the slab, so we write

$$dF_\lambda = -\kappa_\lambda \rho F_\lambda \, dx \qquad (9\text{--}1)$$

In a uniform medium, equation (9–1) may be integrated to yield:

$$F_\lambda(x) = F_\lambda(0) \exp\left(-\kappa_\lambda \rho x\right) \qquad (9\text{--}2)$$

so that the flux diminishes exponentially with depth of penetration. For convenience, astronomers define another measure of this absorption by speaking of *optical depth* τ_λ, where

$$d\tau_\lambda \equiv \kappa_\lambda \rho \, dx \qquad (9\text{--}3)$$

note that τ_λ is dimensionless and equation (9–2) becomes simply

$$F_\lambda(\tau_\lambda) = F_\lambda(0) \exp\left(-\tau_\lambda\right) \qquad (9\text{--}4)$$

For $\kappa_\lambda = 0$, the flux is constant, so that the gas is said to be transparent (*optically thin*) at this wavelength. In general, we speak of an *optically thick* gas as one with optical depth much greater than 1; the base of the photosphere corresponds to this case. Figure 9–4B illustrates how limb darkening arises owing to the photospheric opacity; our earlier statement about the low density (ρ) of the chromosphere implying transparency must now be qualified—κ_λ is also comparatively low in the chromosphere.

(c) H$^-$ Continuous Absorption

The photospheric gas obeys the perfect gas equation, $P = nkT$, where P is the pressure, n is the number density (particles per cubic centimeter), k is Boltzmann's constant, and T is the temperature. Since this pressure supports the photosphere against the gravitational attraction of the body of the Sun, we can deduce the mass density ρ in the photosphere. We find densities far less than those which constitute the "gas under pressure" required for continuum radiation according to Kirchhoff's laws. How, then, does the photospheric continuum radiation arise?

Continuous absorption due to atomic ionization, and emission due to recombination, lead to the wrong wavelength dependence for the observed continuum. By studying the wavelength dependence of solar limb darkening data, it has been found that the chief contributor to the continuous opacity must be H$^-$—*the negative hydrogen ion*. This ion can exist because the single electron of the normal neutral hydrogen atom does not completely "screen" the positive proton. Hence, a second electron may attach itself, albeit fairly loosely; the ionization potential is 0.75 eV compared to 13.54 eV for normal hydrogen. Absorption occurs via the dissociation reaction $(H^- \rightarrow H + e^-)$, and emission takes place when an electron attaches itself to a neutral hydrogen atom $(H + e^- \rightarrow H^-)$. The solar continuum in the infrared and optical is produced by these reactions, which are made significant by the high electron and hydrogen densities in the photosphere.

(d)   The Fraunhofer Absorption Spectrum

(i)   *The Spectral Lines*

In 1814, the German physicist Joseph Fraunhofer (1787–1826) made the first definitive mapping of the vast array of absorption lines seen superimposed upon the photospheric continuum—the *Fraunhofer absorption spectrum*. Not knowing the correct identifications of these features, Fraunhofer designated the strongest lines (starting from the red) with capital letters and the weaker lines with lower-case letters. (See section 8–4 and Figure 8–11 for the definition of line strength.) Today we still refer to the D lines (a resolved doublet of sodium), the *b* lines (magnesium), and the H and K lines (of Ca II). The earliest line identifications included the hydrogen Balmer series and absorption lines of sodium, calcium, and magnesium. Surprising though it may seem, large numbers of faint absorption lines remain unidentified even today; these probably correspond to transitions which are difficult to excite in the laboratory.

Many more solar lines have come under scrutiny since the advent of rocket and satellite spectroscopy, for now we can study that part of the solar ultraviolet spectrum normally blocked by the Earth's atmosphere. The Fraunhofer absorption line spectrum extends to wavelengths as short as 1650 Å. At wavelengths shorter

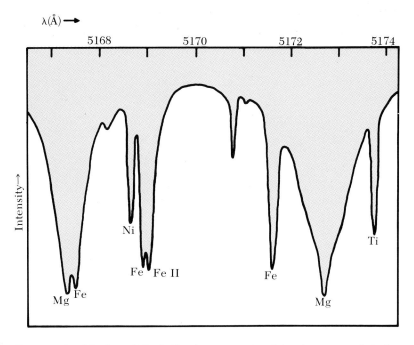

Figure 9–6.   *Solar Spectral Line Profiles.* A narrow region of the solar spectrum including the Mg I lines and several weaker lines.

than the near ultraviolet ($\lambda < 1650$ Å), the solar spectrum is dominated by *emission lines* which are produced in the chromosphere and corona [see sections 9–3(a) and 9–4(c)].

Kirchhoff's laws would lead us to expect that the absorption lines are formed in a region separate from and cooler than the region producing the continuum. However, since the solar continuum is due to a tenuous gas of H^- ions and not a "gas under pressure," the Fraunhofer absorption lines can and do form at the same solar atmospheric level as does the continuum. Furthermore, while the weaker lines originate in the lower photosphere, the stronger ones form in the upper photosphere; in fact, the strongest (the Balmer lines and the H and K lines of Ca II) are formed primarily in the chromosphere.

As already stated, the opacity κ_λ is low in the chromosphere for most wavelengths, for the chromosphere is transparent in the continuum. In strong lines however, the opacity becomes large even at considerable heights in the solar atmosphere for several reasons. These reasons include the abundance of the element, the number of atoms in the lower level of the transition producing the line (which in turn depends on the temperature through the Boltzmann equation; see section 8–4), and the transition probability (the intrinsic atomic parameter that determines the probability that an atom will make that particular transition). The depth of formation differs from line to line and within a given line because of line broadening. Thus, the Fraunhofer absorption lines provide us with a powerful tool to probe different heights of the solar atmosphere.

Recently, high-dispersion spectra have revealed fine structure *wiggles* in most of the Fraunhofer lines (see Figure 9–7). These wavelength oscillations are Doppler shifts due to vertical motions at about 0.4 km sec^{-1} of small-scale (some 1000 km in diameter) structures in the photosphere. The wiggles are approximately

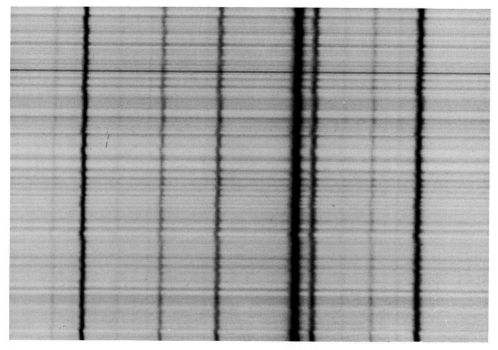

Figure 9–7. *Solar Spectrum Showing "Wiggly Lines."* These wiggles are due to small-scale Doppler motions in the photosphere. (Courtesy of W. Livingston, Kitt Peak National Observatory)

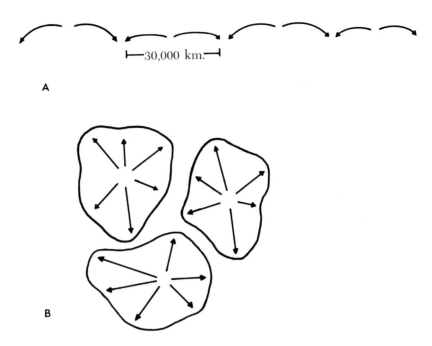

├──30,000 km.──┤

A

B

Figure 9–8. *Schematic Diagram of Supergranulation.* **A**, Side view. **B**, Top view, looking down on solar disk. Magnetic fields and spicules are concentrated at the supergranule boundaries.

periodic in time, and appear to be generated by gas motions induced by the granules at the bottom and continuing up into the chromosphere. In addition, we find horizontal motions parallel to the solar surface with about the same velocities. These arise in larger structures (30,000 km across) called *supergranules*. As indicated in Figure 9–8, gas flows slowly from the center of a supergranule to the edge. Supergranules are apparently related to higher chromospheric structures, and their flowing gases seem to carry magnetic lines of force to their periphery. Hence, the magnetic fields become more localized and concentrated; as we shall see in a later section, "solar activity" is generally associated with strong magnetic fields.

(ii) *Elemental Abundances*

Using the ideas outlined in Chapter 8, we can analyze the Fraunhofer lines to glean information about the photosphere. The characteristic temperatures and pressures are first found, and then the *relative line strengths* (see section 8–4) tell us the *chemical composition*, that is, the relative abundance of the elements. The results are shown in Table 9–1; hydrogen is by far the most abundant element, helium has a number density about 20 times smaller, and all of the heavier elements account for about 1/1000th of the total number density. Note that photospheric temperatures are too low to produce helium absorption lines; hence, the helium abundance has been inferred from (a) its calculated contribution to the total gas pressure, (b) space probe measurements of the solar wind, and (c) observations of solar cosmic rays. The last two techniques yield results exhibiting considerable time variability, so that we cannot yet relate them unambiguously to the photospheric helium abundance.

9–3   THE CHROMOSPHERE

The solar *chromosphere* (see Figure 9–2) extends 8000 to 10,000 km above the photosphere, and its gas density is far less than that of the photosphere. This thin layer is pinkish in hue—as a result of the Balmer (Hα) emission of hydrogen at $\lambda = 6563$ Å—and is usually visible only during a total solar eclipse. Today this region of the solar atmosphere may be examined at leisure outside of eclipse, by using a *coronagraph* which artificially eclipses the solar photosphere with an occulting disk.

(a)   The Flash Spectrum

In the past, the chromospheric spectrum could only be obtained when the Moon covered the entire photosphere during an eclipse; at that instant the photospheric absorption spectrum is replaced by the chromospheric *emission line* spectrum—the so-called *flash spectrum* (see Figure 9–9). Most of the photospheric Fraunhofer absorption lines become emission lines in the chromosphere, and a few extra emission lines (notably those of helium, which require very high temperatures to become excited) appear.

Chromospheric spectra reveal characteristic variations with altitude in the chromosphere. Atomic transitions of low excitation potential, such as some of those of neutral metals (e.g., Fe I), are seen only at the base of the chromosphere. Lines of ionized iron, calcium, and strontium are evident somewhat higher up. The hydrogen Balmer and neutral helium features are seen for many thousands of kilometers above the photosphere. All of the chromospheric lines fade with height, but the strongest

Table 9–1.   Abundances of the Elements in the Solar Atmosphere Relative to Hydrogen by Number*

H	1,000,000
He	50,000†
C	350
N	110
O	670
Ne	28
Na	1.7
Mg	34
Al	2.5
Si	35
P	0.27
S	16
K	0.11
Ca	2.1
Fe	25

* All abundances, except those for He courtesy of G. L. Withbroe.

† There is still considerable uncertainty about this figure, as discussed in the text. The value quoted here is largely based on solar wind and cosmic ray data and may differ from the photospheric abundance. Fluctuations in these data give a range in the He : H ratio of 0.02 to 0.25.

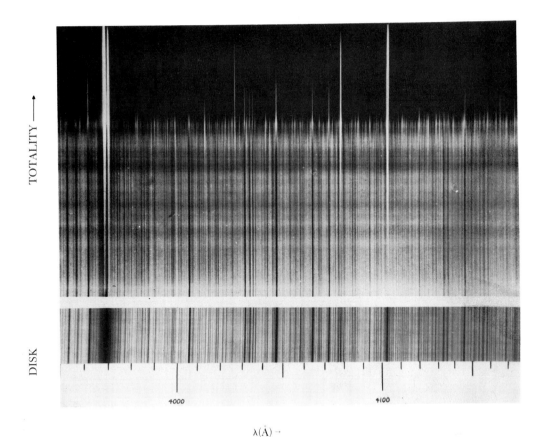

$\lambda(\text{Å}) \rightarrow$

Figure 9–9. *Flash Spectrum.* Obtained by moving the photographic plate during a solar eclipse. Absorption lines arise in the photosphere and these reverse into the chromospheric emission at total eclipse. Several lines already appear in emission before totality. Lines at $\lambda3970$ are due to Ca II and He, at $\lambda4026$ due to He I, at $\lambda4078$ due to Sr II, and at $\lambda4101$ due to Hδ. The disk spectrum is shown separately at the bottom of the photograph. (Lick Observatory)

line of He II ($\lambda4686$) fades most slowly. This decrease in line strengths is due to two major causes: (a) the gas density drops sharply with height, and (b) the temperature increases rapidly with height above the photosphere (see Figure 9–5). The high-excitation-potential lines of helium remain strong, as a result of the high temperatures in the outer chromosphere.*

We mentioned in the preceding paragraphs that the hydrogen Balmer lines are formed predominantly in the chromosphere. Since the chromosphere is hotter than the photosphere, this statement appears to contradict one of Kirchhoff's laws. The explanation of this effect is simply this: the photospheric continuum (and temperatures) is not sufficiently energetic to promote many hydrogen atoms from the ground (Lyman) state to the first excited state. Remember that the Balmer lines must originate from the first excited state. Only in the hotter chromosphere, in accordance with the Boltzmann equation, does the population of second-level hydrogen atoms become significant. Hence, continuum radiation is absorbed in the

---

* Helium (after the Greek *helios*, meaning Sun) was found in chromospheric spectra before it was discovered on Earth. Most helium lines are too weak (*optically thin*) to see in projection on the solar disk, but are readily visible in flash spectra because they are seen against a dark background.

chromosphere to form (a) the Balmer absorption lines (when viewed in projection against the photosphere) and (b) the Balmer emission lines which color the chromosphere (when seen against dark space at the Sun's limb).

(b) Chromospheric Fine Structure

When projected against the solar disk, the chromosphere is optically thin in white light. Only within the narrow bandwidth of certain absorption lines (Hα, and the H and K lines of Ca II) which arise in the chromosphere do we have optical thickness. Radiation at these wavelengths cannot escape from the photosphere, since the chromosphere is essentially opaque here. The chromosphere may be studied at these wavelengths, since the absorption lines are not completely black; the center of each line is darker than the adjacent continuum, but some photons are still emitted in our direction from the chromosphere.

Two types of *monochrometers* (*mono* = one, *chromos* = color) are employed to study a narrow band of wavelengths—the *spectroheliograph* and the *narrow-band polarizing filter*. In a spectroheliograph, sunlight is spread into a spectrum by a spectrograph; a narrow wavelength band of the spectrum is permitted to strike a photographic plate; and the instrument slowly scans the entire solar disk at this selected wavelength. An image of the Sun at that wavelength is built up strip by strip. The primary disadvantage of this technique is its slowness, but almost any wavelength band may be selected. Monochromatic filters are simpler and less expensive, but are less versatile in wavelength selection; however, extremely short exposures of the entire Sun are possible, so that time-lapse motion pictures of transient phenomena can be made.

Monochromatic photographs of the Sun (Figure 9–11) in Hα and Ca II K reveal large bright and dark patches. These are, respectively, the *plages* and *filaments* associated with "solar activity" [see section 9–6(b)]. In addition, there is a distinctive structure over the entire solar disk: the bright *network* (see Figure 9–10) associated with the magnetic fields at the boundaries of supergranules. In the so-called quiet as well as the active areas of the Sun, we see a brightening in Ca II K related to increased magnetic field strengths, but on the quiet Sun these fields are 10 to 30 Gauss, while in active regions they are about 100 Gauss or greater. High resolution Hα filtergrams show an incredible degree of detailed fine structure, much of which still defies interpretation (Figure 9–11).

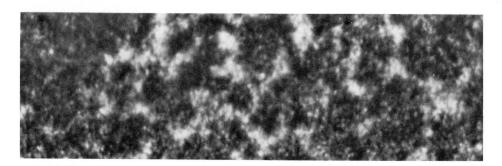

Figure 9–10. *Chromospheric Network Photographed in Ca II K-line.* (Courtesy of N. R. Sheeley and S. Y. Liu, Kitt Peak National Observatory photograph)

To correctly interpret monochromatic pictures of the chromosphere, we must recall the line profiles involved. What do we mean by a *brightening*? As shown in Figure 9–12A, an Hα brightening corresponds to a shallower profile (more emission at the line center. Beyond the limb, of course, Hα is a true emission line.

At the limb of the Sun, tenuous "jets" of glowing gas (500 to 1500 km across) are seen to extend to a distance of ten thousand kilometers upward from the chromosphere (Figure 9–13A). In these *spicules*, which are best observed in Hα radiation, gas is rising at about 25 km sec^{-1}. Although spicules occupy less than one per cent of the Sun's surface area, and have lifetimes of 15 minutes or less, they probably play a significant role in the mass balance of the chromosphere, corona, and solar wind.

As seen in Figure 9–13B, spicules are not distributed uniformly over the solar surface, but form a network pattern. This is, in fact, part of the chromospheric network that constitutes the boundaries of the supergranules; thus, we conclude that spicules only occur in regions of enhanced magnetic fields.

(c) Ultraviolet Spectral Features

Ultraviolet spectral features are expected because of the high chromospheric temperatures; we will consider only hydrogen Lyman α here.

The strong hydrogen Balmer lines of the chromosphere imply an abundance of first-excited-state $(n = 2)$ hydrogen atoms. Absorption from the continuum will excite these atoms to higher levels $(n = 3, 4, \text{etc.})$ producing the Balmer absorption series; however, *most* of the atoms will return immediately to the ground state $(n = 1)$, emitting Ly α photons (at $\lambda 1216$, in the ultraviolet). Hydrogen atoms can be excited to $n = 2$ only by collisions or re-absorption of Ly α photons (self-absorption), since the photospheric continuum is weak in the ultraviolet. Hence, as shown in Figure 9–12B, we see *strong Lyman emission* lines; the central intensity dip is due to self-absorption: Lyman photons emitted low in the chromosphere are absorbed by hydrogen atoms in higher layers.

Spectroheliograms taken in Ly α show essentially the same chromospheric structures as do those made in the Ca II K line.

9–4 THE CORONA

At solar eclipses, the *corona* is seen as a pearly-white halo extending far from the Sun's limb (Figures 9–14 and 9–15); a brighter inner halo hugs the solar limb, and coronal streamers extend far into space. Today, *coronagraphs* (see section 9–3) are used, on Earth and on balloons, rockets, and satellites, to extensively study the tenuous corona.

(a) The Visible Corona

Coronal continuum radiation at optical wavelengths is composed of two parts. Indeed, the corona itself may be subdivided into the *K-corona* (predominating near the Sun) and the *F-corona* (evident farther out).

(*Text continued on page 239.*)

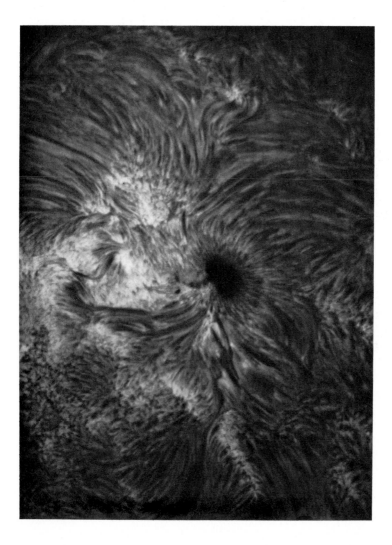

Figure 9–11A.   *Hα Filtergram*, showing detailed fine structure in a region near a sunspot near the center of the disk. (Courtesy of R. B. Dunn, Sacramento Peak Observatory)

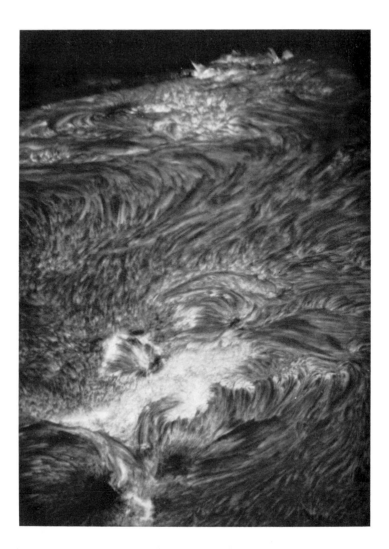

Figure 9–11B.   *Hα filtergram,* showing structure near the Sun's limb. (Courtesy of R. B. Dunn, Sacramento Peak Observatory)

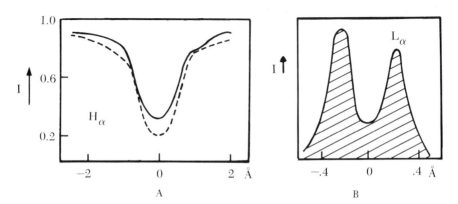

Figure 9–12.   *Hα and Lyman α Line Profiles.* **A,** Hα profile. Dashed line represents the undisturbed Sun; solid lines are bright regions, such as network or plage. **B,** Lyman α profile (see text for explanation).

A

B

Figure 9–13. *Spicules.* **A,** This picture of the Sun's limb was taken in the light of Hα; it shows the top of the chromosphere. Note the high inclination of some of the spicules. **B,** Here spicules are seen as elongated dark features projected against the disk near the limb; they were photographed in the wing of the Hα line. The distribution of spicules is along the chromospheric network, which corresponds to supergranule boundaries. (Courtesy of R. B. Dunn, Sacramento Peak Observatory)

Figure 9–14. *Quiet Sun Corona.* Note the polar plumes and the elongation at the solar equator. (1954 eclipse in Wisconsin; courtesy of J. D. Bahng and K. L. Hallam)

Part of the low-intensity coronal continuum is *identical* in its wavelength dependence to that of the photosphere. This is a result of light scattered by electrons which constitute more than half the particle number density in the K-corona. No photospheric absorption lines are evident in this scattered component, a fact which we attribute to Doppler broadening by rapidly-moving electrons. Since the absorption lines are completely washed out, these electrons must be characterized by extremely high temperatures—average (quiet) coronal temperatures are 1 to 2×10^6 K, that is *millions of degrees Absolute.*

Superimposed on the electron-scattering continuum is the F-corona spectrum, which does show the Fraunhofer absorption lines. This component is due to light scattering from "dust" particles, identical to those grains which pervade interplanetary space. The dust is concentrated in the plane of the ecliptic, for we see the outer extension of the F-corona as the *zodiacal light.* We can separate the F- and K-coronal components because (a) the Fraunhofer lines appear only in the former, and (b) electrons and "dust" differ radically in the way that they polarize the light they scatter.

Figure 9–15. *Active Sun Corona Showing Extensive Streamers.* A neutral filter was used to compensate for the rapid change in brightness from the limb outward. The bright image at the edge of the photograph is Venus. (1966 eclipse in Bolivia; Courtesy of G. Newkirk, High Altitude Observatory)

Solar activity strongly affects the appearance of the K-corona (Figures 9–14 and 9–15). At times of *sunspot maximum* [see section 9–6(a), Sunspots], the corona is very bright and uniform around the solar limb, and bright *coronal streamers* and other condensations associated with active regions are much in evidence. At sunspot *minimum*, the corona extends considerably farther at the solar equator than at the poles, the coronal streamers are concentrated to the equator, and *polar plumes* become visible at the poles.

(b) The Radio Corona

The rapid development of radio astronomy in the last two decades has opened a new window (at "radio" wavelengths, 1 cm to 10 m) on the solar corona and chromosphere. Later in this chapter we will consider the radio bursts associated with solar activity, but here let us discuss the radio properties of the quiet Sun.

Radio photons correspond to very low energies, less than 10^{-4} eV. In the ionized gases of the solar atmosphere, it is the *free electrons* which are responsible for the emission and absorption of such radiation. The interactions involved are called *free-free transitions*: an electron sideswipes an ion or atom, and quantum-mechanically absorbs or emits a low-energy photon while the electron's kinetic energy changes slightly. Note that the electron is *not captured*, but is free both before and after the collision. Both the frequency and strength of interaction increase, the closer the electron passes the scatterer. Hence, the character of the free-free photons depends upon the *gas density*: the denser the gas, the more frequent and energetic will be the interaction; the resulting photons have higher energies or *shorter* wavelengths.

Now we can understand why short-wavelength (1 to 20 cm) radiation characterizes the chromosphere and lower corona, while wavelengths longer than 10 cm arise in the outer corona. At wavelengths longer than about 20 cm, the Sun is observed to be limb *brightened* (into the corona). Hence, we know that the longer the wavelength observed, the *higher* in the corona we are looking. At wavelengths longer than 50 cm, we encounter great radio variability, since the emission from the tenuous electron gas fluctuates wildly as the number density of particles varies.

Just as the H^- ion accounted for the photospheric opacity, so also does the *electron density* lead to the coronal opacity at radio wavelengths. The wavelength-dependence of this electron opacity is the following: the corona is "optically thin" at short wavelengths ($\approx$ 1 cm), so that such radiation can reach us from the chromosphere; at longer wavelengths, it becomes more and more "optically thick," so that the corona is opaque to these wavelengths at greater heights. On the average, we can fit the radio data to the low-energy tail of a Planck blackbody distribution at temperatures of 0.8 to 1.5×10^6 K.

(c)   Line Emission

(i)   *Forbidden Lines*

Superimposed on the visible coronal continuum are some *emission* lines, which were unidentified until about 1942 when W. Grotrian of Germany and B. Edlén of Sweden interpreted them.* The two strongest lines are the "green line" of Fe XIV ($\lambda 5303$) and the "red line" of Fe X ($\lambda 6374$): both of these are forbidden lines.

Two significant obstacles hindered the identification of the coronal emission lines: (a) the fact that the responsible transitions are *forbidden*, and (b) the unexpectedly *high temperatures* of the corona. In quantum mechanics, certain energy levels of an atom are *metastable* because downward transitions from such levels are strongly prohibited by the so-called selection rules. While an ordinary permitted transition takes place in about 10^{-8} sec, these metastable levels may persist for seconds or even days before a forbidden transition occurs. In most laboratory and

---

* They were long called "coronium lines," for they did not fit any known transition.

astrophysical situations, gas densities are so high that collisional de-excitation empties metastable levels very rapidly—there is just not enough time for a forbidden transition to take place. In the near-vacuum of the corona, however, metastable levels populated either by photospheric radiation or by collisions can decay at their leisure, and forbidden emission features are formed.

Exceedingly energetic collisions are required to ionize atoms, such as iron, 9 and 13 times; hence, the coronal gas must be very hot. To produce Fe X, we need a temperature of 1.3×10^6 K, and Fe XIV requires the even higher temperature of 2.3×10^6 K. At times of strong solar activity (e.g., flares), much higher temperatures occur, for the $\lambda 5694$ and $\lambda 5445$ lines of Ca XV (3.6×10^6 K) are seen. We are now sure that all of these temperatures are real; the characteristic range is 10^6 K (quiet corona) to 4×10^6 K (active corona).

(ii)   *Extreme Ultraviolet Lines*

Highly-ionized atoms, such as those present in the solar corona, have lost many of the electrons which "shield" the atomic nucleus. Therefore, the remaining electrons are strongly attracted and tightly bound to the nucleus. Now permitted transitions correspond to very high excitation potentials, and the resulting spectral photons are very energetic—at ultraviolet wavelengths. In fact, the spectral region from 50 to 500 Å (detectable only above the Earth's atmosphere) is dominated by the permitted emission lines from the coronal ions Fe VIII–XVI, Si VII–XII, Mg VIII–X, Ne VIII–IX, and S VIII–XII. This wealth of unambiguous transitions permits us to easily deduce the relative elemental abundances in the corona. The results are consistent with the photospheric abundances.

At wavelengths shorter than 1500 Å, the photospheric continuum becomes undetectably small (see the Planck curve for $\approx$ 6000 K in Figure 8–14). Therefore, the coronal UV emission lines are unobscured by a photospheric background—we needn't occult the solar disk to see them, even in projection on the face of the disk. Thus, we can easily study the structure of the corona at UV wavelengths. Figure 9–16 shows spectroheliograms taken in the "light" of Lyman β (representative of the chromosphere) and Mg X (representative of the corona, since its ionization temperature is over 2.5×10^6 K).

Figure 9–16. *Ultraviolet Solar Images*, constructed from measurements made by Orbiting Solar Observatory IV, can be compared with an Hα photograph (**A**) taken on the same day at the Sacramento Peak Observatory of the Air Force. It shows extensive sunspots on the northern hemisphere of the Sun's surface and lesser activity in the southern hemisphere. The dark patches are filaments. The five ultraviolet images are numbered in order of increasing temperature and therefore height in the solar atmosphere: Lyman continuum at 10,000 degrees (**B**), N III at 100,000 degrees (**C**), O VI at 325,000 degrees (**D**), Mg X at 1.4 million degrees (**E**), and Si XII at 2.25 million degrees (**F**). The last two images respectively correspond to heights of 15,000 kilometers and 200,000 kilometers. The hottest regions are in the palest color; the coolest regions are dark gray. It can be seen that the sunspots exert an influence throughout the solar atmosphere. (From Ultraviolet Astronomy, by Leo Goldberg. Copyright © by Scientific American, Inc. All rights reserved)

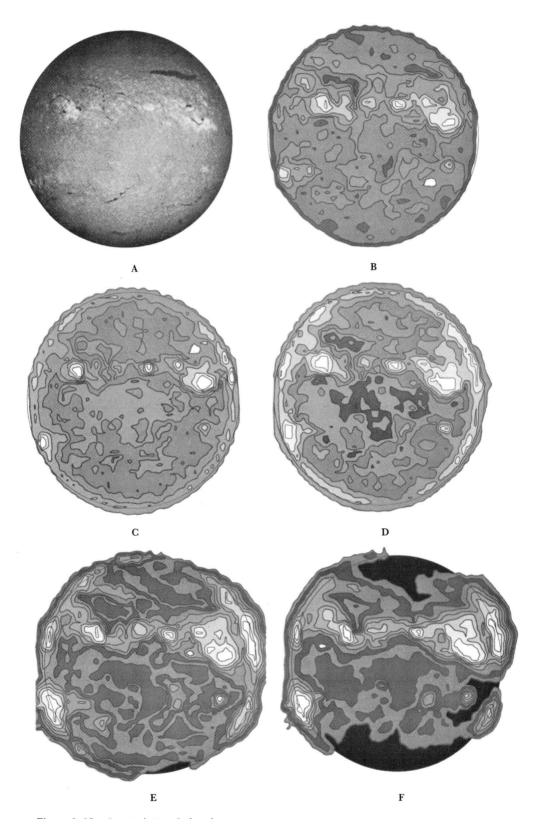

Figure 9–16.   *See opposite page for legend.*

9–5 THE SOLAR WIND

Interplanetary space beyond the solar corona is not empty. The high coronal temperatures imply a substantial pressure tending to blow the corona away from the Sun. The Sun's gravitational attraction on this gas is insufficient to retain it, so that there is a steadily-flowing *solar wind* leaving the Sun. This flowing gas, composed of approximately equal numbers of electrons and protons, is termed a *plasma* (it is electrically neutral on a large scale); as it pervades the Solar System it is called the *interplanetary medium*. The thermal conductivity of the plasma is very high, so that high temperatures prevail over great distances from the Sun. Hence, the wind accelerates as it expands (large-scale speeds near 300 km sec^{-1} at $30R_\odot$, and 400 km sec^{-1} at $215R_\odot = 1$ AU), and the particle density decreases to an average of about five electrons and protons per cubic centimeter at 1 AU. These "quiet Sun" characteristics of the solar wind have been measured by interplanetary space probes.

A plasma couples tightly to lines of magnetic force; in fact, the magnetic field is essentially "frozen into" the gas. Therefore, the solar wind drags the extensions of solar magnetic fields into interplanetary space. The large-scale solar magnetic fields are directly related to the interplanetary fields, via a *sectored structure* (Figure 9–17). The latitude band within 30° of the solar equator may frequently be divided into extended longitude regions of one magnetic polarity or the other. The radially-

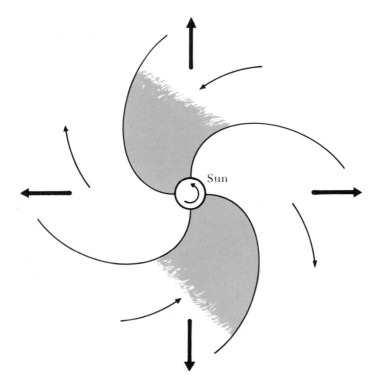

Figure 9–17. *Spiral Nature of the Interplanetary Magnetic Field.* These fields are extensions of the solar photospheric fields. Thin arrows represent the direction of magnetic fields, broad arrows the motion of solar wind particles. Two in-sectors are shown shaded.

outflowing solar wind conveys these fields away from the Sun in sectors; the sector boundaries are distorted into spirals, since the Sun rotates away from the receding gas and magnetic field. Hence, a polarized solar region may have moved beyond the limb before we record the corresponding sector arriving at the Earth.

Solar activity perturbs the simple picture which we have just presented. For instance, at the Earth's orbit the proton density varies from 0.4 to 80 cm^{-3}, and the speed ranges from 300 to 700 km sec^{-1}. There is, however, surprisingly little change in the average behavior over the solar cycle [see section 9–6(a)]. Some fluctuations are apparently due to individual flares; whereas the normal solar wind is composed of "low-energy" ($\sim 10^3$ eV) protons and electrons, solar flares eject clouds of high-energy protons* (10^7 to 10^{10} eV). These clouds rush through the solar wind, altering its speed and density locally, and distorting the magnetic field structure. Not only are these clouds dangerous to unshielded astronauts, but they are also responsible for magnetic disturbances in the Earth's magnetosphere (*magnetic storms*).

9–6 SOLAR ACTIVITY

(a) The Solar Cycle

The Sun is near enough that we can easily observe transient phenomena in its atmosphere. Such phenomena are manifestations of *solar activity*, and they are interrelated through the solar rotation and magnetic field in the *solar cycle*. In the remainder of this chapter we will discuss the *active* Sun, using the generic term, *active region*, to specify an area with sunspots, prominences, plages, and flares (see Table 9–2 for a summary of solar activity).

Sunspots. Prior to 1610, it was believed that the Sun was a spotlessly perfect radiant sphere; in that year Galileo Galilei published his *Sidereus Nuncius* (*Sidereal Messenger*), wherein he reported the existence of dark spots—*sunspots*—on the Sun's surface. Sunspots are photospheric phenomena (see Figure 9–2), which appear darker than the surrounding photosphere (at about 5700 K) because they are cooler (sunspot continuum temperatures are about 3800 K, and sunspot excitation temperatures are about 3900 K). The complex structure of a sunspot is illustrated in Figure 9–19. The darkest central part (with the temperatures just mentioned) is termed the *umbra*; this is usually surrounded by the lighter *penumbra* with its radial filamentary structure. Small sunspots develop from *pores*, larger-than-usual dark areas between bright granules. While most pores and small spots soon disintegrate, some grow into true sunspots of gigantic proportions. The largest have umbral diameters of 30,000 km, and penumbral diameters more than twice as large.

The most important characteristic of a sunspot is its *magnetic field*. Typical field strengths are near 1000 Gauss, but fields as strong as 4000 Gauss have been measured.† These fields inhibit the convective transport of energy to the photo-

---

* *Solar cosmic rays*; see section 9–6 (c, iii).

† These fields are deduced from the observed Zeeman splitting of spectral lines (see Chapter 8).

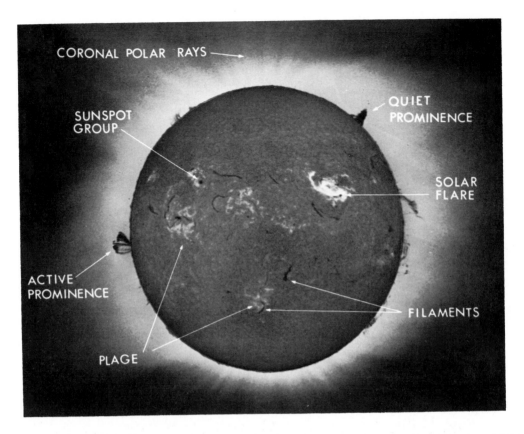

Figure 9–18. *Composite Photograph of the Sun.* The white-light corona is superimposed on Hα prominence and disk photographs. Note that the prominence photograph is a negative, while the disk and corona photographs are positive. (Lockheed Solar Observatory)

sphere, so that the sunspot is *cooler* than its surroundings. Related to the magnetic field is a remarkable horizontal flow of gas in the sunspot penumbra: gas moves out along the lower filaments and inward along the higher filaments (at speeds up to 6 km sec^{-1}). This phenomenon is termed the *Evershed effect*, after the man who first noted it.

A given sunspot has an associated *magnetic polarity*. Lines of magnetic force diverge from a north (positive) magnetic pole, and converge at a south (negative) pole; we are familiar with this characteristic of bar magnets, horseshoe magnets, and even our Earth. But a magnetic pole cannot exist in isolation, since magnetic lines of force must be complete (Maxwell's equations). Therefore, *two* sunspots of complementary polarity are generally found together in a *bipolar* spot group (see the following paragraphs). As usual, there are exceptions to this "rule": sometimes the second magnetic region (still revealed by Zeeman splitting) is so diffuse that only one lone sunspot is seen; at other times, large complicated groups of many sunspots appear—such a group may become the nucleus of a large *active region* [see section 9–6(b)] on the solar disk.

Sunspot Numbers.   Counts of the number of sunspots visible at any given time have been recorded since Galileo's time. Since sunspots tend to group together,

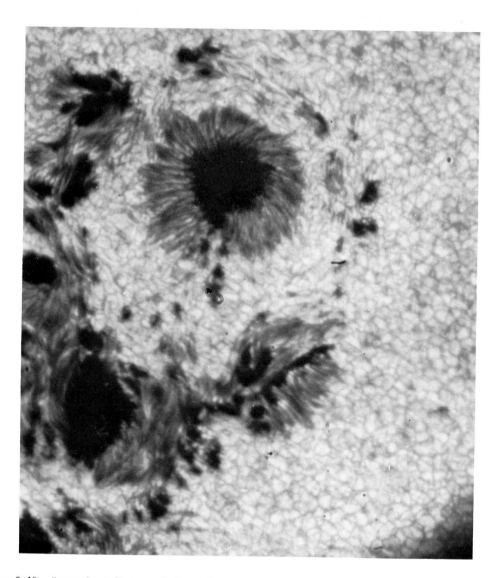

Figure 9–19. *Sunspot Group.* Photograph of part of a sunspot group that includes a classical-type sunspot with almost circular umbra (black) and radial penumbra. Other spots of the group are quite distorted, some having a penumbra only on one side of the umbra. Small dark spots without any penumbra are pores. Note the presence of granulation right up to the spots and penumbral filaments. (Project Stratoscope of Princeton University)

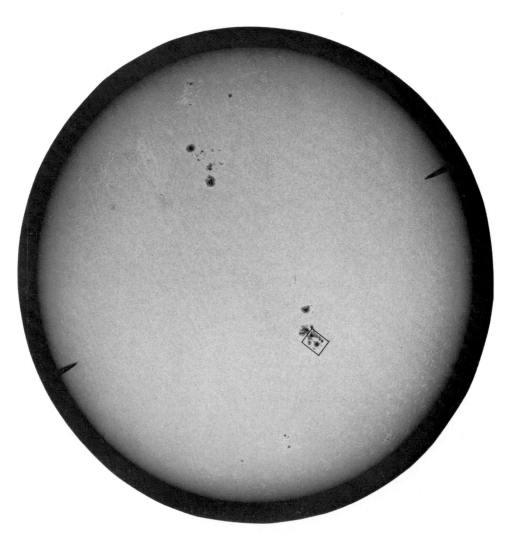

Figure 9–20.   *Disk of Sun in White Light.* Shows the same group illustrated in Figure 9–19 in its entirety, as well as other spots on the disk that day. Note also the limb darkening and the appearance of bright photospheric faculae near the limb. (Hale Observatories)

a meaningful indication of sunspot activity is the *Zurich sunspot number R*:

$$R = K(10g + f)$$

where f is the number of individual spots, g is the number of groups, and K is a normalizing factor which relates different observations to those at the Zurich Observatory (where $K = 1$). In Figure 9–21, the sunspot number is plotted versus time. A cyclic phenomenon is clearly taking place, for successive sunspot maxima (or minima) occur every 11 years on the average (there may be a variation of as much as two or three years in this number, from cycle to cycle). We say that a new cycle *begins* when the number is a *minimum*. (As a result of Figure 9–21, some scientists still speak of the *11-year solar cycle*, but we shall see later in this chapter that the period is actually 22 years!)

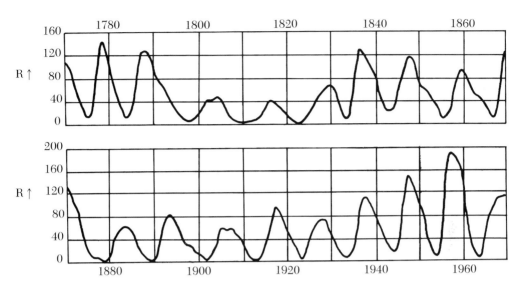

Figure 9–21. *Sunspot Numbers.* The Zurich sunspot number R is plotted as a function of time. The average period is 11 years.

The Butterfly Diagram. The distribution of sunspots in solar *latitude* varies in a characteristic way during the 11-year sunspot number cycle. As seen in Figure 9–22—the famous *Butterfly Diagram*—sunspots tend to reside at high latitudes ($\pm 35°$) at the start of a cycle; most spots are near $\pm 15°$ at maximum; and the few spots at the end of the cycle cluster near $\pm 8°$. Very few sunspots are ever found at latitudes greater than $\pm 40°$. The lifetime of a sunspot ranges from a few days (small spots) to months (large spots), so that the Butterfly Diagram is *not* due to individual sunspots migrating to lower latitudes. In fact, a sunspot "dies" at the same latitude where it was "born" (a characteristic which permits us to determine the solar rotation; see Solar Rotation below). What takes place is this: as the cycle progresses, new spots appear at ever lower latitudes. Note that the first high-latitude spots of a cycle appear even before the last low-latitude spots of the previous cycle have vanished.

The Law of Sunspot Polarity. Since most sunspot groups tend to be magnetically bipolar, it is useful to refer to *preceding* spots and *following* spots (in the sense of solar rotation). The Sun rotates eastward (as does the Earth), so that a preceding spot lies *west* of a following spot *as seen from the Earth*. After George E. Hale discovered the magnetic character of sunspots in 1908, it was realized that all bipolar groups in one solar hemisphere have the same polarity, while those in the other hemisphere have the *opposite* polarity. For example, in the present solar cycle (cycle 20, which began in 1964) preceding spots in the northern solar hemisphere have negative (south) polarity, while preceding spots in the southern hemisphere have positive (north) polarity. Moreover, *the sense of polarities reverses with each cycle*, so that northern hemisphere leading spots were positive during cycle 19. Hence, we now speak of the *22-year solar cycle*.

Solar Rotation. By observing sunspots, Galileo determined that the Sun's surface rotates eastward (synodically) in about one month. Today, the same method

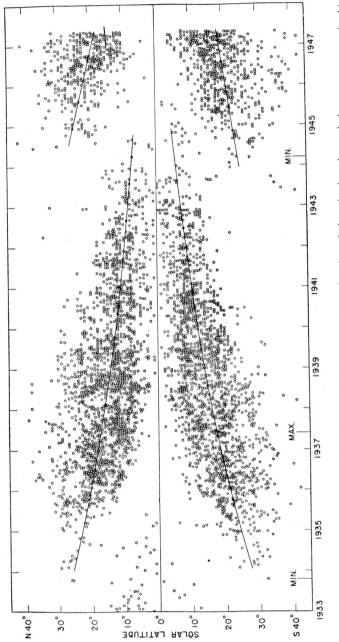

Figure 9–22. *Butterfly Diagram.* The latitude distribution of sunspots as a function of phase in the solar cycle becomes apparent in this representation. (From G. Abetti: *The Sun.* Second Ed. London, Faber and Faber, 1969)

is used in the sunspot zone (other methods, such as Doppler shifts, are necessary above latitude $\pm 40°$), and we know that the Sun rotates *differentially*. That is, the rotation period is shorter at the solar equator ($\approx 25^{\text{d}}$) than at higher latitudes ($\approx 27^{\text{d}}$ at 40°, and 30^{d} at 70°). (We are already familiar with differential rotation in the atmospheres of the Jovian planets; Chapter 7.) The average sidereal period adopted for the sunspot zone is 25.4 days, with a corresponding synodic period (appropriate in discussions of solar-terrestrial events) of 27.3 days.

(b) Active Regions

As the sunspot number increases, so also does solar activity. With each sunspot group is associated a large *active region* (several hundred thousands of kilometers in extent), with characteristics which we shall now discuss (see Table 9–2 for a summary).

(i) *Bipolar Magnetic Regions and Plages*

The most significant property of active regions is their magnetic fields: thousands of Gauss in sunspots, hundreds overall, and only a few Gauss near the periphery. Even when the sunspot group is a tangle of different polarities, the enveloping region is usually bipolar in character; hence, we refer to *bipolar magnetic regions* (BMR).

A *magnetograph* is used to map the magnetic field structure of BMRs (see Figure 9–23). A magnetic field Zeeman-splits a spectral feature into several components, each with a characteristic optical polarization. By comparing the splitting and intensity of these components with a magnetograph, based on the different polarizations, a map of the magnetic field strength and direction is produced.

In white light, photospheric *faculae* are brightenings which mark active regions. The enhanced brightness is due to greater temperatures and densities than the nearby regions of the photosphere, similar to the chromospheric plages to

Table 9–2. Summary of Solar Activity

Photosphere

Sunspots:	(a) strong magnetic fields.
	(b) lower temperature than photosphere.
Faculae:*	denser, hotter, and brighter than photosphere (in white light).
Bipolar magnetic regions (BMR):*	medium to weak magnetic fields.

Chromosphere

Plages:*	(a) brighter than chromosphere in Hα and Ca II lines.
	(b) denser and hotter than chromosphere.
Prominences (filaments):	(a) chromospheric material in corona.
	(b) exhibit motions associated with magnetic fields.
Flares:†	brief brightenings in plages (in Hα and Ca II lines).

Corona

Condensations:*	(a) white light features due to increased electron density.
	(b) increased emission in forbidden and UV lines.
	(c) associated with slowly-varying radio emission.
Radio bursts due to fast electrons trapped in high corona.†	
Particle emission (solar cosmic rays) and enhanced solar wind.†	

* Phenomena related to plages at different heights in the solar atmosphere.
† Flare-related phenomena.

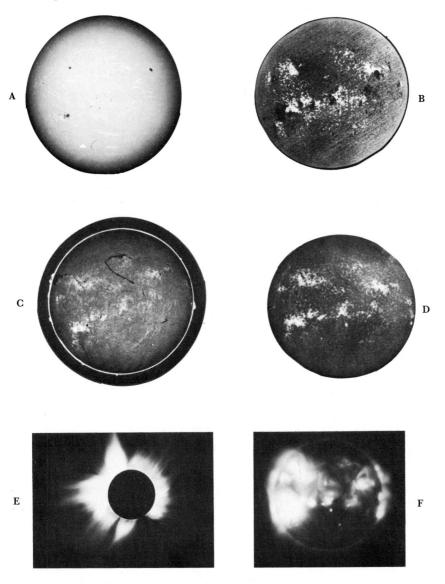

Figure 9–23.   *The Sun on March 7, 1970.* **A**, White light photograph. (Courtesy of Culgoora Solar Observatory.) **B**, Magnetogram. (Courtesy of W. Livingston, Kitt Peak National Observatory.) **C**, Hα filtergram. (Courtesy of P. S. McIntosh, National Oceanic and Atmospheric Laboratories.) **D**, Ca II spectroheliogram. (Courtesy of Sacramento Peak Observatory, Air Force Cambridge Research Laboratories.) **E**, Corona at eclipse. (Courtesy of G. Newkirk, High Altitude Observatory.) **F**, X-ray. (Courtesy of American Science and Engineering)

which the faculae are related. Limb darkening renders faculae visible near the solar limb, though they are invisible near the center of the disk.

Above active regions in the photosphere, we find the bright *plages* in the chromosphere. Plages are regions of higher density and temperature than the surrounding chromosphere, and are due to the magnetic fields of the active regions. These features appear on spectroheliograms taken in the light of Hα and the spectral lines of Ca II; in many respects, they look like concentrations and intensifications of the chromospheric network (see Figures 9–16 and 9–23).

In the corona, active regions manifest themselves again in the higher densities and temperatures of *coronal streamers* and condensations of the white-light corona. Coronal line emission (e.g., the green line and some extreme UV lines of Fe XIV) is stronger over plage regions than elsewhere, and enhanced radio emission arises from increased electron densities. This radio emission characterizes the long-lived active regions, so that it is known as the "slowly-varying component."

(ii)   *Prominences and Other Displays*

Spectacular markers of active regions are the *prominences*, which appear as long dark *filaments* when seen projected on the solar disk. Though visible in white light at a total solar eclipse, these displays are best recorded in Hα or Ca II lines at the limb of the Sun.

Prominences may be thought of as streams of chromospheric gas, occupying coronal regions tens of thousands of kilometers above the chromosphere. This description leads to the following dilemma: how can gas at about 10,000 K exist in the million-degree-hot corona? Magnetic fields are probably responsible for the phenomenon, but we don't know the answer yet. Two characteristic types of prominences are illustrated in Figure 9–24: (a) *quiescent* prominences, and (b) active prominences. Quiescent prominences last for weeks, and look like curtains with gas slowly descending from the corona into the chromosphere; they tend to lie along the "neutral line" separating the two poles of a BMR. Most active prominences only survive a few hours. Among the most active are the *loop* prominences (Figure 9–24), which are closely associated with solar flares and which survive only an hour or so, during which time gas streams down the magnetic field lines joining the BMR poles together.

(c)   **Solar Flares**

Among the most puzzling, spectacular, and energetic phenomena associated with active regions are the *solar flares*. Although these transient outbursts liberate tremendous quantities of energy, we still do not know where in the solar atmosphere (which level) they originate. Flares radiate at many frequencies, from the X-rays and gamma-rays to long-wavelength radio waves; in addition, they emit high-energy particles called *solar cosmic rays* (protons, electrons, and atomic nuclei; see Solar Cosmic Rays in this section). Flare X-rays disrupt terrestrial radio communication by disturbing the Earth's ionosphere. The high-energy particle clouds, which are lethal to unprotected astronauts, reach the Earth in 30 minutes; clouds of low-energy particles and disturbances in the solar wind require from six to 24 hours to transit from Sun to Earth.

(i)   *Optical Manifestations*

Flares usually appear in plages as brightenings in Hα. In fact, the Hα line becomes an *emission* feature (Figure 9–25), reaching maximum brightness within five minutes and decaying in about 20 minutes (three hours for the largest flares). Flares vary in size from 10,000 km (small) to over 300,000 km (largest); in general, the larger the flare, the brighter it is and the longer-lived. Although flares occur quite randomly, at the peak of the solar cycle the *average* occurrence of small flares is hourly and of truly large flares, monthly.

Figure 9–24A.   *Quiescent Prominence.* Notice the details of fine structure; sequences of photographs show the material to be moving downward. (Sacramento Peak Observatory)

Frequently a fast-moving "blast-wave" is produced by a flare. Travelling at a speed of about 1000 km sec^{-1} across the solar disk, this disturbance can perturb prominences and even cause distant filaments to disappear. Remarkably, the disrupted filaments reappear virtually unchanged hours or days later. Other manifestations of flares are spike-shaped *surges*, high temperature loop prominences, or fast *spray prominences* (with speeds of ejection up to 1000 km sec^{-1}; see Figure 9–27).

(ii)   *X-Ray and Radio Bursts*

In some flares, both X-rays and centimeter radio waves occur in conjunction; they probably originate in the upper chromosphere or corona (see Figure 9–28). Both consist of two components: (a) the *slow* component lasting about one-half hour, and (b) the *sudden* component (or *burst*) lasting a few minutes. The burst phenomenon is *nonthermal*, with X-rays more energetic than 20 keV ($\lambda < 1$ Å) and radio emission corresponding to 8×10^7 K. The less-energetic slow components

(*Text continued on page 257.*)

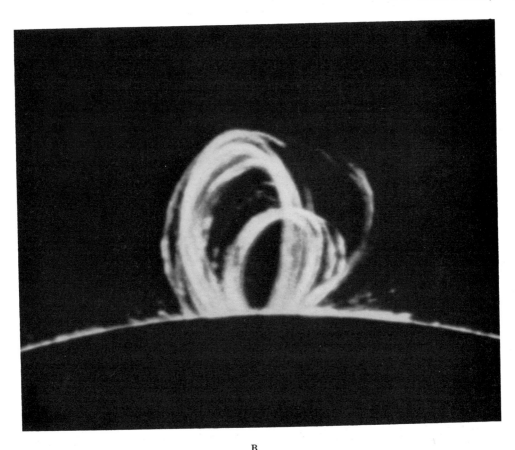

B

Figure 9–24B. *Loop Prominence.* The gas is concentrated along lines of magnetic force. (Sacramento Peak Observatory)

Figure 9–25. *Hα Profiles in Flares of Different Brightness.* All of these were fairly bright flares. Lesser flares have profiles similar to, though shallower than, plage profiles. Dashed curves represent the normal Hα profile from the quiet Sun.

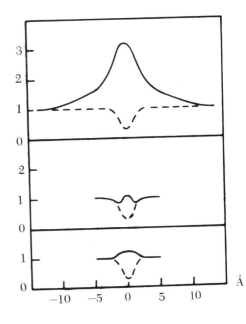

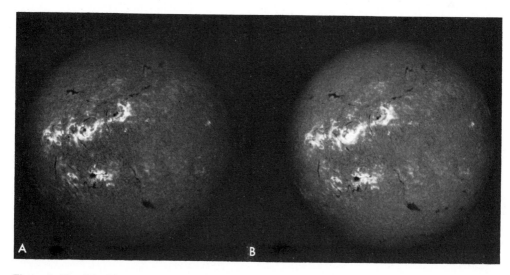

Figure 9–26. *Solar Flare Photographed in Hα.* (June 11, 1969.) **A**, Beginning of flare. **B**, Flare maximum. (Sacramento Peak Observatory)

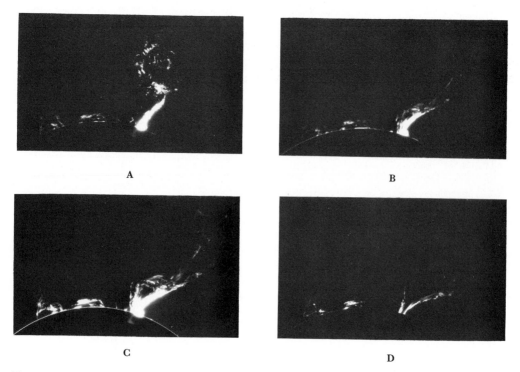

Figure 9–27. *Flare-Associated Spray Prominence.* (March 12, 1969.) **A**, $17^h 46^m$; **B**, $17^h 56^m$; **C**, $18^h 01^m$; **D**, $18^h 14^m$ U.T. (Institute for Astronomy, University of Hawaii)

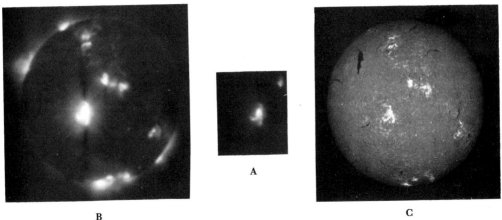

Figure 9–28.   *X-ray Photograph of the Sun.* (June 8, 1968). A small flare occurs near the center of the disk. X-ray wavelengths are **A**, 3.5 to 14 Å, and **B**, 44 to 60 Å. An Hα photograph is shown for comparison in **C** (ESSA). (American Science and Engineering, Inc.)

result from a heating of the corona to four million Kelvins; enhanced densities also stimulate emission in UV and forbidden lines.

Higher in the corona, the flare produces nonthermal radio bursts at meter wavelengths. These disturbances travel through the corona at speeds up to 0.3c. It appears that energetic electrons are trapped by coronal magnetic fields of a few Gauss, leading to the radio emission. Our knowledge of the radio bursts is still rather uncertain.

(iii)   *Solar Cosmic Rays*

Solar flares strongly affect the atomic particles leaving the Sun. The "blast wave" propagates through the solar wind at 1500 km sec^{-1}, disturbing the solar wind flow at the Earth (see section 9–5). Protons, electrons, and atomic nuclei are accelerated to high energies in flares—we call these *solar cosmic rays*. Most of the corpuscles observed are protons, for the electrons lose much of their energy in exciting radio bursts in the corona, and the solar abundance of other nuclei is low. *Alpha particles* (helium nuclei) are the second most abundant nuclei after the protons.

Solar particle energies range from keV (10^3 electron Volts) to about 20 GeV (1 GeV = one Giga-electron-Volt = 10^9 eV), with the bulk of the particles in the MeV (10^6 eV) range. For example, after one large flare, a flux of 3×10^4 protons cm^{-2} sec^{-1} was observed at the Earth in the 10 to 20 MeV range. The highest-energy particles arrive at the Earth within 30 minutes of the Hα flare maximum, followed by the peak number of particles an hour later, with the low-energy cosmic rays bringing up the rear hours later.

About half the flare energy (2×10^{32} ergs for the largest flares!) is in the Hα, half in the shock wave, and only one per cent in solar cosmic rays. However, the origin of the solar cosmic rays is still an unresolved problem. It seems clear that magnetic fields are involved: the particles seem to come from regions with a sharp field gradient (change with distance), and this gradient is lower after the flare. Nevertheless, no acceptable theory has appeared to explain how the magnetic field can dissipate so rapidly, with little evidence for any previous build-up.

(d)   Babcock's Model of the Solar Cycle

In 1960, H. W. Babcock proposed a model of the solar cycle, which seems to satisfy most of the observations. This model, illustrated in Figure 9–29, produces a 22-year solar cycle by coupling the Sun's magnetic fields and differential rotation.

Prior to the start of solar activity, the Sun is taken to have a weak dipolar magnetic field, with the lines of force running along meridians about $0.1R_\odot$ below the surface. Each line completes itself by emerging near the poles. The Sun's differential rotation draws each line out along the equator, wrapping it around the Sun many times. As the density of lines increases, so also does the associated "magnetic pressure"; inside the *flux tubes* which contain the lines, the gas pressure must decrease (to have pressure equilibrium), but then the tubes are lighter than their surroundings—they experience *magnetic buoyancy*. The field strength further increases as gas motions (e.g., convection) twist the lines.

The amplification of the magnetic field is maximum at higher latitudes ($\approx \pm 35°$), so that a *critical* field strength is first attained there. At this point, a *flux loop* rises to the solar surface and appears as a BMR with the strongest fields in the sunspots. Supergranules probably facilitate the emergence of a flux tube, since spots generally originate at the boundaries of supergranules. The *polarity law* is a consequence of the continuity of the field lines from north to south: since the following spot is nearer to the Sun's poles both in latitude and along the flux tube, its polarity is opposite that of the nearest pole.

At lower latitudes, the critical field strength is attained at later times; hence, the Butterfly Diagram. Sunspot maximum corresponds to the time of greatest field strength over the largest latitude band; eventually, lines become mixed and recombine near the equator, so that fewer spots occur later in the cycle. Supergranulation pushes the flux tubes about, dissipating the magnetic fields. The differential rotation causes the following (F) magnetic regions to move toward the poles, while the preceding (P) regions move toward the equator. At the equator, regions of opposite polarity meet from different hemispheres, and annihilate. The F regions converge on the poles, where their opposite polarity first neutralizes and then reverses the polar fields. After 11 years, the polarity of the Sun's overall dipole field is reversed, and the stage is set for the second half of the 22-year cycle.

Problems

9–1.   Show that the Sun is not at rest in the center of the Solar System by calculating the distance from the Sun's center to the center-of-mass of the Sun-Jupiter system.

9–2.   Find the Doppler width of a spectral line at 5000 Å, formed in the Sun's photosphere ($T \approx 5400$ K).

9–3.   (a) Given that the photosphere is at a temperature of 6000 K, would you expect collisional or radiative excitations to be more important in exciting hydrogen atoms to the second ($n = 2$) level?

(b) Would you expect Lyman α to appear in emission or absorption?

9–4.   Consider the following two lines of similar excitation potential: Fe I $\lambda 4144$ and Fe II $\lambda 4173$. Explain in general terms (with reference to the Boltzmann and/or Saha equations) why $\lambda 4144$ is the stronger of the two in the photosphere and the weaker in the chromospheric flash spectrum.

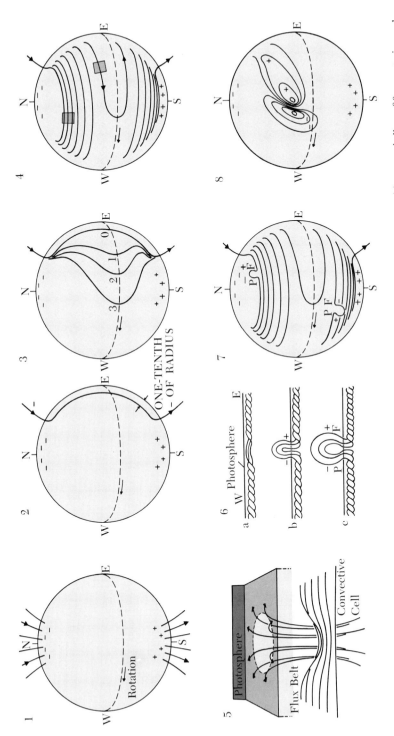

Figure 9–29. *The Babcock Model of the Solar Cycle.* At the initial stage (1) the Sun is similar to a dipole magnet, with magnetic lines of force entering and emerging at latitudes 60° and lying at about one-tenth of a solar radius beneath the surface at lower latitudes (2). Differential rotation draws out the lines of magnetic force along the equator (3), until they are tightly wound (4), but with the density of lines of force different at different latitudes (dark gray areas). Rising convection cells twist the lines of force into ropes (5). Kinks may develop in the lines of force (6), causing the magnetic pressure to increase and the flux ropes to rise and form bipolar magnetic regions (BMR) with preceding (P) and following (F) fields (7). Differential solar rotation and "random walk" type of motion on the solar surface causes the F fields to migrate toward the poles (8), first neutralizing and then reversing the polar fields. (After W. Livingston)

9–5. From our description of the chromospheric and coronal *radio* opacities, describe how you would determine the motion of a solar radio burst through the solar atmosphere.

9–6. How can one determine the temperature structure of a sunspot, using only its continuous spectrum?

9–7. How would you unambiguously assign a sunspot to the old or new cycle, near sunspot minimum?

9–8. Some prominences are said to have speeds greater than the speed of escape from the Sun at the chromosphere. What is the critical speed?

9–9. Make a table, listing all of the atmospheric phenomena of our Sun by level (i.e., photosphere, chromosphere, etc.).

9–10. Using the data given in this chapter, compute the *average rate of mass loss* of our Sun ($M_\odot$ yr^{-1}).

Reading List

Abetti, Giorgio, (Sidgwick, J. B., translator): *The Sun*. Revised Ed. New York, The Macmillan Company, 1957; London, Faber and Faber, 1963.

Aller, Lawrence H.: *Astrophysics: The Atmospheres of the Sun and Stars*. Second Ed. New York, Ronald Press, 1963.

Beckers, Jacques M.: "Solar Spicules." *Annual Review of Astronomy and Astrophysics*, Vol. 10, pp. 73–100 (1972).

Billings, Donald E.: *A Guide to the Solar Corona*. New York, Academic Press, 1966.

Brandt, John C.: *The Physics and Astronomy of the Sun and Stars*. New York, McGraw-Hill Book Company, 1966.

Brandt, John C.: *Introduction to the Solar Wind*. San Francisco, W. H. Freeman and Company, 1970.

Bray, R. J., and Loughhead, R. E.: *Sunspots*. London, Chapman and Hall Ltd., 1964.

Fichtel, C. E., and McDonald, F. B.: "Energetic Particles from the Sun." *Annual Review of Astronomy and Astrophysics*, Vol. 5, pp. 351–398 (1967).

Kuiper, Gerard P., (ed.): *The Sun*. Chicago, The University of Chicago Press, 1953.

Kundu, Mukul R.: *Solar Radio Astronomy*. New York, Interscience Publishers, 1965.

Meadows, A. J.: *Early Solar Physics*. First Ed. New York, Pergamon Press, 1970.

Menzel, Donald H.: *Our Sun*. Cambridge, Massachusetts, Harvard University Press, 1959.

Menzel, Donald H., Whipple, Fred L., and De Vaucouleurs, Gerard: *Survey of the Universe*. Englewood Cliffs, New Jersey, Prentice-Hall, 1970.

Noyes, Robert W.: "Ultraviolet Studies of the Solar Atmosphere." *Annual Review of Astronomy and Astrophysics*, Vol. 9, pp. 209–236 (1971).

Sawyer, Constance: "Statistics of Solar Active Regions." *Annual Review of Astronomy and Astrophysics*, Vol. 6, pp. 115–134 (1968).

Smith, Henry J., and Smith, Elske v. P.: *Solar Flares*. New York, The Macmillan Company, 1963.

Tandberg-Hanssen, Einar: *Solar Activity*. Waltham, Massachusetts, Blaisdell Publishing Company, 1967.

Wild, J. P., and Smerd, S. F.: "Radio Bursts from the Solar Corona." *Annual Review of Astronomy and Astrophysics*, Vol. 10, pp. 159–196 (1972).

Xanthakis, John N., (ed.): *Solar Physics*. New York, Interscience Publishers, 1967.

Zheleznyakov, V. V. (Massey, H. S. H., translator; Hey, J. S., ed.): *Radio Emission of the Sun and Planets*. First Ed. New York, Pergamon Press, 1970.

Zirin, Harold: *The Solar Atmosphere*. Waltham, Massachusetts, Blaisdell Publishing Company, 1966.

Chapter 10
Stellar Distances
and Magnitudes

Having discussed our Solar System and Sun, we now look far beyond to the *stars*. In this chapter, we will consider some of the methods by which the distances to the stars are determined, and we will quantify stellar brightnesses in terms of *magnitudes*.

10-1 THE DISTANCES TO STARS

Within the Solar System, we can determine absolute distances by using Newtonian celestial mechanics (space probe trajectories, and Kepler's laws) and radar astronomy. But even the nearest stars are so distant, in terms of familiar measures such as the AU, that new methods must be employed to determine their distances.

(a) Trigonometric Parallax

The ancients felt secure in their assumption of a geocentric Solar System, since they could detect no stellar movements on the celestial sphere. If the Earth were in motion, then the nearer stars should move *relative to the more distant stars* to reflect this motion. This *parallax* effect, which is familiar to us in everyday life, led people to be sceptical of the Copernican heliocentric model of the Solar System. It was not until 1838, when F. Bessel detected the parallax of the star 61 Cygni,   **261** and F. Struve that of Vega, that Copernicus' theory was incontrovertibly vindi-cated.* The Earth does move about the Sun.

The parallactic displacement of a star on the celestial sphere, due to the Earth's orbital motion, permits us to determine the distance to the star by the method of *trigonometric parallax* (see Figure 10–1). Using the Earth's orbit, we define the trigonometric parallax of the star as the angle (π) subtended at the star by 1 AU at the Sun; if the star is at rest with respect to the Sun, this is half the maximum apparent annual angular displacement of the star as seen from the Earth. Letting a denote the Sun-Earth distance, and d the Sun-star distance, we have:

$$\pi \text{ (radians)} = a/d \qquad \qquad (10–1)$$

Recall that there are 2π radians in a circle (360°), so that one radian equals $57°17'44''.81$ or $206,264''.81$. Hence, if we agree to measure all angles in arc-seconds and all distances in *parsecs* (1 parsec = 206,265 AU), equation (10–1) simply becomes:

$$\boxed{\pi'' = 1/d} \qquad \qquad (10–2)$$

(Note that 1 parsec $\equiv$ 1 pc $= 3.086 \times 10^{18}$ cm $= 3.26$ light-years, where a *light-year* is the distance light travels in one year.)

The measurement and interpretation of stellar parallaxes is a branch of *astrometry*, and the work is exacting and time-consuming. Consider that the *nearest* star, Alpha Centauri, at a distance of 1.3 pc, has a parallax of only $0''.76$ (see Table A2–5); all other stars have smaller parallaxes. Today, stellar parallaxes can be determined with a *probable error* of order $\pm 0''.005$, which means that the quoted parallax $0''.100 \pm 0''.005$ has a 50 per cent probability of actually being between $0''.095$ and $0''.105$. Hence, trigonometric parallaxes are believable only to distances of about 100 pc ($\pi'' = 0''.01$), which is miniscule compared to the 10 kpc (1 kpc = one kiloparsec $= 10^3$ pc) distance to the center of our Galaxy. Indeed, only about 6800 stellar parallaxes have been measured, whereas there are about 10^{11} stars in our Galaxy. Nevertheless, this method is important, because it is an unambiguous distance indicator.

The trigonometric parallax of a star is determined by photographing a

---

* Note that J. Bradley's discovery of stellar aberration in 1729 also proves that the Earth is in motion.

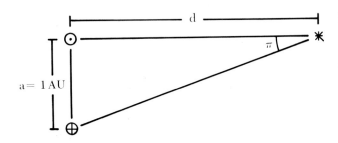

Figure 10–1. *Trigonometric Parallax.* The distance of the Earth from the Sun is 1 AU, and the angle subtended by that distance at the star is π in seconds of arc. The distance of the star from the Sun is d in parsecs.

given star field from several (about five) selected points in the Earth's orbit. The *comparison stars* are selected to be distant "background" stars of nearly the same apparent brightness as the star being measured. Corrections are made for atmospheric refraction and dispersion, and for detectable motions of the background stars; any real space motion (*proper motion* here) of the star relative to the Sun is then extracted. What remains is the small annual parallactic motion; it is readily recognized, since it is cyclic in time. Since a *seeing* resolution of $0''.25$ is considered exceptional, it may seem strange that a stellar position can be determined to $\pm 0''.01$ in one measurement; this is possible because we are determining the *center* of the fuzzy stellar image. At times, these complicating effects may lead to a negative value for the parallax—although this is strictly impossible, we must not discard the value or our final average will be biased toward high values.

(b)   Other Geometrical Methods

To reach distances greater than 100 pc using geometrical methods, we can utilize (a) the Sun's motion through the nearby stars and (b) the motions of *star clusters* which are not too far away. Both of these methods depend upon stellar motions; since this topic is extensively discussed in Chapter 14 (Stellar Motions), we will only summarize the techniques here.

The motion of the Sun among the nearby stars—the *solar motion*—is 20 km sec^{-1} (4.1 AU yr^{-1}) toward the constellation Hercules [$\alpha = 18^\mathrm{h}4^\mathrm{m}$, $\delta = +30°$ (Epoch 1950)]. This baseline grows year by year, so that over an interval of ten years we could measure stellar distances to about 2000 pc *if the nearby stars were stationary in space*. But all stars do move (like the Sun), so that only average parallaxes for groups of stars are possible. By assuming that the peculiar motions (defined in Chapter 14) of the stars in a large sample (preferably of one spectral type; see Chapter 12) average to zero, we can deduce the *mean parallax* (or *secular parallax*) of that sample. Moreover, since the solar motion affects only that component of a star's proper motion parallel to the solar motion, we may use the other (perpendicular) component to find *statistical parallaxes* (see Chapter 14). Note that we have been forced to sacrifice accuracy in order to attain greater distances.

The *moving cluster* method leads to precise individual stellar distances greater than 100 pc; unfortunately, there are very few such clusters (the Hyades cluster in the constellation Taurus is the most famous) where the method is really applicable. A star cluster consists of many stars moving as a group through space, for the stars are bound together gravitationally. If the cluster subtends an appreciable angle on the celestial sphere, the individual stellar proper motions appear to converge to (or diverge from) a single point on the sky—in direct analogy to the meteor radiants discussed in Chapter 5. By measuring the average radial velocity of the cluster, and using trigonometry, we can easily determine the distance to each star in the cluster (again, see Chapter 14).

(c)   Luminosity Distances

Finally, we come to distance determinations based upon stellar brightnesses. We must rely upon these rather indirect methods to truly probe our Galaxy (and other galaxies). Each method yields only *relative* distances, until we calibrate the stars involved. The calibration is accomplished when a nearby representative star (or cluster) is found, for which the absolute distance and brightness can be determined by means of (e.g.) trigonometric parallax or the moving cluster method. Since radiant flux decreases as the square of the distance from the source (see Chapter 8), the absolute brightness follows from the observed apparent brightness once the distance is known.

The first method is that of *spectroscopic parallax*. Intrinsic stellar brightness is closely correlated with spectral type (in the H-R diagram; see Chapter 12). Once we know the spectrum and apparent brightness of a star, we may determine its distance via this correlation.

The second method is that of *periodic variability*. Variable (pulsating) stars of the *RR Lyrae* and *Cepheid* classes brighten and dim periodically, and their periods are coupled with their mean absolute brightnesses. The distance to such a star follows straightforwardly, and if the star is associated with a star cluster (or even another galaxy) the distance to the cluster (or galaxy) is then known.

The last method is termed *main sequence fitting*. As we shall see in Chapter 12, when we plot the absolute brightnesses of the stars in a cluster versus their colors, most of the stars define a *main sequence*—a line on the so-called H-R diagram. For a cluster with an unknown distance, we can plot the *apparent* brightnesses versus colors (unaffected by distance) for the member stars; the resulting main sequence can then be *shifted* to match the main sequence of a calibrated cluster, and the amount of shift-in-brightness tells us the distance to our test cluster. This method is closely related to that of spectroscopic parallaxes.

The subject of stellar distances is of fundamental importance to astronomy and astrophysics. For this reason, we have anticipated later chapters in this book with this cursory exposition. Let us now proceed to discuss stellar brightnesses quantitatively, and link them with stellar distances.

10–2   THE STELLAR MAGNITUDE SCALE

The first stellar brightness scale—the *magnitude* scale—was defined by Hipparchus and refined by Ptolemy. In this qualitative scheme, naked-eye stars fall into six categories: the *brightest* (see Table A2–6) are of *first* magnitude, and the *faintest* of *sixth* magnitude. *Note that the brighter the star, the smaller the value of the magnitude.* In 1856, N. R. Pogson verified W. Herschel's finding that a first magnitude star is 100 times brighter than a sixth magnitude star, and the scale was quantified. Since five magnitudes correspond to a factor of 100 in brightness, one magnitude difference corresponds to a factor of $(100)^{1/5} = 2.512$.* The magnitude scale has been extended to positive magnitudes larger than $+6.0$ to include faint stars (the 200-inch telescope on Mount Palomar can reach to magnitude $+23.5$), and to negative magnitudes for very bright objects (the star Sirius is magnitude -1.4). It is common practice to use the superscript notation, e.g., $+1^{m}\!.0$ to denote an object of magnitude plus one.

---

* This definition reflects the operation of human vision, which converts equal *ratios* of actual intensity into equal *intervals* of perceived (psychological) intensity.

Astronomers find it convenient to work with logarithms (see the Mathematical Appendix) rather than with exponents in making the conversion from brightness ratio to magnitude, and vice versa. Consider two stars of magnitude m and n, with respective apparent brightnesses l_m and l_n. The ratio of their brightnesses (l_n/l_m) corresponds to the magnitude difference $(m - n)$. Since one magnitude difference means a brightness ratio of $(100)^{1/5}$, $(m - n)$ magnitudes refer to a ratio of $\left[(100)^{1/5}\right]^{m-n} = (100)^{(m-n)/5}$, or

$$l_n/l_m = (100)^{(m-n)/5} \tag{10-3}$$

Taking the logarithm (to the base 10) of both sides of equation (10–3) yields*

$$\log (l_n/l_m) = \left(\frac{m - n}{5}\right) \log (100) = 0.4(m - n) \tag{10-4}$$

or

$$\boxed{m - n = 2.5 \log (l_n/l_m)} \tag{10-5}$$

Equation (10–5) defines *apparent magnitude*; note that $m > n$ when $l_n > l_m$ (i.e., brighter objects have numerically smaller magnitudes).

The reader should strive to understand and be able to use equation (10–5). To assist in this process, we present the following worked examples:

(a) The variable star RR Lyrae ranges from $7^m\!.1$ to $7^m\!.8$—a magnitude amplitude of $0^m\!.7$. To find the relative increase in brightness from minimum to maximum we use

$$\log (l_{max}/l_{min}) = 0.4 \times 0.7 = 0.28$$

so that

$$l_{max}/l_{min} \doteq 10^{+0.28} = 1.93$$

This star is almost twice as bright at maximum light as at minimum.

(b) A binary system consists of two stars a and b, with a brightness ratio of 2; however, we see them unresolved as a single "star" of magnitude $+5^m\!.0$. We would like to find the magnitude of each star. The magnitude *difference* is

$$m_b - m_a = 2.5 \log (l_a/l_b) = 2.5 \log (2) = 0^m\!.75$$

Since we are dealing with brightness ratios, it is *not* permissible to put $m_a + m_b = +5^m\!.0$. The sum of the luminosities $(l_a + l_b)$ corresponds to a fifth magnitude star; let us compare this to a 100-fold brighter star, of magnitude $0^m\!.0$ and luminosity l_0:

$$m_{(a+b)} - m_0 = 2.5 \log \left(\frac{l_0}{l_a + l_b}\right)$$

---

* We have used: $\log (x^a) = a \log (x)$ and $\log (10^a) = a \log (10) = a$.

or

$$5.0 - 0.0 = 2.5 \log (100) = 5$$

But $l_a = 2l_b$, so that $l_b = (l_a + l_b)/3$; therefore:

$$m_b - m_0 = 2.5 \log (l_0/l_b) = 2.5 \log (300) = 2.5 \times 2.477 = 6.18$$

Hence, the magnitude of the fainter star is $6^m.18$, and from our earlier result on the magnitude difference, that of the brighter star is $5^m.43$.

10–3 ABSOLUTE MAGNITUDE AND DISTANCE MODULUS

Our discussion in the previous section concerns stars as we see them, that is, their *apparent magnitudes*. But we frequently want to know the intrinsic brightness of a star. It is evident that an intrinsically bright star will appear dim if it is far enough away, and a dim star may look very bright if it is close enough. Our Sun is a case in point: if it were at the distance of the closest star (α Centauri), it would appear slightly fainter to us than that star does. Hence, *distance* is the link between apparent and absolute brightnesses.

The intrinsic brightness of a star is defined by its *absolute magnitude*: the magnitude that would be observed if the star were placed at a distance of 10 parsecs. By convention, absolute magnitude is always capitalized, M, and apparent magnitude is written lower-case, m. The inverse-square law of radiative flux links the apparent brightness l of a star at a distance d to the intrinsic brightness L it would have if it were at a distance $D = 10$ pc, via:

$$(L/l) = (d/D)^2 = (d/10)^2$$

If the absolute magnitude M corresponds to L, and the apparent magnitude m corresponds to l, then equation (10–5) becomes

$$m - M = 2.5 \log (L/l) = 2.5 \log \left[(d/10)^2\right] = 5 \log (d/10)$$

Expanding this expression, we have the useful alternative forms:

$$\boxed{m - M = 5 \log (d) - 5} \tag{10–6}$$

$$\boxed{M = m + 5 - 5 \log (d)} \tag{10–7}$$

and

$$\boxed{M = m + 5 + 5 \log (\pi'')} \tag{10–8}$$

Here d is in parsecs, and π'' is the parallax angle in arc-seconds. The quantity

$(m - M)$ is called the *distance modulus*, for it is directly related to the star's distance in equation (10–6). In many applications we refer only to the distance moduli of different objects rather than converting back to distances in parsecs.

10–4 MAGNITUDES AT DIFFERENT WAVELENGTHS

(a) Magnitude Systems

Detectors of electromagnetic radiation (e.g., the photographic plate, the photoelectric photometer, the human eye, etc.) are sensitive only over given wavelength bands. Hence, a given measurement samples but part of the radiation arriving from a star. Since the intensity of starlight varies with wavelength, the magnitude of a star depends upon the wavelength or color at which we observe. Originally, photographic plates were sensitive only to blue light, and the term *photographic magnitude* (m_{pg}) still refers to magnitudes centered around 4200 Å (in the blue). Similarly, since the human eye is most sensitive to green and yellow, *visual magnitude* (m_v), or the photographic equivalent *photovisual magnitude* (m_{pv}), pertains to the wavelength region around 5400 Å.

Today we can measure magnitudes in the red and infrared, as well as in the ultraviolet and extreme (satellite and rocket) ultraviolet, by using filters in conjunction with the wide spectral sensitivity of photoelectric photometers. Hence, many different magnitude systems (color combinations) are possible. A widely-used magnitude system (which is adequate for this discussion) is the *UBV system*: a combination of ultraviolet (U), blue (B), and visual (V) magnitudes. These three bands are centered, respectively, at $\lambda 3500$, $\lambda 4300$, and $\lambda 5500$; each wavelength band is about 1000 Å wide (see Figure 10–2). In this system, apparent magnitudes are simply denoted by B or V, while the corresponding absolute magnitudes are subscripted, M_b or M_v.

Figure 10–2. *The UBV System.* The response of a photometer in the UBV system to a source with equal energy at all wavelengths. (After H. L. Johnson)

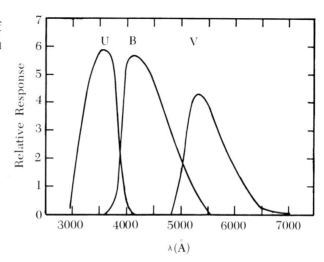

(b)   The Color Index

A quantitative measure of the *color of a star* is given by its *color index, CI,* which is defined as the difference between the magnitudes at two different effective wavelengths. For example,

$$CI = m_{pg} - m_v = M_{pg} - M_v \tag{10-9}$$

where the last equality follows from equation (10–7) because we are talking about one star. Note (a) that the long-wavelength magnitude is always subtracted from the short-wavelength magnitude, and (b) that a star's color index does not depend upon distance. Similarly, the quantities $(B - V)$, $(U - B)$, and even $(R - I)$ (red minus infrared) are also color indices.

Because stars have different temperatures, their spectral energy (Planck) curves peak at different wavelengths (see Chapter 8 and Figure 10–3); therefore, hot stars are blue, while cool stars are red. Using the color index $(B - V)$, we see that a blue star (20,000 K) has a negative color index, since the star is brighter in the blue (smaller B magnitude) than at longer wavelengths (larger V magnitude). A red star (5000 K) has a positive color index, since it is brighter in V than in B.

In Chapter 12 we will discuss the details of spectral classification; there we will see that a star's *spectral type* is a function of the stellar temperature. Hence, color index is very closely related to spectral type, and careful photometry can give us a good approximation to the spectral class of a star. Since we will have occasion to refer to specific spectral types in the following paragraphs, we give the spectral sequence from hot to cool stars (40,000 K to 3000 K) here: O, B, A, F, G, K, and M. Each spectral class is further subdivided into ten subclasses 0 to 9 (e.g., ... B8, B9, A0, A1, ...).

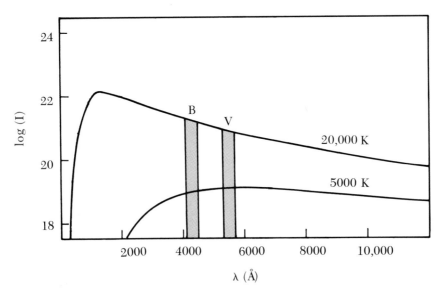

Figure 10–3.   *The Meaning of Color Index.* Two spectral energy (Planck) curves are shown, one for 20,000 K and the other for 5000 K. The pass-bands for the B and V filters are indicated by the shaded areas. The color index is B − V in magnitudes, not intensity; hence, it will be negative for the hot star and positive for the cool star.

Magnitude systems, and the color indices derived therefrom, are observer-dependent, since the percentage of the incident stellar energy measured depends both upon the wavelength (see Figure 10–3) and the particular instruments used. So that all observations may be compared on the same basis, all systems are established with the color index of spectral type A0 stars (about 10,000 K) equal to zero, a convention started with the Harvard photographic and photovisual systems. Earlier spectral types (the O and B stars) have a negative color index, while the index is positive for the cooler later types.

In the preceding discussion, we have assumed that the color indices are intrinsic to stars. However, interstellar space is not a pure vacuum, but is pervaded by "dust grains" which absorb and scatter starlight so that the observed light appears *redder* than when it was emitted (see Chapter 17). This *interstellar reddening* is wavelength-dependent, and it therefore affects color indices. We term the difference between the observed and the intrinsic color indices the *color excess* (see Chapter 17). We mention this here in order to warn the reader that a star's spectral type is not uniquely determined by a *single* color index measurement between two wavelength bands—three-color (or more) photometry is necessary to separate out the effects of interstellar reddening.

(c) Bolometric Magnitudes and Stellar Luminosities

The spectral energy distribution of a star is *sampled* by the color magnitudes we have been discussing. Of greater significance for the overall structure of the star, however, is the total rate of energy output (ergs sec^{-1}) at all wavelengths. If we place ourselves outside of the Earth's obscuring atmosphere,* the radiative flux per unit wavelength ℓ_λ (erg cm^{-2} sec^{-1} Å^{-1}) permits us to define the total *bolometric flux* (erg cm^{-2} sec^{-1}; referred to as *brightness* earlier in the chapter):

$$\ell_{\text{bol}} = \int_0^\infty \ell_\lambda \, d\lambda \qquad (10\text{--}10)$$

The *apparent bolometric magnitude* of the star, m_{bol}, follows from equation (10–5) as:

$$m_{\text{bol}} = -2.5 \log (\ell_{\text{bol}}) + \text{constant} \qquad (10\text{--}11)$$

where the constant is an arbitrary zero-point. The *absolute bolometric magnitude* of the star, M_{bol}, is the bolometric magnitude if the star were set at the standard distance of 10 parsecs. In the notation of equation (10–10), the *visual flux* ℓ_v (for example) is just

$$\ell_v = \int_0^\infty \ell_\lambda S_\lambda \, d\lambda \qquad (10\text{--}12)$$

---

* It is fairly straightforward to correct for *atmospheric extinction*, that is, scattering and absorption. Current practice is to express magnitudes after such correction, as though they were observed above the Earth's atmosphere.

where S_λ expresses the spectral sensitivity of the visual photometric system (the narrow V passband of Figure 10–2). In analogy to equation (10–11), we can then define the *visual magnitude* as

$$m_v = -2.5 \log (\ell_v) + \text{constant} \tag{10–13}$$

Our Sun is the *only* star for which ℓ_λ has been accurately observed. Indeed, ℓ_{bol} is related to the *solar constant*: the total solar radiative flux received at the Earth's orbit outside our atmosphere (1.37×10^6 erg cm^{-2} sec^{-1} or 1.97 calories cm^{-2} min^{-1}). The *solar luminosity* $L_\odot$ (3.90×10^{33} erg sec^{-1}) follows from the solar constant, in the following manner. Using the inverse-square law, we find the radiative flux at the Sun's surface ($R_\odot$). $L_\odot$ is just $4\pi R_\odot^2$ times this flux. The solar energy distribution curve may be approximated by a Planck blackbody curve (Chapter 8) at the *effective temperature* T_{eff} (to account for the effects of spectral emission and absorption lines). Then the Stefan-Boltzmann law (Chapter 8) implies

$$\boxed{L_\odot = 4\pi R_\odot{}^2 \sigma T_{\text{eff}}^4} \ \text{erg sec}^{-1} \tag{10–14}$$

where σ is the Stefan-Boltzmann constant.

If we know the absolute bolometric magnitude of a star, we can use equation (10–5) to find that star's luminosity:

$$M_{\text{bol}}(\odot) - M_{\text{bol}}(*) = 2.5 \log (L_*/L_\odot) \tag{10–15a}$$

Since $M_{\text{bol}}(\odot) = +4^m\!.7$, this becomes

$$\log (L_*/L_\odot) = 1.89 - 0.4 M_{\text{bol}}(*) \tag{10–15b}$$

Usually $M_{\text{bol}}(*)$ is *not* directly observed (although this is now becoming possible with space probes and satellites), but L_* may be deduced by studying the star's spectrum (see Chapter 12); then the absolute bolometric magnitude follows from equation (10–15a).

In practice, we use the *bolometric correction*, *BC*, which is the difference between the bolometric and the visual magnitudes, to determine a star's bolometric magnitude. For example,

$$BC = m_{\text{bol}} - m_v = M_{\text{bol}} - M_v$$
$$= 2.5 \log (\ell_v/\ell_{\text{bol}}) \tag{10–16}$$

Bolometric corrections are *inferred* from ground-based observations by using theoretical stellar models; these corrections have recently been checked and improved with the UV data from the Orbiting Astronomical Observatory satellites. In the (B, V) magnitude system, the bolometric correction is a *minimum* for F5 to F7 stars ($T_{\text{eff}} = 6500$ K); $BC = -0^m\!.07$ for our Sun (a G2 star). (Table A2–7 gives, among other data, the bolometric corrections for stars of different spectral types.)

For F5 stars, the spectral energy curve peaks in the V wavelength band, so that the greatest percentage of the star's energy is detected. For all other spectral types, a smaller percentage of the total radiative energy is measured in the V band; hence, their bolometric corrections are larger (in absolute value) than that for F5 stars. Since only *part* of the total radiation is ever included in visual magnitudes, BC is always *negative*.

In this short chapter, many new concepts have been presented to the reader in rapid succession. Each of these concepts is of great importance in astronomy. We therefore urge a careful re-reading of the chapter; in addition, many problems are presented in the following section to exercise and extend the reader's understanding of this material.

Problems

10–1. Astronomers living on Jupiter would define their astronomical unit in terms of the orbit of Jupiter. If they defined parsec in the same manner as we do, how many Jovian astronomical units would such a parsec contain? How many Earth AU would equal a Jovian parsec? How many Earth parsecs in a Jovian parsec?

10–2. A variable star changes in brightness by a factor 4. What is the change in magnitude?

10–3. What is the combined apparent magnitude of a binary system consisting of two stars of apparent magnitudes $3^m.0$ and $4^m.0$?

10–4. If a star has an apparent magnitude $m = -0^m.4$ and a parallax of $0''.3$, what is
(a) the distance modulus?
(b) the absolute magnitude?

10–5. What is the distance (in parsecs) of a star whose absolute magnitude is $+6^m.0$ and whose apparent magnitude is $+16^m.0$?

10–6. What are the absolute magnitudes of the following stars:
(a) $m = 5^m.0$, distance $d = 100$ pc?
(b) $m = 10^m.0$, $d = 1$ pc (is there such a star?)?
(c) $m = 6^m.5$, $d = 250$ pc?
(d) $m = -3^m.0$, $d = 5$ pc?
(e) $m = -1^m.0$, $d = 500$ pc?
(f) $m = 6^m.5$, parallax $\pi'' = 0''.004$?

10–7. What would the expression for absolute magnitude be, in terms of apparent magnitude and distance, if absolute magnitude were defined as the magnitude a star would have at 100 pc?

10–8. The Sun has an apparent magnitude $m = -26^m.5$.
(a) Calculate its absolute magnitude.
(b) Calculate its magnitude at the distance of α Centauri (1.3 pc).

10–9. Using the data from Table A2–5, how much brighter is the Sun than Sirius, as we see them from the Earth? How much brighter is Sirius than the Sun on an absolute basis?

10–10. A certain globular cluster has a total of 10^4 stars; 100 of them have $M_v = 0^m.0$ and the rest have $M_v = +5^m.0$. What is the integrated visual magnitude of the cluster?

10–11. The V magnitudes of two stars are both observed to be $7^m.5$, but their blue magnitudes are $B_1 = 7^m.2$, $B_2 = 8^m.7$, respectively.

(a) What is the color index of each star?

(b) Which star is the bluer and by what factor (in brightness) is it bluer than the red star?

10–12. What is the color index of a star at a distance of 150 pc with $m_v = 7^m.55$ and $M_B = 2^m.00$?

10–13. What is the absolute bolometric magnitude of a star of luminosity 10^{40} ergs sec^{-1}?

10–14. Given the expressions for the luminosity of a star (equation 10–14) and its bolometric magnitude in terms of that of the Sun (equation 10–15a), find an expression for the bolometric magnitude of the star as a function of its temperature and radius. The effective temperature of the Sun is 5780 K.

10–15. The bolometric correction for a star is $-0^m.4$, and its apparent visual magnitude is $+3^m.5$.

(a) Find the apparent bolometric magnitude of the star.

(b) Find the apparent luminosity of the star.

Reading List

Abell, George: *Exploration of the Universe.* Second Ed. New York, Holt, Rinehart and Winston, 1969.

Armitage, Angus: *William Herschel.* New York, Doubleday Co., Inc., 1963.

Brandt, John C.: *The Sun and Stars.* New York, McGraw-Hill Book Company, 1966.

Chiu, Hong-Yee, Warasila, Robert L., and Remo, John L., (eds.): *Stellar Astronomy.* Vol. 1. New York, Gordon and Breach, 1969.

Doig, Peter: *An Outline of Stellar Astronomy.* London, Hutchinson, 1948.

Hynek, J. A., (ed.): *Astrophysics.* New York, McGraw-Hill Book Company, 1951.

Motz, Lloyd, and Duveen, Anneta: *Essentials of Astronomy.* Belmont, California, Wadsworth Publishing Company, Inc., 1966.

Oke, J. B.: "Absolute Spectral Energy Distributions in Stars." *Annual Review of Astronomy and Astrophysics*, Vol. 3, pp. 23–46 (1965).

Strand, Kaj Aa., (ed.): *Basic Astronomical Data.* Chicago, University of Chicago Press, 1965.

Strömgren, Bengt: "Spectral Classification Through Photoelectric Narrow-Band Photometry." *Annual Review of Astronomy and Astrophysics*, Vol. 4, pp. 433–472 (1966).

Struve, Otto, Lynds, Beverly, and Pillams, Helen: *Elementary Astronomy.* New York, Oxford University Press, 1959.

Swihart, Thomas L.: *Astrophysics and Stellar Astronomy.* New York, John Wiley and Sons, 1968.

Van de Kamp, Peter: *Basic Astronomy.* New York, The Macmillan Co., 1952.

Van de Kamp, Peter: *Principles of Astrometry.* San Francisco, W. H. Freeman and Company, 1967.

Vasilevskis, S.: "The Accuracy of Trigonometric Parallaxes of Stars." *Annual Review of Astronomy and Astrophysics*, Vol. 4, pp. 57–76 (1966).

Wyatt, Stanley P.: *Principles of Astronomy.* Second Ed. Boston, Allyn and Bacon, Inc., 1971.

Chapter 11
Binary Stars

In Chapter 8, we saw that the area of the entire celestial sphere is 4π steradians,* which equals 5.35×10^{11} square arc-seconds. Since few of the $\approx 10^{11}$ stars in our Galaxy (see Chapter 13) are visible, even in large telescopes, the random chance that two stars will appear within a few arc-seconds of one another (we call these *optical doubles*) is very small. Yet many close double, and even multiple, star systems are observed. The first such binary system to be seen was that of Mizar (in the Big Dipper) by J. B. Riccioli in 1650. By 1821, William Herschel had catalogued 800 doubles; today, more than 65,000 *visual binaries* are known.

The stars in a multiple system are physically related; they orbit one another under the influence of their mutual gravitational attraction. That Newton's law of universal gravitation is applicable beyond our Solar System became evident when W. Herschel detected the orbital motion of the Castor star pair in 1804. As we shall see in the following section, known physical laws can be coupled with suitable observations of binary systems to tell us many important stellar characteristics: (a) masses, (b) radii, (c) densities, (d) surface temperatures and luminosities, and even (e) rotation rates. Lest the reader get the idea that such stellar systems are rare, we note here that *most* stars in the neighborhood of our Sun (well over 50 per cent) belong to multiple systems!

For physical and observational reasons, we speak of several different types of binary systems:

Optical binary: Two stars which are not physically associated, but which appear close together on the sky owing to projection effects. Their uncorrelated space motions soon reveal that they are not members of a binary system.

---

* Recall that a steradian is a *square* radian, where one radian equals 206,264″.81. **273**

Visual binary : A system which can be resolved into two stars at the telescope; the mutual orbital motions of these stars are observed to have periods ranging from about one year to thousands of years.

Astrometric binary : Only one star is seen (telescopically), but its oscillatory motion on the celestial sphere reveals that it is accompanied by an unseen *companion*; both bodies are orbiting about their mutual center-of-mass.

Spectroscopic binary : An unresolved system whose duplicity is revealed by periodic oscillations of the lines in its spectrum. In some cases, two sets of spectral features are seen (one for each star) oscillating with opposite phases; in other cases, one of the stars is too dim to be seen, so that only one set of oscillating spectral lines is recorded. Typical orbital periods here range from hours to a few months.

Spectrum binary : An unresolved system whose spectral features do not reveal orbital motion, but where there are clearly two different spectra superimposed. We infer that the two members of a binary system are producing the observed composite spectrum.

Eclipsing binary : A binary system whose two stars periodically eclipse one another, leading to periodic changes in the apparent brightness of the system. Such systems may also be visual, astrometric, or spectroscopic binaries.

11–1 VISUAL BINARIES

As a result of the Earth's turbulent atmosphere, the "seeing image" of a star is a fuzzy spot seldom less than 0″.50 in diameter. The two stars of a binary system are easily resolved (telescopically) as a *visual binary* if their centers are separated by more than 0″.50. The members of a visual binary *must* be well separated in angle at some point in their orbital motion; otherwise, the duplicity will not be resolved. Hence, the observed orbital periods are necessarily long (years to hundreds or thousands of years).

(a) The Determination of Stellar Masses

As shown in Figure 11–1, a single observation of a visual binary is specified by giving the apparent angular *separation* (in arc-seconds) and the *position angle* (angle

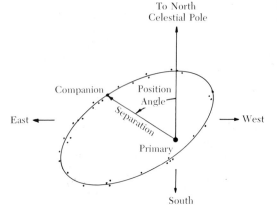

Figure 11–1. *Apparent Relative Orbit of a Visual Binary.*

measured eastward from north, in degrees) of the fainter star (the *companion*) relative to the brighter star (the *primary*). As time passes, these points trace out the *apparent relative orbit* of the binary on the celestial sphere. In 1827, Felix Savary was the first to show (for ξ Ursae Majoris) that this apparent orbit is an *ellipse*.

In Chapter 3, we saw that two gravitating bodies orbit one another—as well as their center-of-mass—in accordance with Kepler's laws. Therefore, the orbit is an ellipse, and the orbital motion satisfies the law of equal areas and the harmonic law. But we do not generally see the true orbit, for the orbital plane of a binary system may be inclined at any angle to the plane of the sky.* It is fortunate, and straightforward to show, that (a) the harmonic law holds for the apparent orbit, (b) the law of equal areas holds (but with a different constant of proportionality) for the apparent orbit, and (c) the elliptical true orbit always projects into an elliptical apparent orbit. The foci of the apparent orbit *do not* correspond to the true foci (in particular, the primary does not lie at one focus of the apparent ellipse). By measuring the displacement of the primary from the apparent focus, we can determine the inclination of the orbit to the celestial sphere; the true eccentricity and the true semi-major axis a'' (in arc-seconds) then follow immediately.

Having determined the true orbit of the visual binary, we may now apply Kepler's third law (the harmonic law) to deduce the *masses* of the member stars. The general form of the third law is:

$$(M_1 + M_2)P^2 = A^3 \qquad (11–1)$$

where mass M is measured in solar masses $(M_\odot)$, the orbital period P is measured in years, and the orbital semi-major axis A is measured in AU. Although we may observe P directly, A follows from a'' only when we know the distance (or parallax π''; see Chapter 10) to the visual binary. Geometrically, we have (see Figure 10–1):

$$A = a''/\pi'' \qquad (11–2)$$

so that equation (11–1) may be written in terms of observables as

$$(M_1 + M_2)P^2 = (a''/\pi'')^3 \qquad (11–3)$$

An accurate value for the *sum* of the stellar masses follows from equation (11–3). To determine the individual masses, we must find the relative distance of each star from the center-of-mass of the system, since

$$M_1 A_1 = M_2 A_2 \quad (\text{where } A_1 + A_2 = A) \qquad (11–4)$$

On the celestial sphere, the center-of-mass travels in a straight path with respect to the background stars, and the binary components weave periodically about this path (see Figure 11–2). By eliminating the center-of-mass motion, and correcting for the orbital inclination, we obtain a_1'' and a_2'', and therefore (a_1''/a_2'') which equals

---

* We say that the *inclination* is 0° when the two planes coincide (i.e., we see the true orbit), and 90° when the orbit is seen edge-on.

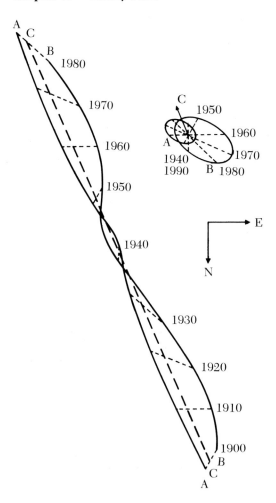

Figure 11–2. *The Motions of Sirius A and B.* At the left we see the apparent motions through the background stars of the primary (Sirius A) and the companion (Sirius B), relative to their center-of-mass C. The inset at the upper right shows the deduced orbital paths of this binary system. (From P. van de Kamp, in S. Flugge, (ed.): *Encyclopedia of Physics.* Berlin, Springer-Verlag, 1958)

(A_1/A_2). Using this method in 1844, F. Bessel discovered that the bright star Sirius was an astrometric binary; later, the dim solar-mass companion (Sirius B) was photographed and found to represent a new class of stars—the white dwarfs (see Chapters 12 and 15).

Let us illustrate how binary star masses are found, by means of a hypothetical example. A visual binary is observed to have a maximum separation of $3''.0$ and a trigonometric parallax of $0''.10$; the apparent orbit is completed in 30 years, and the primary coincides with the focus of that orbit. Since we are seeing the true orbit (why?), the sum of the stellar masses is $30M_\odot$ [from equation (11–3)]:

$$(M_1 + M_2) = (3.0/0.1)^3/(30)^2 = 30$$

The companion is always observed to be five times farther from the center-of-mass than the primary, so that $(A_1/A_2) = 1/5$, and thence [equation (11–4)]:

$$M_1 \text{ (primary)} = 25M_\odot, \quad M_2 \text{ (companion)} = 5M_\odot$$

(b)   The Mass-Luminosity Relation

Just as the Earth's orbit (via Kepler's third law) leads us to the Sun's mass, so also have we deduced binary stellar masses. Since it was necessary to know the distance to the binary system in order to establish these masses, we need only observe the apparent magnitude of each star to find its absolute magnitude or *luminosity* [see equation (10–7)].*

When the observed masses and luminosities for approximately 100 stars in binary systems are graphed, we obtain the correlation shown in Figure 11–3—this result is called the *Mass-Luminosity Relation* (or M-L law). In 1924, Arthur S. Eddington conjectured that the mass and luminosity of normal, main-sequence stars (see Chapters 12 and 15 for details) should be related, in the form:

$$(L/L_\odot) = (M/M_\odot)^\alpha \tag{11–5}$$

His first crude theoretical models indicated a value for the exponent $\alpha = 3$, while later more-refined models predicted $\alpha = 5.5$. To test his conjecture, Eddington assembled stellar data into a diagram similar to Figure 11–3. On such a log-log plot, equation (11–5) graphs as a straight line. Hence, main-sequence stars do seem to conform to equation (11–5), although the exponent varies from (a) $\alpha \approx 3$ for bright and massive stars, through (b) $\alpha \approx 4$ for solar-type stars, to (c) $\alpha \approx 2$ for dim red dwarfs of low mass. Today, astrophysical theories of stellar structure crudely explain these results in terms of (a) the different internal structures of stars of different mass, and (b) the opacities of stellar atmospheres at different tempera-

---

* The absolute color magnitude gives the *bolometric magnitude* M_{bol}, when the bolometric correction BC is applied; the stellar luminosity follows directly from M_{bol}.

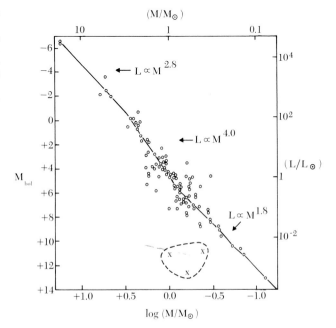

Figure 11–3. *The Mass-Luminosity Relation.* Here the deduced masses of 90 binary stars (○) and three white dwarfs (×) are plotted against their absolute bolometric magnitude M_{bol} (or luminosity L). The drawn-in line is to assist the eye. [From K. Aa. Strand, (ed.): *Basic Astronomical Data.* Chicago, University of Chicago Press, 1963.]

tures. Note that the M-L law does *not* apply to highly-evolved stars, such as red giants (with extended atmospheres) and white dwarfs (with degenerate matter; note the x's in Figure 11–3); it holds only for main-sequence stars. While most stellar masses lie in the narrow range $0.05 \lesssim (M/M_\odot) \lesssim 50$, stellar luminosities cover the vast span $10^{-4} \lesssim (L/L_\odot) \lesssim 10^6$!

(c) Dynamical Parallaxes

The M-L law permits us to determine the distances to main-sequence stars by means of *dynamical parallaxes* (a method not mentioned in Chapter 10). Equation (11–3) is a dynamical relation, which we may invert to define dynamical parallax, via:

$$\pi'' = a''/[(M_1 + M_2)P^2]^{1/3} \qquad (11\text{–}6)$$

In general, a'' and P may be observed for a visual binary, but the sum of the stellar masses is not known. Hence, we proceed by trial and iteration, in the following fashion. A reasonable first guess is $(M_1 + M_2) = 1M_\odot$; we now have a provisional value for the parallax π''. From the apparent magnitudes of the stars, we compute their provisional absolute magnitudes [using equation (10–7)] and luminosities L. Consulting Figure 11–3, we read off new values for the stellar masses, and hence, their sum $(M_1 + M_2)$. We continue in this manner, iterating until all values are stable and consistent.

Let us attempt this procedure with the extraordinary visual binary considered at the end of section 11–1(a).* The apparent bolometric magnitudes of the stars are observed to be $-7\overset{m}{.}5$ (primary) and $-2\overset{m}{.}0$ (companion). Our first iteration yields $\pi'' = 0\overset{''}{.}31$ and $(M_1 + M_2) \approx 12M_\odot$—not too bad. The second iteration gives $\pi'' = 0\overset{''}{.}14$ and $(M_1 + M_2) \approx 25M_\odot$; by the fourth iteration we have the correct distance (10 parsecs), as accurately as the M-L law permits.

Dynamical parallaxes are limited to relatively *nearby* binary systems, where trigonometric parallaxes usually suffice. This follows from equation (11–6), for it is difficult to accurately measure $a'' \lesssim 1\overset{''}{.}0$ and $P \gtrsim 100$ years. Since most stellar masses lie in the range $10^{-1} \lesssim (M/M_\odot) \lesssim 10$, we have $\pi'' \gtrsim 0\overset{''}{.}02$ or $d \lesssim 50$ parsecs.

11–2 SPECTROSCOPIC BINARIES

If a binary system cannot be optically resolved at the telescope, its duplicity may be evident in its spectrum. Although orbital motion may not be detectable, we know that we are dealing with a *spectrum binary* when two different sets of line features are seen superimposed in the spectrum. A more useful, and interesting, case is the *spectroscopic binary*: here, two stars orbit their center-of-mass closely ($\lesssim 1$ AU) and rapidly ($P \approx$ hours to a few months), and the orbital inclination is

---

* We have some hope that the large total mass involved, $30M_\odot$, will not be catastrophic, as a result of the cube root in the denominator of equation 11–6!

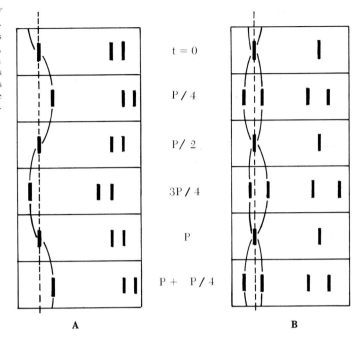

Figure 11–4. *The Spectra of Spectroscopic Binaries.* **A,** A *single-line* spectroscopic binary exhibits only one set of spectral features, which oscillate in wavelength with time. **B,** Two sets of spectral lines oscillate out of phase (sometimes overlapping, as at $t = P$) in the spectrum of a *double-line* spectroscopic binary.

$t = 0$

$P / 4$

$P / 2$

$3P / 4$

P

$P + P / 4$

A

B

not $0°$. The first spectroscopic binary discovered—a subsystem of the Mizar multiple system—was found in 1889 by E. C. Pickering.

The spectrum of a spectroscopic binary (see Figure 11–4) exhibits lines which oscillate periodically in wavelength. If the companion is so dim that its spectral features are not detected, we have a *single-line* spectroscopic binary (Figure 11–4A); two stars of more nearly equal luminosity produce two sets of spectral features which oscillate in opposite senses (in wavelength)—we call this system a *double-line* spectroscopic binary (Figure 11–4B). About 1000 spectroscopic binaries are known, and good orbits have been determined for approximately 400 [see section 11–3(b)].

(a)   The Velocity Curve

To obtain useful information from the spectrum of a spectroscopic binary, we must interpret the behavior of the spectral lines. Since the two stars orbit in a plane inclined (at angle i) to the celestial sphere, that component of their velocity in the line-of-sight *Doppler shifts* their spectral features. (Note that no Doppler shift can occur, as a result of orbital motion alone, when $i = 0°$; the system may then appear as a spectrum binary.) In addition, the center-of-mass of the system is in motion with respect to the Sun, so that the entire spectrum may be Doppler-shifted by some constant amount.

From Chapter 8 [equation (8–13)], the Doppler-shift formula is

$$\Delta\lambda/\lambda_e \equiv (\lambda_o - \lambda_e)/\lambda_e = v_r/c \qquad (11\text{–}7)$$

where λ_e is the *emitted* wavelength (i.e., laboratory wavelength) of a spectral feature, λ_o is the *observed* wavelength, v_r is the radial speed ($+$ for recession, $-$ for approach) of the star, and $c = 3 \times 10^5$ km sec^{-1} is the speed of light. Because of the finite width of spectral lines, we are limited at visible wavelengths (e.g., 5000 Å) to a shift resolution of $\Delta\lambda \gtrsim 0.01$ Å; hence, the radial speed must be $v_r \gtrsim 1$ km sec^{-1} to be detectable. Thus, the periods of observable spectroscopic binaries are necessarily short.

When we convert [using equation (11–7)] the Doppler shifts to radial velocities, and plot the results as a function of time, we obtain the *velocity curve*. Figure 11–5 exhibits the simplest case: circular stellar orbits at the inclination $i = 90°$ (edge-on); the two curves (one for each star) are sinusoidal, and oscillate with exactly opposite phases about the center-of-mass velocity in a period P. In this case, we easily find the distances to the center-of-mass:*

$$R = VP/2\pi \quad \text{and} \quad r = vP/2\pi \tag{11–8}$$

the ratio of stellar masses:

$$M/m = r/R = v/V$$

the relative semi-major axis ($A = R + r$); and from equation (11–1) the sum of the stellar masses:

$$(M + m) = A^3/P^2$$

Therefore, the individual stellar masses follow, and the dynamical characteristics of this spectroscopic binary are completely determined.

In general, this simple picture does not occur. If the case shown in Figure 11–5 were a single-line spectroscopic binary (e.g., only the primary spectrum is seen), we could determine only R and the "mass-function" $m^3/(M + m)^2$ (see the next section for details); a reasonable value for M might be obtained from the primary spectral type—then the system could be approximately deciphered. A greater difficulty is that, unless the system is also an *eclipsing* binary (see section 11–3), we have no idea what the orbital inclination i is. If the velocity curve is purely sinusoidal, we know only that we are dealing with a circular orbit tilted at *some* angle i to the celestial sphere. The amplitudes of the velocity curves give us (via trigonometry) the *observed* (denoted by tildas) circular speeds:

$$\tilde{V} = V \sin i \quad \text{and} \quad \tilde{v} = v \sin i$$

hence, we may determine the mass ratio exactly, since

$$M/m = r/R = v/V = \tilde{v}/\tilde{V}$$

but only the *lower limit*, $A \sin i$, to the semi-major axis is accessible.

If the relative orbit is not circular, but has an eccentricity e, the velocity curves are distorted from pure sinusoids, as shown in Figure 11–6. The double-line curves are mirror images of one another, but with differing amplitudes—an orbital

---

* For example, in one period P, the primary (M) traverses the circumference $2\pi R$ at constant speed V; hence, $VP = 2\pi R$ and equation (11–8) follows.

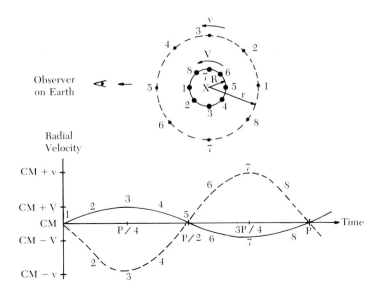

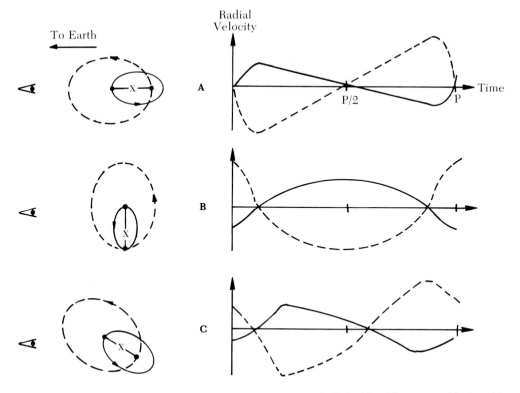

Figure 11–5. *A Simple Velocity Curve.* At the top, the primary (●) and companion
(•) orbit the center-of-mass (X) in circular orbits of inclination
$i = 90°$. At the bottom, the center-of-mass is seen to be receding
at a constant speed (CM), and the stars move at speeds V (primary)
and v (companion) with respect to the center-of-mass. Both stars
complete one orbit in the period P, so that $v > V$.

Figure 11–6. *Velocity Curves For Elliptical Orbits.* The two stars trace elliptical orbits *of the same eccentricity*, but with
semi-major axes in the ratio M/m. Here X is the center-of-mass, and the orientation Ω of the major
axes (with respect to the line-of-sight) is varied; $i = 90°$ in all three cases. **A**, $\Omega = 0°$; skewed
velocity curve due to fast periastron motion and slow apastron motion. **B**, $\Omega = 90°$; the largest
and smallest radial velocities are attained. **C**, $\Omega \approx 45°$; an intermediate case.

inclination i merely reduces *all* radial velocities by the same factor of $\sin i$. The periodicity and characteristic shapes of these curves allow us to find P, e, and Ω (the orientation of the major axis with respect to the line-of-sight) immediately. When $i = 90°$, the relative semi-major axis A and the stellar masses, M and m, may be obtained.

(b)   The Mass-Function

The strong tidal interactions between the component stars in short-period spectroscopic binaries $(P \lesssim 10 \text{ days})$ *circularize* their orbits.* Therefore, let us consider a circular (or small eccentricity, $e \ll 1$) relative orbit at inclination i—what can we say about the masses of the stars? Since we can obtain P, $\tilde{R}$, and $\tilde{r}$ (thence, $\tilde{A} = \tilde{R} + \tilde{r}$) from the velocity curve of a double-line binary, Kepler's harmonic law gives us

$$(M + m) \sin^3 i = \tilde{A}^3/P^2$$

Recall that (M/m) was also found in this case. If we see only the primary (M) in a single-line binary, we can find only the *mass-function* $f(M, m)$, via

$$(M + m)P^2 = A^3 = (R + r)^3 = R^3(1 + r/R)^3$$
$$= R^3(1 + M/m)^3 = \tilde{R}^3(M + m)^3/m^3 \sin^3 i$$

or

$$\boxed{f(M, m) \equiv m^3 \sin^3 i/(M + m)^2 = \tilde{R}^3/P^2} \tag{11-9}$$

$\tilde{R}$ and P are observed, and we have used the relation $MR = mr$.

If the orbital inclination i is unknown, of what use is the mass-function? We cannot evaluate individual stellar masses, but, by combining many data, *statistical* masses may be obtained. If the orbital planes are *randomly* distributed in i, then the mean value of $\sin^3 i$ is 0.59; however, we are more likely to detect spectroscopic binaries with $i \approx 90°$ (i.e., almost edge-on), so we "correct" for this *observational selection effect* by assigning a somewhat larger value to the mean of $\sin^3 i$, say $\approx 2/3 \approx 0.66$. In addition, we have spectral information on the visible components, which can suggest appropriate masses.

11–3   ECLIPSING BINARIES

When the inclination of a binary orbit is close to $90°$, each of the stars can eclipse the other periodically—we call these *eclipsing binaries*. Thousands of such systems are known; most are also spectroscopic binaries, while very few are visual binaries. Figure 11–7 shows why this effect occurs. For a relative orbit of radius ρ, tilted an angle ϕ to the line-of-sight ($\phi = 90° - i$), an eclipse can occur only when

---

* This process appears to require about 10^8 years.

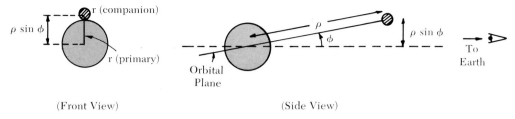

Figure 11–7. *Eclipse Geometry.*

$\rho \sin \phi < [r \, (\text{primary}) + r \, (\text{companion})]$, where r is the stellar radius. Therefore, small orbits are favored; since these have short periods and high orbital velocities, they imply spectroscopic binaries.

(a) Interpreting the Light Curve

In 1669, G. Montonari noted that Algol (β Persei) varies in apparent brightness with a period around three days; John Goodricke attributed this behavior to a stellar eclipse in 1783. Around 1890, H. Vogel found Algol to be a single-line spectroscopic binary with a period of $2^d20^h49^m$. This case illustrates that eclipsing binaries are most readily detectable by their periodically-varying brightnesses. If we plot the apparent brightness of such a binary as a function of time, we obtain the *light curve* (shown in the following figures), which generally exhibits two brightness minima of different depths (corresponding to the two possible eclipses per period).

When the stars are not in eclipse, the observed brightness is essentially *constant* in time.* The deeper minimum—*primary eclipse*—occurs when the hotter star passes behind the cooler star; the other, *secondary eclipse* is shallower. Several types of eclipses are possible: (a) when $i \equiv 90°$, both the *total* eclipse (smaller star behind larger star) and the *annular* eclipse (smaller star in front) are termed *central*; (b) when $\rho \cos i < [r \, (\text{primary}) - r \, (\text{companion})]$, we still have total and annular eclipses; and (c) only *partial* eclipses take place in the event that $[r \, (\text{primary}) - r \, (\text{companion})] < \rho \cos i < [r \, (\text{primary}) + r \, (\text{companion})]$. Note that in every case, the exact same stellar *area* is covered both at primary minimum and at secondary minimum.

Consider the light curve associated with the central eclipses shown in Figure 11–8; here the relative stellar orbit is circular. There are four points (in time during one eclipse) where the limbs of the two stars are tangent; we speak of *first contact* (t_1) when the eclipse begins, *second contact* (t_2) when brightness minimum is reached, *third contact* (t_3) when the smaller star begins to leave the disk of the larger star, and *fourth contact* (t_4) when the eclipse ends. Both primary and secondary minima are *flat*, and they occur exactly half an orbital period apart. If we denote the stellar radii by r_ℓ (larger star) and r_s (smaller star), and the relative orbital speed of the smaller star by v, the geometry of Figure 11–8 implies:

$$2r_s = v(t_2 - t_1) = v(t_4 - t_3) \tag{11–10a}$$

* See, however, section 11–3(c).

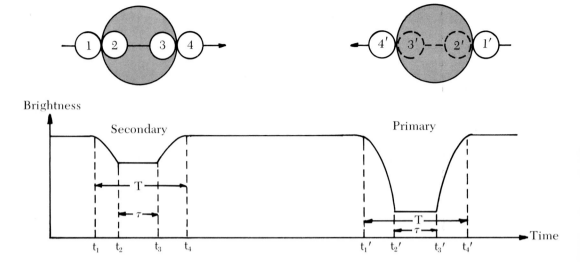

Figure 11–8. *Central Eclipses For Circular Orbit.* Here the smaller star is the hotter. The four *contacts* are explained in the text. The primary and secondary eclipses occur exactly one-half period apart, and for both eclipses the eclipse duration $(1 \to 4)$ and the time at flat minimum $(2 \to 3)$ are the same.

$$2(r_s + r_\ell) = v(t_4 - t_1) \tag{11–10b}$$

But the radius of the relative circular orbit a is clearly:

$$a = vP/2\pi \tag{11–11}$$

where P is the orbital period. By combining equations (11–10) and (11–11), we see that we can determine only the stellar radii *relative to the orbital radius:*

$$\left. \begin{aligned} (r_s/a) &= \pi(t_2 - t_1)/P \\ (r_\ell/a) &= \pi(t_4 - t_2)/P \end{aligned} \right\} \tag{11–12}$$

Without consulting the stellar spectra, we may also determine the ratio of the *effective surface temperatures* of the two stars. Let these effective (blackbody) surface temperatures be denoted by T_ℓ and T_s. Surface brightness (erg cm^{-2} sec^{-1}) is equal to σT_{eff}^4, by the Stefan-Boltzmann law (Chapter 8); since the same stellar area (πr_s^2) is covered at each eclipse minimum, the relative depths of the two eclipse minima (brightness lost in erg sec^{-1}) give us $(T_\ell/T_s)^4$ directly—the *hotter* star is eclipsed at primary minimum.

When the eclipses are *partial* for a circular orbit (see Figure 11–9), both eclipses are still of equal duration (though briefer than for central eclipse), and the brightness minima are *not* flat. Since the two eclipses still occur exactly half an orbital period apart, we know that the orbit is circular.* In this case, it is possible

---

* For elliptical orbits, different times elapse from primary to secondary eclipse and from secondary to primary eclipse. Also, in general, the eclipse durations are not equal. These features permit us to determine the eccentricity e, orientation Ω, and inclination i of the orbit.

to determine (a) the orbital inclination i, (b) the relative stellar radii (r_s/a) and (r_ℓ/a), and (c) the relative effective surface temperatures (T_ℓ/T_s).

(b) Eclipsing-Spectroscopic Binaries

We have seen that light curves yield only *relative* results. It is, indeed, fortunate that most eclipsing binaries are also spectroscopic binaries, for we may determine speeds in km sec^{-1} from their velocity curves. Thence, from equations (11–10) and (11–11), we find absolute values (in km) for a, r_s, and r_ℓ. Since the orbital inclination i follows from the light curve, we can evaluate sin i, and determine the stellar masses, M and m, in grams. By plotting these results, we obtain the mass-radius correlation (for main-sequence stars) shown in Figure 11–10. Mean stellar densities ρ (gm cm^{-3}) may then be computed; for example:

$$\rho_\ell = 3M/4\pi r_\ell{}^3$$

Knowing the stellar radii, we may find the ratio of stellar *luminosities* (from the effective temperature ratio) and the absolute magnitude of the binary; the apparent magnitude of the system then tells us the distance to the binary, and the luminosity of each star. Table 11–1 summarizes the various data which we can obtain from binary stars.

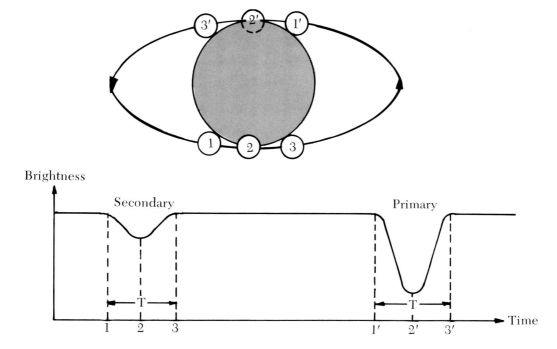

Figure 11–9. *Partial Eclipses For Inclined Circular Orbit.* Again the smaller star is hotter, and the eclipses occur exactly half an orbital period apart. Note that the eclipse minima are *not* flat.

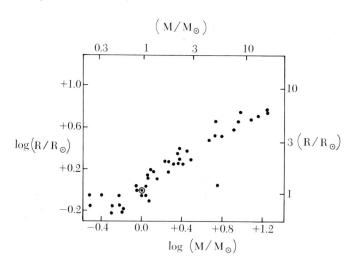

Figure 11–10. *The Main-Sequence Mass-Radius Relation.* Note the strong correlation between stellar masses and radii—more massive stars are larger. Our Sun is denoted by ⊙. [From Kaj Aa. Strand, (ed.): *Basic Astronomical Data.* Chicago, University of Chicago Press, 1963]

(c) Some Useful Complications

The simple light curves which we have considered do not tell the whole story of eclipsing binaries. Much useful information, including knowledge about stellar interiors, may be gleaned from the unusual light curves of more complicated systems. The various parts of Figure 11–11 illustrate the phenomena which will concern us here.

From our experience with the Sun (Chapter 9), we expect that stellar disks will exhibit *limb-darkening* (Figure 11–11A). This feature rounds off the edges of the brightness eclipses.

In tight binary orbits, the hotter star will heat that part of the cooler star's atmosphere which is nearest (on the line between the stellar centers). This hotter gas is more luminous, and leads to the *reflection effect:* just before and after secondary minimum, the system appears brighter than we would otherwise expect (Figure 11–11B).

Table 11–1. Stellar Data From Binary Systems

Type of Binary	Observations Performed or Needed	Parameters Determined
Visual	(a) Apparent magnitudes and π''	Stellar luminosities
	(b) P, a'', and π''	Semi-major axis (a)
		Mass sum $(M + m)$
	(c) Motion relative to CM	M and m
Spectroscopic	(a) Single-line velocity curve	Mass function $f(M, m)$
	(b) Double-line velocity curve	Mass ratio (M/m)
		$(M + m) \sin^3 i$
		$A \sin i$
Eclipsing	(a) Shape of light curve eclipses	Orbital inclination (i)
		Relative stellar radii $(r_{\ell,s}/a)$
	(b) Relative times between eclipses	Orbital eccentricity (e)
	(c) Light loss at eclipse minima	Surface temperature ratio (T_ℓ/T_s)
Eclipsing-Spectroscopic	(a) Light and velocity curves	Absolute dimensions (a, r_s, r_ℓ)
		e and i
		M and m (also densities)
	(b) Same + apparent magnitude	Distance to binary
		Stellar luminosities
		Surface temperatures (T_ℓ, T_s)

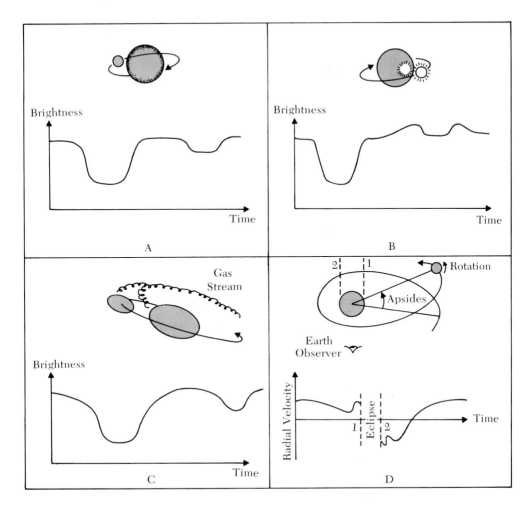

Figure 11–11. *More-Complicated Binaries.* **A,** Limb-darkening. **B,** Reflection effect. **C,** Ellipticity effect and gas streaming. **D,** Stellar rotation causes apsidal motion, and bumps at 1 and 2 in the velocity curve (bracketing the eclipse of the rotating star).

In close binaries, the fluid (gas) bodies of the stars are distorted into prolate spheroids (like footballs) oriented along their line of centers, as a result of the strong tidal forces. This gives rise to the *ellipticity effect* (Figure 11–11C), whereby the observed brightness varies continuously and not just during eclipses. Such systems are usually in circular orbit. If the distortion is great enough, *gas streams* may be drawn from the stellar atmospheres into the space between and around the stars—the system β Lyrae is a famous example. This gas is most readily noticeable in spectra, for bright emission lines can be seen.

Stellar rotation shows up in two ways. The velocity curve exhibits anomalous bumps (Figure 11–11D) just before and just after each eclipse—in one case we see the receding limb of the rotating star (since the rest is already eclipsed) and in the other the approaching limb. In the light curve, stellar rotation appears in the rotation of the line of apsides (i.e., semi-major axis) of elliptical orbits (Figure 11–11D). A rotating star is oblate (flattened), so that its gravitational attraction is *not* that of a point mass; two such stars in a binary interact to change the orientation of their semi-major axes with time.

11–4 INTERFEROMETRIC STELLAR DIAMETERS

Finally, for the sake of completeness, let us briefly mention some other methods by which stellar *diameters* may be determined. All of these methods are basically *interferometric;* they depend upon the constructive and destructive interference of the light waves from a star (see section 8–1).

To see why such indirect procedures are necessary (except, of course, for our Sun), consider the following: at a distance of 1 pc, a star with a diameter of 1 AU (radius $= 108 R_\odot$) subtends an angle of $1''.0$ on the celestial sphere. But this is just the same as the size of the "seeing image" (as a result of the Earth's turbulent atmosphere) seen at a telescope, so that the star's size is unresolvable—besides that, no star is as close as 1 pc.

Within a band $10°$ wide, centered on the ecliptic, stars may be occulted by our Moon. In such a *lunar occultation*, the star does not disappear instantaneously, but fades away in a few seconds. Electromagnetic wave fronts from the star are progressively screened out as time goes on, but the unobstructed portions interfere (actually, *diffract*) to produce a characteristic intensity-versus-time pattern at the Earth. This pattern depends directly upon the angular size of the star; if we know the stellar distance, we can thus deduce the stellar diameter.

In the 1920s, A. E. Michelson invented and used a *stellar interferometer* to measure the angular diameters of large, nearby stars. In this device, widely-separated (many meters) mirrors deflect the starlight to an ordinary focussing telescope, where the wave fronts from different parts of the star produce a characteristic interference pattern (see Figure 11–12). This pattern depends upon the angle between the wave fronts from opposite limbs of the star; this intersection angle increases as the stellar angular diameter increases. The largest star measured ($750 R_\odot$) was Betelgeuse (α Orionis) with an angular diameter of $0''.042$; the smallest was Arcturus (α Boötis) at $0''.020$ ($23 R_\odot$).

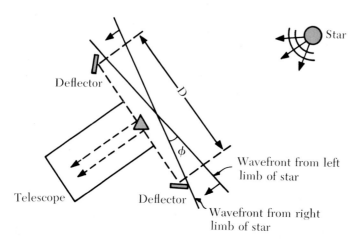

Figure 11–12. *Michelson's Stellar Interferometer.* The widely-spaced (distance $= D$) deflector mirrors send the starlight to interfere in the telescope. Light waves from opposite limbs of the star arrive simultaneously, intersecting one another at the angle ϕ, which is also the angular diameter of the star as seen from the Earth.

Recently, R. H. Brown and R. W. Twiss have developed the *intensity interferometer*, which is analogous to Michelson's device. Here two widely-separated (hundreds of meters!) phototubes convert starlight into an electrical signal, and the two separate signals are mixed electronically to produce an intensity interference pattern. Examples of their results are Sirius (α Canis Majoris) with an angular diameter of $0\rlap{.}{''}00585$ ($1.76R_\odot$), and α Gruis ($0\rlap{.}{''}00098 \rightarrow 2.07R_\odot$). So far, approximately 25 stellar diameters have been measured using interferometers.

The latest technique is a variation on *holography*. To produce a hologram in the laboratory, we illuminate an object and a photographic plate with the coherent light from a laser (via mirrors). The light scattered from the object interferes with the direct laser light at the plate, and an interference pattern is recorded. To produce an image from the plate, we illuminate it with the same laser, and the recorded pattern diffracts this light into a recognizable (three-dimensional!) apparent image. For a star, we take a *very fast* telescopic photograph, so that the Earth's atmosphere is "stationary" during the exposure. Thus, we circumvent the fluctuations which smear the starlight into the "seeing image," and instead obtain a hologram due to the interfering light waves from different parts of the stellar disk. The star's angular diameter is deduced from the hologram pattern.

Note that these interferometric techniques may be (and have been) applied to close binary systems, in order to deduce angular semi-major axes (a'') which are unresolved by the usual visual-binary procedures.

Problems

11–1. (a) From the facts given in section 11–1(a), show that Kepler's harmonic law holds for the apparent orbit of a visual binary.
(b) Using the celestial mechanics you learned in Chapter 3, show that Kepler's law of equal areas holds in the apparent orbit.

11–2. Demonstrate the correctness of equation (11–2). Use a diagram, if necessary.

11–3. What is the sum of the stellar masses in a visual binary of period 40 years, maximum separation $5\rlap{.}{''}0$, and parallax $0\rlap{.}{''}3$? Assume that the orbital inclination is zero.

11–4. Find the distance (in parsecs) to a visual binary, which consists of main-sequence stars of absolute bolometric magnitudes, $M_{\rm bol} = +5\rlap{.}^{m}0$ and $+2\rlap{.}^{m}0$, respectively; the mean angular separation is $0\rlap{.}{''}05$ and the observed orbital period is ten years. What assumptions have you made to arrive at your answer?

11–5. Show that binary systems with small orbits have high orbital speeds.

11–6. The velocity curves of a double-line spectroscopic binary are observed to be sinusoidal, with amplitudes of 20 km sec^{-1} and 60 km sec^{-1}, and a period of 1.5 years.
(a) What is the orbital eccentricity e?
(b) Which star is the more massive, and what is the ratio of stellar masses?
(c) If the orbital inclination is 90°, find the relative semi-major axis (in AU) and the individual stellar masses (in solar masses).

11–7. An eclipsing binary has a period of $2^{\rm d} 22^{\rm h}$, the duration of each eclipse is $18^{\rm h}$, and totality lasts $4\rlap{.}^{\rm h}7$.

(a) Find the stellar radii, in terms of the circular orbit radius a.

(b) If spectroscopic data indicate a relative orbital speed of 200 km sec^{-1}, what are the absolute stellar radii (in km and $R_\odot$)?

11–8. One component of an eclipsing binary has a surface temperature of 15,000 K, and the other 5000 K. The cooler star is a giant with a radius four times that of the hot star.

(a) What is the ratio of the stellar luminosities?

(b) Which star is eclipsed at primary minimum?

(c) Is primary minimum a total or an annular eclipse?

(d) Primary minimum is how many times deeper than secondary minimum (in erg sec^{-1})?

11–9. The Al V (main sequence) star Sirius A has a radius of $1.8\ R_\odot$ and $M_{bol} = 1\overset{m}{.}4$; the radius of its white dwarf companion, Sirius B, is $0.01 R_\odot$ and $M_{bol} = +11\overset{m}{.}5$.

(a) What is the ratio of their luminosities?

(b) What is the ratio of their effective temperatures?

(c) If they orbited at $i = 90°$, which star is eclipsed at primary minimum?

(d) If your photometer can measure magnitudes to an accuracy of $\pm0\overset{m}{.}001$, would you be able to detect the hypothetical primary eclipse? (Use: $\log_{10}(1 + x) = x/2.3$).

11–10. Derive an expression giving the stellar angular diameter θ in milli-arc-seconds $(10^{-3}$ arc-second), when the actual stellar diameter (in $R_\odot$) and distance from us (in pc) are known.

Reading List

Aitken, Robert G.: *The Binary Stars*. New York, Dover Publications, 1964.

Baker, Robert H., and Fredrick, Laurence W.: *Astronomy*. Ninth Ed. New York, Van Nostrand Reinhold Company, 1971.

Batten, Alan H.: "On the Interpretation of Statistics of Double Stars." *Annual Review of Astronomy and Astrophysics*, Vol. 5, pp. 25–44 (1967).

Binnendijk, Leendert: *Properties of Double Stars*. Philadelphia, University of Pennsylvania Press, 1960.

Eggen, O. J.: "Masses of Visual Binary Stars." *Annual Review of Astronomy and Astrophysics*, Vol. 5, pp. 105–138 (1967).

Kopal, Zdeněk: *Close Binary Systems*. New York, John Wiley & Sons, 1959.

Popper, Daniel M.: "Determination of Masses of Eclipsing Binary Stars." *Annual Review of Astronomy and Astrophysics*, Vol. 5, pp. 85–104 (1967).

Strand, Kaj Aa., (ed.): *Basic Astronomical Data*. Chicago, University of Chicago Press, 1963.

Van de Kamp, Peter: *Elements of Astromechanics*. San Francisco, W. H. Freeman and Co., 1964.

Van de Kamp, Peter: "The Nearby Stars." *Annual Review of Astronomy and Astrophysics*, Vol. 9, pp. 103–126 (1971).

Chapter 12
Spectral Classification and the Hertzsprung-Russell Diagram

We end Part 3 with this chapter on the wealth of information which can be discerned by studying *stellar spectra*. To correlate all of these data from our most important "window on the stars," has required extended and herculean efforts on the part of astronomers and astrophysicists. For ease of assimilation, the material is presented in the following fashion: first, we discuss stellar atmospheres, for this is where the stellar spectra originate. Then, we tell the story of the spectral observations—how they have been made, correlated, and interpreted. Finally, we describe that famous synthesis—the Hertzsprung-Russell diagram—and some of its implications. This discussion will lead us to an understanding of the stars themselves (in Chapter 15).

12–1 STELLAR ATMOSPHERES

As we saw in Chapter 9, the spectral energy distribution of the starlight reaching us from a star is determined in that star's *atmosphere*—the region from which radiation can escape essentially unimpeded. To understand stellar spectra, let us discuss a *model* (theoretical) stellar atmosphere and investigate the characteristics which determine the spectral features.

(a) Physical Characteristics

We are interested in the stellar photosphere, that thin gaseous layer shielding the stellar interior from our view. This region has a height which is small compared with the stellar radius R, so we will regard it as a *uniform* shell of gas. The physical properties of this shell may be approximately specified by the average values of its *pressure P*, *temperature T*, and *composition μ* (elemental abundances).

We will make the reasonable assumption that the number density n (number per cm^3) of gas particles (molecules, atoms, ions, and electrons) is high enough that *thermodynamic equilibrium* obtains. This means that particle collisions are sufficiently frequent that both the Boltzmann and the Saha equations hold, and the gas obeys the *perfect gas law*:

$$P = nkT \qquad (12\text{–}1)$$

where k is Boltzmann's constant. The particle number density n is related to both the *mass density* ρ (gm cm^{-3}) and the composition (or *mean molecular weight*) μ via:

$$\rho = \mu m_H n \qquad (12\text{–}2)$$

where $m_H = 1.67 \times 10^{-24}$ gm is the mass of a hydrogen atom. For a star of pure atomic hydrogen, $\mu = 1$. If the hydrogen is ionized, $\mu = 1/2$, since there are as many electrons as protons (hydrogen nuclei), and electrons are far less massive than protons. In general, stellar gases are ionized, and $1/\mu \cong 2X + (3/4)Y + (1/2)Z \approx 2$, where X is the *mass fraction* of hydrogen, Y is that of helium, and Z that of *all heavier elements*. (See Problem 12–11.)

We also assume a *steady-state* atmosphere: although the individual gas particles move about rapidly, on the macroscopic scale nothing changes with time (e.g., no mass motions). This implies *hydrostatic equilibrium*, whereby a typical volume of gas experiences no net force. But the only forces acting are pressure and gravitation. Consider a small volume of photospheric gas (of area A and height dr) a distance r from the star's center (Figure 12–1). The pressure P acts on its base and the pressure $P + dP$ on its top, while the gravitational force exerted inward is $GM\rho(A\,dr)/R^2$ since M is the star's mass. To achieve a static balance, we have

$$[(P + dP) - P]A = -GM\rho(A\,dr)/R^2$$

or

$$dP/dr = -(GM/R^2)\rho = -g\rho \qquad (12\text{–}3)$$

where g is the gravitational acceleration (cm sec^{-2}) or *gravity* at the photosphere. Note that the pressure *decreases continuously* as we move outward through the star.

By parameterizing the pressure in terms of optical depth τ (see Chapter 9), instead of radius r, where $d\tau = -\kappa\rho\,dr$, equation (12–3) takes the useful form

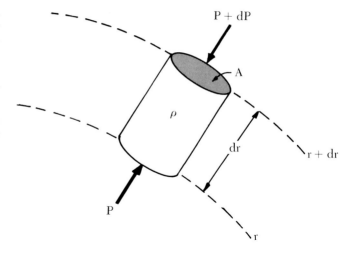

Figure 12–1. *Hydrostatic Equilibrium.* In the photosphere of a star of radius R and mass M, at a distance r from the star's center, resides a volume of gas. Its area is A, its height dr, and its mass density ρ. The pressure P pushes out on its base, the pressure $P + dP$ pushes in on its top, and gravity $(g = GM/R^2)$ trys to accelerate it inward. If the gas is static, all forces must balance exactly.

$$dP/d\tau = +g/\kappa \qquad\qquad (12\text{–}4)$$

In our uniform approximation, then, we may integrate equation (12–4) to

$$P = (g/\kappa)\tau \qquad\qquad (12\text{–}5)$$

so that the atmospheric gas pressure depends upon the stellar surface gravity g and the atmospheric opacity κ $(\mathrm{cm}^2\ \mathrm{gm}^{-1})$.* (We will take $\tau = 1$ as the atmospheric level where spectral features are formed.)

(b) Color Temperatures

From the pressure and composition of the stellar atmosphere, we now turn to the temperature T. We have already noted (Chapters 8 and 9) that the continuous spectrum or *continuum* from a star may usually be well approximated by the Planck blackbody spectral energy distribution. For a given star, the continuum may be used to define an effective or color temperature, via the appropriate Planck curve. We use the word *color* because of Wien's displacement law:

$$\lambda_{\max} T = 2.898 \times 10^7\ \text{Å (K)} \qquad\qquad (12\text{–}6)$$

which states that the peak intensity of the Planck curve occurs at a wavelength $(\lambda_{\max})$ which varies inversely with the Planck temperature T. Hence, most of the radiation intensity occurs at wavelengths near $\lambda_{\max}$, and the star has the *color* appropriate to $\lambda_{\max}$. This implies that hot stars (smaller $\lambda_{\max}$) should be *bluer* than

---

* Photon pressure may augment the perfect gas pressure of particles (e.g., in hot stars). Photons carry *momentum*, as well as energy, so that their scattering (opacity) produces a pressure. (See Chapter 15.)

cool stars (larger $\lambda_{max} \to$ redder). Let us also note here that the hotter a star is, the greater will be its *surface brightness* (or luminous flux in erg cm^{-2} sec^{-1}), in accordance with the Stefan-Boltzmann law:

$$\text{blackbody radiative flux} = \sigma T^4 \qquad (12–7)$$

where $\sigma = 5.67 \times 10^{-5}$ erg cm^{-2} sec^{-1} (K)$^{-4}$.

A word of caution is in order. The effective Planck temperature of a star is usually not identical to its spectral line temperature, for spectral line formation depletes radiation from the continuum. This effect is called *line blanketing* (see Figure 12–2), and it becomes important when the numbers and strengths of spectral lines are large. We will return to this crucial topic in our discussion of subdwarfs. When spectral features are not numerous, we can detect the continuum between them, and obtain a reasonably accurate value for the star's effective surface temperature.

(c)   Spectral Line Formation

In Chapter 8, we saw that spectral absorption features are formed when the molecules, atoms, and ions of a gas absorb continuum photons, and re-emit fewer of these photons toward the observer. The composition of the gas determines which species are available to absorb the photons, while the temperature and pressure determine which spectral features are actually formed. For example, molecular features can originate only in a cool gas, for molecules are easily dissociated; neutral atoms and their spectral lines will predominate at intermediate temperatures, but at high temperatures, all species are ionized, so that only the spectral features arising from ions should be seen.

Let us put these statements on a more quantitative basis. Take a cubic centimeter of gas, in which the number of particles of each elemental species is specified by the composition μ. Now consider the particles of a particular element (e.g., hydrogen). Continuum photons at discrete* wavelengths are absorbed when

---

* *Continuous* absorption occurs when the particles are ionized (further), corresponding to photon wavelengths short (smaller λ) of the various *series limits*.

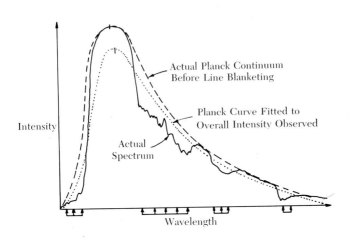

Intensity

Actual Planck Continuum Before Line Blanketing

Planck Curve Fitted to Overall Intensity Observed

Actual Spectrum

Wavelength

Figure 12–2. *Line Blanketing Effect.* The color temperature of this star should correspond to the original continuum Planck curve (dashed), but numerous absorption lines (arrows) blanket the continuum to produce the observed spectrum (solid line). If we fit a Planck curve to the overall stellar intensity, or to a wavelength band which is blanketed, we will deduce a color temperature which is too low (dotted curve).

the particles (neutral atoms or ions) are excited from one atomic energy level to another—the *strength* of each absorption feature is essentially proportional to the number of particles populating a given energy level (see *equivalent width* and the *curve of growth* in Chapter 8). The relative number of atoms in energy levels B (N_B) and A (N_A), where $B > A$, is determined by the *Boltzmann excitation-equilibrium equation:*

$$N_B/N_A \propto \exp\left[(E_A - E_B)/kT\right] \tag{12–8}$$

here E is the energy of the level and T is the gas temperature. At a given temperature T, the higher levels are *less* populated than the lower levels. But the relative number of *ions* in a given ionization stage i ($i = 1 +$ the number of electrons lost by an atom) is determined by both the temperature T and the electron number density N_e ($\#$ cm^{-3}), via the *Saha ionization-equilibrium equation:*

$$\left(\frac{N_{i+1}}{N_i}\right) \propto \frac{(kT)^{3/2}}{N_e} \exp\left(-\chi_i/kT\right) \tag{12–9}$$

here χ_i is the ionization potential from stage i to stage $i + 1$.

Frequently, it is useful to express equations (12–8) and (12–9) in logarithmic form; that is, we take the base-10 logarithm of both sides of each equation. Noting that $\log_{10}(e) = 0.4343$, and expressing T in degrees Absolute (K) and all energies in electron Volts, we have, from equation (12–8):

$$\log(N_B/N_A) = -\frac{5040}{T}|E_A - E_B| + \text{constant} \tag{12–10}$$

and, from equation (12–9),

$$\log(N_{i+1}/N_i) = \frac{3}{2}\log T - \frac{5040}{T}\chi_i - \log N_e + \text{constant}' \tag{12–11}$$

In this way the dependence upon each parameter is clearly manifested. Since the electrons also constitute a perfect gas, we may write equation (12–11) in terms of the *electron pressure* P_e (where $P_e \propto N_eT$) as:

$$\log(N_{i+1}/N_i) = \frac{5}{2}\log T - \frac{5040}{T}\chi_i - \log P_e + \text{constant}'' \tag{12–12}$$

Both the Boltzmann and the Saha equations must be combined, if we are to deduce the population of every energy level—and thence, the spectral line strengths. In Chapter 8 (see Figure 8–13), we saw that the temperature behavior of the combination is the following: at low temperatures, the atoms are all neutral and in their ground states. As the temperature rises, the higher energy levels of the neutral atoms become more populated until $T \approx \chi_I/k$, when single-ionization produces a significant number of ions (depleting the number of neutrals). At still higher temperatures, these ions predominate, but when $T \approx \chi_{II}/k$, double-ionization (two electrons lost) becomes important. After running up the entire ladder of ionization stages, the temperature is so high that only bare nuclei and free electrons remain—the atoms are *fully* ionized—and no more spectral line absorption can take place. The reader should remember that this sequence is *strongly* dependent upon the exact energy level structure of each atomic species, so that the spectral features produced at any temperature *uniquely* characterize the species.

By means of a verbal flow chart, we may express the spectral line behavior to be expected of a given stellar atmosphere. Denoting the strength of a spectral feature by its equivalent width w, we have the functional dependence:

$$w = w(n, T) \tag{12–13}$$

since the rate of absorption is determined by the number density n of atoms and the intensity of the continuum (T). But the equation of state of the gas [equation (12–1)] and the composition of the gas [12–2] imply

$$n = n(P, T, \mu) \tag{12–14}$$

Hydrostatic equilibrium [equations (12–3) and (12–4)] determines the pressure:

$$P = P(M, R, \kappa) \tag{12–15}$$

Now the opacity κ is clearly a function of the number density n, the gas composition μ, and the ionization-excitation state determined by T:

$$\kappa = \kappa(n, T, \mu) \tag{12–16}$$

while the star's luminosity L depends upon its temperature T and radius R, via

$$L = 4\pi R^2 \sigma T^4 \tag{12–17}$$

By suitably combining the dependences of equations (12–14) to (12–17) we see that equation (12–13) becomes, in general:

$$\boxed{w = w(L, T, M, \mu)} \tag{12–18}$$

For stars of a given composition μ (see Stellar Populations, in section 12–3), equation (12–18) reduces to

$$w = w(L, T, M) \tag{12–19}$$

and if we consider, for example, only main-sequence stars with a unique mass-luminosity relation (Chapter 10), we have

$$\boxed{w = w(L, T)} \tag{12–20}$$

Equation (12–20) is taken as the theoretical justification for seeking temperature sequences and two-dimensional Hertzsprung-Russell diagrams in the remainder of this chapter; equation (12–18) covers those cases in which stellar luminosity is not related to stellar mass, and where composition differences are important.

12–2 CLASSIFYING STELLAR SPECTRA

(a) Observations

A star's color and apparent magnitude may be crudely guessed by visual observations through a telescope. Today, however, photographic and photoelectric techniques are employed, for they permit us to (a) reach to much fainter apparent magnitudes (more stars), (b) precisely determine stellar colors, apparent magnitudes, and distances, and (c) obtain stellar spectra.

A single stellar spectrum is produced when starlight is focussed by a telescope onto a *spectrometer* or *spectrograph*, where it is *dispersed* (spread out) in wavelength and recorded photographically. If the star is bright, we may obtain a *high-dispersion* spectrum; that is, few Ångstroms per millimeter on the spectrogram, for there is enough radiation that it may be spread broadly and thinly (the solar spectrum of Figure 12–3 is a good example). At high dispersion, a wealth of detail may be

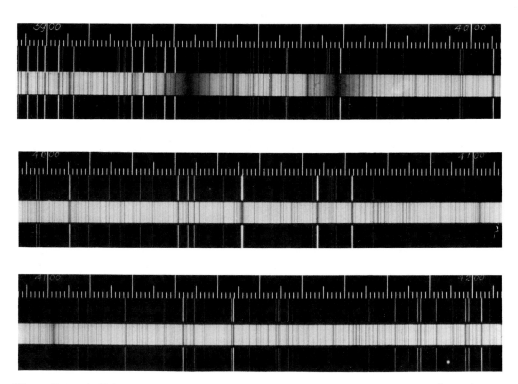

Figure 12–3. *A High-Dispersion Spectrum.* This is a small segment of a high-dispersion (0.5 Å mm^{-1}) solar spectrum. Note the wealth of detail. Such high dispersion is possible only because of the tremendous intensity of sunlight at the Earth. (Hale Observatories)

discerned in the spectrum, but the method is slow (only one stellar spectrum at a time) and we are strictly limited to *bright* stars (a selection effect!). *Low-dispersion* spectra (hundreds of Å mm^{-1}) permit us to reach apparent magnitudes greater than $+15^m_.0$; however, much information is lost on these condensed spectrograms, and single spectra still imply a slow rate of data-gathering.

To maximize the amount of spectral information obtained in a single observation, we now use the *objective prism* (or slitless) technique. A large prism is placed in front of the object lens (or mirror) of the telescope, producing a low- or medium-dispersion spectrum of *every star in the field of view* at the focal plane (i.e., photographic plate). Figure 12–4 shows the spectacular and informative result of this procedure. More than *a half million* stellar spectra, down to a limiting apparent magnitude of $11^m_.0$ to $14^m_.0$, have been obtained and studied using objective prism spectrograms.

If we are interested only in the apparent magnitude and *color* of a star, we need not obtain the entire stellar spectrum. Instead, we can filter the starlight at the telescope, and measure *color magnitudes* in some system such as (U, B, V) or Bengt Strömgren's four-color ($ubvy$) scheme (see Chapter 10). Note that two possibilities exist: (a) the entire field of view may be photographed in each wavelength band

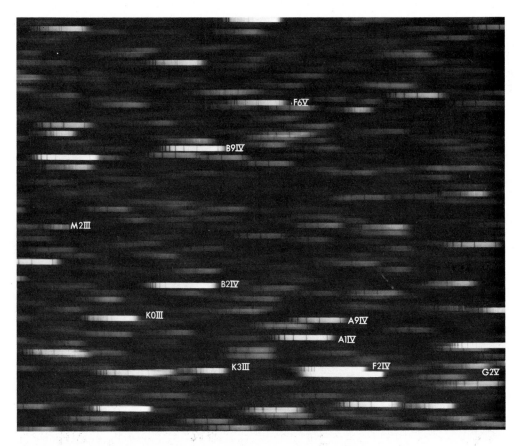

Figure 12–4.   *Objective Prism Spectrogram.* The field of view is near the Milky Way in the constellation Puppis. Many stellar spectra are shown; several have their spectral types indicated. (Department of Astronomy, University of Michigan at Cerro Tololo Observatory)

(many stars), or (b) a given star may be precisely studied, using photoelectric *photometry* (the stellar image is color-filtered and focussed onto a photo-electric device, which produces an electrical signal directly proportional to the star's apparent brightness). From such data, we can determine the star's color index *CI*, and hence (ignoring effects of interstellar reddening, as discussed in Chapter 17) effective temperature and the apparent magnitude. This technique is fast and accurate ($\pm 0\overset{m}{.}001$), and today it even gives us some idea of the important *spectral features* of a star. Note that the effective stellar temperature is deduced by *sampling* the star's spectrum at several wavelengths, and fitting the data to a Planck curve; deviations from a blackbody spectrum (e.g., line blanketing) will necessarily introduce errors into this temperature determination.

(b)   The Spectral Line Sequence

At first glance, the spectra of different stars seem to bear no relation to one another. In 1863, however, Angelo Secchi found that he could (crudely) order the spectra and define different *spectral types*. Alternative ordering schemes appeared in the ensuing years, but around 1920 the Harvard Observatory group of A. J. Cannon and E. C. Pickering introduced the definitive schematization which we use today— the *Henry Draper spectral classification system* (so called because about 400,000 stars were so classified in various volumes of the *Henry Draper Catalogue* by 1949).

At first, the HD scheme was based upon the strengths of the hydrogen Balmer absorption lines in stellar spectra, and the spectral ordering was alphabetical (A through P). The A stars had the strongest Balmer lines, while the P stars had the weakest. Some letters were eventually dropped, and the ordering was re-arranged to correspond to a sequence of decreasing temperatures (recall the effects of the Boltzmann and Saha equations): OBAFGKM. This ordering is now fixed, and it may be remembered by the useful mnemonic "Oh, Be A Fine Girl, Kiss Me!" Stars nearer the beginning of the spectral sequence (closer to O) are called *early-type* stars, while those closer to the M end are referred to as *late-type*. Each spectral type is subdivided into ten parts from 0 (early) to 9 (late); for example, we have . . . F8 F9 G0 G1 G2 . . . G9 K0 . . . . In this scheme, our Sun is designated type G2.

Figure 12–5 shows exemplary stellar spectra arranged in the order of the HD scheme; note how the conspicuous spectral features strengthen and diminish in a characteristic way as we move through the spectral types. In Table 12–1, we have summarized those spectral characteristics which define each spectral type. Study this table (and Figure 12–5) with great care.

(c)   The Temperature Sequence

The HD spectral sequence is a *temperature sequence*, but we must carefully qualify this statement. There are many different kinds of, and ways to determine, temperature. In Figure 12–6, the strengths of various spectral features are plotted against excitation-ionization (or Boltzmann-Saha) temperature, and we see that the spectral sequence does correlate with this temperature.

Theoretically, therefore, the Planck *color* temperature should similarly correlate with spectral type. From the spectra of intermediate-type stars (A to M), we find that the (continuum) color temperature *does* so correlate, but there are difficulties at both the early and late ends of the sequence. For O and B stars, the continuum peaks in the far ultraviolet, where it is undetectable by ground-based observations. Hence, we have been forced to extrapolate (by means of theoretical models) the visible tail of the spectrum to its hidden peak. Recently, however, direct satellite observations in the far ultraviolet have been performed on the Orbiting Astronomical Observatories, and we are beginning to understand the hot O and B stars. For the cool M stars, not only does the Planck curve peak in the far infrared (where few observations have been made), but numerous molecular bands also blanket the spectra of these dim stars.

In practice, we measure a star's *color index*, $CI = B - V$, in the hope that we may thereby determine the effective stellar temperature (see Chapter 10). If the stellar continuum is Planckian, this procedure clearly gives us a unique temperature (see the color index-blackbody temperature plot in Figure 12–7). But observational uncertainties and physical effects do lead to problems: (a) for the very hot (O and B) stars, the color index is small and negative, and Figure 12–7 shows that small uncertainties in CI lead to very large uncertainties in T; (b) for the very cool (M)

Table 12–1. The Henry Draper Spectral Sequence

Spectral Type	Principal Characteristics	Spectral Criteria
O	Hottest blue stars Relatively few lines He II dominates	Strong He II lines—in absorption, sometimes emission. He I lines weak, but increasing in strength from O5 to O9. Hydrogen Balmer lines prominent, but weak compared to later types. Lines of Si IV, O III, N III, and C III.
B	Hot blue stars More lines He I dominates	He I lines dominate, with maximum strength at B2; He II lines virtually absent. Hydrogen lines strengthening from B0 to B9. Also Mg II and Si II lines.
A	Blue stars Ionized metal lines Hydrogen dominates	The hydrogen lines reach maximum strength at A0. Lines of ionized metals (Fe II, Si II, Mg II) at maximum strength near A5. Ca II lines strengthening. The lines of neutral metals are appearing weakly.
F	White stars Hydrogen lines declining Neutral metal lines increasing	The hydrogen lines are weakening rapidly, while the H and K lines of Ca II strengthen. Neutral metal (Fe I and Cr I) lines gaining on ionized metal lines by late F.
G	Yellow stars Many metal lines Ca II lines dominate	The hydrogen lines are very weak. The Ca II H and K lines reach maximum strength near G2. Neutral metal (Fe I, Mn I, Ca I) lines strengthening, while ionized metal lines diminish. The molecular G-band of CH becomes strong.
K	Reddish stars Molecular bands appear Neutral metal lines dominate	The hydrogen lines are almost gone. The Ca lines are strong. Neutral metal lines are very prominent. By late K the molecular bands of TiO begin to appear.
M	Coolest red stars Neutral metal lines strong Molecular bands dominate	The neutral metal lines are very strong. Molecular bands are prominent, with the TiO bands dominating the spectrum by M5. Vanadium oxide (VO) bands appear.

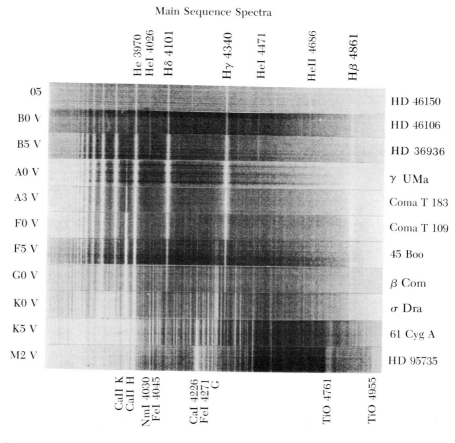

Figure 12–5. *The Spectral Sequence.* Spectra of representative main-sequence (luminosity class V) stars illustrating the changes over the spectral class sequence. (Kitt Peak National Observatory)

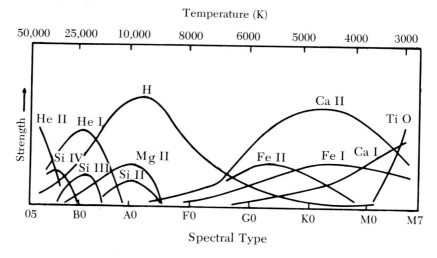

Figure 12–6. *The Spectral Type—Temperature Correlation.* The strengths of the absorption features for various species are plotted against excitation-ionization temperature. This behavior follows from the combined Boltzmann-Saha equations. The HD spectral criteria then determine the sequence of spectral types. (After L. H. Aller)

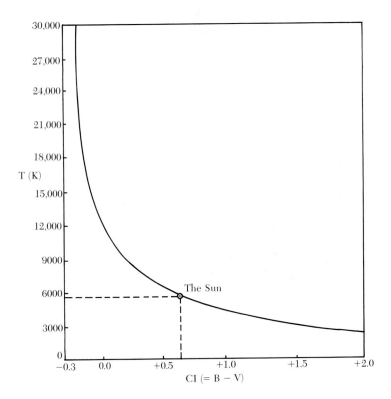

Figure 12–7. *Planck Blackbody Temperature versus Color Index.* Using the UBV color magnitude system, the color index (B − V) is computed for Planck distributions at different temperatures *T*. The results are shown here. (See also Figures 10–2 and 10–3.)

stars, the color index is large and positive, but these faint stars have not been adequately observed, so that *CI* is not well determined for them; and (c) any instrumental deficiencies, calibration errors, or unknown blanketing in the *B* or *V* bands will affect the value of *CI*—and thence, the deduced *T*.

(d) Luminosity Classifications

So far, we have been discussing *one-dimensional* temperature sequences. As shown in section 12–1, this description can represent stars *only* if they all have the *same* mass (*M*), radius (*R*), and composition (*μ*). But Chapter 10 showed us that stars clearly differ in their absolute magnitudes (luminosities, *L*)—so a *two-dimensional* (*L*, *T*) representation is absolutely necessary [see equation (12–20)]. In section 12–3, we will see that 90 per cent of the stars in the solar neighborhood do define a single band—the *main sequence*—on such an (*L*, *T*) plot; but many stars do *not* reside on the main sequence!

As early as 1897, A. C. Maury and E. C. Pickering at Harvard recognized distinctly different spectra for stars of a *given* color temperature; they appended the notation a, b, or c to indicate that certain spectral features were successively *sharper* (that is, *narrower*) than for main-sequence stars. In 1905–1907, E. Hertzsprung confirmed that the c-stars are *much more luminous* than the corresponding main-sequence stars. Between 1914 and 1935, the Mount Wilson (MW) luminosity classification—originated by W.S. Adams and A. Kohlschütter—appeared, which ordered stellar spectra (of the *same temperature!*) according to the anomalous strengths or weaknesses of certain spectral features. After 1937, W. W. Morgan and P. C. Keenan at Yerkes Observatory introduced the currently used *MK luminosity classification* scheme, which defines six stellar luminosity classes (and their subclasses) directly in terms of observational criteria. Let us describe the MK scheme in detail.

Six MK *luminosity classes* are differentiated, with finer subdivisions indicated by an appended a or b, as illustrated in Table 12–2 (note the spectral criteria). If we make a two-dimensional plot of absolute visual magnitude (M_v) versus spectral type, these classes appear as line-segments (Figure 12–8). In this scheme, our Sun is a G2 V star (that is, yellow main-sequence), and its *radius* ($R_\odot$) is much smaller than those of the giants (I to IV) of spectral type G2. Typical designations are: B1 III (β Centauri), A3 V (Fomalhaut), F0 Ib (Canopus), K1 IVa (37 Librae), and M5 V (Barnard's Star).*

In section 12–3, we shall see that Figure 12–8 is an absolute magnitude-spectral type Hertzsprung-Russell diagram. Here, however, let us concentrate on the *physical* interpretation of the different luminosity classes. For a given spectral type, the equivalent terminology—*luminosity effect, surface gravity effect* or *pressure effect*—

---

* Note that the (a, b) subdivisions are still rarely used.

Table 12–2. The Morgan-Keenan Luminosity Classes

Class	I	II	III	IV	V	VI
Subclasses*	(Ia, Iab, Ib)	(IIa, IIab, IIb)	(IIIa, IIIab, IIIb)	(IVa, IVab, IVb)	(Va, Vab, Vb)	(VI)
Name	Supergiant	Bright Giant	Giant	Subgiant	Dwarf†	Subdwarf

Spectral Type	**Spectral Criteria‡**
O	No criteria earlier than O9—note the class convergence in Figure 12–8. Near O9, the ratio of equivalent widths of lines of He I, He II, C III, O III, and Si IV.
B	The ratios of lines of He I, N II, and Si III—especially near B2. After B3, the absorption line strengths of hydrogen Balmer—especially Hδ and Hγ.
A	Until A3, the same Balmer strengths. After A2, the ratios of lines of Fe II, Mg II, and Ti II. For late-type A, the three O I lines in the infrared $\lambda 7771$ to $\lambda 7775$.
F	Balmer hydrogen strengths ineffective after F5. Ratios of lines of hydrogen Balmer, Fe I, and Ca I around F5. In general, the line strength ratios of hydrogen Balmer to Sr II lines.
G and K	Strength of molecular G-band of CH. Enhancement of Hδ and Hγ. Relative line intensities of Sr II and Fe I lines. The strong blue molecular band of CN, and its other absorption bands. Line ratios of Fe I, Fe II, Mn I, Ca I.
M	Line ratios of Fe I, Cr I, Hδ, Sr II, and Y II; also Fe II, Ni I, Ti I, K I, and Ca I. Infrared CN bands.

* In a given spectral type, luminosity *decreases* along the sequence a, ab, b.

† These are the *main-sequence* stars.

‡ The luminosity classes are discerned by studying these spectral features, which depend upon T, and hence, spectral type.

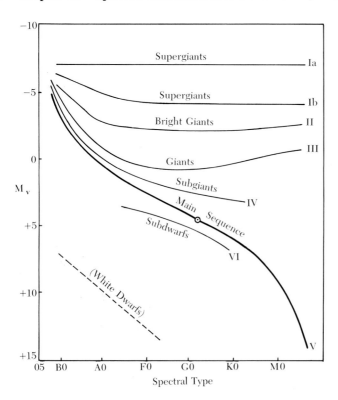

Figure 12–8. *The MK Luminosity Classes.* Note particularly class V—the main-sequence dwarfs. The absolute visual magnitude of class Ia is uncertain to $\pm 0^{m}.5$. Note the convergence of the classes (I–V) at early spectral types.

distinguishes the luminosity classes. Since spectral type "corresponds" to temperature T, equation (12–17) tells us that classes I to IV represent stars with *radii much larger than those of main-sequence stars (dwarfs)*. For example, a G2 supergiant is about $12^{m}.5$ brighter than our Sun; equation (10–5) then implies a luminosity ratio of 10^5, or a supergiant radius of about $300R_{\odot}$! Since stellar masses don't exceed $100M_{\odot}$, this supergiant must be about one million times (10^6) *less dense* than our Sun, on the average. From equation (12–3), we see that the *surface gravity g* of the supergiant is about $10^{-4}g_{\odot}$, so that [equation (12–5)] the photospheric gas *pressure* and electron number densities are also about 10,000 times lower than in the Sun's photosphere. Hence, the spectral features of the supergiant must be different from those of the Sun—in accordance with the Saha equation (12–9)—even though both stars exhibit essentially the *same* color temperature. The pressure effect is somewhat less important than the temperature effect, for it only appears in the equation linearly, while the temperature enters exponentially. Hence, a giant star will exhibit almost the same spectrum as does a main-sequence star of the same spectral type, as long as the giant's surface temperature is *lowered* slightly to compensate for its lower electron density (by the Saha equation, the ratio N_{i+1}/N_i will remain the same if both N_e and T decrease appropriately). Even in this case, however, the spectral lines of the giant will be sharper than those of the main-sequence star, since the giant features suffer much less *pressure broadening* (see section 8–5).

To give the reader some idea of the observed characteristics of stars of different luminosity classes, we present the following data in Table A2–7: absolute visual magnitudes (M_v), color indices ($B - V$), effective surface temperatures (T_{eff}), bolometric corrections (BC), stellar radii ($R/R_{\odot}$), and stellar masses ($M/M_{\odot}$). Study the values and *trends* in this table, in conjunction with Figure 12–8.

(e) Elemental Abundance Effects

The reader may have noticed that we have not yet considered the effects of stellar composition (μ) upon spectral classification, nor have we discussed the subdwarfs (luminosity class VI). The reason for this is simple: the vast majority of the stars in the solar neighborhood have the *same* composition*—they belong to the so-called Population I [see section 12-3(c)]. The subdwarfs (Population II → low mental abundance, $Z \approx 0.001$) reside below the main-sequence because of their metal-deficiency. Fewer heavy elements imply *less* line blanketing, so that these stars *appear* to be hotter and bluer (A to F) than they actually are (probably G to K). Let us now consider those rare "anomalous" cases of Population I, where elemental abundance effects do appear in the spectra (Figure 12-9 is a schematic summary of this discussion).

Wolf-Rayet Stars. In 1867, C. Wolf and G. Rayet discovered three O stars with anomalously strong and wide *emission* lines. Today, only about 200 of these hot (T up to 10^5 K), bright (absolute magnitudes in the range $-4{.}^{m}5$ to $-6{.}^{m}5$) stars are known—the so-called *Wolf-Rayet* or *W stars*. As we shall see in Chapter 16, the exceedingly wide (tens of Ångstroms) emission features of ionized He, C, N, and O seen in their spectra are interpreted as arising in an atmospheric envelope expanding from the star at about 2000 km sec^{-1}. Two abundance sub-branches are distinguished: (a) the *WC stars* with an apparent overabundance of carbon (spectral features of carbon and oxygen, up to C IV and O VI, are prominent), and (b) the *WN stars* with an apparent excess of nitrogen (N III to N V lines dominate). It is believed by some that these branches are caused by excitation-ionization effects, and not by abundance differences.

Hot Emission-Line Stars. In spectral types O, B, and A we find the *Of, Be, and Ae stars* with bright emission lines of hydrogen. Similarly to the W stars, these stars are thought to be slowly losing mass in the form of expanding atmospheric envelopes (where the emission lines arise); see Chapter 16.

Peculiar A Stars. In the spectra of *peculiar A* or *Ap stars*, the lines of ionized Si, Cr, Sr, Eu, and other rare earths are selectively enhanced. In many cases, this enhancement is time-varying (the so-called *spectrum variables*), and it appears to be associated with strong *magnetic fields* (10^3 to 10^4 Gauss) at the stellar surface.

---

* $X \approx 0.70$, $Y \approx 0.30$, $Z \approx 0.01$ → high "metal" (Z) abundance.

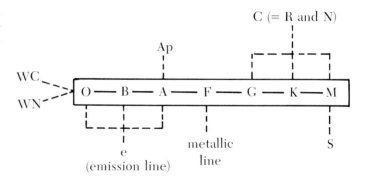

Figure 12-9. *Abundance Anomalies.* A schematization, showing the boxed main sequence of normal Population I composition. We see the hot Wolf-Rayet (WC, WN) stars, the hot emission-line (e) stars, the peculiar A (Ap) stars, the metallic-line F stars, the cool carbon (C) stars, and the heavy-metal-oxide (S) stars.

Metallic-Line F Stars.   These F stars are similar to the Ap stars in their abundance anomalies, but little else can be said at present.

Carbon Stars.   Mixed in with ordinary G, K, and M stars (a temperature range from 4600 K to 3100 K) we find the *carbon* or *C stars*, which appear to be overabundant in carbon relative to oxygen. In the early Harvard classification, these were subdivided into (a) the hotter R stars distinguished by the Swan bands of C_2 and the bands of cyanogen, CN; and (b) the cooler N stars exhibiting C_2, CN, and CH bands with little TiO evident. In addition, we can say only that these stars are *rare* and *giants*.

Heavy-Metal-Oxide Stars.   Finally, among the M stars we find a significant number of *S stars*, which are known to be giants. These stars are distinguished spectroscopically by their enhanced CN absorption bands, but more importantly, by the presence of the molecular bands of the heavy-metal-oxides ZrO, LaO, and YO instead of TiO.

12–3   HERTZSPRUNG-RUSSELL (H-R) DIAGRAMS

Look back at Figure 12–8. In 1911, Ejnar Hertzsprung plotted the first such two-dimensional diagram for observed stars, followed (independently) in 1913 by Henry Norris Russell; today, this plot is called a *Hertzsprung-Russell* or *H-R diagram*. As will soon become clear to the reader, this innocuous-looking diagram represents one of the greatest observational syntheses in astronomy and astrophysics!

(a)   Magnitude versus Spectral Type

In analogy to Figure 12–8, the first H-R diagrams considered stars in the solar neighborhood, and plotted *absolute magnitude* (M) versus *spectral type* (Sp). Figure 12–10 shows this type of plot for stars with well-determined distances within about 15 pc. Note (a) the well-defined main sequence (class V dwarfs), (b) the class III giants, (c) the smattering of supergiants at high luminosities across all spectral types, and (d) the three white dwarfs in the lower left-hand corner. The shaded region shows where Population II stars (such as F to K subdwarfs) could creep into the plot, if they were not assiduously eliminated (by observational selection, as indicated here). Also notice the scarcity of giant stars, and the preponderance (about 90 per cent) of dwarfs—with ever-increasing numbers toward *later* spectral types. By counting the number of stars shown, and dividing by the volume considered, we find a number density of ≈ 0.015 stars pc^{-3} in the solar neighborhood; the average interstellar distance is therefore ≈ 4 pc.

Another H-R diagram of this type, which vastly overemphasizes the relative number of giant stars, is shown in Figure 12–11. Here the *brightest* stars are plotted, so that distances are rather poorly determined (note the broad scatter of the main sequence); the more luminous giant stars are sampled to much greater distances than the dwarfs (a selection effect).

Figure 12–10. *Solar Neighborhood Magnitude-Spectral Type Plot.* Approximately 200 stars with accurately-known distances ($\lesssim 15$ pc) are shown. Compare with Figure 12–8. The shaded region shows where Population II members (e.g., F–K subdwarfs) might encroach upon the diagram. [From K. Aa. Strand, (ed.): *Basic Astronomical Data.* Chicago, University of Chicago Press, 1963]

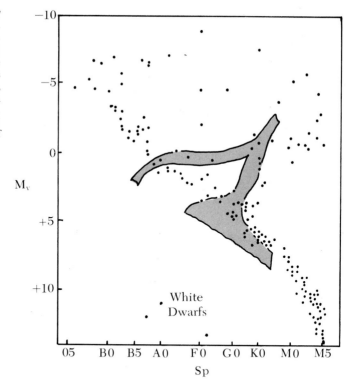

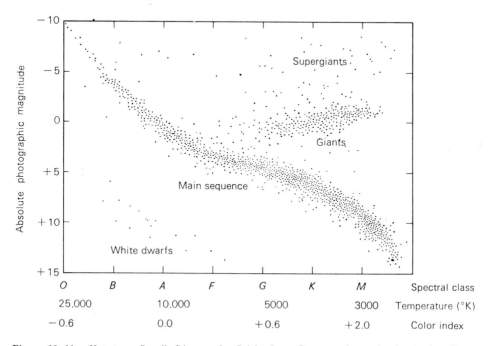

Figure 12–11. *Hertzsprung-Russell Diagram for Bright Stars.* Because observational selection favors intrinsically bright stars, the relative numbers of the various stars are incorrect for unit volume. (From G. O. Abell: *Exploration of the Universe.* Second Ed. New York, Holt, Rinehart, and Winston, Inc., 1969. Reprinted by permission of Holt, Rinehart, and Winston, Inc.)

(b)   Magnitude versus Color

Since stellar colors and spectral types are correlated approximately, we may construct a plot of absolute magnitude versus color—the so-called *color-magnitude H-R diagram*. The relative ease and convenience with which color indices (e.g., $B - V$) may be determined for vast numbers of stars dictates the present predominance of the color $(B - V)$-magnitude (M) plots. The resulting diagrams are very similar to the magnitude-spectral type H-R diagrams considered in section 12–2; let us see what information we can glean from them.

(c)   Stellar Populations

We mentioned earlier that there are two stellar populations:* the young metal-rich Population I and the old metal-deficient Population II. How has this been found to be the case? Let us consider a *stellar cluster* which, by its appearance or the common motion through space of its member stars, is known to be a self-gravitating group of stars all formed at approximately the *same time*. In addition, the distance to every cluster member is approximately the same, so that a plot of *apparent* magnitude versus color *is* an H-R diagram (why?). Figure 12–12 shows such

---

* Actually, these two populations represent the *extremes* of a composition (μ) spectrum. Today, as many as four other populations are recognized *between* these extremes. (See Chapter 13.)

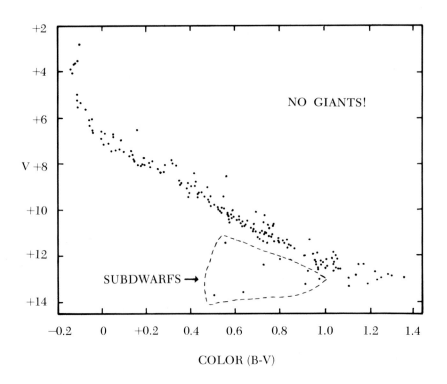

Figure 12–12.   *Color-Magnitude Diagram for the Pleiades.* Note that the main sequence extends much farther into the blue than in the case of the Hyades (Figure 12–13). (After H. L. Johnson and R. I. Mitchell)

a color-magnitude (*CM*) diagram for the Pleiades, a young *open* (or *galactic*) *cluster* of hot blue stars in the constellation Taurus. Notice (a) the well-defined main sequence, (b) the *absence* of giants (luminosity classes I to IV), (c) the curving-up of the early end of the main sequence, and (d) the few subdwarfs (erroneously included because of the errors inherent in determining which stars *actually belong* to the cluster). The stellar spectra exhibit *high* metal abundance ($Z \approx 0.01$), so we assign the stars to Population I. Figure 12–13 shows the color-magnitude (again, apparent magnitude) diagram of a somewhat older (about 100 million years) open cluster, called the Hyades cluster. Again we see the main sequence, but with no stars above $(B - V) \approx +0.2$; in addition, there are several stars in the *giant* region in this Population I cluster.

In our Galaxy, about 120 *globular clusters* have been discovered. A globular cluster is an extremely compact and spherically-symmetric "ball" of stars (up to 500,000 stars in some!). These clusters are found at great distances from the central plane of our Galaxy. Figure 12–14 is the color-magnitude diagram of the globular cluster M3 (the third object listed in C. Messier's catalog of the year 1781). The stellar spectra reveal an exceedingly *low* metal abundance ($Z \lesssim 0.001$), so we assign the stars to Population II. In Figure 12–14, notice (a) the main sequence running from $(B - V) = +0.8$ up to the turn-off at $(B - V) \approx +0.4$, (b) the heavily-populated giant branch, and (c) the high-luminosity branch running toward the left. (The reader should compare this figure with the shaded regions in Figure 12–10.)

Therefore, we have two well-defined stellar populations, exemplified by these clusters. We interpret the variations in terms of differences in chemical composition. Population I stars (open clusters) have a high metal abundance compared to Population II (globular clusters). As we shall see in Chapter 15, one can also understand the color-magnitude diagrams in terms of stellar evolution in the solar neighborhood (Figure 12–11): the main-sequence represents young (Population I)

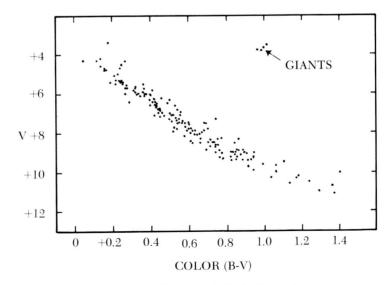

Figure 12–13. *Color-Magnitude Diagram for the Hyades.* Note the later main-sequence turnoff and the giants. (After H. L. Johnson)

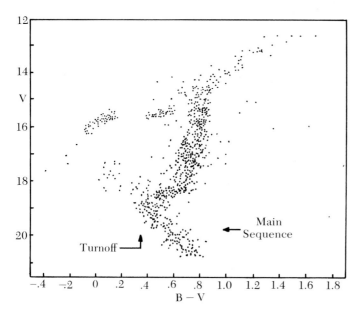

Figure 12–14. *Color-Magnitude Diagram for the Globular Cluster M3.* (After A. R. Sandage and H. L. Johnson)

stars, the giants are older, more-evolved stars (Population I and II), and the white dwarfs are stars at the *end-point* of stellar evolution. The subdwarfs are Population II interlopers passing through the solar neighborhood. Note that Population II is invariably associated with *great age* (old stars).

(d)   Distance Determinations

We end this chapter by discussing two methods by which reasonably accurate *stellar distances* may be determined using H-R diagrams. Both methods depend upon an accurate calibration of the *absolute* magnitude H-R diagram; we will use Figures 12–8 and 12–11 as our standards. [The *moving-cluster method* (see Chapters 10 and 14) gives the best calibration.]

For individual stars (or clusters in which only one star is observed thoroughly), we employ the *method of spectroscopic parallaxes*. From a star's spectrum, we determine both its spectral type and luminosity class. These data *fix* a position in the H-R diagram, from which we can read off the star's absolute magnitude. From the observed apparent magnitude, we compute the distance modulus [equation (10–6)] and thence the stellar distance. Observational errors and the scatter of the H-R diagram imply a magnitude uncertainty of about $\pm 1^m\!.0$, so that the distance is known to within 50 per cent. Spectroscopic parallaxes permit us to probe the vast distances of our Galaxy, as well as nearby galaxies, and so render our cosmic viewpoint less parochial.

Greater accuracy in distance determination is possible by using the *main-sequence fitting technique* for an entire *cluster* of stars. Here we plot the color-*apparent* magnitude diagram of the test cluster on semitransparent paper, and shift this plot up and down (in magnitude) on a calibrated H-R diagram, until the two main sequences overlap at the same spectral types. The difference between the test cluster apparent magnitudes and the calibrated absolute magnitudes, $(m - M)$, is *the same* for every star in the cluster; this number is the distance modulus of the test cluster. For example, M3 has a distance modulus of $+15$, so its distance is about 10 kpc. By using many stars, we can cancel out random errors and achieve good accuracy ($\pm 0\overset{m}{.}2$) in determining the distance modulus. However, some care must be exercised when using this technique on globular clusters, for these clusters have been found to have slightly differing compositions which shift their main sequences significantly. More sophisticated techniques are now available to overcome this troublesome effect.

Problems

12–1.   The absorption spectra of four stars exhibit the following characteristics; what are the approximate spectral types of these stars:
(a) The strongest features are titanium oxide bands.
(b) The strongest lines are those of ionized helium.
(c) The hydrogen Balmer lines are very strong, and some lines of ionized metals are present.
(d) There are moderately-strong hydrogen lines, and lines of neutral and ionized metals are seen, but the Ca II H and K lines are the strongest in the spectrum.

12–2.   To *approximately* which spectral types may the following stars be assigned, if their continuous spectra peak (are of maximum intensity) at the wavelengths:
(a) 500 Å,
(b) 3000 Å,
(c) 6000 Å,
(d) 9000 Å,
(e) 12,000 Å,
(f) 15,000 Å?
(Hint: Plot λ_{max} from Wien's law as a function of spectral class for the main sequence.)

12–3.   If the parallax of a main-sequence star is in error by 25 per cent, how far and in what direction will this star be misplaced from the main sequence in an H-R diagram?

12–4.   Which parameter in the Saha ionization-equilibrium equation is *most* important in explaining the spectral differences between
(a) giants and dwarfs of spectral type G;
(b) B and A dwarfs?

12–5.   (a) What is the ratio of the surface gravities of a K0 V star and a K0 I star?
(b) If both stars had the same atmospheric temperatures and

opacities, what would be the approximate ratio of strengths of the Ca II K lines in their spectra (the ionization potential of calcium is 6.1 eV)?

(c) What assumptions did you make to answer part (b)?

(d) What is the ratio of the mean densities of these two stars?

(e) If the atmospheric electron densities of these stars were directly proportional to their mean densities, would your answer to (b) remain unchanged? Why?

12–6.   Explain clearly, in conjunction with a diagram, the meanings of

(a) line blending;

(b) line blanketing;

(c) line broadening.

12–7.   (a) What distinguishes subdwarfs from ordinary Population I main-sequence stars (give two characteristics)?

(b) Why is the H-R diagram of stars in the solar neighborhood (e.g., within 500 pc) *not* an unambiguous two-dimensional plot?

12–8.   Why is an absolute magnitude-spectral type H-R diagram quite different (in principle) from an apparent magnitude-(B − V) H-R diagram?

12–9.   In determining distances via the main-sequence fitting technique, why must we refrain from comparing the observed H-R diagram of a galactic cluster with the "calibrated" H-R diagram of M3?

12–10.   (a) It is sometimes said that the spectral type of a star depends only upon that star's luminosity and surface temperature; under what conditions is this statement approximately true?

(b) Give three generic examples which violate the statement that stellar masses are uniquely determined by their colors.

12–11.   In section 12–1(a), we wrote the mean molecular weight μ for a fully-ionized gas as: $1/\mu = 2X + (3/4)Y + (1/2)Z$, where, for example,

$$X \equiv \text{hydrogen mass fraction} = \left(\frac{\text{mass density of hydrogen}}{\text{mass density of } all \text{ constituents}} \right).$$

Derive this relation for μ, indicating the assumptions and approximations used at each step in the derivation.

12–12.   The number 5040 appears in equations (12–10) and (12–11).

(a) Show where this number comes from (i.e., how it is derived).

(b) What are the units of this number in these equations?

Reading List

Abetti, Giorgio, (Barocas, V., translator): *Stars and Planets*. London, Faber and Faber Ltd., 1966.

Aller, Laurence H.: *Astrophysics*. Vol. I and II. New York, The Ronald Press Company, 1954, 1963.

Blaauw, Adriaan, and Schmidt, Maarten, (eds.): *Galactic Structure*. Chicago, University of Chicago Press, 1965.

Curtiss, R. H.: "Classification and Description of Stellar Spectra." *Handbuch der Astrophysik*, Vol. 5, Part 1, pp. 1–107 (1932).

Dufay, Jean, (Gingerich, O., translator): *Introduction to Astrophysics: The Stars*. New York, Dover Publications, 1964.

Gingerich, Owen, (ed.): *Theory and Observation of Normal Stellar Atmospheres*. Cambridge, Massachusetts, The M.I.T. Press, 1969.

Motz, Lloyd, and Duveen, Anneta: *Essentials of Astronomy*. Belmont, California, Wadsworth Publishing Company, 1966.

Page, Thornton, and Page, Lou Williams, (eds.): *Starlight*. New York, The Macmillan Company, 1967.

Rose, William K.: *Survey of Modern Astrophysics*. New York, Holt, Rinehart and Winston, 1973.

Schwarzschild, Martin: *Structure and Evolution of the Stars*. Princeton, Princeton University Press, 1958.

Strand, Kaj Aa., (ed.): *Basic Astronomical Data*. Chicago, University of Chicago Press, 1965.

Struve, Otto: *Stellar Evolution*. Princeton, Princeton University Press, 1950.

Swihart, Thomas L.: *Astrophysics and Stellar Astronomy*. New York, John Wiley & Sons, 1968.

Part 4
The Structure and Content of our Galaxy

Chapter 13
Our Galaxy:
An Introductory
View

We have looked at our Solar System, and touched upon some of the basic characteristics of the stars. Now it is time to expand our horizons one step further (not the last step, by any means; see Part V), and consider that magnificent entity of which we constitute but an insignificant part—*our Galaxy*. In this chapter we introduce the reader to the Galaxy, presenting a broad picture so that Chapters 14–17 may be considered in the proper context. In Chapter 18, this context will be reiterated, and our discussion of the Galaxy summarized.

13–1 THE SHAPE OF THE GALAXY

(a) Observational Evidence

The city-dweller of today is often surprised, upon going into the country, to discover the crystal clarity of the night sky and the sharpness of its features. There he rediscovers what has been known since men first gazed at the heavens: that an irregular band of diffuse light about 20° broad encircles the celestial sphere approximately along a great circle. We call this phenomenon *the Milky Way*. The wide-angle grasp of the naked eye has been mimicked and improved upon by the use of specially-designed telescopic cameras, as shown in Figure 13–1 (see also

Figure 13–1. *A Wide-Angle View of the Milky Way.* The center of the Galaxy appears as a bright bulge in Sagittarius with the dark "lane" overlying it. This photograph includes the entire Southern Milky Way from Carina to Aquila. The vanes supporting the camera, which is located far above the mirror, obstruct part of the field of view. (Courtesy of A. D. Code and T. E. Houck)

Figure 13–6). Here the brightest part of the Milky Way is seen in the constellation Sagittarius,* and dark "lanes" of obscuration are evident along the midline of the Milky Way (see Chapter 17). Diametrically opposite on the celestial sphere, in the constellations Perseus and Orion, the Milky Way is rather indistinct and unspectacular.

How are we to understand the Milky Way? In the early decades of the 20th century, the giant new telescopes on Mt. Wilson and Mt. Palomar revealed that the luminous clouds of the Milky Way are *star clouds* (see Figure 13–2); that is, these vague blobs of light were resolved into collections of *billions* of stars. In section 13–2 we shall see how the stars of our *Galaxy* (from the Greek *galaktikos*, meaning milky-

---

* Visible to northern hemisphere observers during the summer.

Figure 13–2. *Star Clouds in Sagittarius.* The contrast between obscured and unobscured areas is very evident in this rich star field. (Hale Observatories)

white) are distributed in space, so that we perceive the Milky Way. In addition, in 1781 the French astronomer Charles Messier published a catalog listing 107 "fuzzy" (i.e., nonstellar-appearing) objects seen while searching for comets; some of these were later found to be glowing clouds of gas in our Galaxy (Chapter 17) and distant globular clusters, but the others—the so-called *nebulae* (from the Latin singular *nebula*, meaning cloud)—remained an enigma until well into the 20th century. Today, we understand these nebulae to be *galaxies*, separate extragalactic entities similar to our own Galaxy (see Chapter 19). Figure 13–3 shows one of the nearest of these objects, the Andromeda galaxy (Messier 31), which turns out to be surprisingly similar to our Galaxy. Better yet, consider Figure 13–4, where two (spiral) galaxies are shown: M74 (also designated NGC 628; seen face-on) and NGC 891 (seen edge-on). Note the amazing similarity between Figures 13–1 and 13–4B; our Galaxy is apparently a highly-flattened (like a pancake) stellar system, which is being viewed (from the Sun) edge-on from a point far from its center. The reader will appreciate that the galactic center lies in the direction of the *central bulge*, that luminous swelling evident in Figures 13–1, 13–3, and 13–4.

(b) The Galactic Coordinate System

To better understand the Galaxy, we will disregard the geocentric coordinates based on the celestial equator and celestial poles, and the heliocentric ecliptic

Figure 13–3. *The Andromeda Galaxy, M31.* A spiral galaxy similar to our own. Note the two elliptical companions. (Lick Observatory)

system, and define a *galactic coordinate system* (see also Chapter 2). Here the center-line of the Milky Way (more accurately, the mass-centroid of the galactic plane) defines a great circle called the *galactic equator*, along which *galactic longitude* ℓ is measured in degrees (0° to 360°) eastward from the direction to the center of the Galaxy in Sagittarius. *Galactic latitude* b is the angular distance on the celestial sphere (in degrees from 0° to ±90°) either north or south from the galactic equator. Hence, the galactic anticenter is at ($\ell = 180°$, $b = 0°$), the north galactic pole (NGP) at $b = +90°$, and the south galactic pole (SGP) at $b = -90°$. In terms of right ascension (α) and declination (δ)—in the geocentric celestial equatorial system—the (epoch 1950) coordinates of the NGP are (12^h49^m, $+27°.4$); hence, the galactic equator is inclined about 63°.5 to the celestial equator and is almost perpendicular to the ecliptic. Figure 13–5 shows how these three important angular-coordinate systems are related: (i) the celestial equatorial (α, δ) system, (ii) the ecliptic (λ, β) system, and (iii) the galactic (ℓ, b) system. Note, in particular, the location of the galactic center ($\ell = 0°$, $b = 0°$) at ($\alpha = 17^h42^m.4$, $\delta = -28°55'$; epoch 1950). In Figure 13–6 we illustrate the galactic coordinate system with a mosaic of photographs covering the entire celestial sphere, in which the Milky Way is aligned horizontally along $b = 0°$.

Prior to 1958, a galactic coordinate system—now denoted (ℓ^I, b^I)—was in use. Its equatorial plane, defined from observations of stars, was not very

Figure 13–4.   *Two Spiral Galaxies.* **A**, A face-on view of Messier 74. (NGC 628; Hale Observatories). **B**, An edge-on view of NGC 891. (Hale Observatories)

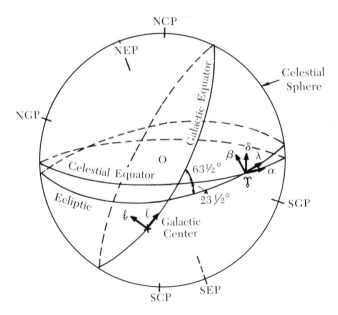

Figure 13–5. *Geocentric, Heliocentric, and Galactic Coordinates.* The observer O maps on the celestial sphere three systems: (1) celestial equatorial (α, δ) with the north celestial pole (NCP), south celestial pole (SCP), and origin at the first point in Aries (Υ); (2) ecliptic (λ, β) inclined 23°5 to celestial equator, with north and south ecliptic poles (NEP, SEP) and origin at Υ; and (3) galactic (ℓ, ℓ) with galactic equator inclined 63°5 to celestial equator, north and south galactic poles (NGP, SGP), and origin at galactic center.

accurate and differed by about 1°5 from that of the current system. In the (ℓ^I, ℓ^I) system, galactic longitude (ℓ^I) was measured in arc-degrees *eastward* along the galactic equator from the point (in the constellation Aquila) where the galactic equator crossed the celestial equator *northward*. The galactic center was at about $\ell^I \approx 327°$. The new system—denoted (ℓ^{II}, ℓ^{II}) originally—was defined in 1958; since 1968, the superscripts have been dropped. Since the Sun's position (above or below the galactic plane) was not accurately known, and since the radio center of our Galaxy—the source Sagittarius A—was not as precisely defined in 1958 as it is today, the actual position of the galactic center may lie as much as 4′ from $(\ell, \ell) = (0°, 0°)$.

13–2 THE DISTRIBUTION OF STARS

The most obvious way to determine the size and shape of our Galaxy is to investigate the *spatial distribution* of the stars we can detect. Here we will learn how to "count" the stars, we will see how misconceptions can arise and *have arisen* in interpreting the data, and we will uncover the stellar indications of the Galaxy's size and its spiral-arm features.

(a) The Star Count Method

(i) *"Gauging" the Stars*

It is evident from Figure 13–1 that our Galaxy is probably a flattened stellar system, being viewed from a point within the system, but near its edge. If this is true, then we should see fewer stars per unit area on the celestial sphere (a) as we look at greater angles from the galactic center in the galactic plane, and (b) as we look toward greater galactic latitudes. Let us describe a quantitative

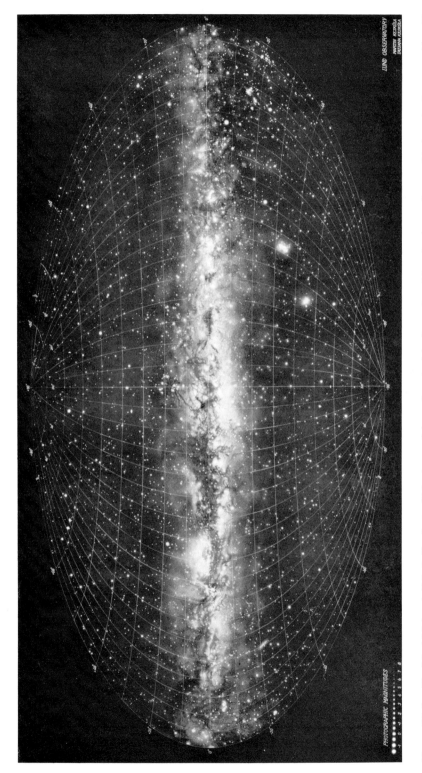

Figure 13–6. *The Celestial Sphere in Galactic Coordinates.* This picture of the entire celestial sphere, plotted in galactic coordinates with the galactic equator ($b = 0°$) horizontal, is a mosaic of many photographs. (Lund Observatory; Courtesy of Sky and Telescope)

procedure—called *star gauging* or *the method of star counts*—for observationally determining the distribution of stars in space.

Consider the solid-angular area $d\Omega$ (e.g., units = square degrees) on the celestial sphere. At a distance r from the observer, this solid angle corresponds to the area dA, where (see section 8–1)

$$dA = r^2 \, d\Omega \tag{13-1}$$

Since the solid angle of a sphere is 4π, the surface area of a sphere of radius r is therefore $4\pi r^2$. If we now move out the distance dr farther, we have bounded the volume

$$dV = dA \cdot dr = r^2 \, dr \, d\Omega \tag{13-2}$$

The number of stars in this volume $N(r)$ is

$$N(r) = n(r) \, dV = n(r)r^2 \, dr \, d\Omega \tag{13-3}$$

where $n(r)$ is the *number density* of stars (number per unit volume) at r. If we are to study a large number of stars, however, it is much easier to consider their *apparent magnitudes* than their distances. Let us assume that we are considering only stars of the same *absolute* magnitude M (selected by spectral type, for example). Then we know that apparent magnitude and distance are related via [equation (10–6)]:

$$(m + 1) - M = 5 \log (r_{m+1}) - 5 \tag{13-4a}$$

$$m - M = 5 \log (r_m) - 5 \tag{13-4b}$$

Subtracting equation (13–4b) from equation (13–4a), we straightforwardly find

$$(r_{m+1}/r_m) = 10^{1/5} = 1.585 \tag{13-5}$$

so that a star appears one magnitude fainter as we increase its distance by a factor of 1.585. Now we may return to equation (13–3), and find the *ratio* of the numbers of stars, $N(m + 1)/N(m)$, in the volumes corresponding to apparent magnitudes $(m + 1)$ and m, respectively:

$$N(m + 1)/N(m) = n(m + 1)r_{m+1}^2 \, dr \, d\Omega / n(m)r_m^2 \, dr \, d\Omega$$

or using equation (13–5):

$$\boxed{\frac{N(m + 1)}{N(m)} = \left(\frac{r_{m+1}}{r_m}\right)^2 \frac{n(m + 1)}{n(m)} = 2.512 \frac{n(m + 1)}{n(m)}} \tag{13-6}$$

Equation (13–6) expresses the result of *differential star counting*, since we are comparing small (differential) but adjacent volumes in the same direction. If the density of stars is *uniform*, $n(m + 1) = n(m)$, then we should count $10^{0.4} = 2.512$ times as many stars at $(m + 1)$ as at m. The effects of a nonuniform stellar distribution are clear from equation (13–6). In practice, we count stars only out to some *limiting magnitude*, m_{max}, determined by observational difficulties and the desired completeness of the sample.

(ii)   *The Herschel/Kapteyn "Universe"*

Because of the vast number of stars in our Galaxy, it has never been possible to count them over the entire celestial sphere. In 1785, William Herschel first "gauged" the stars by the method which we have just described; he sampled 683 regions of the sky, counting the stars in each region. He found far fewer stars at high galactic latitudes than in the galactic plane (the *latitude effect*), and also noted a rapid decrease in the density of stars at great distances from the Sun in the galactic plane. This work seemed to vindicate the "island universe" theories of Thomas Wright (1750) and Immanuel Kant (1755), in which our Sun is at the *center* of a flattened system of stars.*

In 1922, the Dutch astronomer J. C. Kapteyn arrived at a similar conclusion, although the stellar system (centered on the Sun) was now of an oblate spheroidal shape about 12 kpc in diameter. He counted the stars in 206 "selected areas," each about 4 square degrees in angular area, on the celestial sphere. Figure 13–7 shows the model of our Galaxy put forward by Kapteyn; Herschel's earlier picture closely resembled this.

(iii)   *Interstellar Absorption*

When we look at the Milky Way, it is clear that we are not located at the *center* of the Galaxy. But the star-count data of Herschel and Kapteyn indicated that our Sun was at the center. This paradox was resolved in 1930 by the work of R. J. Trumpler, who, while studying star clusters, noted that many (those near the galactic plane) appeared anomalously dim for their observed angular sizes. By equation (10–6), these dim clusters would be very distant, but then their linear sizes (in parsecs) would be unbelievably large. All of these problems disappear when we assume, as did Trumpler, that *interstellar space is not a perfect vacuum*, but is filled

---

* In fact, this stellar system was thought to be the *entire* Universe!

Figure 13–7. *The Herschel/Kapteyn "Universe."* This is a schematic representation of the results of W. Herschel (1785) and J. C. Kapteyn (1922), from their star counts. The oblate spheroidal (i.e., flattened) system of stars is *centered* on our Sun.

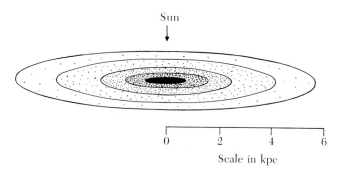

Sun

Scale in kpc

with obscuring material ("dust") which is concentrated toward the galactic plane. Although this assumption justifies itself, the reader should note the dark lanes of obscuration defining the galactic plane in Figures 13–1, 13–4, and 13–6.

Starlight is scattered and absorbed by this material, so we refer to the phenomenon as *interstellar absorption* (see Chapter 17 for details). Since the stars observed at high galactic latitudes (i.e., perpendicular to the plane of the Galaxy, at the Sun's position) do not strongly exhibit the characteristic effects of interstellar absorption, we can believe that the disk of the Galaxy is about 1 kpc thick at the Sun (see Figure 13–7). At low galactic latitudes (near the galactic plane), however, the obscuration is very important. Let us characterize this absorption by a medium of uniform density and opacity (see section 9–2); then equation (9–2) implies an exponential diminution of starlight with distance travelled through the medium. Since stellar magnitude is proportional to the logarithm of the observed brightness [equation (10–5)], we can see that the increase in apparent magnitude due to interstellar absorption is proportional to the distance to the star. Near the galactic equator, the absorption is about $1^m\!.0$ per kiloparsec; in very dense interstellar clouds, it is certainly much higher. Therefore, a star one kiloparsec distant in the galactic plane has an apparent distance modulus of $+11^m\!.0$, though its actual distance modulus is $+10^m\!.0$; this star *appears* to be about 1.6 kpc distant. A star at 5 kpc suffers $5^m\!.0$ of absorption, so that it is not only exceedingly dim, but its distance is overestimated by a factor of 12 if we are unaware of interstellar absorption.

We can now understand why the Galaxy appeared small and heliocentric in the Herschel/Kapteyn interpretation. The interstellar obscuration in the galactic plane was the culprit. By including absorption in the analysis of the star-count data, we find that we are near the periphery of a very large (about 30 kpc diameter) stellar system, and the density of stars increases rapidly toward the galactic center.

(iv)   *Luminosity Functions*

In addition to star counts, we can characterize the stars in our region of the Galaxy by their *luminosity function:* the number of stars per unit spatial volume with a "given" absolute magnitude M. We find a star's absolute magnitude by studying its spectrum, and placing it appropriately on a calibrated H-R diagram.

As a crude, and incorrect, first attempt we can consider the fifty *brightest* visible stars, plotting a *histogram* of the numbers occurring in the absolute magnitude intervals M to $M + 1/2$ (Figure 13–8). It is not unexpected that those stars which appear the brightest are also intrinsically bright (note the preponderance at negative M), but Figure 13–8 is biased, since very luminous stars can be seen to great distances. A truer picture of the actual luminosity function is given by the histogram of the fifty *nearest* stars in Figure 13–9; note the overwhelming predominance of intrinsically faint stars ($10^m\!.0 \lesssim M \lesssim 15^m\!.0$), and the dearth of very luminous stars. A stellar sample complete to a limiting magnitude of $M \approx +17^m\!.0$ is shown in Figure 13–10. We clearly see that most stars are very dim ($M \approx +16^m\!.0$), with few being brighter than our Sun ($M_{pg} = +5^m\!.4$) which is by no means an exceptionally luminous star. We can say that faint dwarfs are in the overwhelming majority, with only a sparse sprinkling of giant stars.

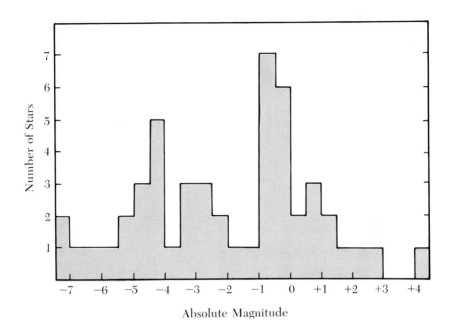

Figure 13–8. *"Luminosity Function" For Fifty Brightest Stars.* This histogram is not a true luminosity function, since a given region of space was not sampled. Such bright stars can be seen to very great distances.

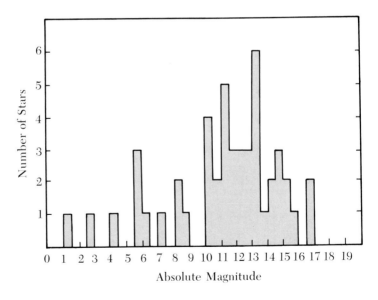

Figure 13–9. *Luminosity Function of Fifty Nearest Stars.* This true luminosity function begins to show the overwhelming preponderance of intrinsically faint stars in the solar neighborhood.

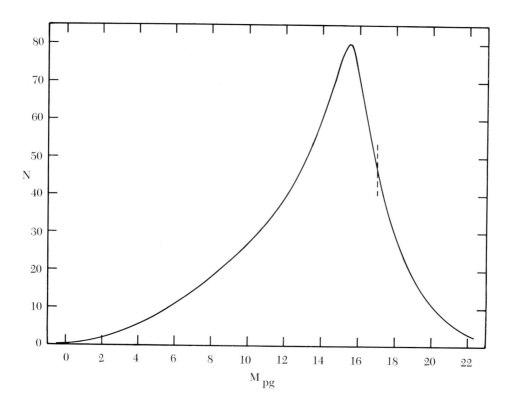

Figure 13–10. *Luminosity Function For Stars Within 10 Parsecs.* The data are believed to be complete to $M = 17^m$; therefore, the drop beyond $15\overset{m}{.}7$ is considered real. The decrease in numbers beyond $M = 18^m$ may not really be as rapid as indicated because of observational limitations. (Data from W. Luyten)

(b) Bright Stars and Clusters

Bright stars and stellar clusters are discernible to very great distances. For example, if we survey to a limiting magnitude of $m = +15\overset{m}{.}0$, B stars of absolute magnitude $M = -5\overset{m}{.}0$ can be seen to a distance of about 100 kpc; interstellar absorption in the galactic plane will, however, limit us to an actual distance of 5 kpc. O and B stars are very young, and they tend to occur in small clusters called *associations;* moreover, these hot, luminous stars can ionize the hydrogen gas in which they are imbedded, producing a bright spherical *H II region* of ionized hydrogen (see Chapter 17). Such associations and H II regions give an indication of the *spiral-arm* structure in the galactic disk (see Figure 18–7). In Chapter 18, we will find that the interstellar hydrogen gas (H I) itself clearly delineates these spiral features.

Recall that a *globular cluster* is a dense spherical cluster of about 10^5 to 10^6 stars; about 120 such globular clusters are known to be associated with our Galaxy, and they are very luminous ($M \approx -4^m$ to -10^m). Since the nearest globular cluster is 3 kpc distant, it is clear that we have here a good probe of galactic distances. The distance to a given globular cluster may be found (a) by the main-sequence fitting technique, (b) by the apparent magnitudes of certain well-known types of stars in the cluster, and (c) by the apparent angular diameter of the cluster. When the distance and direction of globular clusters are plotted, we find that they define a *spherical* system of radius ≈ 15 kpc, with its center 10 ± 1 kpc away in Sagittarius. Harlow Shapley discovered this effect in 1917, and correctly concluded that the Sun is far from the galactic center (the center of the globular system); the accepted current figure puts the Sun at about 10 kpc* from the center, and gives the galactic disk a radius of 15 kpc. This interpretation receives strong support from the observation of a globular cluster system, of similar size and shape, centered upon the middle of the Andromeda galaxy (see Figure 13–3).

13–3 STELLAR POPULATIONS

In Chapter 12, we mentioned that stars could be assigned to different *stellar populations*, characterized by their observed metal abundances (i.e., Z values). In our Galaxy, we find that the stars in the globular clusters are extremely metal-deficient ($Z \lesssim 0.001$)—we call these stars *extreme Population II;* most of the individual stars seen far from the galactic plane (in what is called the *galactic halo*) are of this population. Within about 500 pc of the galactic plane, the spatial density of stars has increased so markedly that we speak of the pancake-shaped *galactic disk* with its *bulge* at the galactic center. Here the metal abundance is much greater than in Population II stars, and it *increases* as we approach the galactic plane—we call these the *disk Population* stars. Both Population II and disk Population stars are very old (billions of years), and their brightest representatives are *red giants*; however, in the galactic plane (especially in the spiral arms) we find the young, luminous, blue stars of *Population I* (greatest metal abundance; $Z \gtrsim 0.01$). Population I stars are seen in *open (galactic) clusters*, O and B associations, and near concentrations of interstellar dust and gas. It was W. Baade in 1944 who first distinguished between these various stellar populations.

Hence, the abundance of heavy elements increases continuously from the halo to the spiral arms buried within the disk. In Chapter 15, we shall see that hydrogen and helium are processed into heavier elements in the centers of stars, and returned to the interstellar medium upon the demise of these stars. Therefore, it appears that our Galaxy began as a spherical cloud of H and He, the metal-poor Population II stars formed, the cloud was enriched with metals as it collapsed toward the galactic plane, and only in the denser regions of the spiral arms is star formation now producing the metal-rich stars of Population I.

This picture seems to be a good representation of the Andromeda galaxy, as well as other spiral galaxies. When such a galaxy is photographed in red light, the globular clusters and the red giants of the disk Population are emphasized; in blue light, the hot young stars of Population I are seen tracing out their spiral-arm patterns (see Figure 13-11).

13–4 GALACTIC DYNAMICS/SPIRAL FEATURES

We have described the stellar content of our Galaxy, but what about the *dynamics* of the system? (Stellar motions are considered in detail in Chapter 14.)

---

* There are strong indications today that this distance may be somewhat smaller, say 8 kpc.

A   **B**

Figure 13–11.   *Stellar Populations.* The spiral galaxy M51 is shown in blue light (**A**), and in red light (**B**). Hot young Population I stars delineate the spiral features in the blue photograph, while the old disk Population and Population II stars form the basic galactic disk and halo in the red photograph. (From Fritz Zwicky; courtesy of Hale Observatories)

Since theoretical models based upon Newtonian gravitation account well for the observed properties of globular clusters, which contain almost a million stars, we will assume that our Galaxy is a system bound together by the mutual gravitational attraction of its stars. The spherical part of the system (i.e., the halo with its hosts of globular clusters) is then analogous to a spherical stellar cluster. The flattened disk, however, clearly reminds us of a *rapidly-rotating* entity; the spiral arms reinforce this impression.

Every star associated with our Galaxy moves within the gravitational attraction produced by all of the other stars. Let us assume *circular* stellar orbits about the galactic center in the galactic plane. In particular, our Sun is in such an orbit, with a radius of 10 kpc. In Chapters 14 and 18, it is shown that (a) near the galactic center, the orbital angular speed ω (radians per second) is approximately constant—we term this *rigid-body rotation*—, while (b) near the Sun, the circumgalactic speed *decreases* with distance from the center of the Galaxy and the Sun's speed is about 250 km sec^{-1} towards $\ell = 90°$. Now it is easy to show (see the problems at the end of this chapter) that a uniform sphere of stars will rotate slowly as a rigid body. At the Sun's position in the Galaxy, however, we are far beyond this central core of stars, and we may approximate the solar motion as a circular Keplerian orbit about a massive central body of mass M_G. Since the centripetal acceleration maintaining this circular orbit is produced by the gravitational attraction between the core (M_G) and the Sun ($M_\odot$), we have:

$$v_\odot{}^2/R = GM_G/R^2 \qquad (13\text{–}7)$$

where R is the distance to the galactic center, $v_\odot$ is the Sun's circular speed, and G is the constant of universal gravitation. Equation (13–7) may be evaluated to give the *effective mass of our Galaxy*:

$$
\begin{aligned}
M_G &= v_\odot{}^2 R/G \\
&= (2.5 \times 10^7 \text{ cm sec}^{-1})^2 (3.1 \times 10^{22} \text{ cm})/(6.7 \times 10^{-8} \text{ cm}^3 \text{ gm}^{-1} \text{ sec}^{-2}) \\
&= 2.9 \times 10^{44} \text{ gm} = 1.5 \times 10^{11} M_\odot \qquad (13\text{–}8)
\end{aligned}
$$

Hence, our Galaxy consists of approximately 100 billion stars; the most massive galaxies known have masses near $10^{13} M_\odot$. From equation (13–7), we also note that the circular speed decreases with distance from the galactic center as:

$$v \propto R^{-1/2}$$

this behavior is in reasonable accord with the observations.

Today we believe that the spiral arms of our Galaxy are also produced by dynamical effects. If this were not the case, these features would *wind up* and disappear (as shown in Figure 13–12) in only a few revolutions of the Galaxy. Now

Figure 13–12. *Spiral Arms Winding up.* Here we show six equispaced "stars" orbiting the galactic center with Keplerian speeds. Within two cycles of rotation ($t = 2T$), these nondynamical spiral arms are so tightly wound as to be undetectable. **A,** Initial line of stars. **B,** Reasonable spiral pattern half a revolution later. **C,** A tight spiral after one full revolution ($t = T$). **D,** Indecipherably wound following two revolutions. Since spiral arms are observed in large numbers of galaxies, and therefore must represent a long-lived or self-perpetuating characteristic, such winding up is *not* believed to occur (see text).

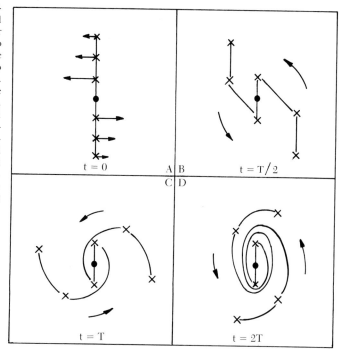

the Galaxy is about 10^{10} years old, and it rotates once every 250 million years ($\approx 2\pi R/v_\odot$), so that it would be very fortuitous if we observed spiral structure which was *not* a steady dynamical feature of the galactic disk. A spiral arm is now thought to be a manifestation of a rotating *density wave* in the galactic disk (see Chapter 18). The disk is initially unstable to density perturbations, which can grow and gravitationally attract material along spiral paths; but these waves rotate only half as fast as the disk, so that material passes through the density pattern in the direction of galactic rotation. W. W. Roberts, Jr. has recently shown that star formation can be triggered as the dust and gas permeating the disk are compressed at the spiral-arm feature; hence, we are beginning to understand why hot young Population I stars delineate spiral structure so well.

13–5   A MODEL OF THE GALAXY

Let us close this chapter by elaborating a schematic model of our Galaxy, which ties together all of the observations and considerations which have been presented above. Figure 13–13 depicts both a side-view and a top-view of the model.

The entire Galaxy is contained within its spherical *halo*, with a diameter of approximately 30 kpc. Here reside (a) the few old Population II halo stars, and (b) about 120 globular clusters. These objects define the spherical envelope, while

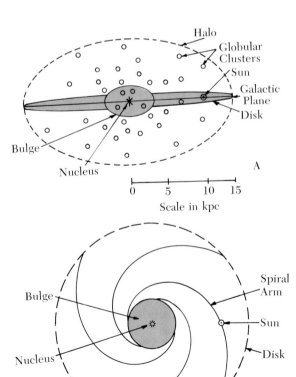

Figure 13–13. *A Model of Our Galaxy.* **A,** Side view, showing spheroidal halo with globular clusters, and oblate bulge, disk, and plane. **B,** Top view, showing spiral pattern in disk, with nucleus at center of bulge. Note the Sun's position (⊙) in the plane near the periphery.

moving along highly eccentric orbits about the galactic center with orbital periods near 1 to 3 × 10^8 years—note the close analogy of this system to the spherical comet cloud centered upon our Sun (Chapter 5).

Bisecting the spherical halo is the circular *galactic plane*. As we approach the galactic plane, the spatial density of stars increases, and their metal abundance rises. We have now reached the oblate spheroidal *galactic disk*, with a perpendicular thickness of about 1 kpc and the large *central bulge* at its center. Most of the mass of the Galaxy ($\approx 10^{11} M_\odot$) resides here, in disk Population stars which move in low-eccentricity orbits about the galactic center.

At the very center of the Galaxy is the small (<100 pc in diameter) massive (about $10^8 M_\odot$) *nucleus*. In other spiral galaxies (e.g., Andromeda) the nucleus has a star-like appearance, but in our Galaxy it is hidden from our view by the interstellar obscuration in the galactic plane. Nevertheless, in 1945, J. Stebbins and A. E. Whitford detected an elliptical region (of extent 8° × 4°) at the galactic center at the infrared wavelength of 10,300 Å; in 1966, E. E. Becklin and G. Neugebauer at the California Institute of Technology, observing at wavelengths near 2 microns ($\approx$ 20,000 Å), determined the nucleus to be a core 30″ in diameter (1.5 pc in linear dimensions!) with several nearby sources in a sphere of radius 10 pc. The galactic nucleus is undoubtedly an extremely compact stellar cluster, wherein events of great violence (explosions) are and have been taking place.

At the midline of the disk is the *galactic plane*, with its veneer of *spiral arms* in a region only about 500 pc thick (consider how thin this really is; the ratio of diameter to thickness is about 60:1). Newborn stars of Population I signal the presence of the spiral density waves propagating around the galactic plane, as a result of their high luminosities and short lifetimes. Myriad clouds of hydrogen gas and dust occupy the galactic plane, with some tendency to collect near the spiral features. The *interarm* regions are not void, but have their quota of disk Population and old Population I stars.

Our Sun, a Population I star about 5 billion years old, is located 10 kpc from the galactic center, in association with one of the spiral arms. Although it is not exactly in the galactic plane, it is only a negligible distance above it. Carrying the Solar System with it, the Sun orbits the galactic center once every 250 million years in a path of very low eccentricity. As we gaze out upon the "celestial flywheel" which is our Galaxy from the Earth, we see the neighboring stars participating with us in the galactic rotation towards $\ell = 90°$, and the breathtaking band of the Milky Way encircling the sky from its center in Sagittarius.

Problems

13–1.   Using a simple diagram, explain why our Galaxy appears as the Milky Way in the night sky.

13–2.   The Andromeda galaxy is situated at the galactic latitude $\ell = -20°$, and its plane is inclined about 80° to the celestial sphere. What is the inclination of our Galaxy as seen from Andromeda?

13–3.   Using Figure 13–5, describe the various aspects of the Milky Way on evenings of different seasons for an observer in London, England.

Use the altitude-azimuth horizon system of coordinates, and mention the tilt of the Milky Way to the horizon and the observability of Sagittarius.

13–4. The differential star counts of section 13–2(a, i) may be extended to *integral* star counts, if we count *all* of the stars in a certain line-of-sight out to some apparent magnitude *m*.

Assume that stars are uniformly distributed in space.

(a) Show that for the integral count ratio, $\overline{N}(m + 1)/\overline{N}(m) = 3.98$.

(b) What simplifying assumption has been made in part (a)?

(c) Make a schematic diagram, plotting $\log [\overline{N}(m)]$ versus *m*, and show what effect each of the following would produce:

(i) a uniformly-distributed interstellar obscuration of $1\overset{m}{.}0$ per kpc,

(ii) a strongly-absorbing interstellar "dust" cloud localized at the apparent magnitude m_{cloud}.

13–5. What stellar population would you expect to find (and why):

(a) in the nucleus of our Galaxy?

(b) in the spiral arms of the Andromeda galaxy?

(c) in the Pleiades star cluster?

(d) in intergalactic space (i.e., beyond the halo)?

(e) in the galactic bulge?

13–6. (a) A globular cluster is in elliptical orbit ($e = 0.9$) about the center of our Galaxy, reaching *apogalacticon* (furthest distance from the center) at the distance 20 kpc. What is the *perigalacticon* (nearest) distance, and how long will this cluster require to complete one orbit?

(b) What is the approximate speed of escape from the Galaxy, in the solar neighborhood, if the Sun's circular orbital speed about the galactic center is 250 km sec^{-1}?

13–7. Consider the center of our Galaxy to be a spherical star cluster of radius *R* and uniform mass density ρ.

(a) What is the total mass (M) of this cluster?

(b) What is the mass contained within the sphere of radius $r < R$?

(c) In terms of *M*, what is the angular speed ω of a star in circular orbit at a distance r ($< R$) from the cluster's center?

Note that $\omega = v/r$, where *v* is the circular orbital speed of the star (e.g., in km sec^{-1}).

(d) Using an analogy, explain why we refer to the result of part (c) as *rigid-body rotation*.

13–8. Assume that the galactic disk may be approximated by a plane-parallel slab, 500 pc thick and with the galactic plane at its midline. If the Sun is located in the galactic plane,

(a) how many magnitudes of absorption are there at $\ell = 90°$?

(b) how many magnitudes of absorption at the general galactic latitude ℓ?

(c) explain why the region $\ell \lesssim 10°$ is called the "zone of avoidance" (i.e., essentially total obscuration) in terms of apparent magnitudes.

13–9. In the direction perpendicular to the galactic plane, approximately how thick (in kpc) is the galactic bulge? (Hint: Use Figure 13–6.)

13–10. In this chapter we mentioned the provocative suggestion that stars are formed when disk materials (gas and "dust") catch up with a

density wave. Assuming that the newborn stars continue about the galactic center with the circular speed appropriate to their distance:
(a) how far will an O star move from the spiral arm of its birth in one million years?
(b) why are we surprised to find our Sun "in" a spiral arm (or near one, at least)?

Reading List

Becker, W., and Contopoulos, G., (eds.): *The Spiral Structure of Our Galaxy.* IAU Symposium No. 38, Basel, Switzerland. Dordrecht, Holland, D. Reidel Publishing Co., 1970.

Blaauw, Adriaan, and Schmidt, Maarten, (eds.): *Galactic Structure.* Chicago, The University of Chicago Press, 1965.

Bok, Bart J.: *The Distribution of the Stars in Space.* Chicago. The University of Chicago Press, 1937.

Bok, Bart J., and Bok, Priscilla F.: *The Milky Way.* Third Ed. Cambridge, Massachusetts, Harvard University Press, 1957.

Goldberg, L., and Aller, L. H.: *Atoms, Stars, and Nebulae.* New York, McGraw-Hill Book Company, 1943.

Kurth, Rudolf: *Introduction to Stellar Statistics.* London, Pergamon Press, 1967.

Lequeux, J.: *Structure and Evolution of Galaxies.* New York, Gordon and Breach Science Publishers, 1969.

Mihalas, D., and Routly, P. M.: *Galactic Astronomy.* San Francisco, W. H. Freeman and Company, 1968.

O'Connell, D., (ed.): *Stellar Populations.* New York, Interscience Publishers, 1958.

Page, Thornton, and Page, Lou Williams, (eds.): *Stars and Clouds of the Milky Way.* New York, The Macmillan Company, 1968.

Shapley, Harlow: *The Inner Metagalaxy.* New Haven, Yale University Press, 1957.

Chapter 14
Stellar Motions

In Chapter 13, we saw that our Solar System resides in a rotating spiral galaxy, known as the Milky Way Galaxy, and that our Sun is one of its multitudinous stars. In this chapter, we study the observable motions of the stars in our Galaxy (especially those in the solar neighborhood), verifying the galactic rotation and evaluating its characteristics quantitatively. We begin by discussing the observational properties of stellar motions. Then we consider the random deviations of these motions from the average galactic motion. Finally, the characteristic rotational velocity of our Sun about the center of the Galaxy and differential galactic rotation are investigated.

Before we begin, a word of clarification concerning coordinates and reference systems is in order. In Chapters 2 and 13, we referred to *geocentric* (Earth-centered) coordinate systems, such as the system of *right ascension and declination*, and to the *galactic coordinate system*. It should be readily apparent that *dynamical* investigations are best referred to a *heliocentric* (Sun-centered) coordinate system (see Chapters 3 and 5). In practice, therefore, astronomers differentiate between *apparent* celestial positions (i.e., the geocentric positions of objects determined from Earth-bound observations) and *true* celestial positions (i.e., the heliocentric positions relative to our Sun). True stellar positions (e.g., heliocentric right ascension and declination) differ from apparent positions, since the latter exhibit *periodic* displacements, such as aberration orbits and geocentric parallax orbits, as a result of the Earth's orbital motion around the Sun. It is a straightforward procedure to correct terrestrial observations to a heliocentric basis,* since the Earth's orbital elements are well

---

* In more advanced discussions of celestial mechanics, the center-of-mass of our Solar System, called the *barycenter*, is taken as the origin of the *barycentric coordinate systems*. In direct analogy with our discussion of the Earth-Moon system (see Chapter 6), we find that all members of the Solar System, *including the Sun*, orbit about this barycenter. The barycenter is located approximately 10^6 km ($1.1R_\odot$) from the center of the Sun.

known. Henceforth, we will speak only of true or heliocentric celestial positions. In the second section of this chapter we shall introduce yet another dynamical reference system, called the *Local Standard of Rest*, which is based upon the rotational dynamics of our Galaxy. The reader should be careful to discern which of the various coordinate and reference systems is being used at each point in the following discussion.

14–1   COMPONENTS OF STELLAR MOTIONS

Stellar motions are manifestations of the velocities of stars through space. These velocities are *vectors*, which may be resolved into two perpendicular components (as we saw in Chapter 3): the radial velocity along the line-of-sight, and the tangential velocity in the plane of the sky or celestial sphere.

(a)   Radial Speed

The *radial speed* of a star is its speed of approach or recession; therefore, it is easily obtained from the Doppler shift of the stellar spectral lines. To determine this Doppler shift, a laboratory comparison spectrum is photographed adjacent to the stellar spectrum, and the relative positions of lines in the two spectra are measured (see Figure 14–1). The *comparison spectrum* is an emission line spectrum of an element for which the wavelengths of the spectral lines are well established, such as iron (Fe) or neon (Ne). We must first determine the nature of the stellar spectral absorption or emission lines—for instance, the hydrogen Balmer series is usually easily recognized—then we may compare the observed wavelength of each line λ_o with the emitted-laboratory or rest-wavelength λ_e. The measured Doppler shift, $\Delta\lambda = |\lambda_o - \lambda_e|$, permits us to deduce the radial speed of the star from the Doppler formula [equation (8–13)]:

$$v_r = \left(\frac{\Delta\lambda}{\lambda_e}\right)c \qquad (14\text{–}1)$$

where c is the speed of light (see Chapter 8). When $\lambda_o > \lambda_e$, the spectrum is *redshifted*, and v_r is the radial speed of *recession*; when $\lambda_o < \lambda_e$, the spectrum is *blueshifted*, and the star is *approaching* us. To refer these motions to the Sun, we must correct for the component of the Earth's orbital velocity (speed $= 30$ km sec^{-1}) along the line-of-sight to the star.

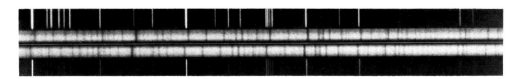

Figure 14–1.   *Spectrograms Showing Radial Velocity.* Two spectrograms of the spectroscopic binary α Geminorum show the difference in the velocity shifts at two different times. (Lick Observatory)

The radial speed may be determined for *any* star for which a spectrum may be obtained. This means that the distance to the star is irrelevant, for it is only the apparent magnitude of the star which determines whether it is bright enough to produce a spectrum when observed through a telescope-spectrograph combination. Hence, radial speeds may be obtained for very distant bright stars. From equation (14–1), we see that we cannot measure radial speeds smaller than 1 km sec^{-1}, since $\Delta\lambda$ usually cannot be discerned to accuracies better than ± 0.01 Å at visible wavelengths.

(b) Proper Motion

The motion of a star in the plane of the celestial sphere is called the *proper motion μ*, and it is usually expressed in seconds of arc per year ($''$ yr^{-1}). For a given speed perpendicular to the line-of-sight, the proper motion will be greater the closer the star is to us. Hence, we see an obvious analogy with trigonometric parallax (Chapter 10). Conversely, very distant stars will exhibit no measurable proper motion, and they may therefore be used as *reference* or *background stars*. Most proper motions are exceedingly small, with the largest known being that of Barnard's star, an exceptional 10$''$ yr^{-1} (see Figure 14–2). Compared to the cyclic nature of parallax orbits, proper motions have the distinct advantage of being *cumulative*: measurements may be made many years apart, so that small annual angular displacements can accumulate to an easily measured amount! An accuracy of $\pm 0\overset{''}{.}003$ yr^{-1} is attainable when observations are spaced over decades, but great care must be exercised in the selection of standards of reference. As a result of galactic rotation (see section 14–4), the ideal reference objects lie *beyond* the Galaxy; in fact, the present fundamental reference system is in the process of being based upon the external galaxies. Since observations of proper motions are made many years apart, precautions similar to those taken for parallax observations are necessary. In particular, corrections must be made for stellar parallax and for the aberration of starlight. In addition, *fundamental* reference systems are required over the entire celestial sphere, and these are not easily established. Nevertheless, proper motion measurements are well worth their trouble, for they are of paramount importance to our knowledge of the structure of the Galaxy.

(c) Tangential Speed

The proper motion of a star is due to that star's *tangential speed, v_t*. This is its linear speed in the direction perpendicular to the line-of-sight. To convert the angular measure of proper motion into a linear speed (km sec^{-1}) transverse to the line-of-sight, we must know the distance d to the star, for

$$v_t = d \sin \mu \cong d\mu \qquad (14\text{–}2)$$

The last part of equation (14–2) follows because μ is very small (less than 5×10^{-5} radian yr^{-1}). In this equation we must be careful to remember our units, for d is normally given in parsecs and μ in $''$ yr^{-1}. Equation (14–2) gives v_t in (pc yr^{-1}) when μ is in (radians yr^{-1}), but v_t is found in (AU yr^{-1}) when μ is in ($''$ yr^{-1});

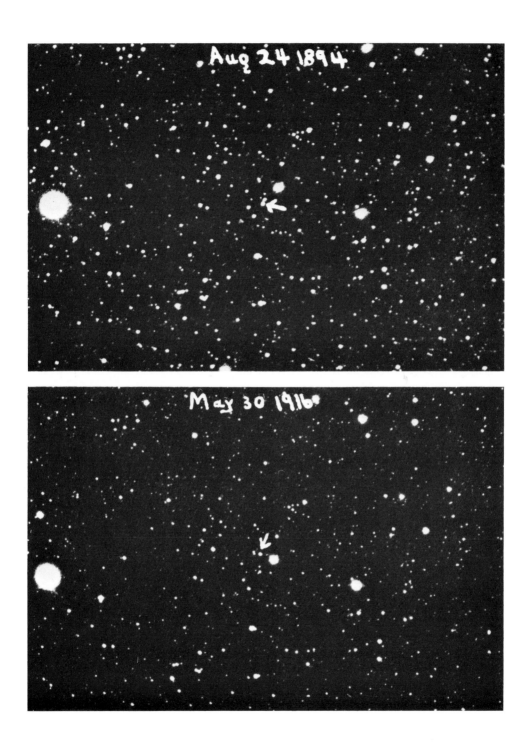

Figure 14–2. *Proper Motion of Barnard's Star.* Here a "first epoch" plate (August 24, 1894) is compared with a "second epoch" plate of the same region of the sky, taken 22 years later (May 30, 1916). Barnard's star (indicated by the arrows) moves $10''$ yr^{-1}. (Yerkes Observatory)

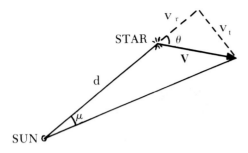

Figure 14–3. *The Space Velocity Resolved into Components.* A star (*) is a distance d from the Sun ($\odot$), and moves with space velocity **V**. This space velocity is resolved into a radial component v_r and a tangential component v_t (causing the proper motion μ). The angle between the space velocity and the line-of-sight is θ.

we use d in parsecs in either case. By applying the appropriate conversion factors (AU → km, and yr → sec), we find that

$$v_t = 4.74\ \mu''d = 4.74\ (\mu''/\pi'') \quad \text{km sec}^{-1} \qquad (14\text{–}3)$$

where the distance d is given in parsecs, the parallax π'' in seconds of arc, and the proper motion μ'' in $''$ yr^{-1}. For example, a star whose distance is 100 pc and whose proper motion is $0''.1$ yr^{-1} has a tangential speed of 47.4 km sec^{-1}.

(d) Space Motion

By referring to Figure 14–3, we see how the *space velocity* **V** of a star with respect to the Sun has been decomposed into two perpendicular components: (1) the radial velocity, whose magnitude is v_r (the radial speed), and (2) the tangential velocity, with magnitude v_t (the tangential speed). From the Pythagorean theorem (and the law of vector addition in Chapter 3) we find

$$V^2 = v_r^2 + v_t^2 \qquad (14\text{–}4)$$

Once again we caution the reader that all speeds in equation (14–4) must be in the same units; this implies [from equation (14–3)] that we must know the distance to the star. The angle which the true or space velocity makes with the line-of-sight is θ, and it is readily found from

$$\tan \theta = v_t/v_r$$

14–2 THE LOCAL STANDARD OF REST AND PECULIAR MOTIONS

(a) The Local Standard of Rest and Solar Motion

In our Galaxy, there are two physically-preferred frames of reference. The first, to which we will refer implicitly throughout Part 4, is the *galactocentric* system: centered at the nucleus of the Galaxy, its reference plane is the galactic plane and its reference axis is the galactic rotation axis. We are primarily interested in the second system here—the so-called *Local Standard of Rest* (LSR).

The LSR may be defined in two, nonequivalent, ways; in practice, the two frames of reference are essentially the same. The *dynamical LSR* is that reference frame, instantaneously centered upon the Sun, which moves in a *circular* orbit about the galactic center at the circular speed appropriate to its position in the Galaxy. Therefore, all stars in the solar neighborhood which are in circular galactic orbits are essentially *at rest* in the dynamical LSR. Any deviations from circular motion, *in the solar neighborhood*, will appear as peculiar stellar motions [see section 14–2(b)] *with respect to the dynamical LSR*. On the other hand, the *kinematical LSR* is that reference frame, instantaneously centered upon the Sun, in which the *space velocities* of all the stars in the solar neighborhood *average* to zero. Hence, the kinematical LSR is the frame of the mean motion of the local stars about the center of the Galaxy. Note that the kinematical LSR is *not unique*, for we have not specified the stars or the spatial volume which we are to velocity-average. Nevertheless, the two different LSRs are practically equivalent if we consider only the region within about 100 pc of the Sun, since the vast majority of the nearby stars are in almost prefect circular orbits about the galactic center.

The Sun's galactic orbit is not perfectly circular. Therefore, with respect to the LSR there is a *solar motion* of 20 ± 1 km sec^{-1} towards the constellation Hercules ($\alpha = 18^{\mathrm{h}}4^{\mathrm{m}} \pm 7^{\mathrm{m}}$, $\delta = +30° \pm 1°$). We say that, on the celestial sphere, the Sun is moving toward the *solar apex* and away from the *solar antapex*. The nature and extent of the solar motion were first demonstrated by William Herschel in 1783 using statistical methods; let us paraphrase his analysis in modern terminology.

The stars in the solar neighborhood exhibit *peculiar motions* [see section 14–2(b)] with respect to the LSR; that is, they swarm about like sluggish bees. If the Sun were at rest in the LSR, the average of these peculiar velocities would also be zero with respect to the Sun. But if the Sun moves with respect to the LSR, each star will (in addition to its peculiar motion) *reflect* this solar motion to an extent dependent upon that star's position on the celestial sphere (see Figure 14–4).

Figure 14–4. *The Reflected Solar Motion.* This diagram shows how the solar motion, with respect to the LSR, affects eight stars on the celestial sphere. In addition to their peculiar motions, the stars *reflect* the solar motion in their radial (v_r) and proper (μ) motions. Note the apex and antapex.

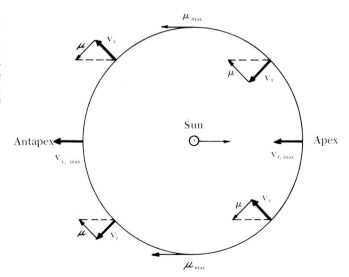

Stars on the great circle 90° from both the apex and antapex will, *on the average,* exhibit the largest proper motions toward the antapex; hence, proper-motion averages (out to about 50 pc) will tell us the locations of the apex and antapex. We may determine the speed of the solar motion by averaging the radial speeds of stars near the apex, and also near the antapex; the average radial speed of approach will be greatest at the apex (another way to locate the apex), and of recession greatest at the antapex (another way to locate the antapex). Finally, note that the average proper motion vanishes at apex and antapex. Since the stars sampled for radial speed can be much more distant than those sampled for proper motion, a different apex location will result from each sample; in practice, the two locations are essentially identical.

(b) Stellar Peculiar Motions

The solar motion was just found by *averaging* stellar peculiar motions *with respect to the Sun.* To find the peculiar motion of a given star *with respect to the LSR,* we must now subtract the reflected solar motion from the *observed* stellar peculiar motion components.* Figure 14–5 illustrates the geometrical basis for these corrections. Let λ be the angle on the celestial sphere between the solar apex and the star.

---

* Remember that the effects of the Earth's orbital motion have already been eliminated.

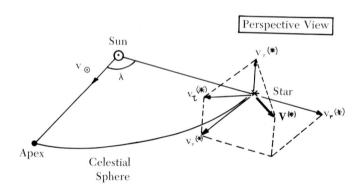

Figure 14–5. *Stellar Peculiar Motion Components.* At the top, a perspective of the star's (∗) space velocity **V**, its radial component v_r, and its tangential component v_t (further decomposed into the tau v_τ and upsilon v_υ components). At the bottom, the solar motion $v_\odot$ is split into radial ($v_\odot \cos \lambda$) and upsilon ($v_\odot \sin \lambda$) components.

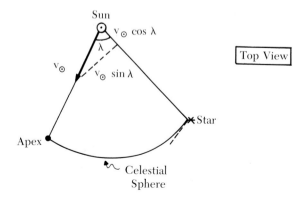

Then the star's peculiar radial speed with respect to the LSR is simply:

$$v_r(*) = v_r \text{ (obs)} + v_\odot \cos \lambda \tag{14–5}$$

where v_r (obs) is the heliocentric radial speed and $v_\odot = 20$ km sec^{-1} is the speed of the solar motion. Only at $\lambda = 90°$ will we observe the true stellar peculiar radial speed.

On the other hand, the tangential motion is a more complicated affair. By "tangential" we mean that component in the plane of the celestial sphere—perpendicular to the line-of-sight to the star—that is, the tangential speed v_t (obs) given by the heliocentric proper motion μ'' in equation (14–3). Now the direction to the solar apex and the line-of-sight to the star define a plane which intersects the celestial sphere in a great circle (joining the apex to the star). The star's tangential velocity may be decomposed into two mutually perpendicular components: the upsilon component $[v_v(*)]$ in this plane, and the tau component $[v_\tau(*)]$ perpendicular to the plane, where

$$v_t{}^2 = v_v{}^2 + v_\tau{}^2$$

The tau component is clearly unaffected by the solar motion, so that

$$v_\tau(*) = v_\tau \text{ (obs)} = 4.74\left(\frac{\tau'' \text{ (obs)}}{\pi''}\right) \tag{14–6a}$$

However, the upsilon component is influenced by a component of the solar motion:

$$v_v(*) = v_v \text{ (obs)} + v_\odot \sin \lambda = 4.74\left(\frac{v'' \text{ (obs)}}{\pi''}\right) + v_\odot \sin \lambda \tag{14–6b}$$

In equations (14–6), the proper motions (in $''$ yr^{-1}) in the tau and upsilon directions are denoted by τ'' and v'', respectively; the speeds v must be specified in km sec^{-1}. If equations (14–6) are written entirely in terms of angular proper motions, we have:

$$\tau''(*) = \tau'' \text{ (obs)} \tag{14–6a'}$$

$$v''(*) = v'' \text{ (obs)} + \left(\frac{v_\odot \sin \lambda}{4.74}\right)\pi'' \tag{14–6b'}$$

From equations (14–5) and (14–6), we may deduce the star's peculiar speed relative to the LSR as:

$$V^2(*) = v_r{}^2(*) + v_t{}^2(*) = v_r{}^2(*) + v_\tau{}^2(*) + v_v{}^2(*) \tag{14–7}$$

The tangential component depends upon the stellar parallax (equations (14–6), so that stellar peculiar motions relative to the LSR may be determined only for *nearby* stars (within approximately 100 pc).

(c) **Peculiar Motion Parallaxes**

Three statistical methods of determining stellar parallax (π'') are evident from our previous discussions.

Inverting equation $(14\text{--}6b')$, we may write:

$$\pi'' = \left(\frac{4.74}{v_\odot \sin \lambda} \right) [v''(*) - v''(\text{obs})] \qquad (14\text{--}8)$$

For a single star, equation $(14\text{--}8)$ tells us nothing, since we do not know $v''(*)$—the upsilon component of that star's peculiar proper motion. However, for a large number of stars in a particular region of the sky (λ), we may frequently assume that their $v''(*)$'s are *random*,* so that they average (indicated by a *bar* on top) to zero:

$$\overline{\pi}'' = -\left(\frac{4.74}{v_\odot \sin \lambda} \right) \overline{v''(\text{obs})} \qquad (14\text{--}9)$$

If the region spans a significant *range* of angular distances λ from the solar apex, we must be more subtle. Let us write equation $(14\text{--}8)$ as

$$v_i''(\text{obs}) = v_i''(*) - \left(\frac{v_\odot}{4.74} \right) \pi_i'' \sin \lambda_i \qquad (14\text{--}10)$$

for the ith star being considered. Multiplying through on both sides by $\sin \lambda_i$, and summing over all the stars (i.e., summing over all i), we find:

$$\overline{v''(\text{obs}) \sin \lambda} = -\left(\frac{v_\odot}{4.74} \right) \overline{\pi''(\sin^2 \lambda)}$$

or

$$\overline{\pi}'' = -\left(\frac{4.74}{v_\odot} \right) \left[\frac{\overline{v''(\text{obs}) \sin \lambda}}{\overline{\sin^2 \lambda}} \right] \qquad (14\text{--}11)$$

Equation $(14\text{--}11)$ resulted only because of two extremely important facts: (a) $\overline{v''(*) \sin \lambda}$ vanishes, since the random projections of $v''(*)$ are also random with a mean of zero (draw a diagram to convince yourself of this); and (b) $\overline{\pi''(\sin^2 \lambda)} = \overline{\pi}''\overline{(\sin^2 \lambda)}$ since both π_i'' and $\sin^2 \lambda_i$ are random and *uncorrelated*. Equations $(14\text{--}9)$ and $(14\text{--}11)$ give us the *mean parallax* of the group of stars being considered.[†]

The quantity actually computed in equation $(14\text{--}11)$ is

$$H'' = \left(\frac{v_\odot}{4.74} \right) \pi'' \approx 4.2\pi'' \qquad (14\text{--}12)$$

---

* Note section 14–3, where this assumption is completely violated.

† More advanced texts refer to this as the *least-squares* mean parallax.

which we term the *secular parallax*. It is caused by the solar motion through the LSR, at the rate of about 4.2 AU yr^{-1}; here we have a continuously-growing baseline for parallax determinations, with a larger annual value than that provided by the Earth's orbital motion. A star with no tangential velocity, at $\lambda = 90°$, would be observed to exhibit this secular parallax each year.

Finally, in direct analogy to the mean parallaxes determined from the upsilon (υ) components, we may use the tau (τ) components—which are unaffected by solar motion—to find *statistical parallaxes*. Assuming that our group of stars has random peculiar motions with respect to the LSR, we clearly have [from equation (14–6a)]:

$$\overline{|v_r(*)|} = \overline{|v_\tau(*)|} = \frac{4.74}{\pi''} \overline{|\tau'' \text{ (obs)}|} \tag{14–13}$$

where the vertical bars denote *absolute values* (positive, without respect to the actual sign). Therefore,

$$\boxed{\overline{\pi''} = \frac{4.74}{\overline{|v_r(*)|}} \overline{|\tau'' \text{ (obs)}|}} \tag{14–14}$$

and the $v_r(*)$'s may be computed directly from equation (14–5).

Table 14–1 summarizes this somewhat complicated discussion of parallaxes. Once we have found the average parallax $\overline{\pi''}$ for a group of stars, the average stellar apparent magnitude $\overline{m}$ gives us the *average* stellar absolute magnitude $\overline{M}$, via:

$$\overline{M} = \overline{m} + 5 \log \overline{\pi''} + 5 \tag{14–15}$$

Equation (14–15) leads to a believable result, only if we select stars (a) of a particular spectral type and luminosity classification, and (b) within a narrow range of apparent magnitudes. The *method of mean parallaxes* has been applied to determine the absolute magnitudes of A stars, RR Lyrae stars, and Cepheid variables (concerning the latter two, see Chapter 16). Note that, in the *method of statistical parallaxes*,

Table 14–1.   Statistically-Determined Stellar Parallaxes

Type of Parallax		Equation	Physical Basis
Annual		π''	The Earth's solar orbit
	Mean	$\overline{\pi''_v} = -\left(\dfrac{4.74}{v_\odot}\right)\overline{\dfrac{\upsilon'' \text{ (obs) } \sin \lambda}{(\sin^2 \lambda)}}$	The solar motion and upsilon components of proper motion
	Secular	$H'' = \left(\dfrac{v_\odot}{4.74}\right)\pi''$	Reflected solar motion 90° from the solar apex
Statistical		$\overline{\pi''_\tau} = \dfrac{4.74\,\overline{|\tau'' \text{ (obs)}|}}{\overline{|v_r(*)|}}$	Radial speeds and tau components of proper motion

we must choose stars at approximately the same distance for both the radial and proper motions; otherwise, the proper motion sample will be biased towards nearer stars compared to the radial velocity sample.

14–3 MOVING CLUSTERS

We briefly mentioned *moving stellar clusters* in Chapter 10. Recall that such a cluster is a gravitationally-bound group of stars travelling through the Galaxy. Hence, all the *cluster members* exhibit the *same* peculiar motion relative to the LSR,* and we can identify the members of a given cluster by this property. Random observational errors (a) permit a small number of noncluster-members to slip into our chosen sample, and (b) limit us—via equation (14–3)—to those clusters nearer than about 500 pc.

Consider a small, or distant, cluster which subtends a *small* solid angle on the celestial sphere. Its radial speed v_r and angular proper motion μ''—with respect to the LSR—may be obtained by studying a *single* star (and correcting for the reflected solar motion). To avoid errors, due to the possible inclusion of noncluster-members, we usually take averages over several stars. But the *direction* of the cluster's space velocity is unknown (it is too distant for proper motion measurements), so that we *cannot* find the cluster distance from equation (14–3). Some other distance criterion must be used in this case (see the following paragraphs).

On the other hand, the situation is much better when we have a cluster (usually nearby) which subtends a *large* solid angle on the sky. Then (see Figures 14–6 and 14–7), since all cluster members are moving in the same spatial direction, their proper motions appear to converge toward (or diverge from) a single point on the celestial sphere—we call this the *convergent* (or *divergent*) *point*. This phenomenon is directly analogous to the *radiant* of a meteor shower (see Chapter 5), where perspective effects cause an incoming parallel stream of meteoroids to appear to radiate from a single point on the sky. Again, we know the radial speed v_r of the moving cluster, once we have determined it for a single member star, but now we also know the direction of the cluster's space motion. When we extend the proper

---

* The stellar motions are *completely correlated*, not random.

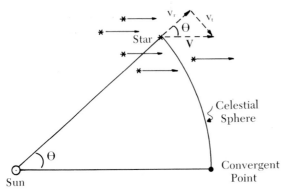

Figure 14–6. *Moving Cluster Geometry.* The cluster of stars (*) is moving with space velocity **V**, and the proper motions appear to point toward the convergent point. The angle θ from the convergent point to a member star is the angle between the space velocity and our line-of-sight to the star. Therefore, from trigonometry: $v_r = V\cos\theta$ and $v_t = V\sin\theta$.

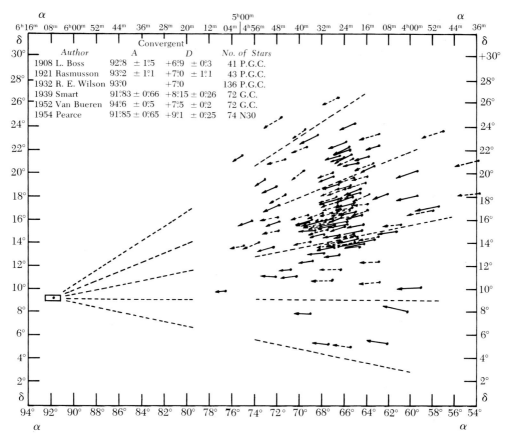

Figure 14–7. *The Hyades.* This most famous moving cluster (in the constellation Taurus) contains about 200 identified members, subtends 25° × 25° on the sky, and converges towards α = 6ʰ8ᵐ, δ = +9°. (From O. Struve, B. Lynds, and H. Pillans: *Elementary Astronomy.* New York, Oxford University Press, 1959.)

motions μ'' of the member stars on the celestial sphere, they intersect at, for example, the convergent point of the cluster (see Figure 14–7). Figure 14–6 shows that the angular distance θ from the convergent point to a member star is the same as the angle between the line-of-sight to that star and the star's space velocity **V**. Hence, we immediately have:

$$V = v_r/\cos\theta \qquad (14\text{--}16)$$

But we know that $v_t = V\sin\theta$, so, using equations (14–3) and (14–16), it gives us the *cluster distance*, via its parallax:

$$\pi'' = 4.74\mu''/v_r\tan\theta \qquad (14\text{--}17)$$

In contrast to the average parallaxes discussed in section 14–2(c), these *moving cluster parallaxes* apply to *individual* stars; therein lies their extraordinary power.

Those moving groups which have received the most attention are: (a) the

Hyades cluster (about 200 members) in the constellation Taurus (see Figure 14–7), (b) the Sirius group (which includes most of the bright stars in the Big Dipper) in Ursa Major, and (c) the Coma group in Coma Berenices. In 1869, R. A. Proctor was the first to note the moving-cluster character of the Hyades and Sirius groups. In 1908, Lewis Boss developed the convergent-point method for moving clusters, applied it to the Hyades, and found a distance of 41 pc (the current best value is 40 ± 1 pc).

Once we have obtained an accurate cluster distance, we can make an *absolute* calibration of the H-R diagram of that cluster (see Chapters 10 and 12); the Hyades distance (obtained by the moving-cluster method) now provides the *fundamental distance scale* upon which all distances greater than about 100 pc* (even to the "limits" of our Universe!) are based. By using the Hyades H-R diagram and the main-sequence fitting technique (Chapters 10 and 12), the scale has been extended to the Praesepe ("Beehive") cluster in Cancer at 158 pc and to the Double Cluster (*h* and χ Persei) in Perseus at 2290 pc—with an accuracy of about 10 per cent in the distances. In later chapters (particularly Chapters 16 and 19), we will see how this distance scale may be further extended to extragalactic objects and the Universe.

To give the reader some feeling for the effects of observational errors, let us consider the uncertainties in the Hyades distance scale. Using $d = 1/\pi''$ in equation (14–17), we have:

$$d = v_r \tan \theta / 4.74\mu'' \tag{14–18}$$

Taking the natural logarithm ($\ln = \log_e$) of both sides gives

$$\ln d = \ln v_r + \ln (\tan \theta) - \ln (\mu'') + \text{constant};$$

then differentiating yields:

$$\frac{\Delta d}{d} = \frac{\Delta v_r}{v_r} + \frac{2 \, \Delta \theta}{\sin (2\theta)} - \frac{\Delta \mu''}{\mu''} \tag{14–19a}$$

Since the uncertainties (Δ) in equation (14–19) are assumed to be *random*, their sign is not important, so that we will use:

$$\left| \frac{\Delta d}{d} \right| = \frac{1}{3} \left[\left| \frac{\Delta v_r}{v_r} \right| + 2 \left| \frac{\Delta \theta}{\sin (2\theta)} \right| + \left| \frac{\Delta \mu''}{\mu''} \right| \right] \tag{14–19b}$$

Now v_r is about 20 km sec^{-1}, and we have already shown that Δv_r is ± 1 km sec^{-1}, so that $|\Delta v_r/v_r| \approx 0.05$; from Figure 14–7 we have $\theta \approx 25°$ and $\Delta \theta \approx \pm 1° = \pm 0.017$ radians, so that $2 |\Delta \theta / \sin (2\theta)| \approx 0.05$; finally, $\mu'' \approx 0''.1$ and $\Delta \mu'' \approx 0''.005$, so that $|\Delta \mu''/\mu''| \approx 0.05$. Therefore, equation (14–19b) tells us that $|\Delta d/d| \approx 0.05$, or the Hyades distance is uncertain by approximately 5 per cent.

---

* Distances less than 50 pc are *based* on trigonometric parallaxes.

14–4 GALACTIC ROTATION

(a) Differential Galactic Rotation and Oort's Formulae

In section 14–2, we said that stars in the solar neighborhood orbiting the galactic center in perfect circular orbits would be *at rest* in the LSR. But this statement implies *rigid-body rotation* (ω = a *constant* angular speed about the galactic center) of our region of the Galaxy; the particles comprising a rigid body remain at *fixed* distances from one another, while every particle moves around the center of the body in the *same* period. Now $\omega = v/r$, where v is the circular orbital speed (e.g., km sec^{-1}) and r is the orbital radius (e.g., kpc); hence, for a rigid body, we have $v \propto r$. Since our Sun orbits beyond the bulk of the mass of the Galaxy, we know that the solar orbit must be approximately Keplerian—but then $v \propto r^{-1/2}$ and $\omega \propto r^{-3/2} \neq$ constant. We say that there is *differential galactic rotation*, when the orbital angular speed is a function of distance from the galactic center, $\omega = \omega(r)$. Let us now follow Jan Oort's brilliant work of 1927, and derive the effects of such differential galactic rotation (particularly in the solar neighborhood). At the end, we will see that these effects are indeed observable.

We first assume *circular* galactic orbits *in* the galactic plane, and confine our attention to this case. Consider Figure 14–8. Here we have defined: (a) R as, for example, the star's distance from the galactic center, R_0 as the Sun's distance from the center, and d as the Sun-star distance; (b) Θ as the star's circular orbital speed and Θ_0 as the orbital speed of the LSR; (c) ℓ as the galactic longitude of the star and α as the angle between the line-of-sight to the star and its orbital velocity; and (d) ω as the star's galactic *angular* speed and ω_0 as the LSR's.

Figure 14–8. *Galactic Rotation/The Oort Geometry.* Here the Sun, the galactic center, and a star define the galactic plane; all orbital motions are circular. The various symbols are defined in the text.

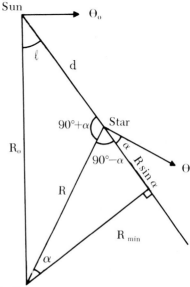

We begin with v_r, the star's radial speed with respect to the LSR. From Figure 14–8, we see that:

$$v_r = \Theta \cos \alpha - \Theta_0 \sin \ell \qquad (14\text{–}20)$$

while the law of sines (see the Mathematical Appendix) gives us:

$$\frac{\sin \ell}{R} = \frac{\sin (90° + \alpha)}{R_0} \equiv \frac{\cos \alpha}{R_0} \qquad (14\text{–}21)$$

But $\omega = \Theta/R$ (and $\omega_0 = \Theta_0/R_0$), so using equations (14–20) and (14–21):

$$\boxed{v_r = R_0(\omega - \omega_0) \sin \ell} \qquad (14\text{–}22)$$

For a rigid rotation, $\omega \equiv \omega_0$, so that $v_r \equiv 0$ (as we stated in the preceding paragraph); differential galactic rotation implies a finite radial speed for the star.

How about v_t, the star's tangential speed with respect to the LSR? Again, from Figure 14–8:

$$v_t = \Theta \sin \alpha - \Theta_0 \cos \ell \qquad (14\text{–}23)$$

now the law of sines gives

$$\frac{\sin \ell}{R} = \frac{\sin (90° - \ell - \alpha)}{d} \equiv \frac{\cos (\ell + \alpha)}{d} \equiv \frac{\cos \alpha \cos \ell - \sin \alpha \sin \ell}{d} \qquad (14\text{–}24)$$

where the last equality follows from the identity: $\cos (x + y) = \cos x \cos y - \sin x \sin y$. We may solve equation (14–24) for $\sin \alpha$, and use equation (14–21) for $\cos \alpha$, to put equation (14–23) in the form:

$$\boxed{v_t = R_0(\omega - \omega_0) \cos \ell - d\omega} \qquad (14\text{–}25)$$

Now rigid rotation $(\omega = \omega_0)$ implies: $v_t = -d\omega_0$, which simply means that the LSR coordinate system is *inertial*—its axes always point in the same directions *relative to all other galaxies and the Universe*, and are not "glued" or fixed to the rigid body.

The *Oort Formulae*, equations (14–22) and (14–25), are completely general;* we will use them extensively in section 14–4(c) and in Chapters 17 and 18. Let us here, however, specialize to the solar neighborhood, or equivalently where $d \ll R_0$.

---

* For concentric, circular, coplanar orbits.

Near the Sun, ω is approximately equal to ω_0, and in fact (to first approximation*):

$$(\omega - \omega_0) \cong (d\omega/dR)_{R_0}(R - R_0) \tag{14–26}$$

where $(d\omega/dR)_{R_0}$ is the rate-of-change of the orbital angular speed with respect to distance, evaluated at the distance $R = R_0$. Historically, we use the *Oort constant* A defined as

$$A \equiv -(R_0/2)(d\omega/dR)_{R_0} \tag{14–27}$$

to write the radial speed [equation (14–22)] as:

$$v_r = -2A(R - R_0)\sin \ell \tag{14–28}$$

But the reader can see that, for small d, Figure 14–8 implies $(R_0 - R) \cong d\cos \ell$, so that equation (14–28) takes the final form:

$$\boxed{v_r = Ad\sin 2\ell} \tag{14–29}$$

where the identity, $\sin 2\ell = 2\sin \ell \cos \ell$, has been used. The tangential speed equation (14–25), in the same approximation,† becomes:

$$\boxed{v_t = d[A\cos 2\ell + B]} \tag{14–30}$$

where the *Oort constant B* is given by

$$B \equiv A - \omega_0 \tag{14–31}$$

and where the identity, $\cos^2 \ell = (1/2)(1 + \cos 2\ell)$, has been employed.

   If the radial and tangential speeds, given by equations (14–29) and (14–30), are plotted *versus* galactic longitude (ℓ), the resulting curves are termed *double-sinusoids* (sinusoidal with a period of 180°, not 360°). Figure 14–9 shows observed stellar motions plotted in this way—it is clear that our Galaxy *is rotating differentially*. Figure 14–10 illustrates the physical cause for the results shown in Figure 14–9: since A is positive (why?), the angular speed $\omega(R)$ is *decreasing* with increasing R

---

   * If $f(x)$ is a smoothly-varying curve (function of x), then we may find the value of this curve *near* some point $x = x_0$ by using the *Taylor expansion:*

$$f(x) = f(x_0) + (df/dx)_{x_0}(x - x_0) + (1/2)(d^2f/dx^2)_{x_0}(x - x_0)^2 + \cdots$$

Now $(df/dx)_{x_0}$ is the curve's *slope* at x_0, and $(d^2f/dx^2)_{x_0}$ is *curvature* or the rate-of-change of the slope with x; recall that $(x - x_0)$ is small. The first two terms in this expansion, with $(df/dx)_{x_0} = $ constant, excellently approximate $f(x)$ when the curvature is negligible.

   † That is, equation (14–26) and the relation $(R_0 - R) \cong d\cos \ell$ are assumed, and terms like d^2 and smaller are neglected.

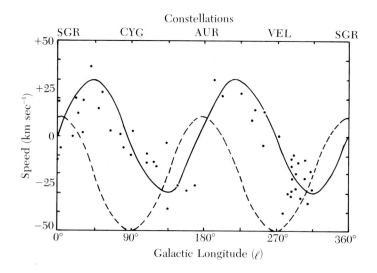

Figure 14–9. *Verification of Differential Galactic Rotation.* Here the radial velocities of Cepheids, in the ranges $|\ell| < 5°$ and 1 kpc $\lesssim$ $d \lesssim$ 2 kpc, are plotted *versus* galactic longitude (ℓ). These motions are with respect to the LSR. The *solid* double-sinusoid is the theoretical curve; the dashed curve is what we would expect for *tangential* speeds. (After S. P. Wyatt)

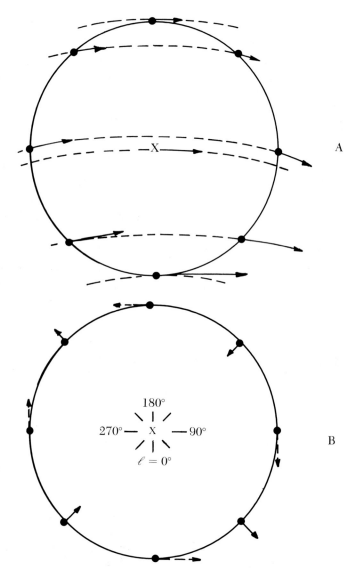

Figure 14–10. *Effects of Differential Galactic Rotation.* **A**, Stars and the LSR (**X**) orbit the galactic center with smaller speeds at greater distances. **B**, With respect to the LSR, we see characteristic variations (with ℓ) of radial speed (solid) and tangential speed (dashed).

in the solar neighborhood; in fact, it is decreasing rapidly enough such that the following effects are observable: (a) stars with $R < R_0$ are moving about the galactic center more rapidly than the LSR, and (b) stars with $R > R_0$ are orbiting at slower speeds than the LSR. Hence, both sets of stars appear to be moving toward larger values of ℓ; those closer in are passing us by, while those farther out are being left behind by the LSR. Note the close analogy with the planetary motions as seen from the Earth, where the inferior planets outpace us while the superior planets are bypassed by the Earth. The resulting apparent stellar motions with respect to the LSR are shown in Figure 14–10, decomposed into radial and tangential parts; compare this with Figure 14–9.

(b)   **Characterizing the LSR**

In terms of our Galaxy, the LSR is specified by its distance R_0 from the center and its circular orbital speed Θ_0. How can we determine these parameters? The stellar effects of differential galactic rotation in Figure 14–9, for example, may be evaluated using equations (14–29) and (14–30), to find Oort's constants A and B. Since radial velocity data permit us to reach great distances, the reasonably accurate value of $A = +15$ km sec^{-1} kpc^{-1} has been found from B stars and Cepheid variables (see Chapter 16). Proper motion data is inevitably limited to nearby stars, but secular parallax studies give us the more uncertain value of $B = -10$ km sec^{-1} kpc^{-1}. Using equation (14–31), we therefore have $\Theta_0/R_0 \equiv A - B = +25$ km sec^{-1} kpc^{-1}. Clearly, an independent determination of either R_0 or Θ_0 is needed, before the other unknown can be found.

In the next section (and especially in Chapter 18), we will see that a good value for the combination AR_0 is obtainable from radio astronomical observations of the neutral atomic hydrogen (H I) in our Galaxy. Using the previously-determined A, we find $R_0 \cong 10$ kpc; therefore, $\Theta_0 \cong 250$ km sec^{-1}.

By assuming that the systems of (a) globular clusters and (b) RR Lyrae stars (see Chapter 16) are symmetrically distributed about the galactic center, another independent value of $R_0 \approx 10$ kpc (uncertain) has been obtained.

If we could somehow determine $\omega(R)$ for our Galaxy, and thence $(d\omega/dR)_{R_0}$, equation (14–27) would give us R_0; unfortunately, we must already know R_0 if we are to obtain $\omega(R)$.

An orbital speed, $\Theta_0 \approx 300$ km sec^{-1}, has been determined by studying the radial velocities of nearby galaxies (those close enough that the Hubble expansion is not important; see Chapter 19) and assuming that the LSR is moving only with respect to the average frame of rest of these galaxies. That is, the galactic center is assumed to be the "local standard of rest" of these galaxies. It is not too surprising that the accuracy of this result is difficult to ascertain.

Table 14–2 lists the currently adopted values for the parameters which we

Table 14–2.   Parameters of the LSR

Oort Constants	$\{$	A	$+15$ km sec^{-1} kpc^{-1}
		B	-10 km sec^{-1} kpc^{-1}
Angular Speed		ω_0	$+25$ km sec^{-1} kpc^{-1}
Distance to Galactic Center		R_0	10 kpc
Circular Orbital Speed		Θ_0	250 km sec^{-1}

have just discussed. These are based primarily upon studies within our own Galaxy, and they may change slightly in the future.

To better understand differential galactic rotation, let us *derive* the Oort constants for the LSR in Keplerian orbit about the mass $M_G = 1.5 \times 10^{11} M_\odot$. Then

$$\omega^2 R = GM_G/R^2$$

or

$$\omega(R) = (GM_G/R^3)^{1/2} \tag{14-32}$$

Differentiating equation (14–32) with respect to R gives:

$$(d\omega/dR) = -(3/2)(GM_G)^{1/2}R^{-5/2} = -3\omega/2R$$

so that Oort's first constant is [by equation (14–27)]:

$$A = +(3/4)\omega_0 = +19 \text{ km sec}^{-1} \text{ kpc}^{-1} \tag{14-33a}$$

Finally, from equation (14–31):

$$B = A - \omega_0 = -(1/4)\omega_0 = -6.4 \text{ km sec}^{-1} \text{ kpc}^{-1} \tag{14-33b}$$

Note that $R_0 = 10$ kpc was used in equation (14–32) to obtain equations (14–33).

(c)   The Rotation Curve of Our Galaxy

As a result of differential galactic rotation, $\omega = \omega(R)$, so that $\Theta = \Theta(R)$—the latter relation is called the *orbital velocity* or *rotation curve* of our Galaxy. The rotation curves for other (nearby) galaxies are easily found by measuring the radial velocities of H I features in these galaxies (see Chapter 19), but for our own Galaxy the problem is more intricate.

Let us return to Figure 14–8, and notice that the *maximum* radial speed $v_{r,\text{max}}$ observed at a given galactic longitude ℓ occurs when the line-of-sight passes *closest* (R_{min}) to the galactic center. Here the line-of-sight is tangent to the orbit, and from Figure 14–8 we have:

$$R_{\text{min}} = R_0 \sin \ell \tag{14-34}$$

Now equation (14–22) is easily inverted to the general forms:

$$\Theta(R_{\text{min}}) = v_{r,\text{max}} + \Theta_0 \sin \ell \tag{14-35a}$$

and

$$\omega(R_{\text{min}}) = \omega_0 + (v_{r,\text{max}}/R_0 \sin \ell) \tag{14-35b}$$

where $v_{r,\text{max}}$ is *observed* and R_{min} may be found from equation (14–34). Note that, in equations (14–35), at least two of the following must be known before we can determine the rotation curve: R_0, Θ_0, and ω_0. If we confine our attention to the radial velocities of, for example, Cepheids nearby ($d \ll R_0$), then equations (14–28) and (14–34) lead to

$$v_{r,\text{max}} = 2AR_0 \sin \ell (1 - \sin \ell) \qquad (14\text{--}36)$$

and we obtain the combination (AR_0) by observing at galactic longitudes near (but smaller than) 90° and near (but larger than) 270°.*

Equation (14–36) is useful for Cepheid observations to distances of 1 to 2 kpc, but the interstellar obscuration (see Chapters 13 and 17) severely impedes stellar studies at greater distances. Fortunately, H I can be "seen" by radio techniques to vastly greater distances, and then we must use equations (14–35). Since the hydrogen clouds tend to delineate spiral features, there are only a few places where the line-of-sight is tangent to a spiral arm, and we may evaluate $v_{r,\text{max}}$ from these H I observations. By combining both the optical and radio data, however, Maarten Schmidt has deduced the rotation curve of our Galaxy shown in Figure 14–11. Here we see rigid-body rotation near the galactic center, the effects of the galactic disk farther out, and essentially Keplerian motion beyond R_0.

---

* These provisos follow because the line-of-sight can be tangent only to orbits *interior* to R_0.

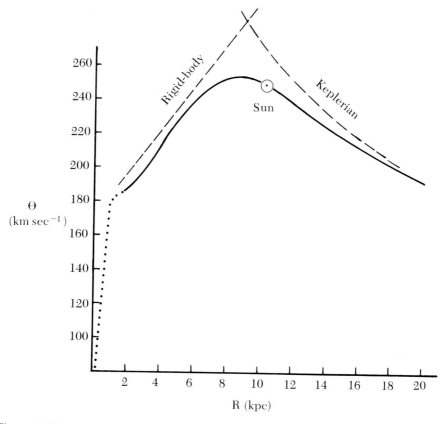

Figure 14–11. *The Rotation Curve of Our Galaxy.* M. Schmidt used both optical and radio data to compute the solid curve; the dotted line is an extrapolation to the galactic center. Dashed lines represent "rigid" rotation and "Keplerian" motion.

(d)   High-Velocity Stars

Let us now return to the solar neighborhood to complete our study of stellar motions. Nearby stars are observed to be moving in every possible direction with respect to the LSR (consider the solar motion, discussed in the preceding sections); these stars are in *eccentric orbits* about the galactic center, so they cannot be at rest in the LSR. Let us decompose a star's motion, *with respect to the galactic center*, into three mutually-perpendicular components: (i) Π, the speed radially-outward (towards $\ell = 180°$), (ii) Θ, the speed towards $\ell = 90°$ as before, and (iii) Z, the speed perpendicular to the galactic plane (positive towards $b = +90°$). Now the stellar motion components, *with respect to the LSR*, are clearly $(\Pi, \Theta - \Theta_0, Z)$; for example, in this notation the solar motion is $(-10.4, +14.8, +7.3)$ km sec^{-1}. Therefore, the Sun is moving inward toward the galactic center, forward toward Cygnus, and is rising out of the galactic plane.

In Figure 14–12, we schematically show the velocity distributions of several types of stars *in the galactic plane* $(\Pi, \Theta - \Theta_0)$. Here we can clearly see (a) that young (early-type) stars are almost at rest in the LSR, (b) that older (late-type) stars are moving faster with respect to the LSR, and (c) that very old Population II stars are in rapid motion relative to the LSR. The Z-components of the stellar velocities behave similarly, so that the older a star is, the more rapidly and the farther it moves out of the galactic plane. This phenomenon is clearly associated with the origin of the stars: young stars are born in spiral arms and move in nearly circular orbits in the galactic plane, whereas the older disk and Population II stars were born far from the galactic plane (and even in the galactic halo).

It is possible to compute velocity curves for stellar orbits of different eccen-

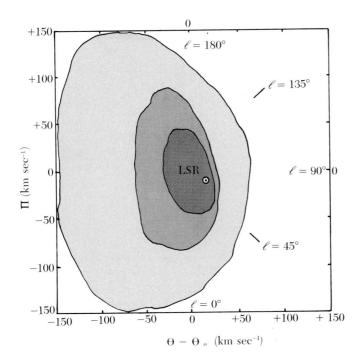

Figure 14–12. *Stellar Motions in the Galactic Plane.* For stars in the solar neighborhood, the observed velocities in the plane of the Galaxy are plotted (schematically) with respect to the LSR. Note the Sun ($\odot$). The region of *darkest* shading is where young A stars are found; the *intermediate* shading corresponds to older K giants; and the large *lightly*-shaded area is occupied by very old disk and Population II stars. Note the *absence* of stars with $\Theta - \Theta_0 \gtrsim$ 65 km sec^{-1} towards $\ell = 90°$! [From A. Blaauw and M. Schmidt, (eds.): *Galactic Structure.* Chicago: University of Chicago Press, 1965]

tricities and semi-major axes, and to plot these curves on Figure 14–12. This procedure gives us the famous *Bottlinger Diagram*, from which we find (a) that young stars and our Sun are in nearly circular galactic orbits at $R \approx R_0$, and (b) that very old stars (including subdwarfs) and the globular clusters are in highly eccentric orbits with $R_0/2 \lesssim R \lesssim R_0$ (here R represents the semi-major axis).

Two points of extraordinary interest are contained in Figure 14–12. First, the velocity region for A stars—called the *velocity ellipse,* for obvious reasons—is *not* aligned perpendicular to $\ell = 0°$, but exhibits a *vertex deviation* of about 15°. This phenomenon is still unexplained, though it may be related to the conditions under which these young stars were formed.

Secondly, the stellar velocities seem to avoid the direction of galactic rotation ($\ell = 90°$), and no star moves faster than about 65 km sec^{-1} in this direction. This effect is most noticeable for the so-called *high-velocity stars*, with in-plane speeds greater than 65 km sec^{-1} relative to the LSR. Two separate phenomena are involved here. Apparently, a star whose total speed is greater than 315 km sec^{-1}, relative to the galactic center, will *escape from the Galaxy;* hence, the deficiency of stars with $\Theta - \Theta_0 \gtrsim 65$ km sec^{-1}. (This escape speed is significantly less than that computed assuming Keplerian orbits, owing to the distribution of mass in the galactic disk.) The high-velocity stars arise, not because they are moving extremely rapidly through the Galaxy, but because they are moving much more slowly than the LSR. Figure 14–13 shows how such a star, moving slowly near apogalacticon in a highly eccentric orbit, appears to be moving rapidly with respect to the LSR. For example, the LSR (moving at $\Theta_0 = 250$ km sec^{-1}) seems to be receding at 100 km sec^{-1} from a star at apogalacticon (in the solar neighborhood) with $\Theta = 150$ km sec^{-1}. Such high-velocity objects also include the RR Lyrae stars, and the recently discovered *high-latitude* (large ℓ) *high-velocity* H I *clouds.*

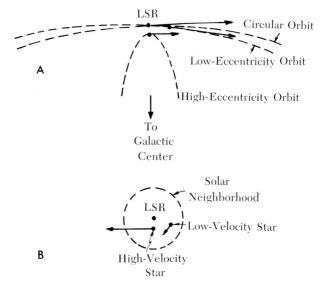

Figure 14–13.   *High-Velocity Objects.* **A,** Three galactic orbits are shown: the circular LSR orbit, the low-eccentricity orbit of a nearby A star, and the high-eccentricity orbit of a nearby globular cluster at apogalacticon. **B,** Relative to the LSR, these same objects in the solar neighborhood appear as a low-velocity A star and a high-velocity globular cluster.

Problems

14–1.   The Fe I emission lines (at 4415.1 Å and 4442.3 Å) in a comparison spectrum are located 15.00 mm and 15.43 mm, respectively, from an arbitrary reference point. If a stellar Ca I line (of rest-wavelength 4425.4 Å) is measured to be at 15.27 mm,
(a) what is the observed *wavelength* of the Ca I line?
(b) what is the *radial velocity* of this star?

14–2.   By making the appropriate conversions of *units*, show that equation (14–3) follows from equation (14–2).

14–3.   A star, located 90° from the solar antapex on the celestial sphere, is at rest in the LSR a distance 10 pc from the Sun. As seen from the Sun,
(a) by what angle (in arc-seconds) will this star appear to move on the celestial sphere in ten years?
(b) in what direction will the star appear to move?

14–4.   The star δ Tauri is a member of the Taurus moving group. It is observed to have a proper motion of $0''.115$ yr^{-1}, a radial velocity of $+38.6$ km sec^{-1}, and to lie $29°.1$ from the convergent point of the group.
(a) What is this star's parallax?
(b) What is its distance in parsecs?
(c) Another star belonging to the same group lies only 20° from the convergent point; what are *its* proper motion and radial velocity?

14–5.   Refer again to the data given in Problem 14–4. Assume that a probable error of $\pm 0''.005$ yr^{-1} is associated with the proper motion of δ Tauri. If we independently measure the trigonometric parallax of this star (with a probable error of $+0''.005$), what are the uncertainties (in parsecs) in the distances determined from these two separate parallaxes?

14–6.   Show, step by step, a detailed derivation of equations (14–22) and (14–25)—the famous Oort Formulae.

14–7.   In analogy to Problem 14–6, show a detailed derivation of equation (14–30).

14–8.   Assume that the mass of our Galaxy is 1.5×10^{11} solar masses, and that it is *all* concentrated in a point at the galactic center.
(a) Plot a *rotation curve* (Θ versus R), with appropriate units and exemplary values along each axis, for this hypothetical case.
(b) Indicate the rotation *period* at $R = 5$, 10, and 20 kpc.
(c) What is the speed of escape from $R = 10$ kpc?

14–9.   In analogy to Problems 14–6 and 14–7, give an explicit derivation of the two equations (14–35).

14–10.   A star, in a Keplerian galactic orbit of eccentricity 0.8 and semi-major axis 7 kpc, moves through the solar neighborhood on its outward journey in the galactic plane. What is the velocity of this star with respect to the LSR?

Reading List

Blaauw, Adriaan, and Schmidt, Maarten, (ed.): *Galactic Structure*. Chicago, The University of Chicago Press, 1965.

Mihalas, Dimitri, and Routly, Paul M.: *Galactic Astronomy*. San Francisco, W. H. Freeman and Company, 1968.

Page, Thornton, and Page, Lou Williams, (ed.): *Stars and Clouds of the Milky Way*. New York, The Macmillan Company, 1968.

Podobed, V. V. (Vyssotsky, A. N., translator): *Fundamental Astrometry*. Chicago, The University of Chicago Press, 1965.

Smart, W. M.: *Stellar Kinematics*. New York, John Wiley & Sons, 1968.

Staal, Julius D. W.: *Focus On Stars*. London, George Newnes Ltd., 1963.

Strand, Kaj Aa., (ed.): *Basic Astronomical Data*. Chicago, The University of Chicago Press, 1963.

Swihart, Thomas L.: *Astrophysics and Stellar Astronomy*. New York, John Wiley & Sons, 1968.

Trumpler, Robert J., and Weaver, Harold F.: *Statistical Astronomy*. Los Angeles, University of California Press, 1953.

Van de Kamp, Peter: *Principles of Astrometry*. San Francisco, W. H. Freeman and Company, 1967.

Woolard, Edgar W., and Clemence, Gerald M.: *Spherical Astronomy*. New York, Academic Press, 1966.

Chapter 15
The Structure and Evolution of Stars

Just as cells are the building blocks of living organisms, so also may we consider *stars* the basic entities in space. The observational characteristics of stars were our primary concern in Chapters 9 through 14; here we will discuss the *physical laws* which govern the structure and evolution of these bodies. By suitably combining these laws into theoretical *stellar models*, we can understand (a) the "equilibrium" configurations of stars, (b) their evolution in time, and (c) the astrophysical basis for cluster H-R diagrams. (The reader should recall Chapters 9 and 12, where we actually began this development with the physics of stellar atmospheres.)

We will concentrate upon normal stars, which are in the vast majority, and return later (Chapter 16) to the rare peculiar stars with their fascinating variability. This is not to say that these are two distinctly different classes of objects, for a normal star generally exhibits several *brief* phases of peculiarity during its lifetime. If we lived a billion years we would see this, but at this instant of cosmic time we happen to catch most stars in their sedate states—only a few are in active or violent phases now.

By its very nature, a normal star is so hot that it must be *entirely gaseous*. Let us see how this comes about, by beginning this chapter with the physical precepts applicable to stellar-sized masses.

15–1   THE PHYSICAL LAWS OF STELLAR STRUCTURE

(a)   Hydrostatic Equilibrium

A star is a massive object, held together by its *self-gravitation* and supported against collapse by its *internal pressures*. The simplest stellar model is a static, spherically symmetric "ball" of matter; that is, no parameter of physical interest depends upon time or angle, but only upon the radial distance r from the star's center. *Hydrostatic equilibrium* ensues, since the inward gravitational attraction exactly balances (i.e., no net force) the outward pressure forces at every point (r) within the star. As we approach the center of the star, the pressure must steadily increase to counterbalance the increasing weight of the material which lies above.

Using Figure 15–1, we can express these considerations quantitatively. At the radius r is a *thin spherical shell* of thickness dr ($\ll r$). Since the surface area of this shell is $4\pi r^2$, its volume is $4\pi r^2\, dr$. The mass density (gm cm^{-3}) of the material in the shell is denoted by $\rho(r)$, so that the shell's mass is $4\pi r^2 \rho(r)\, dr$. From Chapter 3, we know that the mass beyond r has no gravitational influence upon the shell, while the entire mass inside r—called $M(r)$—attracts the shell as would a point of mass $M(r)$ at the star's center. Therefore, the inwardly directed gravitational force upon the shell is

$$\frac{GM(r)\left[4\pi r^2 \rho(r)\, dr\right]}{r^2} \qquad \textbf{(15–1a)}$$

where $G = 6.670 \times 10^{-8}$ dyne cm^2 gm^{-2} is the Newtonian constant of universal gravitation. From Figure 15–1, we see that the pressure $P(r)$ (force per unit area) pushes outward on the shell, while the pressure $P + dP$ pushes inward, so that the net *outward* pressure *force* is

$$-4\pi r^2\, dP \qquad \textbf{(15–1b)}$$

Figure 15–1.   *Hydrostatic Equilibrium.* The spherical star (total radius = R) is subdivided into *thin* spherical shells of matter, such as the one—of thickness dr—at the radial distance r. The pressure P pushes outward on the shell at r, while the *smaller* pressure $P + dP$ (since dP is *negative*) pushes inward at $r + dr$.

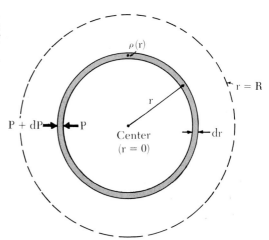

Combining equations (15–1), we have the *equation of hydrostatic equilibrium:*

$$dP/dr = -GM(r)\rho(r)/r^2 \qquad (15\text{–}2)$$

We see that the pressure steadily *decreases* as we move to greater values of r.

There appear to be three independent parameters in equation (15–2): $P(r)$, $\rho(r)$, and $M(r)$. But Figure 15–1 clearly shows that $M(r)$ increases by the amount, $dM(r) = 4\pi r^2 \rho(r)\, dr$, when we add the indicated spherical shell (i.e., as we move from r to $r + dr$). Therefore, $M(r)$ is determined from $\rho(r)$ via:

$$dM/dr = 4\pi r^2 \rho(r) \qquad (15\text{–}3)$$

our second fundamental equation. If the star's outer radius is R, then the *total mass* of the star is

$$M = \int_{r=0}^{r=R} dM(r) = 4\pi \int_0^R \rho(r) r^2\, dr \qquad (15\text{–}4)$$

We have simply added together the masses of all of the onion-like shells in our model star.

Hence, we need know only $\rho(r)$ to determine, first $M(r)$ via equation (15–3), and then $P(r)$—the *pressure profile* within the star—via equation (15–2). Given the star's radius, R, we also find its total mass, M, using equation (15–4).

(i) *The Sun's Central Pressure*

To perfect our physical intuition, let us crudely calculate the necessary pressure at the center of our Sun, using the equation of hydrostatic equilibrium. We know that: $G = 6.670 \times 10^{-8}$ dyne cm^2 gm^{-2}, $M_\odot = 1.989 \times 10^{33}$ gm, and $R_\odot = 6.96 \times 10^{10}$ cm; therefore, the *mean* solar density is $\bar\rho_\odot = 3M_\odot/4\pi R_\odot{}^3 = 1.41$ gm cm^{-3}. Taking the surface pressure as zero, letting $r = dr = R_\odot$, and using $M(r) = M_\odot$, in equation (15–2) quickly yields:

$$P_c \simeq GM_\odot\bar\rho_\odot/R_\odot = 2.7 \times 10^{15} \text{ dyne cm}^{-2}$$

But the atmospheric pressure at the Earth's surface is well known to be 1 atmosphere $= 1.013 \times 10^6$ dyne cm^{-2}, so that the staggering pressure, $P_c \simeq 2.7 \times 10^9$ atmospheres, must exist at the Sun's center to support the weight of its mass.

Our computed mean solar density, $\bar\rho_\odot$, is slightly greater than that of water (1.00 gm cm^{-3}), but since the Sun (and in fact, all stars) is centrally condensed, the actual central density is more like 160 gm cm^{-3}. How, then, are we justified in treating stars as gaseous spheres? In the next section, we will see that the tremendous pressures supporting stars imply very high interior temperatures—high enough to maintain the gaseous state.

(b)   Equations of State

To solve our stellar model, we need the density profile $\rho(r)$. The microscopic constitution, and local state, of the stellar materials must be examined in detail—are we dealing with a gas, a liquid, or a solid? At the end of this chapter, we shall see that (quantum) liquids and solid lattices of atomic nuclei become important when we consider *white dwarfs* and *neutron stars*; in the vast majority of cases, however, we are interested only in gases.

For normal stars, then, we can assume that the material is a *perfect gas*, which obeys the *perfect gas law** (see Chapter 5):

$$P(r) = n(r)kT(r) \tag{15–5}$$

Here the pressure $P(r)$ is related directly to the gas-particle-number density $n(r)$ (particles cm^{-3}), Boltzmann's constant $k = 1.381 \times 10^{-16}$ erg K^{-1}, and the gas temperature $T(r)$. In Chapter 12, we saw that $n(r)$ could be expressed in terms of $\rho(r)$ and the *gas composition* $\mu(r)$:

$$n(r) = \rho(r)/\mu(r)m_{\mathrm{H}} \tag{15–6}$$

where $m_{\mathrm{H}} = 1.67 \times 10^{-24}$ gm is the mass of a hydrogen atom; recall that, in terms of the *mass fractions* of hydrogen (X), helium (Y), and all heavier elements—"metals"—(Z), the composition (or *mean molecular weight*) is

$$\mu = (2X + \tfrac{3}{4}Y + \tfrac{1}{2}Z)^{-1} \approx \tfrac{1}{2} \tag{15–7}$$

Therefore, the perfect-gas equation of state becomes:

$$P(r) = \rho(r)kT(r)/\mu(r)m_{\mathrm{H}} \tag{15–8}$$

As we will see later, $\mu(r)$ is usually specified when we construct our stellar model, so that only $T(r)$ remains to be determined.

(i)   *The Sun's Central Temperature*

By inverting the perfect gas law of equation (15–8), using the P_c and $\bar\rho_\odot$ values found in section 15–1(a, i), and by making the same crude but simplifying approximations, we may estimate the required central temperature T_c of our Sun:

$$T_c \simeq P_c\mu m_{\mathrm{H}}/\bar\rho_\odot k \tag{15–9}$$

Since μ is near 1/2 in our Sun (i.e., mostly hydrogen) we have

$$T_c \simeq 12 \times 10^6 \text{ K} \tag{15–10}$$

---

* In massive stars, the gas pressure is significantly augmented by *photon pressure* (or *radiation pressure*): $P_{\mathrm{rad}}(r) = (a/3)T^4(r)$, where $a = 7.564 \times 10^{-15}$ erg cm^{-3} K^{-4} is the radiation constant.

which is close to the accurate (computer-determined) value of about 15 million kelvins!

It should be abundantly clear that no solid nor liquid can exist at these temperatures, and, in fact, even the gas is dissociated into ions and electrons—a neutral mixture termed a *plasma*. Our earlier statement, that stars are gaseous spheres, is thus seen to be consistent and true. The gaseous state persists throughout the star, since $\rho(r)$ decreases approximately as rapidly as $T(r)$ as we move outward through the star.

(c) Modes of Energy Transport

To determine $T(r)$, we must consider how energy is transported from the stellar interior to the surface, where it is radiated away into space. The star's self-gravitation forces the stellar center to be at a much higher temperature (15 million K for our Sun) than the stellar surface (only about 6000 K for the Sun); the second law of thermodynamics—which prohibits "perpetual motion machines"—then implies that heat energy must flow from the higher-temperature regions to the lower-temperature regions.

At least three *modes of energy transport* suggest themselves: conduction, convection, and radiation. *Conduction* occurs when energetic (i.e., hot) atoms communicate their agitation to nearby cooler atoms via "collisions"; this mode is extremely efficient in solids (especially metals and in degenerate gases [see section 15–3(d, i)], but is of negligible importance in perfect gases owing to their low thermal conductivities. *Convection* is the transport of heat energy by means of mass motions in fluids. As we saw in Chapter 9, when $T(r)$ varies rapidly enough with distance,* the fluid becomes unstable and "boils." This process takes place in limited regions of most stars, with hot fluid masses rising, releasing their heat energy, and sinking again to pick up more energy. The top of such a *convective zone* is observed at the base of the Sun's photosphere (see Chapter 9). Unfortunately, no completely adequate mathematical theory of convective transport has yet been devised.

The third mode, *radiative transport*, is usually the most important means of energy flow in stars. Here high-energy *photons* flow outward from the star's interior, losing energy via scattering and absorption in the hot plasma of the so-called *radiative zone*. At the extreme temperatures of stellar interiors, the most important sources of such *opacity* (see Chapters 9 and 12) are (a) *electron scattering*—the scattering of radiation (photons) by free electrons, and (b) *photo-ionization*—the detachment of electrons from ions using the radiation energy (see Chapter 8). Returning once again to Figure 15–1, we can now crudely derive the *equation of radiative transport*. At the base of our thin shell, the spherical surface is essentially a blackbody emitter at temperature $T(r)$, so that by equation (8–40) we have the outward radiative flux (erg cm^{-2} sec^{-1}), $F(r) = \sigma T^4(r)$, where $\sigma = 5.67 \times 10^{-5}$ erg cm^{-2} sec^{-1} K^{-4} is the Stefan-Boltzmann constant. But at $r + dr$, the temperature is $T + dT$, and the outward flux is $F + dF = \sigma(T + dT)^4 \simeq \sigma(T^4 + 4T^3\, dT)$. Now dT is *negative*, since the exterior of the shell must be cooler than its interior,

---

* That is, there is a steep temperature gradient (dT/dr).

so that the flux absorbed within the shell is:

$$dF = +4\sigma T^3(r)\, dT \qquad (15\text{–}11a)$$

This absorption is due to the *opacity* $\kappa(r)$ of the shell material, and from equation (9–1)

$$dF = -\kappa(r)\rho(r)F(r)\, dr \qquad (15\text{–}11b)$$

Equating equations (15–11), and defining the *luminosity* (erg sec^{-1}) via $L(r) = 4\pi r^2 F(r)$, we see that the total energy flow through our thin spherical shell per second is:

$$L(r) = -\frac{16\pi\sigma r^2 T^3(r)}{\kappa(r)\rho(r)}\left(\frac{dT}{dr}\right)_r \qquad (15\text{–}12)$$

The complete machinery of the theory of radiative transfer introduces an additional factor of (4/3) into equation (15–12), so that the correct equation of radiative transport is:

$$L(r) = -\frac{64\pi\sigma r^2 T^3(r)}{3\kappa(r)\rho(r)}\left(\frac{dT}{dr}\right)_r \qquad (15\text{–}13)$$

Our stellar model is now complete, with equation (15–13), since $\kappa(r)$ depends only upon $\mu(r)$, $T(r)$, and $\rho(r)$; at the star's surface, $L(r)$ equals the observed (bolometric) luminosity of the star.

(i) Solar Luminosity from Radiative Transfer

For our Sun, a good estimate of (dT/dr) is $(-T_c/R_\odot)$, which equals the miniscule gradient -2×10^{-4} K cm^{-1}. Using equation (15–13), we will now attempt to estimate the solar luminosity, $L_\odot$. Setting $r = R_\odot$, $T(r) = T_c$, and $\rho(r) = \bar{\rho}_\odot$, we have:

$$L_\odot \cong -\frac{64\pi\sigma R_\odot^2 T_c^3}{3\kappa\bar{\rho}_\odot}\left(-\frac{T_c}{R_\odot}\right) = \frac{9.5 \times 10^{36}}{\kappa}\ \text{erg sec}^{-1} \qquad (15\text{–}14)$$

where we have yet to determine a reasonable opacity κ (cm^2 gm^{-1}). From its dimensions, κ is clearly the *interaction area* per gas particle multiplied by the number of particles per gram of the stellar material; a gram of completely ionized hydrogen contains 6×10^{23} protons, and the same number of electrons. For electron scattering, the interaction area per electron is approximately 10^{-26} cm^2; for hydrogen photo-ionization, this area per atom is near 10^{-16} cm^2. In the solar interior, the latter opacity source dominates, so that (very approximately) $10^{-2} \ll \kappa \lesssim 10^8$— accurate opacities are extremely difficult to determine. Therefore, our prediction for the solar luminosity falls in the broad range $10^{29} \lesssim L_\odot \ll 10^{39}$ erg sec^{-1}; the "mean" value of 10^{34} erg sec^{-1} is very close to the known value, $L_\odot = 3.90 \times 10^{33}$ erg sec^{-1}.

(d)   Energy Sources

Since the stellar luminosity represents energy loss, no star is perfectly static; our stellar model is, however, an excellent approximation for times which are short compared to the stellar evolution time. We must, therefore, ask (a) how long will the star remain in essentially steady-state, and (b) what *energy source* maintains this period of stability? Geological and paleontological evidence indicates that our Sun has been radiating energy at a fairly steady rate for several billion years; hence, energy generation must take place in stellar interiors.

The rate of energy production per unit mass of stellar material (erg sec^{-1} gm^{-1}) is denoted by $\epsilon(r)$. As we shall see, $\epsilon = 0$, except in stellar cores and in certain localized spherical shells. For our Sun, we may estimate the average value of ϵ, called $\bar{\epsilon}_\odot$, needed to maintain the solar luminosity $L_\odot$ via:

$$\bar{\epsilon}_0 \simeq L_\odot / M_\odot = 2.0 \text{ erg sec}^{-1} \text{ gm}^{-1}$$

Consulting Figure 15–1 once again, we can find how such energy generation within our thin spherical shell augments the stellar luminosity. The luminosity $L(r)$ enters the bottom of the shell, while the *greater* luminosity $L + dL$ leaves the top—owing to the energy produced in the shell's mass $4\pi r^2 \rho(r)\, dr$. The additional luminosity is clearly:

$$dL = 4\pi r^2 \rho(r)\epsilon(r)\, dr \qquad \text{(15–15)}$$

Equation (15–15) expresses the *balance* between net energy loss from the shell, dL, and that generated within the shell. Thus we have an energy or *thermal equilibrium*.

In a quasi-static gaseous star, energy may be generated only by (a) gravitational contraction and/or (b) thermonuclear fusion reactions. Each of these processes is important at some stage in the star's evolution; in 1854 H. von Helmholtz, and in 1861 Lord Kelvin, suggested the first mechanism, (a), while our understanding of the significance of the second, (b), began in 1938 with the pioneering work of H. Bethe and C. F. von Weizsacker. Let us consider these energy sources in detail.

(i)   *Gravitational Contraction*

In Chapter 3, we showed that gravitational potential energy could be transformed into kinetic energy of motion (as when a rock is dropped near the Earth's surface); one form of kinetic energy is *heat* (when the rock hits the ground). An imperceptibly slow contraction of the planet Jupiter was conjectured in Chapter 5 to explain that body's apparent production of heat energy. Consider a very slowly contracting star. The heat energy of its interior provides the pressure—due to the random motions of the gas particles—which supports the star against its own self-gravitation. When the star contracts to a smaller radius, the self-gravitation increases, so that the internal pressures (and, hence, temperatures and heat energy) must also increase to maintain approximate hydrostatic equilibrium. But the gravitational potential energy decreases about twice as fast as the heat energy increases; hence, to conserve the total energy of the system, approximately half of the potential energy change must be radiated into space—the star's luminosity.

This energy-conversion process may be illustrated by a simple analogy (see Chapter 3). A small satellite of mass m moves in a circular orbit of radius r at speed v about a great mass M. From equation (3–30), the satellite's kinetic energy is $mv^2/2$, while from equation (3–31) its gravitational potential energy is $-GMm/r$. Since the centripetal acceleration (v^2/r) maintaining the orbit is provided by the mutual gravitational attractive acceleration, $GM/r^2 = v^2/r$, we see that the kinetic energy $mv^2/2 = GMm/2r$ or *half* the magnitude of the potential energy. If we now move the satellite to a smaller (stable) orbit at $r - dr$, the increase in kinetic energy is certainly only half the decrease in potential energy (which becomes more negative). To conserve total energy (potential plus kinetic), the other half of the potential energy change must be transmitted to the agent which alters the satellite's orbit—in the case of a star, this is the energy which is radiated away.

Let us apply these considerations to our Sun, as did Helmholtz and Kelvin. For each gram of solar material, the average gravitational potential energy which is available for radiation is

$$GM_\odot/2R_\odot = 9.54 \times 10^{14} \text{ erg gm}^{-1} \tag{15–16}$$

Comparing this with $\bar{\epsilon}_\odot$ we see that *Helmholtz-Kelvin contraction* can sustain the Sun—at its present luminosity—for only 15 million years; some other energy source must be sought, if we are to account for billions of years of sunshine. In section 15–3, we will see where gravitational contraction is important in stellar evolution.

(ii)   *Thermonuclear Reactions*

Only after about 1938 did it become clear that the long-term energy source for stars must be *thermonuclear fusion reactions*. In this process, light atomic *nuclei* collide with such violence and frequency—in the high-temperature, high-density stellar interior—that they coalesce into heavier nuclei and release tremendous quantities of energy (an example is the hydrogen bomb, which has unleashed thermonuclear fury on Earth). We say that the lighter elements "burn" to form heavier elements in this process of *nucleosynthesis*.

We discussed the constitution of atomic nuclei in section 8–2, where we saw that the strong nuclear force overcomes the electrostatic repulsion of the positively-charged *protons*, and binds from one to 260 nucleons (protons and *neutrons*) in a region about 10^{-13} cm in diameter. (The reader should review this earlier discussion, especially the comments on *notation*, before proceeding here.) Two nuclei will fuse together to form one larger nucleus if they approach within 10^{-13} cm of one another; but they must be moving fast enough to overcome their mutual electrostatic repulsion—all nuclei have a positive charge. The magnitude of this repulsion, the *Coulomb barrier*, is proportional to the *product* of the nuclear charges, so that the easiest fusion reaction involves two protons (i.e., hydrogen nuclei); such reactions become significant at temperatures around 10 million kelvins. Other nuclei are involved at higher temperatures, as we shall see.

The high abundance of hydrogen makes it an important constituent in stellar nuclear reactions. The next stable nucleus is helium, $_2\mathrm{He}^4$, with atomic weight 4. Since the hydrogen nucleus, the proton, only has atomic weight 1, four protons are required to make one helium nucleus. The atomic weights do not exactly match, for the more exact atomic weight of a proton is 1.0078, and four of them add to 4.0312, while the weight of $_2\mathrm{He}^4$ is 4.0026, leaving a *mass defect* of 0.0286. This is converted to energy by Einstein's equation for the equivalence of mass and energy (derived from his theory of special relativity),

$$E = mc^2 \tag{15–17}$$

where c is the speed of light. Since a unit atomic weight is 1.66×10^{-24} gm, the energy released by the conversion of four $_1\mathrm{H}^1$ nuclei to one $_2\mathrm{He}^4$ nucleus is

$$E = (0.0286) \times (1.66 \times 10^{-24}) \times (9 \times 10^{20})$$
$$= 4.3 \times 10^{-5} \text{ erg}$$

We can also use equation (15–17) to determine the total energy store of the Sun, if we assume that it originally consisted of pure hydrogen, all of which will eventually be converted into helium. The mass defect liberated in the form of energy in this thermonuclear conversion is the fraction $0.0286/4.0312 = 0.0071$ of the available mass of original hydrogen. Since it is only in the central core that the temperature and pressure are sufficiently high to permit nuclear reactions, only about ten per cent of the mass of the Sun is available for energy conversion. Hence, the total thermonuclear energy available in the Sun is

$$E_{\text{total}} = m(_2\mathrm{He}^4 - 4_1\mathrm{H}^1) \times c^2 \times 0.1 M_\odot$$
$$= 1.28 \times 10^{51} \text{ ergs}$$

which, with the present solar luminosity of 3.90×10^{33} ergs sec^{-1}, would last about 10 billion years. The best estimates of the age of the Solar System yield figures around 5 billion years, so there is ample energy from this reaction to sustain our Sun.

Recall that four hydrogen nuclei are required to produce one $_2\mathrm{He}^4$ nucleus. A simultaneous collision fusing four independent particles is extremely improbable, even at the densities prevalent in the centers of stars, so we must seek a *step process* or *reaction chain*. Experimental data pertaining to the types and rates of nuclear reactions probable in stellar interiors are obtained in the laboratory by bombarding target nuclei with projectile particles in high-energy accelerators.

Two different processes lead to the conversion of hydrogen into helium: the proton-proton (p-p) chain and the carbon (CNO) cycle. The p-p chain dominates at temperatures less than 2×10^7 K, while the CNO cycle is prominent at higher temperatures. In the Sun, for instance, both processes take place, but the p-p chain is the most important. The CNO cycle plays a negligible role in stars lower (i.e., later) on the main sequence than the Sun, while for stars earlier (and hotter) than F stars, it predominates.

The *proton-proton chain* consists of the following reactions:

$$\left.\begin{aligned}
_1H^1 + {}_1H^1 &\rightarrow {}_1H^2 + e^+ + \nu \\
_1H^2 + {}_1H^1 &\rightarrow {}_2He^3 + \gamma \\
_2He^3 + {}_2He^3 &\rightarrow {}_2He^4 + {}_1H^1 + {}_1H^1
\end{aligned}\right\} \qquad (15\text{–}18)$$

$_1H^2$ is heavy hydrogen or deuterium, whose nucleus contains one proton and one neutron; e^+ is a positron, ν is a neutrino, and γ is a photon.* Conservation of charge is maintained in the first reaction by the emission of the positron. Note that the first two steps must occur twice before the last can take place, and that a total of six protons is involved even though two are again released in the final step. Other reactions may occur instead of the last step of this chain, for example

$$_2He^3 + {}_2He^4 \rightarrow {}_4Be^7 + \gamma$$

Then there are two possible branches from $_4Be^7$, both resulting in $_2He^4$. All three chains operate simultaneously in a star, but the first, equation (15–18), is the most important.

The neutrinos produced by the proton-proton chain should escape the Sun without interacting with any other solar material. Since 1965, R. Davis has attempted to detect these solar neutrinos by means of an elaborately shielded tank (located far underground in the Homestake mine in South Dakota) filled with 100,000 gallons of tetrachloroethylene (C_2Cl_4). Although neutrinos react only very weakly with matter, they should transmute some of the Cl^{37} to A^{37}. These radioactive argon atoms can then be collected and counted. The neutrino flux measured by Davis is well *below* that predicted by current theoretical solar models. The source of this discrepancy remains an unsolved mystery.

The *carbon (CNO) cycle* also converts hydrogen into helium, but it requires a carbon nucleus as a catalyst:

$$\left.\begin{aligned}
_6C^{12} + {}_1H^1 &\rightarrow {}_7N^{13} + \gamma \\
_7N^{13} &\rightarrow {}_6C^{13} + e^+ + \nu \\
_6C^{13} + {}_1H^1 &\rightarrow {}_7N^{14} + \gamma \\
_7N^{14} + {}_1H^1 &\rightarrow {}_8O^{15} + \gamma \\
_8O^{15} &\rightarrow {}_7N^{15} + e^+ + \nu \\
_7N^{15} + {}_1H^1 &\rightarrow {}_6C^{12} + {}_2He^4
\end{aligned}\right\} \qquad (15\text{–}19)$$

Each step in the chain need occur only once to convert four protons into an *alpha particle* (helium nucleus). The second and fifth steps take place because $_7N^{13}$ and $_8O^{15}$ are unstable isotopes of their respective elements, with half-lives of only a few minutes. *Half-life* is the time in which half of the original quantity of isotope

---

* A *positron* has the same mass as an electron, but its charge is positive. *Neutrinos* have neither mass nor charge, only energy, and are therefore very elusive and difficult to detect.

has disintegrated into its more stable nuclear form. The cycle starts with the reaction between carbon and hydrogen, but ends with the release of an identical carbon nucleus; hence, we say the $_6C^{12}$ acts as a *catalyst*. Although the temperature may be sufficiently high, the CNO cycle cannot operate in a star unless carbon is available. This fact is of prime importance in considering the evolution of Population II stars, which formed with essentially no carbon, as compared to Population I stars, which began with a significant amount of carbon.

Higher temperatures are required for the carbon cycle because of the higher Coulomb barriers of the carbon and nitrogen nuclei, compared with those of protons and helium nuclei. As a result, the temperature dependence of the p-p reaction rate goes approximately as T^4, while it varies as T^{15} for the carbon cycle.

At very high temperatures, around 10^8 K, other reactions will begin transmuting helium into heavier elements: namely, carbon, oxygen, and neon. Three α-particles ($_2He^4$) will form carbon; thus,

$$\left.\begin{array}{l} _2He^4 + {}_2He^4 \rightleftarrows {}_4Be^8 + \gamma \\ _4Be^8 + {}_2He^4 \rightarrow {}_6C^{12} + \gamma \end{array}\right\} \tag{15–20}$$

This is known as the *triple-α process*, the first stage of helium burning. The intermediate product, beryllium-8, is not very stable, and the right-to-left (back) reaction occurs readily. Nevertheless, an equilibrium will be established where some $_4Be^8$ takes part in the second step of the chain. Other reactions follow straightforwardly with the addition of further α-particles, leading to the formation of $_8O^{16}$, $_{10}Ne^{20}$, $_{12}Mg^{24}$, and even heavier elements. Of the light elements other than hydrogen, helium, and carbon, most are rare deep inside stars, because such elements (e.g., deuterium, lithium, beryllium, and boron) quickly combine with protons at temperatures of only a few million degrees to form one or two helium nuclei; for example:

$$_3Li^7 + {}_1H^1 \rightarrow {}_2He^4 + {}_2He^4$$

As we shall see, the triple-α and other helium burning reactions play a major role in the evolution of the stars. Moreover, by reactions similar to these, we now see that it is possible to start with stars of pure hydrogen and eventually produce any of the heavy elements.

15–2   THEORETICAL STELLAR MODELS

(a)   Recapitulating the Parameters

In the preceding section, we discussed the physical principles basic to stellar structure: hydrostatic equilibrium, the perfect gas equation of state, the various modes of energy transport, and the gravitational and thermonuclear sources of stellar energy. These are the *tools* used by astrophysicists in computing *stellar models*—theoretical "stars" in which physical parameters and their rates of change throughout the star are exactly described. To reiterate, these mutually dependent parameters include temperature $T(r)$, mass $M(r)$, density $\rho(r)$, pressure $P(r)$, luminosity $L(r)$, rate of energy production $\epsilon(r)$, and chemical composition in terms of the mean molecular weight $\mu(r)$. Their interdependence is stated by the basic equations of stellar structure:

Hydrostatic equilibrium: $dP/dr = -GM(r)\rho(r)/r^2$ (15–2)

Mass equation: $dM/dr = 4\pi r^2 \rho(r)$ (15–3)

Radiative transport: $\dfrac{dT}{dr} = \dfrac{-3\kappa(r)\rho(r)}{64\pi\sigma r^2 T^3(r)} L(r)$ (15–13)

Energy generation:
(Thermal equilibrium) $dL/dr = 4\pi r^2 \rho(r)\epsilon(r)$ (15–15)

Equation of state: $P(r) = k\rho(r) T(r)/\mu(r)m_{\mathrm{H}}$ (15–8)

These equations describe how the parameters vary through the star *only* if we know their values at some particular points (or shells) in the star, such as at the center and the surface. These values constitute the *boundary conditions*. At the center, for example, where $r = 0$, the boundary conditions for the mass and the luminosity must be $M(r) = 0$ and $L(r) = 0$. So that the theoretical models may bear some relation to *real* stars, we may use *observed* stellar characteristics for the boundary conditions at the surface. Thus, at $r = R$, the radius of the star, $M(R) = M$, $L(R) = L$, $T(R) = T_e$, the effective surface temperature or photospheric temperature, and both $\rho(r)$ and $P(r)$ approach zero.* The properties of the stellar material ultimately depend on the *chemical composition* of the star. This will determine the type of energy generation ϵ and the opacity κ.

Chemical composition as characterized by the mean molecular weight μ is of enormous importance in stellar structure, as may be seen from these basic equations. The equation of state shows the dependence of $P(r)$ upon μ; since the hydrostatic equation indicates that $\rho(r)$ is strongly dependent on $P(r)$, it follows that $\rho(r)$ depends on μ. Note that $\rho(r)$ appears in all the other equations. The difficulty lies in knowing *how* the composition varies through the star. At best, this must be estimated, and the parameters resulting from the calculations must be compared with observed quantities. There is, therefore, an element of trial and error in such calculations.

Many theoretical models have been computed for our Sun, for it is the best-observed star and serves as the observational prototype for all stars. These models differ in (a) the relative abundances of hydrogen, helium, and the heavy elements assumed for the newly-formed Sun, and (b) the degree of mixing of the elements and their participation in thermonuclear reactions in the solar interior. Compositional mixing becomes more thorough, the greater the extent of the *convection zone*. Current solar models which best fit the various observational requirements indicate that a large fraction of the hydrogen at the center of the Sun has already been converted to helium, so that there is a *central helium-enriched core*. Present energy generation, hydrogen burning via the p-p chain, still occurs primarily in

---

* To be more exact, one can use the photospheric density for $\rho(R)$.

the core. Energy transport is *radiative* for most of the interior, but beyond $0.8R_\odot$ the temperature gradient becomes sufficiently steep to maintain *convection*. The solar granulation constitutes direct evidence for such a convection zone, as already discussed in Chapter 9 and illustrated schematically in Figure 9–1.

Models for other stars differ from that for the Sun in the extent of the convection zone and therefore the degree of mixing. In stars more massive than the Sun (i.e., those higher on the main sequence), convection occurs at the central core, and the envelope is entirely radiative.

(b)   The Russell-Vogt Theorem

In 1926, H. N. Russell and H. Vogt independently showed that the structure of a star [hence $M(r)$] is *uniquely* determined by the chemical composition and the mass. The *Russell-Vogt theorem* tells us that the values of the other parameters, such as temperature, pressure, density, luminosity, and distance from the center (r), are completely determined for any point in the star specified by $M(r)$. Moreover, there is only one solution possible, given the total mass and the composition. Actually, because of the interdependence of the several parameters, the Russell-Vogt theorem may be generalized, for any one of the three parameters, M, R, or L, will determine the structure if the chemical composition is also known.

The importance of the theorem arises in interpreting the H-R diagram, for it tells us that stars of the same composition but of differing masses should lie along a smooth line—the greater the mass, the higher the luminosity and the surface temperature. This line, as we shall see, is the zero age main sequence. Stars that lie elsewhere on the H-R diagram, such as the red giants, have chemical compositions different from the main-sequence stars as a result of the gradual accumulation of products of nuclear reactions. Such compositional changes are part of stellar evolution.

15–3   STELLAR EVOLUTION

The study of the changes which take place in stars, as they alter their composition because of thermonuclear reactions, is a part of the subject of *stellar evolution*. The term must be thought of not as strictly analogous to biological evolution, where succeeding generations evolve new characteristics, but rather in the more general sense including the growth or development of individuals.

(a)   The Birth of Stars: Protostars and
Pre-main-sequence Stars

Stars form out of clouds of dust and gas. The *globules*, or small dense interstellar dust clouds sometimes seen projected against star clouds or bright nebulae, may well be precursors of such protostars. A later stage is represented by the *Herbig-Haro objects*—reasonably round, self-luminous objects which have a fuzzy appearance, and which change in appearance over the years. The *T Tauri stars* are believed to be the first stages of truly stellar objects. They are always found very closely associated with interstellar dust and gas, indicative of their formation out of this matter.

The entire process of star formation is still not completely understood, but it seems inevitable that stars are the result of the gravitational contraction of a cloud containing dust and gas. As such a cloud contracts, gravitational potential energy is partially transformed into kinetic or thermal energy, and is partially radiated away.

Calculations by R. B. Larson, P. Bodenheimer, C. Hayashi, and others in the late 1960s show that, as the result of the initial free-fall collapse, the center of the cloud reaches stellar densities very rapidly, within a thousand years or so. This core at first includes only a small fraction of the total mass, but grows as material from the rest of the cloud falls into it. Such growth by *accretion* requires about a million years for a star like the Sun, and less time for more massive stars.

The temperature of the original cloud is believed to be close to 10 K, but of course it rises rapidly as the stellar core forms. With the rise in temperature, particles in and near the core vaporize, molecules (especially H_2) dissociate, and finally atoms become ionized. The infalling dust obscures the core, initially completely absorbing its radiation so that it is unobservable. At later stages, the dust particles absorb the radiation from the core and then re-emit it at *infrared* wavelengths, thus permitting it to escape the cloud. The cloud continues to remain optically opaque until virtually all of the material has fallen into the core.

Pre-main-sequence stars, therefore, should be strong infrared emitters; in fact, a substantial number of infrared sources *are* identified with objects believed to be pre-main-sequence stars. Eventually, the dust thins enough for the star to become visible at optical wavelengths, although it may continue to radiate strongly in the infrared because part of the radiation is still absorbed and then re-emitted by the surrounding dust. Some of the variable stars embedded in dense dust clouds, either reflection nebulae or dark nebulae (see Chapter 17), may be examples of this stage of a pre-main-sequence star. Many of these variables are T Tauri-type stars, which will be described in Chapter 16 (see Figure 15–2). Why they are variable is still not understood, for they are in quasi-hydrostatic equilibrium. Gradually, the infrared radiation becomes less dominant, as the dust either has fallen into the star or is ejected from it.

At this point the star, and it truly is a star now, has a surface temperature ranging from 4000 K to 7000 K, or even higher depending on its mass. The radius is still somewhat greater than it will be when it finally reaches the main sequence. Contraction continues, increasing the temperature at the center until it is sufficiently high to initiate hydrogen burning. The star has now reached the main sequence.

The position of a pre-main-sequence star in the H-R diagram changes in a complex manner during its evolution. An example based on Larson's recent calculations for a solar-mass star is illustrated in Figure 15–3. The evolutionary track shown starts at the stage when most of the radiation from the core escapes the cloud in the form of infrared radiation. Prior to that time the cloud was totally opaque. The time scale is shorter the more massive the star, for the gravitational acceleration will be greater. Because more potential energy is converted into kinetic energy, more energy must be radiated away and the surface temperature will be higher. Hence, the position at which a star arrives on the main sequence depends critically on its mass. Massive stars are hot and bright, while small stars are cool and faint, as is confirmed observationally by the mass-luminosity relation [section 11–1(b)].

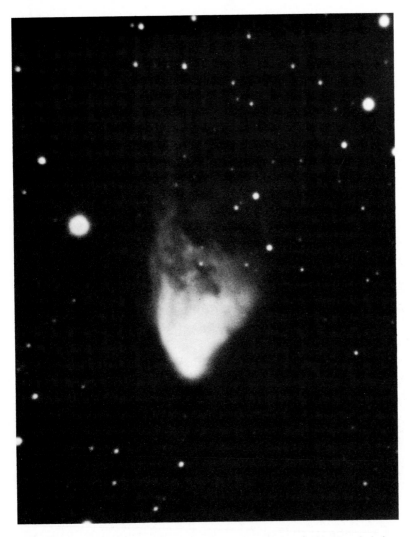

Figure 15–2. *R Monocerotis—a T Tauri-Type Star.* The star is embedded in nebulosity, and both nebula and star are variable. (Hale Observatories)

Stars less massive than $0.05M_\odot$ or perhaps even $0.1M_\odot$ do not attain central temperatures sufficiently high for hydrogen burning to be an important source of energy. Nearly all the energy for such "stars" derives from gravitational contraction. Eventually this contraction halts as the gas becomes degenerate [see section 15–3(d, i)]. The star cools off slowly to become a *black dwarf*.

(b) Evolutionary Tracks for Different Stars

When thermonuclear reactions become effective as a form of energy generation, the internal stellar pressures quickly attain the level necessary for hydrostatic equilibrium. At this time, contraction stops altogether and the star is stable; it is now truly a star, whereas prior to this it was a protostar. We refer to this moment

Figure 15–3. *Possible Evolutionary Track for a Pre-Main-Sequence Star of One Solar Mass.* The numbers next to the curve are the times in years since the formation of the stellar core. The solid circle is the point at which half of the total mass has fallen into the core, and the cross indicates when such accretion is complete and the star becomes optically visible as a pre-main-sequence star. Its final position on the main sequence is indicated by ⊙. The dashed lines indicate constant radius locii. (After R. B. Larson)

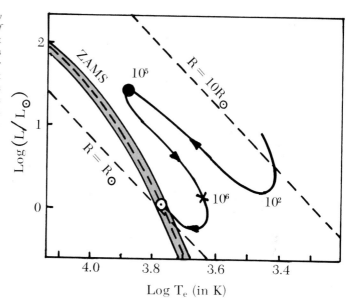

of stellar development as *age zero*. Zero-age stars have surface temperatures and luminosities corresponding to main-sequence stars; in other words, they lie on the main sequence of the H-R diagram. As hydrogen is converted into helium, the composition and therefore the mean molecular weight change, thus altering the structure of the star. These changes are gradual at first, and the star does not deviate greatly from its initial position on the main sequence. To specify the position in the H-R diagram for stars just starting their hydrogen burning (transmutation of hydrogen to helium), we refer to the *zero age main sequence* (*ZAMS*).

The development of stars on and from the zero age main sequence depends primarily on two characteristics: the total mass and the initial composition. In the discussion of the protostars, we saw that the rate of stellar development depends strongly on the *total mass*, and that the more massive stars lie higher on the main sequence. Massive stars cannot remain on the main sequence as long as the smaller-mass stars can; this is readily seen by comparing rates of energy output (luminosities). Early-type stars (O and B stars) expend their energies at far greater rates in proportion to their masses than does the Sun. Conversely, a main-sequence M star is exceedingly sedate and conservative in its energy expenditures.

We also noted earlier that the structure of a star depends on the mean molecular weight or *composition*. Moreover, if there is not enough $_6C^{12}$ available to act as a catalyst for the CNO cycle, that chain of reactions will not occur, however high the temperature is. Stars currently being formed out of the interstellar gas and dust will accrete a significant fraction of their mass (on the order of one per cent) in the form of elements heavier than hydrogen and helium. If they are slightly more massive than the Sun, such stars will then generate energy predominantly by the carbon cycle. These are the bright, young Population I stars in the spiral arms, where the dust and gas are found.

Originally, the Galaxy itself was a cloud of gas, presumably composed primarily of hydrogen, with some helium, but with few heavy elements. As the proto-Galaxy condensed, stars formed out of this hydrogen gas cloud. Because there was no carbon present, however, the carbon cycle could not operate even in the massive stars; therefore, their evolution differed from that of Population I stars. These stars born in the early stages of the formation of the Galaxy are the stars now called old Population II stars.

(i) *The Population I Star of* $5M_\odot$

We can plot a continuous curve on the H-R diagram representing the positions a star occupies at different stages of its evolution; such a plot is called an *evolutionary track*. Let us follow the evolution of a Population I star with a mass five times that of the Sun, as outlined by I. Iben. It is possible that details of the track will change as new nuclear reaction data and model calculations become available, but this description fits most of the observational data and represents the present state of our knowledge. We have chosen a $5M_\odot$ star rather than the Sun as an example, because of its more rapid development; there must presently exist stars that have traveled the entire evolutionary track. Solar-type stars have changed only slightly from their original positions on the zero age main sequence. Table 15–1 summarizes the important characteristics and phases of stellar evolution.

The start of interval A at O in Figure 15–4 represents the main-sequence position of the star in the H-R diagram at age zero, when hydrogen burning started at the center of the star. Initially, only temperatures at the very center are sufficiently high for the CNO cycle to operate, and fresh hydrogen is supplied by convective mixing throughout a central core. Conditions in the core change in a rather complex fashion, both as a result of the depletion of hydrogen in the core and because of

Table 15–1. **Summary of Stellar Evolution** (see Figure 15–4)

	Stage	
< O	Protostar	(a) Rapid gravitational collapse of dust and gas cloud, accompanied by heating of the interior and the ionization of atoms.
	Pre-main-sequence star	(b) Semi-hydrostatic equilibrium; contraction and heating continue.
O	ZAMS star	Hydrogen burning commences.
A, B	Initial evolution on main sequence	Hydrogen is consumed in the core, and some contraction occurs.
C	Evolution off main sequence	Hydrogen is depleted in the core, followed by the establishment of an isothermal helium core and a hydrogen-burning shell.
D, E	Evolution to the right in the H-R diagram	Rapid contraction of the core, expansion of the envelope, and a narrowing of the hydrogen-burning shell.
F, G, H	Red giant	Increased energy output, convective envelope, and the start of helium burning.
I, J	Cepheid?	Contraction of the convective shell, and helium burning in the core as the major energy source.
K, etc.	Supergiant?	Helium-burning shell.

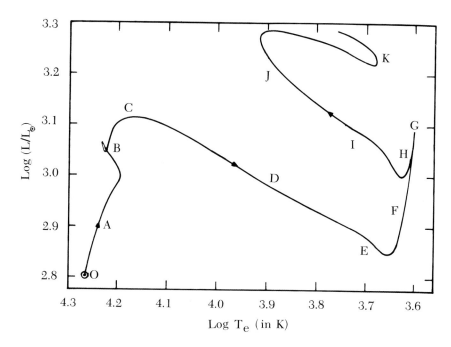

Figure 15–4. *Evolutionary Track for a Star of Mass 5M$_\odot$.* O—Start on zero age main sequence. A—Hydrogen burning in core (6.44 × 10^7 yr). B—Overall contraction phase (2.2 × 10^6 yr). C—Hydrogen burning in thick shell (1.2 × 10^6 yr). D—Shell narrowing phase (8 × 10^5 yr). E—Convective envelope begins to extend inward rapidly—surface abundances begin to change. F—Red giant phase (5 × 10^5 yr). G—Ignition of triple-alpha process. H—First phase of core helium burning (5 × 10^6 yr). I—Disappearance of deep convective envelope—rapid contraction (6 × 10^5 yr). J—Major phase of core helium burning (8 × 10^6 yr). K—Overall contraction with exhaustion of central helium. (After I. Iben)

other nuclear reactions involving oxygen and nitrogen. When only a small fraction of all the material in the core is hydrogen, first the core and then the whole star contracts (stage B).

Eventually all the core hydrogen is exhausted. As a result of the gravitational contraction of the core, material just beyond the core is pulled into higher temperature regions and hydrogen burning starts in a shell surrounding the (primarily) helium core. This shell is fairly thick at first (stage C), but as a larger fraction of the total mass of the star is concentrated into the core, the shell narrows (stage D). Throughout stage C the core is so dense that, to a large extent, it supports the weight of material above it. When too much material is in the core (called the Schönberg-Chandrasekhar limit), this equilibrium situation ends and the core contracts much more rapidly. When this happens, energy generation in the shell is accelerated and the outer envelope expands. Such an expansion must be accompanied by cooling at the surface, and the position of the star in the H-R diagram moves to the right (stages C to E).

During the last part of the expansion phase, convection develops in the envelope and changes the direction of the evolutionary track from decreasing to increasing luminosity (region E) by carrying a greater part of the energy outward to the star's surface. This is the beginning of the *red giant* phase of the star's life (stages F to H). Temperatures in the interior continue to rise owing to the core contraction; finally a point is reached (about 10^8 K) where the triple-α process can begin at the star's center. Helium burning at this stage is short-lived; the star's surface cools and it drops from position G to H.*

Once again gravitational contraction takes over until temperatures are sufficiently high to reignite helium. The balance between helium burning in the core and hydrogen burning in the shell gradually shifts in favor of helium burning, and the surface of the star becomes both hotter and brighter (I, J).

When the star reaches this region of the H-R diagram, the *outer layers* become quite unstable and may alternately expand and contract. The pulsations cause the star to change its luminosity periodically; it is a variable star. The phase just before point I may correspond to a Cepheid variable (see Chapter 16). When the helium is exhausted and the core consists of carbon, contraction again occurs and a helium-burning shell is formed; by now the star may be a supergiant (K). These last phases of stellar evolution are the most uncertain, and the theoretical position of the star in the H-R diagram is subject to change as further work is done in the theory of stellar evolution. This proviso is also true (to a lesser extent) for the core helium-burning stages.

(ii) *Differences Due to Mass*

Stars of other masses follow the same general development as that described for the $5M_{\odot}$ star. The details and especially the time scales differ. Figure 15–5 shows the evolutionary tracks followed across the H-R diagram for stars of several representative masses. Less massive stars, like the Sun, have more extensive hydrogen-burning regions but no convective cores. Less massive stars suffer fewer of the structural changes and subsequent contractions which more massive stars undergo; hence, their tracks are considerably simpler and smoother. In solar-type stars, hydrogen burning extends over a relatively large fraction of the star, both in the core-burning phase (phases A and B in Figure 15–5) and during the shell-burning phase (up to E). As in the more massive stars, the red giant phase (F to H) corresponds to the establishment of a deep convective envelope.

In very massive stars (e.g., $15M_{\odot}$) the temperatures in the core become high enough to start the triple-α process shortly after the star leaves the main sequence. Helium burning in the core starts well before the star has become a red giant (note the location of stage G for the $15M_{\odot}$ star). The star moves horizontally rather than upward across the H-R diagram. Stars may move in and out of the red giant region as well as the unstable atmosphere zone (Cepheid variables) several times; massive stars probably do this more often than less massive ones.

Stars less massive than the Sun enter the red giant region twice. After the first red giant phase they evolve through the instability zone, which in this case corresponds to the RR Lyrae variables. Of the low-mass stars, only old Population II stars, like those in the globular clusters, have had time to develop to this stage.

---

* The sharp peak in the evolutionary track is referred to as the *helium flash*.

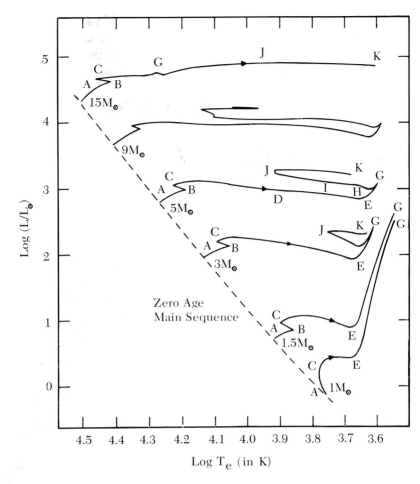

Figure 15–5. *Evolutionary Tracks For Stars of Different Masses.* Note that the luminosity scale has been compressed compared to Figure 15–4. (After I. Iben)

Other differences due to mass will be noted as we discuss the terminal phases of stellar evolution. Some low-mass stars probably go through a planetary nebula phase prior to becoming white dwarfs. More massive stars explode as supernovae and end up as neutron stars. *Very* massive objects *may* end up as black holes.

(c) Interpreting Cluster H-R Diagrams

If a cluster of stars is formed, such that all member stars contract out of the gas cloud more or less simultaneously, the locations of the stars in the H-R diagram will depend upon the *time* elapsed since the initial (formation) event. The times listed in Table 15–2 tell us that 10^8 years after the formation of the cluster, stars with masses greater than about $8M_\odot$ will have evolved beyond the end of the track shown. The $5M_\odot$ stars will have reached stage K, and less massive stars will still lie on the main sequence. The rapidity of the post-main-sequence phase, compared to evolution on and near the main sequence, allows us to "date" a cluster on the basis of its H-R diagram. In the above example of 10^8 years, the cluster has stars on the main sequence up to a luminosity corresponding to 3 solar masses or slightly more, say $L/L_{\odot} = 100$ or absolute magnitude $= 0^{m}0$. Somewhat more massive stars lie a little to the right of the main sequence, and some stars have reached the giant branch, but have not yet gone beyond it. Such a cluster closely resembles M11, as seen in Figure 15–6.

The *turn-off* from the main sequence, therefore, tells us the time elapsed since the stars first arrived on the zero age main sequence. *Ages of clusters* may be found by comparing the turn-off points in the H-R diagram of Figure 15–6 with the scale on the right of the diagram. The observed H-R diagrams of various clusters represent the actual loci of the ends of the evolutionary tracks for member stars to that particular time since formation. These loci are sometimes referred to as *isochrones* (constant-time lines). The fact that computed isochrones closely resemble observed H-R diagrams is one of the major triumphs of modern astrophysics.

The time taken up by the initial (protostar) contraction is sufficiently short that it can usually be neglected, particularly for well-developed clusters. In many clusters, however, the contraction time for massive stars is so much faster than that for light stars that the massive stars will have already started to evolve off the main sequence by the time the low-mass stars reach it. An example of a very young cluster is NGC 2264, which contains a large number of T Tauri stars—that is, the pre-main-sequence stars discussed earlier (Figure 15–7). Here we see massive stars which have already reached the zero age main sequence, as well as less massive stars which are still in protostar form.

Another young cluster is h and χ Persei; this cluster has many massive and luminous stars. The turn-off at the very upper end of the H-R diagram of this cluster (see Figure 15–6) and the appearance of the supergiant branch have usually been attributed to the rapid evolution of these massive stars.

At the other extreme, we have two old open clusters (Population I) whose stars are highly evolved. One of these, M67, is seen from Figure 15–6 to have stars of about $1.25 M_{\odot}$ evolved onto the red giant branch, while stars of $1 M_{\odot}$ are just about to leave the main sequence. The other cluster, NGC 188, is even older, and stars of $1 M_{\odot}$ have already progressed to the red giant stage. One estimate gives the ages for these clusters as 5×10^{9} and 11×10^{9} years respectively.

The H-R diagrams of globular clusters (Population II) shown in Figure 15–8 differ from the H-R diagrams typical to both intermediate Population I clusters (like M11) and the old Population I clusters (like M67). The globular clusters are probably the same age as M67 or somewhat older, but, because they

Table 15–2.   The Time Scales of Stellar Evolutionary Tracks (see Figure 15–5)

Mass $(M/M_{\odot})$	Interval (years)				
	O, A–B	C–E	F–G	G–J	J–K
1	9×10^{9}	1.4×10^{9}	$\gtrsim 10^{9}$		
1.5	1.6×10^{9}	4.5×10^{8}	$\gtrsim 2 \times 10^{8}$		
3	2.3×10^{8}	1.5×10^{7}	4.2×10^{6}	6.6×10^{7}	6×10^{6}
5	6.8×10^{7}	2.1×10^{6}	4.9×10^{5}	1.6×10^{7}	9.3×10^{5}
9	2.2×10^{7}	2.4×10^{5}	6.5×10^{4}	3.8×10^{6}	1.6×10^{5}
15	1.0×10^{7}	7.6×10^{4}		1.5×10^{6}	3.5×10^{4}

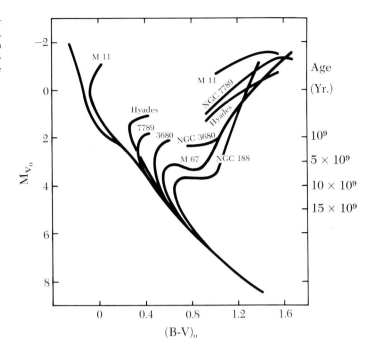

Figure 15–6. *H-R Diagram for Open Clusters.* Schematic H-R diagram as observed for open clusters (Population I) of different ages. (After A. R. Sandage and O. J. Eggen)

started with almost pure hydrogen and helium and virtually no heavy elements, their evolutionary tracks took different forms than for Population I clusters. The present H-R diagrams for globular clusters represent the end points of the evolutionary tracks. Differences from one cluster to another are attributed to differences in the initial chemical composition. Some clusters like M13 and 47 Tucanae were formed with a somewhat higher metal abundance than others like M15. The lower ends of the main sequences seem to fall below the zero age main sequence; whether this is in fact correct still remains uncertain. They can be brought into coincidence by correcting the stellar magnitudes and colors for the effects of spectral absorption lines on the total flux, a phenomenon termed *line blanketing* [see section 12–1(b)].

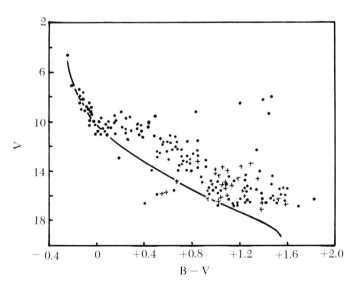

Figure 15–7. *H-R Diagram for NGC 2264.* The solid line represents the zero age main sequence. T Tauri stars are shown as crosses. Note how the bright blue end of the ZAMS is already well populated, but fainter, less massive stars are still in the pre-main-sequence stages. (After M. Walker)

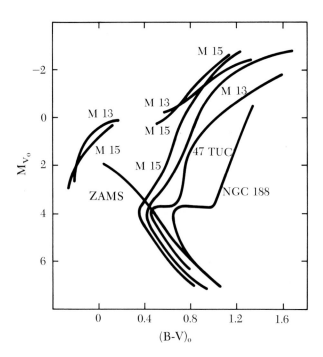

Figure 15–8. *H-R Diagram for Globular Clusters.* The composite diagram for three globular clusters compared with the old galactic cluster NGC 188 and the zero age main sequence (see also Figure 15–6). (After A. R. Sandage)

The reason that clusters of the same age can be separated into Populations I and II lies in the evolution of the Galaxy as a whole. The gas and dust near the galactic plane were contaminated with heavy elements, primarily ejected by rapidly evolved stars (supernovae, discussed in the following section), even at the time the old galactic Population I clusters were formed. The gas far from the plane was still of the original primordial composition, and thus produced Population II-type stars and globular clusters.

(d)   Stellar Demise and Final States

What happens to a star as it evolves beyond the helium-burning stage? Here knowledge is largely replaced by speculation. Nevertheless, we recognize two types of stars as representing the end of stellar evolution: the ubiquitous white dwarfs and the neutron stars (observed as pulsars). A third category is that of the intriguing, but highly speculative, black holes. Our discussion here is limited to the properties of these objects pertaining to stellar evolution. The observational characteristics of the phenomena (planetary nebulae and supernovae) that may lead to their formation are left to the following chapter, as is the discussion of the objects identified as neutron stars—the pulsars.

(i)   *White Dwarfs*

There are several roads to the white dwarf stage, depending on the mass of the star. In the case of a star like our Sun, contraction of the carbon core may not produce temperatures sufficiently high to bring about burning of carbon. The core will, nevertheless, contract to a highly compressed state, and the increasing temperatures will accelerate the helium shell burning rates. The situation is similar to stage C described for the $5M_\odot$ star. As in that case, the outer envelope expands and cools. Now, however, the expansion is so great that the envelope becomes separated from the core as a transparent shell; this is a *planetary nebula* (see Chapter 16, especially Figure 16–7). The core, having lost its envelope, now stands revealed as a hot, very dense "star," a *white dwarf*.

In less massive stars, say less than $0.35M_\odot$, helium burning may not become appreciable because of low central temperatures. Such a star continues to contract gravitationally to become a white dwarf even without ejecting an envelope. Other, more massive stars end as white dwarfs by some kind of mass loss, perhaps by exchange of mass between members of a binary system. A substantial number of white dwarfs are in fact faint companions to larger, more massive stars; Sirius B and 40 Eridani B are well-known examples.

Observationally, *white dwarfs* are stars which lie well below the main sequence, mostly to the left in the H-R diagram; their surface temperatures range from 5000 K to 50,000 K. They are subluminous both for their spectral type and for their mass; we find that they have very small radii, of the order of $0.02R_\odot$ and smaller (i.e., they are the same size as our Earth!). Since their masses may be of the order of a solar mass, this means that they are exceedingly dense. At such densities, the stellar material no longer behaves as an ordinary gas, but becomes so tightly packed that the electrons cannot move completely at random. Hence, the electrons' motions are subject to limitations imposed by the close presence of other electrons. Some electrons may still move at very high velocities, but they cannot change their velocities by collisions as in an ordinary gas; the electrons may change velocity only by exchanging "orbits" with other electrons. The laws of quantum mechanics still hold and the *Pauli exclusion principle* remains valid: only two electrons with opposite spins can have a given energy in a given volume at one time. Because of the close packing, less space is available; therefore, the number of possible velocities or energies permissible for an electron becomes smaller. Such a material is called a *degenerate* electron gas. In such a degenerate gas, the electrons are distributed more or less uniformly throughout the medium, surrounding the nuclei. The nuclei themselves are regularly spaced and become more tightly constrained as the pressures increase, until they are so fixed with respect to each other that they resemble a crystalline lattice. Such a condition more closely resembles a solid than a gas, and we may refer to this as a *quantum solid;* the state just prior to this is called a *quantum liquid*. Central densities in white dwarfs may be as high as 10^7 or 10^8 gm cm^{-3}, compared with about 150 gm cm^{-3} for the Sun!

The cause of these high densities lies in the fact that all available nuclear energy has been expended, and the star contracts gravitationally until stopped by the pressure of the degenerate electron gas. Only stars of mass smaller than about $1.4M_\odot$ (called the Chandrasekhar limit) can be stable white dwarfs because of limitations imposed by the stellar structure, which in turn depends upon both hydrostatic equilibrium and the nature of the degenerate electron gas.

Since there is no free internal energy left after contraction stops, the star gradually cools. This cooling process is exceedingly slow, for the star may take 10^{10} years to cool to 3000 K. Eventually it will fade into a "black dwarf."

(ii)   *Supernovae and Neutron Stars*

Temperatures in the cores of massive stars $(M > 1.4M_\odot)$ easily become high enough for carbon burning to occur (about 6×10^8 K), wherein carbon nuclei react with each other. Such a core will already also contain some of the other products of helium burning, such as oxygen and neon, and these too will become involved in nuclear reactions. Many types of reactions will occur at temperatures near 3×10^9 K. The end result is a core containing a variety of heavy nuclei, in particular an abundance of iron, Fe^{56}.

When this core collapses, the outer layers also fall in. This collapse is sudden and implosive, and brings lighter elements from outer shells to the extreme temperatures of the core. As a consequence, this material undergoes nuclear reactions at a tremendous rate. The collapse also causes a shock wave that blows off the outer parts of the star, releasing accelerated particles that become cosmic rays, and high-energy electrons and other gases. During the last moments of the collapse, nuclear processes synthesize (build up) very heavy elements, especially through neutron capture by heavy nuclei. Much of this heavy-element matter is then ejected from the star into the interstellar medium. The explosion, and the ejection of this matter, is observed as a *supernova*, for enormous amounts of radiation are emitted, so much that the star may briefly rival the luminosity of the entire galaxy of which it is a member.

Supernovae eject a substantial fraction of the original star's mass; thus, they enrich the interstellar medium with heavy elements. They are believed to be one of the chief producers of elements heavier than iron in the Universe. There are, in addition, some stars which apparently synthesize certain heavy elements when in one of the red giant or other late phases of their evolution.

Although the outer part of the star is ejected, the inner core continues to collapse. This stellar remnant contains only a fraction of the former mass. The violence of the collapse carries it beyond the white dwarf stage, compressing the matter to a final density of about 3×10^{14} gm cm^{-3} or more, compared to the 10^8 gm cm^{-3} for white dwarfs. In approaching such densities and pressures, electrons are accelerated to velocities close to the speed of light, and tunnel into the nuclei to combine with the nuclear protons, forming neutrons. At a still higher pressure and density, the individual nuclei disintegrate into their component neutrons and few remaining protons. The star itself is now almost like a giant nucleus. At this point, when the density is about 3×10^{14} gm cm^{-3}, the material becomes virtually incompressible as a degenerate neutron gas. The star is a neutron star, with a diameter of 10 to 30 km, and a mass of the order of 1 to $3M_\odot$. (Remember that the rest of the originally massive star was ejected into space.)

The theory of neutron stars was first considered in the 1930s by L. Landau and R. J. Oppenheimer, but these objects remained hypothetical until the discovery of the pulsars, which could only be explained as neutron stars (see Chapter 16).

(iii)   *Black Holes*

Far more speculative, but perhaps even more fascinating, are the *black holes* which may represent the stage of absolutely final collapse for some objects. Stars with masses from 1 to $10^6 M_\odot$, even $10^{10}M_\odot$, may end up as black holes. Solar-mass stars might become black holes if their final collapse is so rapid that it overrides the explosive stage. Very massive objects cannot become either white dwarfs or neutron stars, but they can (theoretically) collapse completely to such a state that the density approaches infinity.

According to Einstein's general theory of relativity, the gravitational effect from such an object on photons is so strong that light cannot escape; hence, the object is invisible or "black." Anything that approaches such a black hole too closely also becomes "swallowed up" and loses its identity. Indeed, the theory predicts that such a final state, the black hole, is *inevitable* for some massive objects and stars. Only time will tell whether observation confirms this amazing result.

Problems

15–1. Verify that about 6×10^{14} grams of hydrogen are converted into helium in our Sun every second.

15–2. (a) If a star is characterized by $M = 2 \times 10^{35}$ gm and $L = 4 \times 10^{39}$ erg sec^{-1}, how long can it shine at that luminosity if it is 100 per cent hydrogen and converts all of the H to He?
(b) Do a similar calculation for a star of mass 10^{33} gm and luminosity 4×10^{32} erg sec^{-1}.

15–3. (a) Consider the following hypothetical nuclear reaction chain:

$$_1H^1 + {}_6C^{13} \rightarrow (?)$$

$$_2He^4 + (?) \rightarrow (??) + e^+ + v$$

$$(??) \rightarrow (???) + \gamma$$

Using Appendices 3 and 4 at the end of this book, unambiguously identify the nuclear species (?), (??), and (???).
(b) Do the same for the α-decay reaction:

$$_{92}U^{238} \rightarrow (?) + {}_2He^4$$

15–4. In your own words, briefly describe the evolution of the following stars from a cloud of gas and dust to the final demise:
(a) $M = 10M_\odot$,
(b) $M = 0.1M_\odot$.
Clearly indicate which stages of the evolution are highly uncertain.

15–5. Using Figures 15–5 and 15–6, and the data given in Table 15–2, sketch the H-R diagrams for star clusters of ages 10^7 years, 10^8 years, and 10^9 years (these are *constant-time* lines!). Clearly label the axes, and comment upon the significance of your results (e.g., turn-off points).

15–6. Consider stars of $1M_\odot$. Compute the mean mass density $\bar{\rho}$ for the following:
(a) our Sun $(R_\odot = 7 \times 10^5$ km),
(b) a white dwarf $(R = 10^4$ km),
(c) a neutron star $(R = 10$ km).
Now consider a carbon $({}_6C^{12})$ nucleus, of radius $r = 3 \times 10^{-13}$ cm, and compute its mean density; discuss the significance of all these results!

15–7. (a) A white dwarf has an apparent magnitude $m_v = 8^m\!.5$ and parallax

$\pi'' = 0''.2$. Its bolometric correction is $+2^m.1$ and $T_{eff} = 28,000$ K. Find the radius of the star. Compare to the radius of Earth.

(b) A neutron star has $T_{eff} = 5 \times 10^5$ K and a radius of 10 km. Find its luminosity.

(c) A pre-main-sequence cloud starts out with $T = 15$ K and $R = 4 \times 10^4 R_\odot$. Find $L/L_\odot$ and the wavelength of the peak of the Planck curve.

15–8. Calculate the kinetic energy $(mv^2/2)$ for each of the following:

(a) nova outburst which accelerates $10^{-5} M_\odot$ to $v = 10^3$ km sec^{-1},

(b) formation of a planetary nebula in which $1 M_\odot$ is accelerated to $v = 20$ km sec^{-1},

(c) supernova outburst which accelerates $1 M_\odot$ to $v = 4 \times 10^3$ km sec^{-1}.

How long, in years, would it take the Sun to radiate away these energies?

Reading List

Arnett, W. D., et al.: *Nucleosynthesis*. New York, Gordon and Breach Science Publishers, 1968.

Chandrasekhar, S.: *An Introduction to the Study of Stellar Structure*. Chicago, University of Chicago Press, 1939; New York, Dover Publications, 1957.

Chiu, H.-Y.: *Stellar Physics*. Waltham, Blaisdell Publishing Company, 1968.

Chiu, H.-Y., and Muriel, A., (eds.): *Stellar Evolution*. Cambridge, Massachusetts, MIT Press, 1972.

Clayton, Donald D.: *Principles of Stellar Evolution and Nucleosynthesis*. New York, McGraw-Hill Book Company, 1968.

Cox, John P., (and Giuli, R. Thomas): *Principles of Stellar Structure*. Vols. 1 and 2. New York, Gordon and Breach Science Publishers, 1968.

Eddington, Arthur S.: *The Internal Constitution of the Stars*. Cambridge, England, Cambridge University Press, 1926.

Hayashi, Chushiro: "Evolution of Protostars." *Annual Review of Astronomy and Astrophysics*, Vol. 4, pp. 171–192 (1966).

Iben, Icko, Jr.: "Stellar Evolution: Within and Off the Main Sequence." *Annual Review of Astronomy and Astrophysics*, Vol. 5, pp. 571–626 (1967).

Johnson, Martin: *Astronomy of Stellar Energy and Decay*. London, Faber and Faber, Ltd., 1948; New York, Dover Publications, 1959.

Meadows, A. J.: *Stellar Evolution*. New York, Pergamon Press, 1967.

Menzel, Donald H., Bhatnagar, Prabhu L., and Sen, Hari K.: *Stellar Interiors*. New York, John Wiley and Sons, 1963.

Motz, L., and Duveen, A.: *Essentials of Astronomy*. Belmont, California, Wadsworth Publishing Company, 1966.

Ostriker, Jeremiah P.: "Recent Developments in the Theory of Degenerate Dwarfs." *Annual Review of Astronomy and Astrophysics*, Vol. 9, pp. 353–366 (1971).

Page, Thornton, and Page, Lou Williams, (eds.): *The Evolution of Stars*. New York, The Macmillan Company, 1968.

Rose, William K.: *Introduction to Astrophysics*. New York, Holt, Rinehart and Winston, 1973.

Schwarzschild, Martin: *Structure and Evolution of the Stars*. Princeton, New Jersey, Princeton University Press, 1958; New York, Dover Publications, 1965.

Stein, R. F., and Cameron, A. G. W., (eds.): *Stellar Evolution*. New York, Plenum Press, 1966.

Swihart, Thomas L.: *Astrophysics and Stellar Astronomy*. New York, John Wiley and Sons, 1968.

Chapter 16
Variable and
Unusual Stars

In Chapter 15 we discussed the structure and evolution of the majority of the observed stars—the slowly-changing *normal* stars. There we alluded to the existence of rapidly evolving and/or spectacularly changing stars, and mentioned that most normal stars pass through several such phases during their lifetime. In this chapter, we intend to discuss the observational and structural properties of such *variable stars* in detail. This investigation will help to illuminate and clarify many of the intriguingly peculiar phases of normal stellar evolution, and will illustrate the diverse phenomena with which stellar astrophysics must concern itself.

The variable stars of primary interest are the *intrinsic variables*, a term used to differentiate them from the geometrical variables (such as eclipsing binary stars). Such variables may be roughly divided into two categories: (a) the *pulsating stars*, whose atmospheres undergo periodic expansion and contraction, and (b) the *cataclysmic* or *explosive variables*, which exhibit sudden and dramatic changes. The pulsating stars include Cepheids, RR Lyrae stars, the irregular RV Tauri stars, and the long-period or Mira stars. Novae, dwarf novae (U Geminorum stars), and supernovae comprise the cataclysmic variables. Variables which do not fit neatly into either category include flare stars, T Tauri stars, spectrum variables, and magnetic stars. Tables 16–1 to 16–3 list the characteristics of the most important types of variable stars, and Figure 16–1 indicates their generic positions in the H-R diagram.

As we mentioned earlier, a single star may pass through several stages of variability during its evolution. A star of one solar mass or less may initially be a T Tauri star before settling onto the main sequence; then, much later it may pulsate **387**

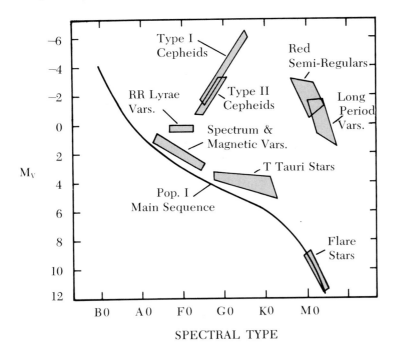

Figure 16–1. *The Positions of Variable Stars in the H-R Diagram.*

as an RR Lyrae star after passing through the giant stage. More massive stars contract to the main sequence very rapidly, so that it is difficult to say whether any have been observed during this contraction phase. In later stages of evolution, however, these massive stars are observed as Cepheid variables, following the giant or supergiant stage. To become a white dwarf, a massive star must lose much of its mass, and it may do this as a red giant, a planetary nebula, or a supergiant. Many stars apparently reach their final stages of evolution by collapsing to the white dwarf or neutron star stage; the recently-discovered pulsars are probably neutron stars formed by supernova explosions.

16–1 THE NOMENCLATURE OF VARIABLE STARS

In Chapter 2 we discussed the names of stars. To designate variable stars (both intrinsic and geometrical), astronomers have devised another system of nomenclature. Each variable star is given a set of capital letters, followed by the genitive name of the constellation in which the variable occurs. The capital letters follow an alphabetical sequence based on the order of discovery: the first variable discovered in a given constellation is designated R, with the rest of the alphabet through Z used for successively discovered variables. After Z, double letters are used in the order: RR, RS–RZ, SS, ST–SZ, . . . , ZZ. Then we return to the beginning of the alphabet (the letter J is omitted to avoid confusion with I): AA, AB–AZ, BB, BC–BZ, . . . , QZ. If there are more variables within a constellation, we resort to numbers starting with V335 (V for variable), since the single and double letters have already accounted for 334 variables. Some examples of variable star designations are (in sequence!): R Monocerotis, T Tauris, RR Lyrae, UV Ceti, AG Pegasi, BF Cygni, V378 Orionis, and V999 Sagitarii.

Until recently, not all novae were included in this system; each was simply designated by the constellation name and the year of occurrence (e.g., Nova Aquilae 1918). Since 1925 A.D., novae have been given variable star designations, such as RR Pictoris and DQ Herculis; today, even the earlier novae have been assigned such designations, so that Nova Aquilae 1918 is also known as V603 Aquilae.

Variable stars which are bright enough to have been given a proper name or a Greek letter designation have not been renamed. Therefore, we have β Canis Majoris, δ Cephei, and Algol, all of which are well-known variables and prototypes of their stellar classes.

16–2   PULSATING STARS

(a)   Observational Data on Pulsating Stars

The vast majority of known variable stars have been detected by observations of the fluctuating stellar luminosity. The most important of the pulsating stars are the *Cepheid variables*; let us consider them as the prototypes of pulsating stars. In distinct contrast to the light curves of most eclipsing stars, the light curves of pulsating stars exhibit *continual* changes in brightness. In addition, the spectrum of a pulsating star varies periodically, corresponding to a change in stellar surface temperature which may range over an entire spectral class. The spectral lines show variable Doppler shifts, from which *radial velocity curves* (for the stellar atmosphere) may be deduced. From the periodic radial velocities, we find that the star alternately expands and contracts; the change in the star's radius follows straightforwardly. Figure 16–2 illustrates the typical periodic variation of all these parameters.

A binary star model was involved in the early attempts to explain the behavior of Cepheids. The observed velocity curves, however, implied that the binary orbit

Table 16–1.   Pulsating Stars

Type	Prototype	$\overline{M}_v$	Spectral Class	Pulsation Period Range	Characteristic Period	Population
Classical Cepheids	δ Cephei	-0.5 to -6	F6 to K2	1^d to 50^d	5^d to 10^d	I
Population II Cepheids (W Virginis)	W Virginis	0 to -3	F2 to G6	2^d to 45^d	12^d to 28^d	II
RR Lyrae Stars	RR Lyrae	$+0.5$ to $+1$	A2 to F6	$1^h.5$ to 24^h	$0^d.5$	II
Long Period Variables (Mira)	o Ceti	$+1$ to -2	M1 to M6	130 to 500^d	270^d	disk
RV Tauri Stars	RV Tauri	-3	G, K	20^d to 150^d	75^d	II
Beta Canis Majoris Stars	β Canis Majoris	-3	B1, B2	4^h to 6^h	5^h	I
Semi-Regular Red Variables	α Herculis	-1 to -3	K, M, R, N, S	100^d to 200^d	100^d	I and II
Dwarf Cepheids	δ Scuti	$+4$ to $+2$	A to F	1^h to 3^h	2^h	I

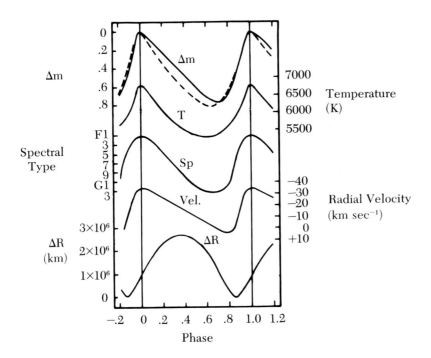

Figure 16–2.   *Profiles of Observables for a Typical Pulsating Star.* Phase-dependent curves are shown for the light variation (Δm)—dashed curve shows change in magnitude if the radius stayed constant; temperature (T); spectral type (Sp); radial velocity (Vel.); and radius (ΔR). (After W. Beckers)

must be smaller than the sum of the stellar radii. Moreover, if the brightness minimum is interpreted as an eclipse minimum, then the radial velocity should be zero; in actuality, the observed radial velocity is a *maximum* at minimum brightness! Therefore, the binary star model had to be rejected.

The pulsational interpretation of Cepheids was shown by A. S. Eddington to require the following relationship between pulsation period P and mean stellar density $\bar{\rho}$:

$$P \propto \bar{\rho}^{-1/2}$$

Observations vindicate this prediction. A *simple* theory of pulsation would predict that a pulsating star should be hottest (and therefore brightest) when it is most highly compressed. Note that the change in stellar radius does not contribute significantly to the change in luminosity. From Figure 16–2, however, we see that the star is actually expanding most rapidly at the time of maximum brightness. This phenomenon is referred to as the *phase lag discrepancy.* Attempts to explain this phase lag have only recently met with apparent success.

In general, the characteristics described in this section pertain to all pulsating stars. The reader is nevertheless urged to study Table 16–1 with care, for there we have summarized the distinguishing features and population characteristics of the several types of pulsating stars. For instance, note that despite their importance as prototypes and distance indicators, Cepheids of Population I are far outnumbered by the long-period variables and the RR Lyrae stars.

Long-period variables exhibit a very large change in *visible magnitude*. This occurs because these stars are cool (about 2000 K), so that most of their radiation lies in the infrared; an increase of even 500 K in surface temperature will shift much more radiation into the visible wavelength band than appears to be warranted by a naive application of Stefan's law (Chapter 8). Although the total amount of radiation emitted by the star increases only slightly, the Planck spectral energy curve slides over into the visible band and dramatically increases the amount of *visible* radiation. In addition, many molecules can exist at these low temperatures, forming a *veil* over the star's surface; as the temperature increases, these molecules are dissociated, and more radiation can penetrate the veil. In fact, these stars may be an important source of interstellar grains (solid particles). The pulsation mechanism discussed below probably does not apply to the long-period variables.

(b) The Pulsation Mechanism

A star pulsates because it is not in hydrostatic equilibrium; the force of gravity acting on the outer mass of the star is not quite balanced by the interior pressure (see Chapter 15). If a star expands as a result of increased gas pressure, the material density (and pressure) decreases until the point of hydrostatic equilibrium is reached and *overshot* (owing to the momentum of the expansion). Then gravity dominates, and the star starts to contract. The momentum of the infalling material carries the contraction beyond the equilibrium point. The pressure is again too high, and the cycle starts anew. Energy is dissipated during such pulsation (analogous to frictional losses), and eventually this loss of energy should result in a *damping* of the pulsations. The prevalence and regularity of pulsating stars imply that the dissipated energy is replenished in some way. The problem of what keeps the pulsations going has been tackled many times over the years since Eddington first considered it.

It now seems that Eddington's original idea of a *valve mechanism* is of paramount importance in explaining stellar pulsations. Only recently have real strides been made in our understanding of the nature of this "valve" and how it leads to the oscillation of a stellar atmosphere. The original suggestion, that the increase of temperature and pressure at the time of compression accelerates the rate of thermonuclear energy generation, is not supported by stellar model calculations. The pulsations are confined to the outer part of the star, where nuclear reactions do not play a role. Thermonuclear reactions probably generate energy at a more or less constant rate even in the interiors of variable stars.

On the other hand, the rate at which the energy is transported outward from the stellar interior can be altered by a "damming" process. We define *opacity* (see also Chapter 9) as the amount of radiative energy absorbed; therefore, a changing opacity will act as a valve. When the stellar atmosphere is transparent, radiation flows freely and the star is bright. Similarly, when the opacity is greatest and radiation is prevented from escaping, the star is faint. If the star is compressed at the time of greatest opacity, the excess radiation is dammed up and exerts pressure on the outer layers of the star; this process provides the energy necessary to continue the pulsations. The atmospheres of pulsating stars have a zone or layer in which the opacity *increases* with the additional availability of radiation. This occurs because singly-ionized helium absorbs the radiation to become doubly ionized. This He^+ ionization region is cooler than the surrounding regions because energy normally used to heat the gas is used to ionize it. S. A. Zhevakin was the first to show that the helium ionization zone contributes to the instability of the stellar atmosphere and thus perpetuates the pulsations. It is possible that zones where ionization of hydrogen and neutral helium occurs also play roles in pulsation mechanisms, but He^+ seems to be the major culprit.

Pulsating stars occupy well-defined areas of the H-R diagram (see Figure 16–1); this observation can be explained in terms of the depth of the He^+ ionization zone. The depth of this region depends upon the structure of the star, which in turn is a function of the star's stage of development. When the zone lies too deep, the valve action is insufficient to overcome damping. When the zone is shallow, the damming action is inefficient and pulsations are not given the necessary impetus. The period-luminosity law (see the following paragraph) is explicable in terms of the position in the H-R diagram where the star becomes unstable to this damming "valve" mechanism.

Complex numerical computations (including nonlinear differential equations) are necessary to follow the behavior of theoretical stellar models of pulsating stars. Such calculations, using the opacity mechanism of the He^+ ionization zone, have been remarkably successful in reproducing the observations; these models have even removed the phase lag discrepancy discussed earlier.

(c) The Period-Luminosity Relation

The fact that the pulsation period of a Cepheid variable is directly related to its luminosity means that Cepheids are among the most powerful tools for determining distances, especially extragalactic distances. This law was first discovered from a study of the variables in the Magellanic Clouds, two small nearby companion galaxies to our Galaxy which are visible in the night sky of the southern hemisphere. To a good approximation, one can consider all stars in each Magellanic Cloud to be at the same distance from us. H. Leavitt, working at Harvard in 1912, found that the brighter the mean apparent luminosity, the longer was the period of the Cepheid variable. Harlow Shapley recognized the importance of this *period-luminosity relation* (P-L Law) and attempted to find the zero-point, for then a knowledge of the period of a Cepheid would immediately indicate its absolute magnitude. This calibration was difficult to perform because of the relative scarcity of Cepheids and their large distances. None are sufficiently near to allow a trigonometric parallax to be determined. Shapley had to depend upon the relatively inaccurate method of statistical parallaxes. His zero-point was then used to find the distances to many other galaxies. These distances were severely revised in the early 1950s, when it became apparent that something was amiss with the distance scale. W. Baade recognized the error during his very careful study of M31 (the Andromeda galaxy), and his revision led to an increase by a factor of two in the distance scale of the Universe.

Further work showed that there are *two* types of Cepheids, each with its own separate, though parallel, P-L law. The revised relation appears in Figure 16–3. The classical Cepheids are the more luminous, are of extreme Population I, and are found in spiral galaxies. Population II Cepheids, also known as W Virginis stars after their prototype, are found in globular clusters and other Population II systems. The present zero-point for the P-L law for classical Cepheids relies heavily upon

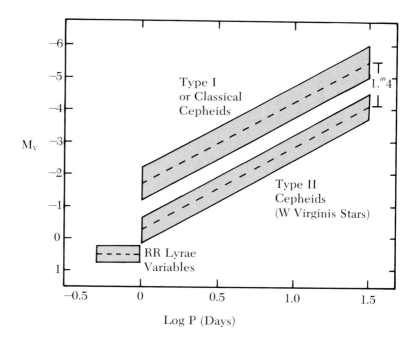

Figure 16–3. *The Period-Luminosity Relation.* This diagram shows the correlation between the absolute visual magnitude (M_v) and the period of pulsation (P) for Cepheid variables. By convention, RR Lyrae variables are included in the same diagram.

five such variables discovered in galactic clusters. The H-R diagrams of these clusters permit us to find distances by using the technique of main-sequence fitting. The P-L law for Population II Cepheids is displaced to fainter magnitudes by about $1^m.4$.

By convention, the RR Lyrae variable stars also appear on the diagram, although they do not show any systematic change in mean luminosity with period. The available evidence indicates that they all have about the same mean absolute magnitude (about $+0^m.5$). All RR Lyrae stars have periods of less than a day. Although they are useful as distance indicators, the RR Lyrae stars can only be used within our own Galaxy and the nearest galaxies owing to their low luminosity.

Cepheids, however, can be seen to far greater distances, for they are highly luminous objects; some are as bright as -6 absolute magnitude. The relatively rare W Virginis stars are fainter than classical Cepheids by $1^m.4$, but they are still brighter than the more numerous RR Lyrae stars. The pulsation periods and mean apparent magnitudes of variables are readily determined from suitably spaced observations. Reference to the period-luminosity law, therefore, immediately yields the *distance modulus* simply and directly. Unfortunately, any corrections for interstellar absorption still remain unknown and must be determined independently (see Chapter 17).

16–3 MISCELLANEOUS VARIABLES

Several types of variable stars do not fit in the category of pulsating stars, but are nevertheless of importance and interest. The T Tauri stars play a significant role in stellar evolution, for these are stars on their way to the main sequence after the protostar stage. Flare stars have arrived on the main sequence, but exhibit stellar activity similar to that observed on the Sun. The stage of evolution of the magnetic and spectrum variables remains uncertain, although they are almost surely in a late phase.

(a)   T Tauri Stars

T Tauri stars were already mentioned in Chapter 15 as stars still in the process of contracting onto the main sequence. At this stage of stellar evolution, these low-mass stars have an extensive convection zone, and surface activity is rampant. Spectral activity includes emission in the hydrogen Balmer lines, and also from ionized calcium and other metals. Some T Tauri spectra also show the forbidden lines typical of gaseous nebulae, indicating the importance of the surrounding nebula (gas cloud). The underlying spectra of these stars are usually of type G, K, or M; a few have spectra resembling earlier types, even A0 or B8. The continuum also varies, so that a total light variation occurs together with the sporadic appearance of emission lines. Large fluctuations occur in the ultraviolet; in fact, the ultraviolet is often excessively strong for the apparent spectral type.

The position of T Tauri stars on the H-R diagram is just above and to the right of the main sequence, just where we expect to find pre-main-sequence stars (see Figure 15–7). T Tauri and related objects radiate very strongly in the infrared region of the spectrum $(>1\mu)$. R Monocerotis (see Figure 15–2), for instance, radiates 90 per cent of its total luminosity in the infrared. As discussed in Chapter 15, the infrared excess of these stars is attributed to a surrounding dust cloud, which absorbs much of the star's short-wavelength radiation and then re-emits it at infrared wavelengths. This dust may be either the remainder of the material from which the protostar formed, or matter ejected from the star as it collapsed.

In some cases, the infrared objects are extended; therefore, they are called *infrared nebulae* rather than stars. This term is used by F. J. Low and his colleagues, who have done much of the pioneering work in this field. These nebulae, although bright in the infrared, are very cold (200 K and less). They may well be clouds of dust surrounding groups of protostars.

(b)   Flare Stars

Solar flares are the most energetic and spectacular aspect of solar activity. There must be flares on other G-type stars, but the amount of energy radiated by even the largest flares is negligible compared to the total stellar radiation. However, on a dwarf M star, which radiates very much less than the Sun, a flare with the energy of a large solar flare would lead to a *twofold* increase in brightness! In fact, several cool main-sequence stars have flared at irregular intervals by brightening several magnitudes in a matter of seconds. The light curves are similar to those of solar flares, for the decline is far slower than the ascent. The emission line spectra include the hydrogen Balmer lines, ionized calcium, and helium. Joint observations by radio and optical observers have verified that some of these flare stars emit radio bursts simultaneously with the flares.

The variations of these stars very probably result from the presence of a convection zone near the stellar surface. The Sun has such a convection zone, and solar activity is to some extent a result thereof. Flare stars are among the few types of known *radio stars*, which also include the pulsars discussed later in this chapter, some infrared stars, and some X-ray stars.

(c)   Magnetic Stars and Spectrum Variables

There exist stars which have the basic characteristics of A-type stars, except for certain spectral peculiarities attributable to abundance abnormalities (see Chapter 12). Some of these stars exhibit *variable spectra*, such that the intensity of certain lines varies approximately periodically. These are the spectrum variables. Many of the peculiar and metallic-lined A stars have strong integrated magnetic fields; these fields range up to 34,000 Gauss, but most of these magnetic stars have fields from hundreds to 3000 Gauss (recall that the strongest magnetic fields on the Sun are about 4000 Gauss, and these fields are restricted to very small areas in sunspots). Most observed stellar magnetic fields are variable, with some undergoing polarity reversals; sometimes the magnetic variability is coupled with spectrum variability. The light variations of magnetic stars and spectrum variables are very small, amounting to about $0^{m}.1$.

One possible explanation for the magnetic variability is that the magnetic axes are tilted with respect to the rotational axes (just as is the case for the Earth). This model is referred to as the *oblique rotator*. Perhaps more simply, both the spectrum and magnetic variations are due to large "spots" on the stellar surface that are periodically brought into view by rotation.

Despite much observational and theoretical work, these peculiar stars are still a mystery, and their place in the evolutionary scheme is a matter of controversy.

16–4   EXTENDED OR EXPANDING STELLAR ATMOSPHERES: MASS LOSS

(a)   The Atmospheric Model

The presence of an extended gaseous envelope around a star is deduced from the line profiles of certain spectral lines. Figure 16–4 illustrates the current model of such a star and the regions contributing to different parts of the profile. The portion of the *shell or extended atmosphere* seen projected against the star's photosphere produces a narrow absorption line, while that part not projected against the disk (the annular region) is seen as an emission line. Normally, the emission is superimposed on the stellar photospheric absorption; the extent to which the under-lying absorption profile is distorted by this emission line depends on the strength of the emission, which in turn is a function of the density of the stellar atmosphere. The atmosphere cannot be too dense, however, for then it would affect the continuum radiation as well as the spectral lines. The widths of the several components of the profile depend upon the motions of the contributing regions. For instance, if the atmosphere is turbulent, both the emission and absorption features will be broad.

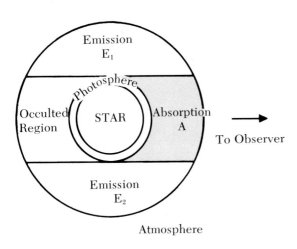

Figure 16–4. *The Model for Extended-Atmosphere Stars.* Radiation from the star's photosphere is absorbed (region A) on its way to the observer, and stimulates emission (regions E$_1$ and E$_2$) in the extended stellar atmosphere.

(b)   Be and Shell Stars

If the star rotates more rapidly than its atmosphere, the underlying stellar absorption profile is widened more than the atmospheric emission line. This results from the greater Doppler shifts of the approaching and receding limbs of the photosphere compared with those of the extended atmosphere (see Figure 16–5A). The absorption due to the projected gas remains narrow because the motion is across the line of sight. The *B emission stars* (*Be*) and *shell stars* both fit this model, but they differ in that the latter contain more material in the envelope. Sometimes net outward velocities are observed, indicative of expansion and possibly loss of material. The mass lost in this manner is negligible compared to the mass of the star, and to the mass which the star must lose if it is to become a white dwarf. Observationally, it is found that normal (main sequence) early-type stars do not seem to lose any mass, while giants and especially supergiants are ejecting material at a substantial rate (see the following paragraphs). The formation of a shell or ring around the B stars may result from the rapid rotation. The rotational velocities may be so high that material is ejected centrifugally; some additional force such as that due to magnetic fields may be necessary to keep this material moving.

(c)   Mass Loss from Giants and Supergiants

In a sense, we should include giants and supergiants in the category of stars with extended atmospheres. For instance, there is evidence of *mass loss* from M-type giants and supergiants, indicating that the stellar atmospheres are expanding. The spectra of many M giants and supergiants have narrow absorption lines, superimposed on and shifted with respect to the broad emission Ca II lines of the star itself (see Figure 16–5B). These features are interpreted as being due to circumstellar material, similar to the shells of the early-type stars discussed earlier. The Doppler velocities of the shifts are of the order of tens of km sec^{-1}. To convert these velocities to rates of mass loss we must know the density of the material ejected. Estimates of the density lead to figures for the mass loss ranging from 10^{-6} to 10^{-8} solar masses per year. The more luminous the star and the later the spectral type, the higher is the rate of mass loss.

Table 16–2.   Extended-Atmosphere Stars

Type	Prototype or Example	$\bar{M}_v$	Spectral Class	Expansion Velocity (km sec^{-1})	Rotation Velocity (km sec^{-1})	Rate of Mass Loss ($M_\odot$ yr^{-1})	Mass ($M_\odot$)	Population	Age or Lifetime (yrs.)	Light and/or Spectral Variations
Be	48 Per	-4	B	—	500 to 600	$\leq 10^{-6}$	~ 10	I	—	slow erratic
Shell Star	{ γ Cas, Pleione }	-4	B	50	500	$\leq 10^{-7}$	~ 10	I	—	slow erratic
Wolf-Rayet	{ HD 66811, HD 68273 }	-4 to -6.8	WN, WC	120 to 2500	—	10^{-6}	4 to 10	I	10^6?	none
P Cygni	P Cyg	-6?	B	130	0	?	—	—	—	—
O and B Supergiants	{ ζ Pup, δ, ε, ι Ori }	-7	O and B	1000 to 1800	—	10^{-6}	10 to 30	I	10^6	none
M Supergiants	{ α Her, α Ori }	-2 to -8	M Ia to II	up to 26	—	5×10^{-6} to 5×10^{-9}	—	I	5×10^6	none
Planetary Nebulae	Ring Nebula in Lyra	0 to $+8$	W stars? O, B	10 to 30	—	10^{-4} to 10^{-5}	≤ 1.4	II	30,000	none
β Lyrae	{ β Lyr, γ Cyg }	-6?	B, B	—, 130	—, 0	10^{-4}, ?	~ 5	—	—, —	—, —

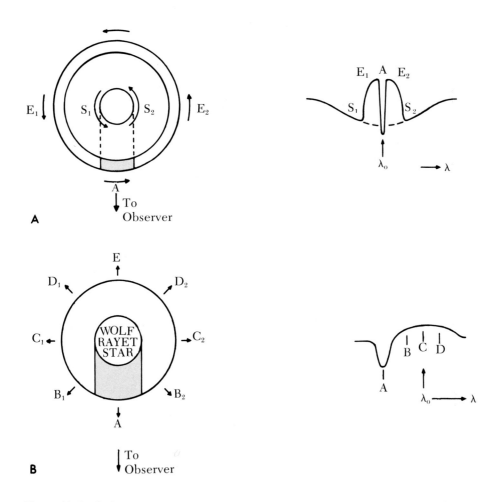

Figure 16–5. *Rotation and Expansion of Extended-Atmosphere Stars.* **A**, *A rotating star.* The line profile shows the absorption A of the undisplaced line center, and the Doppler-shifted emission peaks of E_1 and E_2. S_1 and S_2 are the absorption lines of the star Doppler shifted by the rotation to the blue and red, respectively. **B**, *Expansion of the atmosphere.* The line profile has absorption feature A displaced to the blue because of the atmospheric expansion. The contribution to the emission C is at the undisplaced line center, whereas B and D are Doppler shifted owing to the expansion.

At the other end of the spectral sequence, we find O and B supergiants ejecting mass at high velocities (1000 to 2000 km sec^{-1}). Spectra in the visible region do not show this phenomenon, for there are no suitable spectral lines. The material being ejected from the star is at a very low density; hence, collisional excitation is negligible, and most atoms and ions are in their lowest energy level (ground state). Owing to the high temperature, however, most of the gas is ionized; therefore, radiative excitation leading to an absorption line from the ground state (known as a *resonance line*) must originate from an *ion*. None of the ions produce such resonance lines in the visible spectral region. In the rocket ultraviolet, however, there are several such transitions: from Si III and IV, C III and IV, and N V.

In fact, it was from rocket spectra of early-type supergiants, such as those in Orion, that the ejection of mass was discovered. The ejection velocities are far higher than for the late-type supergiants, but the mass loss is only slightly greater because of differences in the gas densities. Rocket spectra of *main-sequence* O and B stars seem to confirm the absence of expansion in that part of the H-R diagram.

Since both early- and late-type supergiants are losing mass, we might expect similar behavior of the intermediate types, the A through K supergiants. There is indeed some evidence that these also eject matter, but apparently rather more sporadically than those stars already discussed.

(d) Wolf-Rayet Stars

Wolf-Rayet (W) stars are hot early-type stars whose spectra show strong and wide emission lines of He I, He II, C III, C IV, N III, and N V (see also Chapter 12). The broad emission lines often have narrower absorption lines superimposed, but displaced to the blue. These observations fit the interpretation shown in Figure 16-5B, in terms of a stellar atmosphere expanding at about 1000 km sec^{-1}. Nevertheless, the true nature of Wolf-Rayet stars is still far from clear. There remains the peculiar differentiation into the carbon and nitrogen (WC and WN) sequences. It is unclear whether these branches should be attributed to real differences in abundances, and perhaps stage of evolution, or whether they result from a difference in excitation.

The position of W stars in the scheme of evolution also remains to be clarified. That they are young stars of Population I is deduced from their association with OB stars, both in open clusters and in H II regions [see section 17-2(b)]. It is not entirely clear whether they are stars that are evolving onto the main sequence or are in the post-main-sequence state, though most astronomers believe that they have left the main sequence. Many W stars are members of spectroscopic binary systems, usually with O or B stars as companions. In addition, a high proportion of Wolf-Rayet stars have composite spectra, and it has been suggested that all W stars are binaries. Orbital data from spectroscopic binaries yield mass ratios of about 1:3, with the W stars as the less massive members. We can conclude, therefore, that their masses are of the order of 4 to 10 solar masses. B. Paczynski has suggested that the relatively low mass (low for their luminosity) is due to an exchange of matter between two stars of a close binary system. The binary nature is thereby assumed to be a basic characteristic of W stars. This mass loss occurs during the approach to the red giant phase of the initially more-massive star, which must evolve more rapidly than its companion because of its greater mass. The Wolf-Rayet star is the remnant, consisting largely of helium and heavier elements. The alternative interpretation considers W stars to be pre-main-sequence stars, slower to evolve than their O or B companions because of their lower mass.

(e) Planetary Nebulae (See also Chapter 17)

Planetary nebulae, so named because some resolve into disks reminiscent of planets when seen with a telescope, also fit the expanding atmosphere model. In fact, the atmosphere is really a large shell, large enough and of sufficiently low

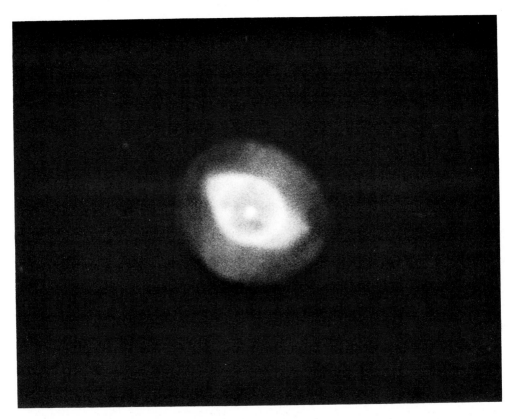

Figure 16–6.   *The Planetary Nebula in Hydra.* (Hale Observatories)

density so that most of the receding portion (E in Figure 16–5B) is visible and the spectral lines are doubled. The velocities of expansion, however, are only some tens of km sec^{-1}, far lower than those for the Wolf-Rayet stars.

Until recently, planetaries were believed to be very rare. The most spectacular stage with rings or a disk is indeed rare, but it has become evident that a substantial fraction of all stars probably go through the planetary nebula stage. The small number of observable planetaries is due to the short duration of this stage, which lasts only 30,000 to 50,000 years. Figure 16–7 illustrates the probable development of a star into a planetary nebula. We see there that the nebula is material ejected from the central star during the contraction which terminates the red giant stage. The extended envelope includes so much of the former star that what is seen as the central star was formerly the core. It does not follow from this model that all red giants necessarily become planetary nebulae, nor that all white dwarfs were once planetaries.

The central star is very hot and therefore radiates strongly in the ultraviolet; hence, the atoms and ions in the envelope *fluoresce* (i.e., absorb ultraviolet radiation to become ionized and emit at longer wavelengths upon recombination and cascade). The central stars generally have spectra that can be classified as type O or Wolf-Rayet, but these stars are not identical to normal O and W stars in either luminosity or mass. Both absolute magnitude and mass are very difficult to establish because all planetaries lie at distances too great for the measurement of their trigonometric

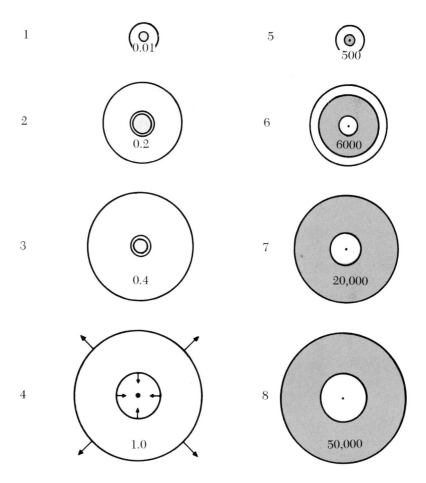

Figure 16–7. *Schematic Evolution of a Planetary Nebula.* (1) A star such as the Sun converts hydrogen into helium inside its dense, hot core (shaded). (2) As hydrogen becomes exhausted, energy generation is confined to a thin layer which surrounds the core. (3) As the evolution proceeds, the core (now mainly helium) contracts, while the outer envelope continues to expand. The star has become a red giant. (4) When the outer envelope is about 1.5×10^8 km in diameter, it escapes into space, as the inner part forms the planetary's central star. (5) A new planetary appears with dense, ionized hydrogen (shaded) surrounding a central star. The neutral hydrogen is invisible. (6) The disk of ionized hydrogen is large enough to see in a telescope. The spectrum contains strong forbidden lines of oxygen and nitrogen. (7) All the hydrogen and most of the other elements have been ionized when the nebula reaches this size. The density and total brightness decline. (8) The nebula is now barely seen against the sky and is about to fade from view. (For each stage the diameter is roughly indicated in astronomical units.) (After L. H. Aller)

parallaxes. We can nevertheless make reasonable estimates on the basis of physical processes operating between the star and nebula. The stellar masses are believed to be of the order of $1 M_\odot$, while estimates of the mass contained within the envelopes range from $0.1 M_\odot$ or less to $0.5 M_\odot$. Apparently many planetary nuclei are *underluminous* for their spectral type, as might be expected for a star in the process of losing its mass as it changes from a giant star to a white dwarf.

If we ask what place the planetary nebulae occupy in the evolutionary scheme, we immediately raise the problem of their population characteristics. The planetaries in our Galaxy are primarily concentrated to the disk and center, like the long-period variables and the RR Lyrae stars. Their galactic orbits are elongated rather than circular, in contrast to the orbits of Population I objects. At least one is known to be a member of a globular cluster—that is, an extreme Population II object. Several planetaries occur in the Magellanic Clouds, but their population characteristics are still ambiguous. It is probably safe to say that most planetary nebulae belong to the intermediate disk population, but isolated cases are found at both extremes, Populations I and II.

16–5   CATACLYSMIC VARIABLES

In this category we include stars which eject matter suddenly and energetically, in contrast to the slow expansion and small mass loss of the planetary nebulae. These eruptions are accompanied by *enormous* changes in luminosity, from a few magnitudes for the dwarf novae to over 20 magnitudes in the case of supernovae (see Table 16–3). It is probable that all cataclysmic variables, except the supernovae, suffer several eruptions in their lifetimes. Repeated outbursts of dwarf novae and recurrent novae have been observed, with the interval between outbursts being a function of the amplitude of the "eruption." If such a relationship extends to "normal" novae, their inter-burst interval must be of the order of 10,000 years.

(a)   Novae

Nova is the Latin word meaning "new"; novae (plural) are stars which suddenly become visible in the sky where no star was seen before. Actually, the star is usually too faint to be visible prior to its eruption, for a nova characteristically increases in brightness some nine to 12 magnitudes from the pre-nova stage to nova maximum. The rise to maximum brilliance is very rapid, as shown by the light curves depicted in Figure 16–8. The initial rise brings the star to within two magnitudes of the maximum in only two or three days, while the final increase in luminosity takes a day for "fast" novae and weeks for very "slow" novae. The decline (decay) from the maximum is far more gradual; the time spent at maximum light is relatively short, generally only a matter of days. Erratic light fluctuations of large amplitude may occur during the decline; occasionally a drop of many magnitudes is observed, followed by a practically complete recovery some weeks later (see Figure 16–8). The length of time from the maximum to the final leveling-off of the light curve (at approximately the pre-nova level) ranges from months for fast novae to years for slow ones.

Table 16–3.   Cataclysmic Variables

Type	Example	M_{max}	M_{min}	Δm	E per Outburst (ergs)	Cycle Time	Mass Ejected per Cycle ($M_\odot$)	Velocity of Ejection (km sec^{-1})	Mass of Star ($M_\odot$)	Duration of Decline
Supernovae I	Tycho's	-20	?		10^{51}	—	≤ 1	10,000	1	—
Supernovae II		-18	$+2.7$	>20	10^{50}	—	?	10,000	≥ 4	—
Novae: Fast	GK Per = Nova Per	-8.5 to -9.2	to $+4.8$	11 to 13	6×10^{44}	$10^{6??}$ yrs	10^{-5} to 10^{-3}	500 to 4000	1 to 5	1/2 to 2 yrs
Slow	DQ Her	-5.5 to -7.4	$+4.8$ to $+8.1$	9 to 11	—	—		100 to 1500	0.02 to 0.3	3 to 9 yrs
Recurrent	T Cr B	-7.8	0	8	10^{44}	18 to 80 yrs	5×10^{-6}	60 to 400	2	<1 yr
Dwarf Novae	{ U Gem, SS Cyg	$+5.5$	$+9.5$	4	6×10^{38}	40 to 100 days	10^{-9}	—	~ 0.4	days to mos.

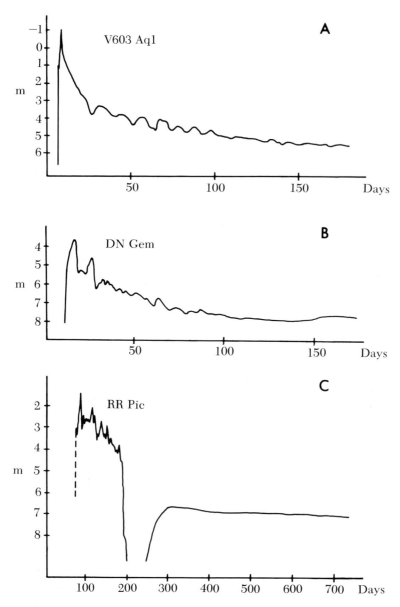

Figure 16–8. *Light Curves of Novae.* Note premaximum halt on all light curves, rapid early decline, transition zone of rapid fluctuations or sharp drop followed by recovery, and slow final decline. **A**, Example of a very fast nova, V 603 Aquilae. **B**, Example of fast nova, DN Geminorum. **C**, Example of slow nova, RR Pictoris. (After D. B. McLaughlin)

Spectra obtained just prior to and at maximum light show that material is ejected from the star at velocities up to 2000 km sec^{-1}. The complex changes of the spectrum during the development of the nova need not concern us here; suffice it to say that *several* layers or shells of ejected material are seen as the outer envelope becomes progressively more transparent and we see down to the next layer. In addition, the expanding material may change character, and its spectrum changes accordingly. At one stage, for instance, most nova spectra include the bright forbidden lines characteristic of emission nebulae—hence, the term "nebular stage." The velocities of ejection differ at different nova stages; some velocities are directly related to changes in the light curve, as if there were ejections subsequent to the initial eruption.

There is only one true nova for which a spectrum was recorded prior to the outburst: Nova Aquilae 1918 (also known as V603 Aquilae). This spectrum is similar to that of a hot blue star without any spectral lines, presumably a blue subdwarf. As far as we can tell, most novae eventually return to such a state. Some recurrent novae have similar (high-temperature) continuous spectra between outbursts. These observations strengthen the suggestion that all novae were originally *blue subdwarfs* and that they return to this stage after the outburst.

Some light fluctuations persist even at minimum light. Superimposed on these rapid and erratic variations, we sometimes find light variations which are best interpreted as stellar eclipses. Such novae clearly must be members of binary star systems; in fact, an increasing amount of evidence implies that *all* novae are members of binaries. This would imply that the presence of a companion is a necessary condition for a star to become a nova. One binary model based upon this idea consists of a red giant, or a star in the process of expanding into the red giant phase, and a hot subdwarf or white dwarf. As the red star extends its atmosphere, gaseous material crosses the *Lagrangian surface*. This is the region where the net gravitational force produced by each star on a small body vanishes. For example, consider an Apollo spacecraft traveling to the Moon; the vehicle passes through a Lagrangian point when the force of the Earth's gravitational attraction equals that of the Moon. There are Lagrangian surfaces for any two-body system, such as binary stars. When material crosses the Lagrangian surface, it passes from the gravitational influence of one body into that of the other, and therefore *falls into* the latter body. Hence, the gas from the red giant's atmosphere escapes from the red giant and falls into the dwarf or forms a gaseous disk. The influx of hydrogen-rich gas onto a degenerate star (see Chapter 15) which has used up most of its hydrogen may cause further nuclear reactions near the stellar surface or in the interior. This new supply of energy heats the outer envelope of the dwarf star so that it becomes non-degenerate, and the stellar atmosphere suddenly expands. In other words, there is a nova explosion. Repeated explosions may occur if more material flows from the red giant to the white dwarf. The length of time between "explosions" is to some extent inversely related to the violence of the ejection (as indicated by the increase in luminosity). Such a relationship seems to apply to the recurrent novae, with intervals of 30 to 50 years between eruptions. The essential differences which characterize the several types of novae are noted in Table 16–3.

Novae represent an advanced stage of evolution of stars that happen to be members of close binary systems. The distribution of novae in our Galaxy corresponds to Population II. The discovery that several X-ray sources are close binary systems (see the following paragraphs), in which the emission of X-rays is attributed to mass exchange, is suggestive of a close link between novae and X-ray sources. No known old nova, however, has as yet been identified with an X-ray source.

The high ejection velocities quoted earlier are based upon measurements of spectral line profiles; these profiles resemble those of Wolf-Rayet stars (see

Figure 16–5B). The ejected gas expands as a shell, and sometimes this shell becomes visible as a nebula surrounding the nova. Over the years, the nebula expands perceptibly, and its *rate* of expansion is measurable as a proper motion in seconds of arc per year (see Chapter 14). Spectra obtained during the same epoch give expansion velocities directly in km sec^{-1}. If we assume that the velocity of expansion is uniform in all directions (i.e., the expansion is *isotropic*), then the observed proper motion must correspond to the same velocity. The geometry of the expansion then allows us to find the *distance* to the nova. From Chapter 14, we have

$$d = \frac{v_r}{4.74\ \mu''}$$

where μ'' is the proper motion (in $''$ yr^{-1}), v_r is the radial velocity (in km sec^{-1}), and the factor 4.74 is a conversion factor which yields d in parsecs. DQ Herculis and Nova Persei (GK Persei; see Figure 16–9) are the best examples of novae with expanding nebulae. The importance of this method of determining their distances is obvious, for it is both direct and unambiguous (see Problem 5). In this way we can establish the absolute magnitudes of novae at maximum and minimum, and from that information we can find the total amount of energy released in the explosion.

Figure 16–9.   *The Expanding Nebula Around Nova Persei.* The nova exploded in 1901; this photograph was made in 1949. (Hale Observatories)

The chief problem which remains is usually the unknown amount of interstellar absorption between us and the nova (see Chapter 17).

(b) Dwarf Novae: U Geminorum Stars

Many orders of magnitude less energetic per outburst, but similar to novae in other respects, are the *U Geminorum stars*, or dwarf novae. Their brightenings occur quite frequently, though erratically. The duration of the high-luminosity phase and the amplitude of the brightness increase are correspondingly less than for novae. The term "dwarf" applies not only to the relative change in luminosity and to the brightness at maximum but also to the magnitude at minimum, which is considerably fainter than for the classical novae.

Dwarf novae may be related to novae, but they are not just miniature versions of the phenomena. A rapidly rotating disk of gas surrounds what is presumably a hot star. This disk may be the seat of the outbursts, but in contrast to regular novae, no expanding envelope is observed; hence, there is no evidence for mass loss.

(c) Supernovae

Supernovae are even more spectacular than novae, for they attain absolute magnitudes in the range from -18 to -20. Although supernovae are relatively rare, the appearance of some in our Galaxy has been noted in historical records. From such records, especially the Chinese chronicles, we find, for example, that the supernova of 1054 A.D. reached an apparent magnitude of -6; this is even brighter than Venus when it is brightest at -4. Much of what we know about supernovae has been gleaned from studies of external galaxies. In the case of small galaxies, the luminosity of a supernova may rival the total brightness of the galaxy. Like ordinary novae, supernovae exhibit a very rapid rise to maximum light, and then drop two or three magnitudes within a month before declining more gradually.

There are at least two types of supernovae, differentiated primarily by their spectra and partly by their light curves: type I supernovae appear in both elliptical and spiral galaxies (see Chapter 19), while type IIs occur only in spirals. One may deduce, therefore, that type I supernovae belong to Population II, and type II are Population I. The type I supernovae attain the brighter maxima (see Table 16–3) and have broad-line spectra at maximum—so broad that whether they are absorption or emission lines is still a matter of debate. Spectra of type II supernovae are continuous at maximum, although they resemble those of ordinary novae at later stages. While an ordinary nova explosion ejects only a small fraction of the stellar mass, a supernova explosion may eject over one solar mass at velocities around 10,000 km sec^{-1}! Here we are dealing not with a surface phenomenon but with something far more fundamental.

As we saw in Chapter 15, it is generally believed that a supernova explosion results from very rapid gravitational collapse, when most of the "normal" thermonuclear reactions leading to nuclear synthesis can no longer occur. During this collapse, gravitational potential energy is converted into incredible amounts of radiation and heat. As a result, heavy nuclei are synthesized and ejected into the interstellar medium, thus "salting" the gas and dust from which stars of a later generation will arise. Some atomic nuclei are accelerated from the stellar surface and pervade the Galaxy as cosmic rays.

We are fairly certain that no supernovae have occurred in our own Galaxy during this century, but we can identify several supernova *remnants* with historically recorded outbursts: for instance, the Crab nebula corresponds to the 1054 A.D. supernova. These remnants are the material ejected by the star, and are now seen as filamentary nebulae which radiate strongly throughout the electromagnetic spectrum. The stars themselves have shrunk in size and brightness, some possibly to the white dwarf stage but most into neutron stars. Such neutron stars have been observed as the *pulsars*—rapidly and periodically fluctuating radio sources.

(d)   The Crab Nebula—A Supernova Remnant

This beautiful object (see Figure 16–10) is the nearest, most intriguing, and most spectacular *supernova remnant*. It is therefore to be expected that the Crab nebula is the object of much observational and theoretical study. Occasionally, scientists are surprised by the revelation of some previously unsuspected characteristic of the Crab, such as the identification of a pulsar with the central star in 1968. Most other supernova remnants share some of the characteristics portrayed by the Crab nebula, so it is also important as a prototype.

The distance to the Crab nebula has been established by the techniques outlined for novae; it is about 2000 pc. We can also use the present rate of expansion of the nebula to extrapolate back to the time of outburst; the result **corroborates** the identification with the 1054 A.D. supernova (note that some acceleration since the initial expansion must be taken into account). The expansion is not perfectly uniform in all directions, so that the central (parent) star was not unambiguously identified until the pulsar was identified optically.

Even in the visible region of the spectrum, the nebula presents varied aspects, depending upon whether it is photographed (a) in the radiation from one of the emission lines (such as $H\alpha$), (b) in the continuum radiation, or (c) through a polarizing filter. The line radiation emanates from clearly-defined filamentary features (see Figure 16–10), where the gas density is enhanced over that of the rest of the nebula. Underlying the filaments, and more concentrated to the central part of the nebula, is the region emitting in the continuum. This region also possesses considerable structure, which can be described as vague wisps or fibers. The appearance and position of these wisps change with time, probably owing to motions of the gas or to compression waves moving through the gas.

An important clue to the nature of the Crab nebula is the fact that the continuum radiation is *strongly polarized* (see Chapter 8). Moreover, this wispy nebula cannot be a hot dense gas such as that which produces a continuous spectrum according to Kirchhoff's laws. Finally, the nebula is a strong radio emitter, for it is identified with the radio source Taurus A (first radio source to be discovered in the constellation Taurus). The wavelength dependence of the radiation in both the visible and radio regions differs greatly from a Planck blackbody curve: it is *nonthermal radiation*. These characteristics led I. S. Shklovsky to propose (in 1953) that *synchrotron radiation* is the source of both the optical and radio continua. When energetic electrons are accelerated by a magnetic field, they spiral around the

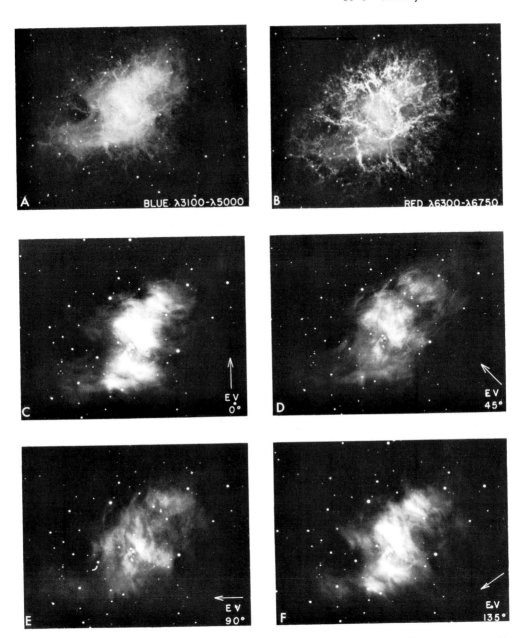

Figure 16–10. *The Crab Nebula.* **A**, Blue continuum radiation. **B**, Hα light. **C** through **F**, Polarized light, with the arrows indicating the direction of the electric vector. (Hale Observatories)

magnetic field lines (see Figure 16–11). This motion causes the electrons to emit strongly polarized continuous radiation, whose intensity at a given frequency depends upon both the magnetic field strength and the energy of the electrons. The higher the mean electron energy, the higher is the frequency at which the intensity is a maximum. Many radio sources have a synchrotron radio spectrum, but few simultaneously emit synchrotron radiation at the higher frequencies corresponding to visible radiation. The Crab nebula is one of these few.

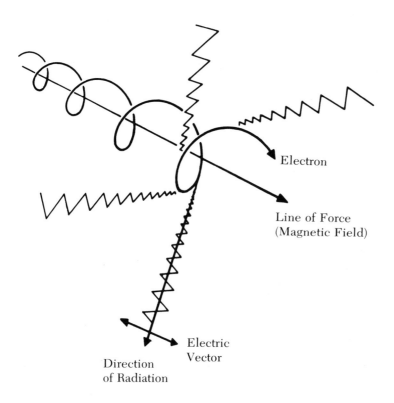

Electron

Line of Force
(Magnetic Field)

Electric
Vector

Direction
of Radiation

Figure 16–11. *The Origin of Synchrotron Radiation.* A relativistic electron spirals a magnetic line of force. The orbital acceleration causes the electron to emit plane–polarized synchrotron radiation.

In fact, synchrotron radiation also accounts for the *X-ray emission* which is observed from the Crab. That the nebula as a whole, and not just the central star, generates X-rays was shown by rocket observations during a lunar occultation of the Crab nebula. The X-ray emission faded gradually during the occultation, a phenomenon characteristic of an extended source rather than a point source. Later data have shown that the central star, a pulsar, is itself also a strong source of X-rays. The intensity as a function of wavelength (or counts per second as a function of X-ray energy, as measured experimentally) fits the synchrotron mechanism model.

The magnetic field in the nebula is estimated as 5×10^{-4} Gauss; this is stronger than the general galactic magnetic field, but is small compared to the magnetic fields of the Earth (about 1 Gauss) and the Sun (in the range $1-10^3$ Gauss). Nevertheless, this field strength is sufficient to produce copious quantities of synchrotron radiation, when the electrons are traveling at *relativistic speeds* (almost the speed of light). But where do these relativistic electrons come from? We know the following (observational) facts about the Crab nebula: (a) we continue to see it approximately 920 years after the outburst, (b) the nebular gases exhibit time-varying (i.e., accelerated) motions, and (c) phenomenal amounts of synchrotron radiation continue to be emitted by the nebula. Taken together, these facts demand the existence of a *strong energy source* somewhere within the nebula. The identity of this source has long been a mystery; now, however, we have an excellent candidate— the *Crab nebula pulsar*.

Although the Crab nebula is representative of supernova remnants, it is spectacular and unique in the intensity of its radiation at all wavelengths, in part because of its proximity and in part because of the recentness of the outburst. Other supernova remnants are seen as faint filamentary nebulosities. For the older remnants, such as the Cygnus loop (see Figure 16–12), these filaments are attributed to the interaction of the expanding debris with the surrounding interstellar gas, which produces a shock wave. Radio emission from such remnants must be due to synchrotron radiation, but X-rays, if present, may be from the hot plasma behind the shock front.

16–6 PULSARS

(a) The Observational Data

In 1967, a large radio telescope was activated in Cambridge, England, to study the *scintillations* of radio sources. Scintillation is the rapid "twinkling" of a radio source due to density fluctuations in the interplanetary plasma (the solar wind) and in the interstellar medium; it is completely analogous to the

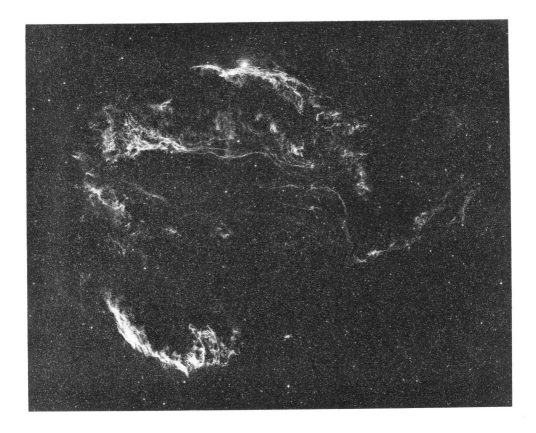

Figure 16–12.   *The Cygnus Loop—a Supernova Remnant.* (Hale Observatories)

twinkling of visible stars (due to density fluctuations in the Earth's atmosphere). Almost immediately, weak precisely-periodic radio signals were detected coming from a fixed position on the celestial sphere. The importance of this accidental discovery was quickly recognized, and after a detailed investigation, A. Hewish and his colleagues announced the existence of the *pulsar* CP 1919 (that is, Cambridge Pulsar at right ascension 19^h19^m). Other radio pulsars were discovered in rapid succession, both at Cambridge and elsewhere, and presently more than 85 of these objects are known.

Pulsars present us with a fascinatingly complex array of observational data. Their primary distinguishing characteristic is a *train of radio pulses*, consisting of pulses approximately 20 to 50 milliseconds (1 millisecond = 10^{-3} sec) in duration separated by time intervals (e.g., 1.3373 sec for CP 1919) which are *constant to 1 part in 10^9!* The periods of known pulsars range from 0.03 to 4 seconds, a factor of 10^2. The incredible precision of the pulse periods implies that they must be referred to the barycenter of our Solar System (see the footnote in Chapter 14), after appropriate corrections are made for (a) the Earth's rotation, (b) the Earth's motion about the center-of-mass of the Earth-Moon system, and (c) the Sun's motion around the barycenter.

Figure 16–13 illustrates some representative pulse trains. The pulses or bursts last only tens of milliseconds, with no detectable radio emission between pulses. Individual pulses may be resolved into 20 to 30 subpulses of submillisecond duration, so that the primary pulses are actually the envelopes of these secondary pulses. Pulse amplitudes vary markedly, a phenomenon usually attributed to scintillation (see first paragraph of this section).

Pulsars are most readily observed at low frequencies; for instance, the first discoveries were made at 81.5 MHz (8.15×10^7 Hertz $\rightarrow$ a wavelength of 3.7 meters). The intensity of the pulses diminishes rapidly as we go to higher frequencies, while the pulses become broader and more regular in shape. Nevertheless, the Crab pulsar has been observed from the radio wavelengths, through the visible and X-ray, to the gamma-ray region of the electromagnetic spectrum. At all of these frequencies, the Crab pulsar emits pulsed radiation with a period of approximately 33 milliseconds. Only the Crab pulsar has been seen optically, and it is identified with a blue star (characterized by a featureless spectrum indicating high temperatures) at the center of the Crab nebula; this star is considered to be the supernova remnant. By correcting for the extended X-ray emission of the nebula, the Crab pulsar's *pulse energies* at radio, visible, and X-ray frequencies have been found to be in the ratio $1:10^2:2 \times 10^4$.

Approximate distances to pulsars, and some properties of the interstellar medium, may be deduced directly from pulsar observations. We have already mentioned the *radio scintillations*, which probe the nonuniformity of the interplanetary and interstellar plasmas. In addition, a given pulse is observed to arrive at the Earth *later* as we look at lower frequencies. This phenomenon is called *frequency drift* or *velocity dispersion*, and it is due to a slowing-down of the pulse velocity by electrons in the line-of-sight to the pulsar (analogous to the index of refraction discussed in Chapter 8, and the lower propagation velocity of light in a material medium). Longer wavelengths are slowed down more, and, from the observations, we may deduce the *mean electron density* in the line-of-sight. Conversely, if we know

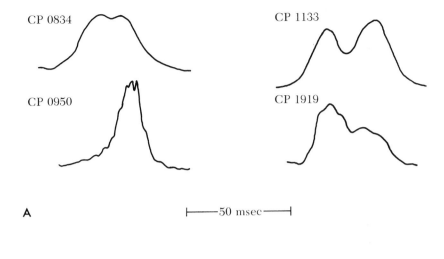

A   ⊢——50 msec——⊣

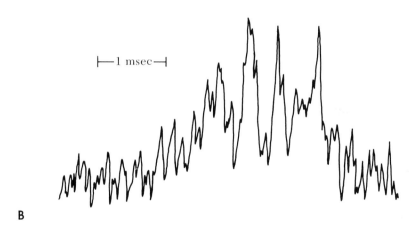

B

Figure 16–13.   *Typical Pulse Profiles of Pulsar Radiation.* **A**, The varieties of mean pulse shapes. **B**, Fine structure within a single pulse from CP 1133. (After A. T. Moffet and R. D. Ekers)

(or can estimate) the mean electron density, the *distance* to the pulsar follows immediately. Combined with the observation that most pulsars lie at low galactic latitudes, these data imply that pulsars are quite "local" (within a few kpc) and lie in the galactic disk. Finally, we know that the plane of polarization of linearly polarized radiation (see Chapter 8) is *rotated*, when the radiation propagates through a *magnetized* plasma; this effect is known as *Faraday rotation*. This phenomenon depends upon (a) the mean electron density, (b) the mean magnetic field strength, (c) the square of the wavelength of the radiation, and (d) the distance traveled through the medium. Since pulsar bursts are strongly linearly polarized, we may immediately find that the mean magnetic field in the galactic disk is about 10^{-6} Gauss.

(b)   Theoretical Pulsar Models

Prior to the discovery of the pulsars, theoreticians had predicted the existence of *neutron stars* (see Chapter 15). A neutron star is actually a gigantic nucleus, composed almost entirely of neutrons and stabilized by its own self-gravitation (and by the pressure of the degenerate neutron gas). These objects have a mass variously estimated from $0.2 M_\odot$ to $3 M_\odot$, a radius near 10 km, and a mean density of 10^{14} gm cm^{-3}. Since neutron stars are believed to be formed in supernova explosions, the discovery of the pulsar in the Crab nebula (a supernova remnant) appeared to vindicate the neutron star hypothesis; all subsequent observations have indeed verified that we are dealing with neutron stars.

The present consensus is that pulsars are *rapidly-rotating, magnetic, neutron stars*, with the magnetic axis inclined to the axis of rotation. The pulses are intimately connected with the magnetic field (which may have a strength near 10^{12} Gauss at the surface of the neutron star). The extraordinary stability of the pulse period is due to the rotation of the neutron star. The energy radiated by the pulsar derives from the energy of rotation, and leads to a slowing-down of the rotation; hence, the pulsar period lengthens. Many pulsars have been observed to have *slowly* increasing periods, a phenomenon which is totally inexplicable if the pulsars are *pulsating* white dwarfs (a theory advanced soon after the discovery of pulsars); in addition, a white dwarf cannot rotate at the observed high frequencies. The total energy radiated by the Crab nebula is more than adequately explained by this rotation mechanism.

Many ingenious mechanisms have been advanced to explain the *origin* of the pulsar bursts. Most theories involve the behavior of high-energy charged particles near the surface of the rapidly-rotating star; these particles interact with the very high magnetic field generally associated with a neutron star. The star is presumed to be enveloped by a kind of magnetosphere. Pulses may arise because of the inclination of the magnetic axis to the rotational axis or because of instabilities in the plasma of the magnetosphere.

The Crab and the Vela pulsars (that is, neutron stars) have been identified with visible supernova remnants. Many pulsars occur where there is no visible supernova remnant, but this does not imply that they were not formed in supernovae; the supernova remnants may have simply faded into undetectability. Both the Crab pulsar and the Vela pulsar have exhibited a fascinating and abrupt *change in period*. In early 1969, the Vela pulsar's slowly increasing period was interrupted by a 134 nanosecond (1 nanosecond = 10^{-9} sec) *decrease* in the period, and the Crab pulsar has also abruptly decreased its period. These phenomena have been interpreted (and "explained") by M. Rudermann as *starquakes*: neutron stars may have a *solid crust*; as the star's rotation slackens, the crust is strained to the "breaking" point and relaxes the tension by means of a starquake. Needless to say, this interpretation is the subject of intense controversy, as are all theories pertaining to pulsars.

16–7   X-RAY SOURCES

Several different types of objects appear to be X-ray sources—that is, objects emitting strongly in the 0.1 to 10 Å wavelength region (1 to 100 keV).

A discussion of solar X-rays was included in Chapter 9. Reference to extragalactic sources will be made in Chapter 19, and we have already discussed earlier in this chapter those sources that are identified as supernova remnants. Rockets and balloons obtained the early data on X-ray sources, but our knowledge expanded greatly with the launching of the X-ray satellite, UHURU, in December 1970. Most X-ray sources have not yet been identified with either optical or radio objects.

The strongest source is Sco X-1 (first X-ray source to be discovered in the constellation Scorpius), identified with a 13th magnitude blue star and representative of several other sources. The X-ray emission is erratically variable by as much as a factor two, changing within minutes or hours. Spectral changes accompany the intensity variations. These changes correspond to a temperature range from 50 to 150 million kelvins, if interpreted as a thermal phenomenon in keeping with the observed exponential spectrum. Such a thermal source may be a region of hot, low-density ionized gas, a plasma, surrounding a central object. X-rays are generated when the electrons and protons in the gas make close encounters and are accelerated by the Coulomb forces. This *thermal bremsstrahlung* is the high-energy analog to the free-free transitions responsible for thermal radio emission [see section 9–4(b)]. The origin of the energy heating the gas is unknown, but many astronomers favor the idea that it involves a close binary surrounded by the streaming plasma.

Some X-ray sources have been unambiguously identified as binaries (e.g., Cen X-3 and Her X-1), since their X-ray intensity exhibits sinusoidal variations. Superposed on the two-day orbital period of Cen X-3 are short pulses, with a period of 4.8 seconds. The model proposed by the American Science and Engineering group (responsible for UHURU) is that the X-ray emission originates from a very small, rapidly pulsating or rotating object. This object revolves about and is periodically occulted by a large massive star.

Today, more than 100 X-ray sources are known. The opening of the X-ray window and the birth of *X-ray astronomy* promise to add immeasurably to our knowledge of the heavens.

Problems

16–1. In a given constellation, the following designations have been assigned to variable stars: V502, SU, II, V956, XY, and AK. List these stars in the order of their discovery.

16–2. In your own words, describe and correlate the fluctuation profiles of the typical pulsating star illustrated in Figure 16–2.

16–3. A Cepheid variable in a hypothetical galaxy is observed to pulsate with a period of ten days, and its mean apparent visual magnitude is 18^m. It is not known whether this is a Population I or Population II Cepheid.
(a) What are the two possible distances to the galaxy (neglect interstellar absorption)?
(b) What is the ratio of these distances?
(c) Would this ratio change if we considered other galaxies? Explain!
(d) Will this ratio differ for Cepheids of different periods?

16–4. If our telescope has a *limiting magnitude* of 22^m, what is the *maximum* distance to which we can see the following stars:

(a) RR Lyrae stars,
(b) Classical Cepheids,
(c) W Virginis stars,
(d) ordinary novae,
(e) dwarf novae,
(f) supernovae?

How do these distances compare with the diameter of our Galaxy? (Hint: Be sure to consult Tables 16–1 to 16–3.)

16–5. The outburst of Nova Aquilae (V603 Aql) occurred in June, 1918, at which time it attained a brightness of $-1\overset{m}{.}1$. Spectra showed Doppler-shifted absorption lines corresponding to a velocity of 1700 km sec^{-1}. By 1926, the star was surrounded by a faint shell 16″ in diameter. Find the distance to Nova Aquilae in parsecs and the absolute magnitude at maximum light.

16–6. If a star becomes a supernova, by what amount does its luminosity change if it was originally of absolute magnitude $M_v = +5\overset{m}{.}0$? $M_v = +2\overset{m}{.}0$? The absolute visual magnitude of a supernova at maximum is about $-15\overset{m}{.}0$.

16–7. Consult Chapter 9, and find the energy output (erg sec^{-1}) of the most energetic solar flares. Referring to Chapter 12, compare this with the typical energy outputs of stars (of radius $1R_\odot$) of the following spectral types:

(a) F,
(b) G,
(c) K,
(d) M.

By what factor is the luminosity of each star increased during such a flare event? Which stars would you consider to be *observable* flare stars?

16–8. Consider a rotating star with an extended atmosphere. If the stellar atmosphere is *both* rotating and expanding, draw the observed profile of a given spectral line (see Figure 16–5) when the atmosphere rotates *more slowly* than the star.

16–9. If the pulse period of a *pulsar* is known to be constant to 1 part in 10^9, then

(a) to what percentage accuracy can we determine the Earth's orbital velocity around the Sun;

(b) how accurately can the Earth's rotation period be determined;

(c) if the pulsar has a mass of $1M_\odot$, and orbits a similar star (binary system) once every five years, how accurately can we determine the semi-major axis of the binary orbit?

Reading List

Blaauw, A., and Schmidt, Maarten, (eds.): *Galactic Structure.* (See especially Chapters 8, 13, 14, 15.) Chicago, The University of Chicago Press, 1965.

Cameron, A. G. W.: "Neutron Stars." *Annual Review of Astronomy and Astrophysics*, Vol. 8, pp. 179–208 (1970).

Campbell, L., and Jacchia, L.: *The Story of Variable Stars.* New York, McGraw-Hill Book Company, 1951.

Christy, Robert F.: "Pulsation Theory." *Annual Review of Astronomy and Astrophysics*, Vol. 4, pp. 352–392 (1966).

Davis, R. D., and Smith, F. G., (eds.): *The Crab Nebula*. IAU Symposium 46. Dordrecht, Holland, D. Reidel Publishing Company, 1971.

Glasby, J. S.: *Variable Stars*. Cambridge, Harvard University Press, 1969.

Hewish, A.: "Pulsars." *Annual Review of Astronomy and Astrophysics*, Vol. 8, pp. 265–296 (1970).

Minkowski, R.: "Supernovae and Supernova Remnants." *Annual Review of Astronomy and Astrophysics*, Vol. 2, pp. 247–266 (1964).

Paczynski, B.: "Evolutionary Processes in Close Binary Systems." *Annual Review of Astronomy and Astrophysics*, Vol. 9, pp. 183–208 (1971).

Smak, Jozef I.: "The Long-Period Variable Stars." *Annual Review of Astronomy and Astrophysics*, Vol. 4, pp. 19–34 (1966).

Sobolev, V. V.: *Moving Envelopes of Stars*. Cambridge, Harvard University Press, 1960.

Chapter 17
The Interstellar Medium

We have studied the general form and motion of our Galaxy, and have seen the varieties of stars which populate it. Now let us consider the physical content of the vast regions of space between the stars—the interstellar medium. Just as our own Solar System is pervaded by gas and plasma (the solar wind), magnetic fields, particles, and rocks, so also is interstellar space filled with gas, dust, magnetic fields, and particles. In this chapter, we shall concentrate upon the dust and gas in the galactic disk, leaving the detailed discussion of the galactic magnetic field and high-energy cosmic rays (relativistic particles) to Chapter 18. Let us begin our present discussion by investigating the characteristics of interstellar dust.

Table 17–1. **Evidence for Interstellar Dust and Gas**

Dust	Gas
Dark nebulae	Interstellar absorption lines
General obscuration	H II regions (bright nebulae)
Reflection nebulae	21-cm hydrogen line
Interstellar reddening	Radio molecular lines
Interstellar polarization	Thermal radio emission
	Nonthermal radio emission

17–1   INTERSTELLAR DUST

(a)   Dark Nebulae and the General Obscuration

Numerous dark patches stud the Milky Way; some are apparent only on photographs, while others, such as the Great Rift in Cygnus and the Coal Sack near the Southern Cross (see Figure 17–1), are immediately evident to the naked eye. These *dark nebulae* are opaque clouds obscuring the light of the stars behind them. In many cases, dark nebulae lie adjacent to or superimposed upon bright nebulae; an example is the famous Horsehead nebula in Orion (Figure 17–2). As we shall see later, the bright nebulae in such instances are usually gaseous, in distinct contrast to the dark nebulae which are dust clouds. Sometimes very small dark regions called *globules* overlie bright nebulae, as may be seen in Figure 17–3. Many astronomers consider these globules to be protostars—dense clouds of dust and gas in the process of condensing and collapsing to form a star (see Chapter 15).

Far more difficult to detect is the *general obscuration* caused by dust distributed more uniformly and thinly than in the dark clouds. We have already mentioned this obscuration in our discussion of observed and expected star counts (Chapter 13). Star counts do not *prove* the existence of obscuration; the data might be interpreted as a real decrease in the number of stars with distance from the Sun. Much more conclusive proof was presented in a study of galactic clusters published by R.

Figure 17–1.   *The Southern Coal Sack: A Dark Nebula* seen projected against the background of the Milky Way. (Harvard College Observatory)

Figure 17–2. *The Horsehead Nebula in Orion.* (Hale Observatories)

Trumpler in 1930. He found that he could classify clusters according to their appearance, using such criteria as (a) the concentration of stars to the center, (b) the range in brightness of the member stars, and (c) the total number of stars contained in the cluster. Furthermore, he found that the diameter of a cluster was correlated with its appearance classification. The observed angular diameters could therefore give a direct measure of the distances to the clusters. Distances which Trumpler derived in this manner were systematically *less than* those based on the magnitudes and spectral types of the member stars (spectroscopic parallaxes). He concluded that only the presence of general "absorption" (diminishing the apparent brightness) could explain the discrepancy, so that the equation for the distance modulus should be rewritten from

$$m - M = 5 \log d - 5 \qquad\qquad (\mathbf{17\text{--}1})$$

to

$$m - M = 5 \log d - 5 + A \qquad\qquad (\mathbf{17\text{--}2})$$

Here A represents the total amount of absorption (in magnitudes) for the distance d from our Sun.

If this general interstellar absorption were truly uniform throughout the Galaxy, A could be expressed as a simple linear function of distance, $A = kd$.

Figure 17–3. *Globules Superimposed on a Bright Nebula.* (Part of Rosette Nebula). (Hale Observatories)

Observations clearly show, however, that the pervasive absorption is not uniform; the total amount of absorption between us and a star or cluster differs with direction in the sky and with the character of the intervening space. This nonuniformity should not surprise us, if we remember the dark nebulae and their irregular appearance.

The absorption effects of a *discrete* dust cloud may be compared with those of the general obscuration by considering *star counts*. In Figure 17–4, we plot the number of stars $N(m)$ seen to a limiting apparent magnitude m. The best results are obtained if we use only stars whose absolute magnitudes are known; for a large sample this is impossible, so that allowances must be made for a range in intrinsic brightness. The curves labelled "unobscured" correspond to a region of the sky which has no interstellar absorption. A dark nebula which is localized produces the following effects (Figure 17–4A): (a) the stars in front of the cloud are unobscured; (b) at the apparent magnitude corresponding to the cloud distance, the curve is shifted toward fainter magnitudes as obscuration takes place; and (c) beyond the cloud the obscured curve runs *parallel* to the unobscured curve, since the same amount of absorption (A) is experienced by all of these more distant stars. In the case of general obscuration (Figure 17–4B), the absorption steadily increases with distance, or $A = A(d)$, and the obscured $N(m)$ curve is displaced to fainter and fainter apparent magnitudes compared with the unobscured curve.

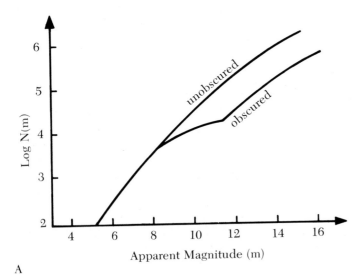

A

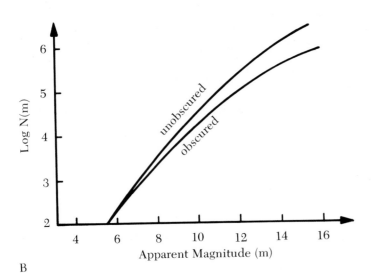

B

Figure 17–4. *Interstellar Obscuration and Star Counts.* **A,** The effect of a discrete dust cloud at $m \cong 10$. **B,** The cumulative effect of general obscuration. Note that the curves marked "unobscured" refer to star fields with *no* interstellar absorption or obscuration.

Note that stars *appear* to be fainter (larger m), and therefore farther away, whenever there is interstellar obscuration.

There are other indications of the prevalence of obscuration in the galactic plane. For instance, the observed absence of (extragalactic) galaxies at low galactic latitudes, termed the *zone of avoidance*, can be attributed only to obscuration in the galactic plane.

Figure 17–5. *The Milky Way in Cygnus Showing the Effect of Obscuration by Dust Clouds.* (Hale Observatories)

(b) Interstellar Reddening

The absorption A in equation (17–2) depends upon whether we are dealing with visual or blue magnitudes. The interstellar dust between us and a star does not dim that star's light identically at all wavelengths; more light is absorbed in the blue than in the red. As a result, the light from the star appears redder than in the absence of the dust—hence, the term *interstellar reddening*. The obscuration is primarily a form of scattering rather than an absorption phenomenon. The reddening arises from *selective scattering*, so that if equal numbers of red and blue photons are incident upon a dust cloud, a greater number of the blue photons is scattered out of the beam. Hence, a proportionately larger number of red photons penetrates through the cloud and reaches the observer (see Figure 17–6).

Reddening must *increase* the color index observed for a star. We define *color excess* as the difference between the observed and intrinsic color indices:

$$CE = CI \text{ (observed)} - CI \text{ (intrinsic)} \tag{17–3}$$

The intrinsic color index depends upon the spectral type of the star, and it can be quickly established if the spectrum is available. It is also possible to find the color

Star

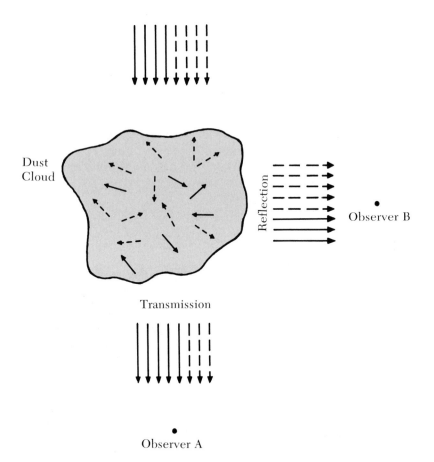

Figure 17–6. *Selective Scattering and Interstellar Reddening.* Equal numbers of red photons (solid lines) and blue photons (dashed lines) leave the star. More blue photons are scattered by the dust than red photons. Therefore, observer B sees a *blue* reflection nebula, while observer A sees *reddened* starlight.

excess without taking a spectrogram, by comparing several color indices, such as blue-visual $(B - V)$ and ultraviolet-blue $(U - B)$.

The wavelength dependence of the interstellar extinction (absorption) is found by comparing the brightness of stars of similar spectral type at a number of wavelengths. By such a comparison of stars reddened by different amounts, astronomers have found that the extinction is proportional to $1/\lambda$ in the visual and photographic regions. Such data also established that in most regions of the Galaxy the absorption in visual magnitudes is three times the color excess; thus

$$A_v = 3(CE) \qquad (17\text{–}4)$$

We may use this value of A in equation $(17\text{–}2)$ to find the true distance to the star, providing the star itself does not have a peculiar spectral distribution or affect the nature of dust grains in its immediate vicinity.

As an example, consider a star with $m_v = 13.0$, $CI = 1.6$, and spectral type = G0 V. Reference to the H-R diagram (Chapter 12) tells us that a star of this spectral type has an intrinsic color index of $+0.6$, and $M_v = +5$. Therefore, from equation (17–3) we have $CE = 1.0$ and from equation (17–4) $A_v \cong 3.0$; now substituting in equation (17–2)

$$m_v - M_v = 5 \log d - 5 + A_v$$

we obtain

$$13.0 - 5.0 = 5 \log d - 5 + 3$$

$$\log d = 2, \quad d = 100 \text{ pc}$$

Therefore, although the star *appears* to be 400 pc distant, its actual distance is only 100 pc. Alternatively, we could say that the *true* distance modulus is

$$m_v - M_d = m_v - M_v - A_v = 5 \log (d/10)$$

$$= 13 - 5 - 3 = +5^m$$

(c) Reflection Nebulae

When a dust cloud lies to one side of a star rather than between the observer and the star, it scatters the light from the star toward the observer. Here we are dealing with the same scattering phenomenon that is responsible for interstellar reddening, but instead of viewing the light which filters *through* the dust, we see the light which is scattered out of the star-to-cloud direction. Figure 17–6 illustrates this situation for an observer (B) viewing a *reflection nebula*. Each of the dust particles scatters a bit of the star light toward us. Since the particles scatter blue light more effectively than red, reflection nebulae appear *bluer* than the incident starlight which they scatter. Another type of bright nebula is an *emission nebula*, which consists primarily of gas excited by a hot star. Such emission nebulae will be discussed later, but they are mentioned here to emphasize the difference between reflection and emission nebulae. Much of the light from emission nebulae arises from hydrogen line radiation, of which the red Hα line is the most prominent in the visible spectral region. Emission nebulae therefore appear *reddish*, while reflection nebulae are blue. Color photographs of nebulae show this difference quite dramatically.

Although we distinguish between these two types of bright nebulae, they are often found in close proximity; a reflection nebula may lie adjacent to an emission nebula, as occurs in the constellation Orion. This illustrates that dust and gas are usually closely mingled when they are found in the interstellar medium (see section 17–3).

(d) Interstellar Polarization

The optical polarization of starlight by the interstellar medium was discovered accidentally by J. S. Hall and W. A. Hiltner as they searched for predicted periodic changes in the degree of polarization of light from an eclipsing binary star. The observed polarization was far larger than anticipated, and remained constant. Light from other stars, including single stars, also turned out to be polarized, up to six or seven per cent (the nature and properties of polarized light are briefly explained in Chapter 8). The amount of polarization is correlated with the interstellar reddening, in the sense that a large degree of polarization is found only for stars with a large color excess. It is evident, therefore, that interstellar dust causes most of the optical polarization of starlight, although some stars have recently been discovered which show intrinsic and variable polarization.

Only *nonspherical* particles can polarize light; hence, the discovery of interstellar polarization presented a totally unexpected clue to the nature of the *interstellar dust grains*. Furthermore, even nonspherical particles cannot polarize light if they are oriented at random. The observations indicate that the grains must be "lined-up" (on the average). Although *not* a real proof, the occurrence of interstellar polarization is a strong argument for the existence of interstellar magnetic fields, for even relatively weak fields will align the particles. The most generally accepted theory visualizes elongated particles (small cylinders) spinning with their short axes aligned along the magnetic field. Therefore, we can use polarization data to map the magnetic field of the Galaxy as seen from the Sun. Polarization will be strong and ordered when the magnetic field is perpendicular to the line of sight, and will be weak and random when we look along the field (i.e., down a magnetic tube). Some observations of interstellar polarization in the southern Milky Way are reproduced in Figure 17–7. The Sagittarius spiral arm crosses the line-of-sight from $\ell = 310°$ to $\ell = 55°$, and we can conclude from the figure that the galactic magnetic field (on the average) lies *along* spiral arms. As discussed in Chapter 18, *radio* interstellar polarization is the only *direct* evidence for this magnetic field, and it confirms the interpretation that the optical polarization is related to the magnetic field.

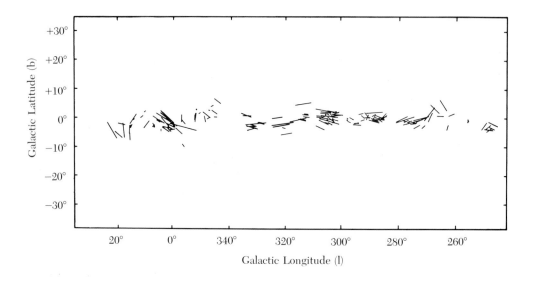

Figure 17–7. *Interstellar Polarization of Starlight.* This diagram depicts the polarization of starlight near the galactic plane, as seen in the southern hemisphere. Each line represents one star, and indicates the amount of polarization (length of line) and the polarization direction.

(e) The Nature of the Interstellar Grains

The observed effects of the interstellar grains—interstellar reddening, extinction, reflection, and polarization—give us some clues to the nature of the particles involved. Although both theorists and experimentalists have worked diligently to decipher the data, no uncontroversial final solution has yet been found. The possibilities which can explain most of the observations include the following:

(a) elongated, perhaps needle-shaped, dirty-ice grains,
(b) grains of graphite (carbon),
(c) particles with graphite cores and ice mantles,
(d) large and complex molecules similar to hydrocarbons,
(e) silicate particles,
(f) diamonds or other forms of carbon.

The strength of the interstellar obscuration and the characteristics of the reddening require *particles* rather than ordinary molecules, atoms, or electrons. Interstellar polarization requires *nonspherical* particles that can be aligned by a magnetic field; therefore, pure ice particles are excluded because they would not be so aligned. Graphite or graphite-core particles could fit, for carbon in the form of graphite readily forms into highly-flattened plates or flakes. This nonspherical attribute is not unique to graphite, however, for ice and ice-like materials also tend to form flat crystals. The idea of diamonds may seem far-fetched, but, in fact, the reddening curve can be reproduced fairly well by a suitable range of sizes of diamonds! There is strong support for silicate particles, alternative (e), or a *mixture* of silicates and either alternative (b) or (c). Part of the evidence in favor of silicates comes from the wavelength dependence in the ultraviolet (rocket and satellite observations) and part from infrared (around 10μ) spectral features.

The correct model must explain the *formation* of the particles; this is a major hurdle, since the chemistry of interstellar space, with its very low densities and temperatures, is still very poorly understood. A truly satisfactory model must show which atoms will stick together and how. Proponents of the graphite or graphite-plus-ice models point out that certain cool stars, such as the R and N stars (also known as *carbon stars*), represent likely sources for the formation of such grains. These carbon stars are variables, which pulsate with periods of about 100 days and undergo changes in surface temperature of hundreds of kelvins. Carbon may condense in the cooling atmosphere and grow into graphite grains, which are then pushed away from the star by radiation pressure to become part of the interstellar medium. These small particles may serve as the nuclei to which other atoms and molecules, such as water, methane, ammonia, and an occasional metal atom, will adhere. The grains will not grow indefinitely; collisions with other grains, with fast-moving atoms, and even with cosmic rays will keep their size small. Such *grain destruction* occurs in collisions of dust clouds, and also when cold clouds become H II regions upon encounters with O or B stars (see the following section). The average diameter of the dust grains, deduced from the observations, is about a tenth of a micron (i.e., 10^3 Å).

The controversy over the nature of the dust grains rages unabated, for both the experimental data and our understanding of these data are far from complete.

Only the future can tell us what we presently do not know about this fascinating component of the interstellar medium.

Closely related to the problem of grain formation and destruction is the dependence of *complex molecule* formation on interstellar grains. As we shall discuss later, a large variety of molecules have been observed, generally in close association with cold, dense, dust clouds. Irregular grains may trap atoms for a sufficient length of time to permit the building up of complicated molecules. There will be both evaporation of molecules from and condensation of molecules onto grains.

17–2 INTERSTELLAR GAS

In addition to dust grains, interstellar space is filled with gas. In this section, we shall investigate the physical and observational properties of the many species of *interstellar gas* which constitute most of the interstellar medium. While the dust between the stars makes its presence known by its broad spectral influence upon the continuous light from distant stars, the interstellar gas produces its own characteristic emission and absorption line spectra. The temperature and density of the gas (e.g., the proximity of a gas cloud to a hot star) determine these characteristic spectral features (see Chapter 8). In general, the gas is essentially *transparent* over a wide spectral range, despite the fact that the total mass of the gas in our Galaxy is very much greater than the total mass of the dust!

If there is so much more gas than dust, then why is it that the dust is responsible for interstellar obscuration? Certainly the number density of dust grains is vastly smaller than the number density of the gas. To answer the original question, we must consider the *absorption coefficient* of each material; this parameter is analogous to the opacity discussed in Chapter 15. Dust particles interact very strongly with visible light, over a broad range of wavelengths, while the interstellar gas cannot interact with visible light (i.e., hydrogen and helium in their ground states cannot absorb visible-light photons). Therefore, at visible wavelengths, the absorption coefficient per particle is astronomically greater for the dust than it is for the gas, and it is the dust which obscures.

(a) Interstellar Optical Absorption Lines

Some stars have absorption lines in their spectra which are quite out of character with their spectral class. For instance, many B stars exhibit sharp, sometimes multiple, lines of Ca II. Some spectroscopic binaries show particular spectral lines which remain *fixed* in wavelength, while the rest of the spectral lines shift periodically to the red or blue in response to the binary stellar motions (see Figure 17–8A). These absorption lines must originate in the interstellar medium, for they certainly cannot be due to the star. *Multiple lines* arise when there are several absorbing clouds along the line-of-sight (see Figure 17–8B). Optical absorption lines, identified as interstellar in origin, include those from: Ca I, Ca II, Ti I, Ti II, Na I, and the molecules CN and CH. Interstellar H_2 and CO have been observed in the rocket ultraviolet.

These absorption lines are sharp because thermal Doppler broadening is negligible at the low temperatures which characterize the interstellar medium.

Figure 17–8. *The Discovery and Appearance of Interstellar Absorption Lines.* **A**, Interstellar lines were discovered when the spectrum of a binary system was found to have un-Doppler-shifted absorption lines. Only one spectrum is observed, that of the shaded star. **B**, Each spectral line has several Doppler-shifted components; each component is attributable to a separate absorbing cloud.

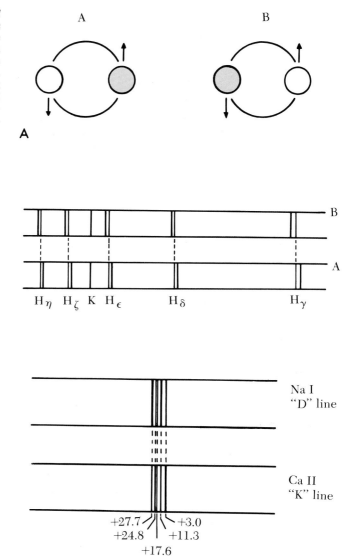

The intensity of a line depends upon the amount of gas lying between the star and the observer; if the gas is distributed uniformly through space, the intensities of interstellar absorption lines depend directly upon the *path length* traversed by the starlight. Although the actual distribution of the gas is not perfectly uniform, this approximation is sufficiently accurate so that we may use the intensity of interstellar lines as approximate distance indicators for stars in the galactic plane. Since the density of the gas is exceedingly low, the path lengths to such stars may be very long (as much as a kiloparsec or two) before a sufficient number of atoms is encountered by the line-of-sight to produce a measurable spectral line.

The low gas density also plays a role in preventing ions from recombining into neutral atoms after photo-ionization. Sufficiently energetic photons and cosmic rays will occasionally encounter and ionize the widespread gas atoms and molecules. In order to recombine, an ion must capture an electron, but at interstellar densities the chance of such a capture is very small.

Some interstellar lines are not sharp; occasionally we find lines which are tens of Ångstroms wide. These broad lines have not yet been definitely identified, but the fact that such lines are even better correlated with interstellar reddening than are normal interstellar lines suggests that they are associated with the dust grains rather than with the gas. Several theories have been proposed to explain these lines; one school of thought advocates that they are formed by complex organic molecules adhering to the surfaces of the dust grains. At present, no theory is generally accepted.

(b)   Emission Nebulae: H II Regions

(i)   *Hydrogen Line Emission*

Among the most spectacular objects to be photographed with telescopes are the *emission nebulae*—clouds of gas caused to shine by the intense radiation from a hot star. Hot O and B stars emit tremendous amounts of ultraviolet radiation; such energetic photons, with wavelengths less than 912 Å, will ionize any hydrogen atoms they encounter. Therefore, if such a hot star is surrounded by a cloud of gas, the hydrogen atoms close to the star will be ionized and we will have an *H II region* or emission nebula. As we move away from the star, we find that the energetic photons are degraded in energy; eventually they can no longer ionize the hydrogen, and the H II region terminates (i.e., neutral hydrogen prevails). Let us now examine the physical processes which determine the structure of an H II region in greater detail.

H II regions are observed to be roughly spherical in shape, with a very sharp outer boundary. It is clear that these objects can serve as approximate distance indicators (in fact, they are presently used to determine extragalactic distances), once we understand the physical basis for their structure. In 1948, B. Strömgren published his now-classic analysis of H II regions; the brief discussion which follows is based upon his work.

The hydrogen gas in interstellar space is extremely dilute and cold. Most of the gas is H I (neutral hydrogen) in the ground state, since collisional excitation is rare. Therefore, only photons whose wavelength is *less than or equal to 912 Å* can ionize the gas to H II (see Figure 8–9); note that 912 Å corresponds to the Lyman continuum limit (ionization potential) of hydrogen. Let us now place a *hot* star in the midst of this cool H I gas; we use a hot star ($T_{eff} \gtrsim 20{,}000$ K) because the Planck spectral curves (see Chapter 8) imply that very little ultraviolet radiation ($\lambda \lesssim 912$ Å) is emitted by cool stars. Therefore, we will consider only the extensive H II regions produced by O and B stars. If the gas density is reasonably uniform, the ultraviolet radiation from the central star ionizes all of the hydrogen in a spherical volume of space; we term this region the *Strömgren sphere*. Equilibrium is established when the rate of recombination (H II + e$^-$ → H I) equals the rate of photo-ionization; the Strömgren sphere is maintained by the continual re-ionization of recombined H I atoms due to the flux of ultraviolet photons from the central star.

The sharp outer boundary of an H II region is due to several factors. As we move to greater distances from the star, the inverse-square law diminishes the flux of ultraviolet photons, and ionization of the recombined H I atoms is no longer

Figure 17–9.   *An H II Region* (Lagoon Nebula). Bright O and B stars excite the surrounding hydrogen gas. Note the superimposed globules, pointing to the close association between gas and dust. (Lick Observatory)

possible. Thus the ratio of H I to H II rises sharply with increasing distance from the star, and the material quickly becomes opaque to the Lyman continuum, giving rise to the sharp boundary. In addition, most of the H II recombines to an *excited* state of the neutral H I; the atom then quickly cascades to the ground state, emitting several *low-energy* ($\lambda > 912$ Å) photons in the process. Since the H I atoms spend so little time in the excited states, practically all of these low-energy photons (as well as the star's photons with $\lambda > 912 \,\text{Å}$) escape from the H II region. Therefore, the H II region *fluoresces* by converting the stellar ultraviolet radiation into lower-energy photons, with the bulk of the radiation escaping as the visible (longer wavelength) Balmer lines. The H II region is therefore spectacularly visible as a bright reddish *emission nebula*—reddish because most of the Balmer radiation is in Hα.

The radius of the Strömgren sphere clearly depends upon the following parameters: (a) the *surface temperature* of the central star, which determines the flux of ultraviolet photons; and (b) the *density* of the hydrogen gas, which determines the rates of photo-ionization and recombination. An additional complication arises if the gas density decreases with distance from the star, for then the size of the nebula may be limited by the gas density alone. Note that the gas within the Strömgren sphere is *hot*, owing to the high kinetic energies of the electrons and ions, while the gas outside the nebula is very cold.

H II regions are observed to radiate many interesting and informative spectral lines and continua. Although some radiation short of the Lyman continuum limit does leak out, the spectrum of the central star (as seen through the gas cloud) exhibits severe depletion at ultraviolet wavelengths. As we have already mentioned, the strongest optical emission lines from the nebula itself are the hydrogen Balmer lines, especially Hα. Recently, radio line emission at centimeter wavelengths has been observed; this is due to very-low-energy electronic transitions between very high excitation levels of H I, such as from level $n = 110$ to $n = 109$ and from 105 to 104. The cold neutral hydrogen which surrounds the emission nebula may be observed by the 21-cm (radio) line which it emits, but we defer our discussion of this important line until later in this chapter (and in Chapter 18). The ionized hydrogen (H II) within the Strömgren sphere has no electrons, so it cannot radiate spectral lines; nevertheless, low-energy (radio) continuum radiation emanates from the H II region, as a result of free-free transitions (discussed in the next section and in Chapter 9). Optical fluorescence lines of helium are also strong in the spectra of emission nebulae; together with the recently-discovered radio recombination lines of helium (arising from transitions between high-excitation levels), these lines permit us to (a) study the excitation mechanisms operating in H II regions, (b) investigate the elemental abundances (especially He/H) of the interstellar medium, and (c) probe the spiral structure of our Galaxy. Forbidden lines of O II, O III, and Ne III appear in these emission nebulae as well as in the planetary nebulae discussed later.

(ii) *Continuous Radio Emission*

We have seen that the ultraviolet photons from a hot star liberate electrons from hydrogen atoms in the gaseous nebula surrounding the star. These electrons then move freely through the gas, sometimes recombining with ions and sometimes exciting atoms or ions,* but more often interacting with ions in a *free-free transition*. Again we have a situation similar to that which occurs in the solar corona (see Chapter 9). A free electron travels past an ion in a hyperbolic orbit of a given energy. This orbit can be altered by the quantum-mechanical absorption or emission of a photon with any amount of energy. When an assembly of electrons and ions

---

* Leading to the emission of forbidden lines, as discussed in section 17–2(d).

(i.e., a plasma) is involved, the individual free-free emissions add up to a continuum; since the characteristic kinetic energies are small, this continuum radiation occurs predominantly at infrared and radio wavelengths. In short, an H II region is a source of radio emission characterised by the mean energy of the electrons (i.e., by the temperature of the gas). To distinguish this emission from synchrotron radiation (see Chapter 16), we use the term *thermal radio emission*.

(c) Supernova Remnants

The most spectacular of the known remnants in our Galaxy is the Crab nebula, already discussed in detail [section 16–5(d)]. Material ejected from supernovae certainly becomes part of the interstellar medium. Moreover, the ejected matter sweeps up any surrounding gas and dust as it expands; this produces a shock wave that excites and ionizes the gas, which then becomes visible as an emission nebula. X-rays emitted by supernovae are also instrumental in ionizing nearby gas. As we have already seen, supernova remnants are radio emitters because of their synchrotron radiation.

Therefore, when a supernova explodes, a vast amount of energy is released, and this energy greatly affects existing interstellar matter. S. P. Maran and his colleagues have suggested that some emission nebulae may actually be fossil Strömgren spheres created by supernova explosions. They are "fossil" because the exciting star is no longer present to keep the gas ionized, but the gas has not yet entirely returned to the neutral state. This hypothesis was proposed to explain the *Gum Nebula* (named after its discoverer, C. S. Gum), an enormous nebula covering 60 by 30 degrees in the sky, corresponding to a linear diameter of 360 pc.

(d) Planetary Nebulae: Nebular Forbidden Lines

Closely related to the diffuse H II regions, *planetary nebulae* differ in that they are more compact and of higher surface brightness. When seen through a telescope, a planetary nebula appears as a round, greenish disk or halo which superficially resembles a planet—hence the name. Closer examination reveals that the nebula is excited by a very hot, central star, similar to the Wolf-Rayet type but not so bright. Gas densities in the nebulae surrounding these stars are higher than in H II regions; hence, collisions between electrons, atoms, and ions occur more frequently. Collisional excitation and de-excitation are therefore significant, so that spectra of planetaries differ in important ways from those of H II regions. For a detailed discussion of the physical origin and evolution of planetary nebulae, consult section 16–4(e).

Although the spectral lines of hydrogen and helium are quite pronounced in the spectra of planetary nebulae, the strongest lines are those of O III, O II, and Ne III. The latter lines do not correspond to ordinary electronic transitions, but instead arise when an electron jumps from an excited metastable state to the ground state. Since quantum mechanics strongly suppresses such transitions, the resulting spectral lines are termed *forbidden lines*. This situation reminds us of the coronal forbidden lines (see Chapter 9), except that in the solar corona the forbidden lines are due to ions which have lost nine or more electrons, while in planetary nebulae, only one or two electrons have been removed. The physical reason for this difference is the great disparity in the temperatures characterizing the two situations: in the solar corona, temperatures of 2 million degrees Absolute prevail, while in planetary nebulae, the temperature is of the order of 10,000 K or less. Forbidden lines occur in each case because of the *low gas density;* when an atom is excited to a metastable level, the chance of collisional de-excitation from that level is slight, so that the atom may remain in that level long enough to make the "forbidden" (i.e., low probability) transition to the ground level. At the gas densities found in stellar atmospheres, collisions with electrons and other atoms occur sufficiently often that an atom will not remain in a metastable state long enough to make the forbidden transition.

The forbidden nebular lines of [O III] were identified in 1927 by I. S. Bowen, when he recognized that the wavelengths λλ 5007 and 4959 correspond to the calculated energy differences between the metastable level and two of the three closely-spaced ground levels of O III (see Figure 17–10A). (The square bracket notation [ ] is used to signify forbidden transitions.) These lines, which

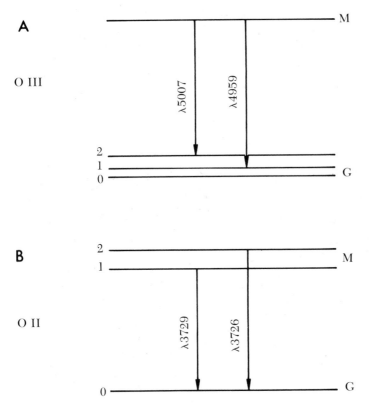

Figure 17–10. *Some Forbidden Transitions of O III and O II.* These are energy level diagrams of two species of ionized oxygen showing the excited metastable state (*M*) and the ground state (*G*). **A,** For O III, *M* is a single level while *G* has three sublevels. Only two forbidden lines (at 4959 Å and 5007 Å) are seen. **B,** For O II, *M* is split into two sublevels while *G* is single. The two possible forbidden lines (at 3726 Å and 3729 Å) are seen.

give planetaries their *greenish* appearance, have never been observed in the laboratory because sufficiently low densities cannot be obtained and collisional excitation and de-excitation prevail. The third transition, from the metastable level M to the lowest ground level G_0, is much more strongly forbidden than the other two; therefore, it does not appear, for the atom can leave the metastable level by either of the other transitions $M \rightarrow G_2$ or $M \rightarrow G_1$. A similar pair of forbidden lines arises from [O II], except that in this case the metastable level is double and the ground level single (see Figure 17–10B).

Ions such as O II, O III, Ne III, and N II act as *cooling agents* in gaseous nebulae. Hydrogen atoms require large amounts of energy to become excited (10.15 eV for the first excited state), but most of the free electrons in the nebula do not have this much kinetic energy. The cooling-agent ions, however, all have energy levels near 2 or 3 eV; when an electron collides with one of these ions, it gives up part of its kinetic energy to excite the ion to one of these low (metastable) levels. Within a minute or two (in contrast to the 10^{-8} sec for ordinary levels), the ion gives up this energy by emitting a forbidden line which escapes the nebula. Collisions must be rare; otherwise, de-excitation would prevent the occurrence of forbidden transitions. On the other hand, collisions will be frequent enough that collisional excitation is fairly common. Therefore, these ions extract energy from the electrons, and since the kinetic energy of the electrons is a measure of the temperature of the nebula, the result is a lower temperature for the planetary nebula or H II region. Although most of the forbidden lines are comparatively stronger in planetaries than in H II regions, this cooling mechanism also plays a very important role in the latter.

(e) Interstellar Radio Lines

(i) *The Neutral Hydrogen Line at 21 cm*

Far from hot stars, the interstellar gas is cold. Here hydrogen, the most abundant of the elements, is neutral and in its ground state. This ground state, however, is actually split into two levels separated by a very small energy difference. The reason for this phenomenon lies in the fact that both the proton and the electron have an "intrinsic spin." Just as we are not entirely justified in thinking of electron orbits in atoms as being analogous to planetary orbits (see Chapter 8), so too we cannot make a rigorous analogy between intrinsic spin and a rotating billiard ball. Nevertheless, we found the Bohr model of the atom a useful conceptual device in Chapter 8; similarly, the spinning-ball model of intrinsic spin may be used to advantage in the present discussion. The reader is probably aware of the fact that a moving charge produces a magnetic field. Since both the proton and electron are *charged particles*, their spin motion generates a dipole magnetic field (like the field of a tiny bar magnet) which we can characterize by the term *magnetic moment*. The magnetic moment of a spinning particle is represented by a *vector*, which is proportional to the vector angular momentum (see Chapter 3) of the particle.

In Figure 17–11 we illustrate two possible ground-state configurations of the neutral hydrogen atom. In one configuration, the magnetic moment vectors of the proton and electron are parallel or *aligned;* since vectors add (see the Mathematical Appendix), there is a lot of magnetic energy present. Just as two parallel bar magnets will repel one another, so too will the proton and electron be less

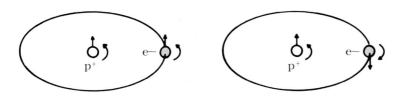

Figure 17–11. *Spin Alignment in the Hydrogen Atom: The Origin of 21-cm Radiation.* The hydrogen atom consists of one proton (p$^+$) and one electron (e$^-$), both of which have *spin*. The *lowest* energy state of the system is shown on the right (spins antiparallel). When the relative spin flips, the atom makes a transition from the left (higher energy) to the right (lower energy), emitting a photon of 21-cm wavelength in the process.

tightly bound together in their mutual orbits. If the magnetic moment vectors are antiparallel or *opposed*, we have the second configuration, which is characterized by less magnetic energy and a more tightly bound orbit. Therefore, the aligned state lies at a slightly *higher* energy than the opposed state; we refer to this effect as the *hyperfine splitting* of the ground state of the hydrogen atom. A spontaneous transition from the higher hyperfine state to the lower one can occur, accompanied by a relative "spin-flip" (from aligned to opposed) and the simultaneous emission of a very-low-energy photon. This emission produces the famous 21-cm radio spectral line of neutral hydrogen; the characteristic frequency of this transition has been accurately determined as 1420.406 MHz.

When hydrogen atoms collide in the interstellar medium, they generally exchange their electrons; this collisional transfer is the chief mode of changing the hyperfine states of these atoms. If the spin of the newly-acquired electron has the same orientation as the old, no change in energy level occurs; otherwise, there will be a change in level (either up or down). In other words, collisions may result either in no change, in excitation, or in de-excitation. A change in either direction takes place about once every 400 years for a given interstellar hydrogen atom. On the other hand, an atom in the excited hyperfine state makes a *spontaneous* downward transition, followed by the emission of a 21-cm line quantum, only once every 11 million years (on the average) because this transition is strongly forbidden. The lifetime for spontaneous emission is unaffected by the numerous collisional excitations and de-excitations (of the hyperfine levels) which occur in the interim.

When we consider distances on the order of a kiloparsec, an enormous number of hydrogen atoms lies along a line-of-sight, in spite of the exceedingly low gas densities in our Galaxy; among these atoms, enough downward radiative transitions occur to produce a detectable 21-cm spectral line. In 1944, H. C. van de Hulst predicted that this line could be observed; the state of technology did not meet the observational requirements until 1951, when the line was first seen.

The 21-cm line of H I has proved to be a very powerful tool in studies of galactic structure. Different regions of the Galaxy move at different velocities with respect to us, and the observed 21-cm line is Doppler-shifted accordingly. Combined with our knowledge of the internal motions of our Galaxy, the observed Doppler-shifted line may be used to estimate the distance to the cloud of hydrogen producing that emission. When we examine the shape or *profile* of the 21-cm line, we find that it often has several peaks; this indicates that the gas is concentrated in discrete regions, such as spiral arms, rather than distributed smoothly throughout the Galaxy. Further details on how we can map the distribution of hydrogen gas in the Galaxy are mentioned in Chapter 18.

(ii)   *Molecular Lines at Radio Wavelengths*

One of the most exciting recent developments in astronomy has been the discovery of *interstellar molecules* of a variety and complexity far greater than anticipated. These range from simple molecules like CO, CN, and OH to such complex organics as formaldehyde (H_2CO) and methanol (CH_3OH). They were found by searching for spectral lines at radio wavelengths. Table 17–2 lists those molecules for which radio lines have been discovered at the time of this writing. These molecules are important because they let us probe dense clouds of gas and dust, some of which must be on their way to collapsing into protostars. Furthermore, study of these molecules must eventually lead to a better understanding of the chemistry of the interstellar medium: How are such molecules formed? How soon after formation are they destroyed? What is their association with interstellar grains? What are the peculiar conditions of their excitation? For many people, however, the most exciting aspect of the molecules concerns their implications with respect to *life* outside the Solar System. H_2O, NH_3, HCN, H_2CO, HC_3N, and HNCO are among the molecules used in laboratory experiments to synthesize amino acids and nucleotides, the "building blocks" of life. The fact that these molecules exist in interstellar space indicates that their formation does not require extraordinary conditions. Whether they could survive the heating accompanying the condensation

Table 17–2.   Interstellar Molecules with Radio Lines

Molecule	Chemical Name	Wavelength (cm)
OH	Hydroxyl	18, 6.3, 5.0, 3.7, 2.2
CO	Carbon Monoxide	2.7
CN	Cyanogen	2.6
CS	Carbon Monosulfide	0.20
SiO	Silicon Monoxide	0.23, 0.34
H_2O	Water	1.35
NH_3	Ammonia	1.2, 1.3
H_2S	Hydrogen Sulfide	0.18
HCN	Hydrogen Cyanide	0.35
OCS	Carbonyl Sulfide	0.27
HCHO	Formaldehyde	6.6, 6.2, 2.1, 1.0, 0.2
HCHS	Thioformaldehyde	9.5
CH_3OH	Methanol	36, 1.2
CH_3CN	Methylcyanide	0.27
HCOOH	Formic Acid	18.3
HNCO	Isocyanic Acid	1.36, 0.34
CH_2NH	Formaldimine	5.8
CH_3C_2H	Methylacetylene	0.35
CH_3HCO	Acetaldehyde	28.1
NH_2HCO	Formamide	6.5
X-ogen	Unknown	0.34
X_2	Unknown	0.33

of protostars and protoplanets is most uncertain, but they may not need to, since they seem to form so readily.

The first molecule to be detected in the radio region was the hydroxyl molecule, OH, with several lines around 18 cm. I. S. Shklovsky had predicted that these lines might be observed as early as 1953, but they were not detected until 1963, after their characteristic frequencies had been firmly established in the laboratory. Four transitions near a wavelength of 18 cm (frequencies of 1612, 1665, 1667, and 1721 MHz) occur because of the splitting of the ground level of the OH molecule. With the detection of ammonia, NH_3, in 1968, the avalanche of newly-discovered molecular lines really started, so that now there are more than 20 molecules known, represented by over 50 lines (see Table 17–2). These lines lie at wavelengths of between 2 mm and 36 cm. The short-wavelength end borders on the infrared, and future work will probably lead to the discovery of new molecular lines at still shorter wavelengths.

Some "radio molecules," such as OH and H_2O, are associated with certain red and infrared stars. All molecules, however, appear to be connected with dust. OH, H_2CO (formaldehyde), and CO lines are fairly widespread, originating in large dust clouds. Some *small* dense dust clouds apparently enable peculiar excitation conditions to exist which cause them to be strong sources of H_2O, OH, and other molecular lines. Many of these dense clouds lie in the direction of and are connected with H II regions; the Orion nebula is a prime example of an H II region containing many dense condensations. The number densities in such clouds are estimated as 10^3 to 10^6 H_2 molecules per cubic centimeter; other molecules are, of course, far less abundant though more readily observed. The cloud temperatures are low, usually 10 K to 30 K, sometimes 100 K, and, for the formaldehyde clouds, as low as 1 K (see the following paragraphs).

Some molecular lines, including those of OH, are seen as *absorption* lines superimposed on the continuum of some bright radio sources, such as H II regions or the center of the Galaxy. Most of the radio molecular lines are *emission* lines due to rotational transitions [see section 8–3(c)]. Emission requires that the molecules must be excited above the ground state by some mechanism.

Several different molecules are often observed in the same source: for example, the clouds associated with such H II regions as the Orion nebula, and clouds in the direction of the galactic center. A comparison of the intensities and Doppler shifts of these lines leads to some surprising results. The molecules OH, NH_3, H_2CO, HCN, and CH_3OH have all been detected in the direction of Sagittarius A, the strongest lines (OH and H_2CO) in absorption, the others in emission. The 21-cm line of H I also appears as a strong absorption line. The strong lines are all multiple, for the line-of-sight intersects several clouds, each with a slightly different radial velocity that produces its own Doppler-shifted component in the profiles. Although the molecular lines have components at the same velocities as the H I line features, the relative strengths differ very greatly. The strongest molecular features do *not* correspond to the strongest H I features (see Figure 17–12). Physical conditions must differ substantially from cloud to cloud. In clouds with strong molecular lines, but weak H I, most of the hydrogen is presumably in the form of H_2 molecules rather than neutral hydrogen atoms, for it is unreasonable to suppose that the overall abundances of the elements vary to that extent.

Figure 17–12. *Profiles of Molecular Lines* in the direction of the galactic center. The vertical line represents zero radial velocity with respect to the Sun. Features that appear strongly in the 21-cm H I line also appear in other lines, especially those of OH and H_2CO, but with very different relative strengths. (Adapted from D. M. Rank, C. H. Townes, and W. J. Welch)

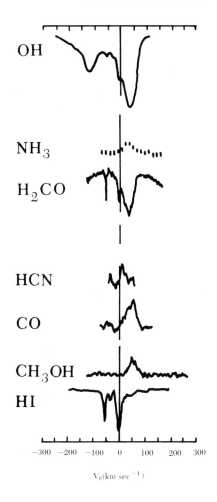

$V_R(km\ sec^{-1})$

OH lines also appear in *emission* from parts of bright H II regions. Some, but by no means all, OH sources show strong H_2O emission as well. The H_2O emission is extraordinarily variable, with intensity changes occurring in periods of months or days. Some previously strong H_2O sources have disappeared entirely. Although also variable, the OH radiation changes far less erratically. Superimposed on H II regions, one finds several groups of OH emission regions separated by several seconds of arc. Within each group, about 1″ in diameter, are several sources, each of dimension about 0″.0005. In fact, one source observed in the Orion nebula corresponds to a linear diameter of 0.4 AU. Such extraordinarily small angles are measured by *long-base-line interferometer techniques;* simultaneous and carefully-synchronized observations are performed at radio observatories in (for example) Sweden and West Virginia. Thus, the baselines are many thousands of kilometers and hence millions of wavelengths long. Although the *sizes* can be determined by these techniques to 0″.0005, the positions thus far are only accurate to some seconds of arc. So it is not known to what extent the OH and H_2O regions are coincident. That they are not *exactly* coincident is shown by the fact that there exist slight differences in Doppler shifts.

The emissions from these small OH and H_2O regions far exceed what would be expected from thermal excitation by collisions. Ridiculous temperatures, such as 10^{13} kelvins, are calculated on the basis of a thermal interpretation of the intensities. Moreover, the relative strengths of the OH lines bear no resemblance to the theoretical predictions, and the lines often exhibit strong polarization. The energy levels of the molecules are apparently subject to population inversion, by which we mean that *more* molecules are in the upper levels than in lower levels; hence, the Boltzmann equation (Chapter 8) is absolutely violated, and thermal equilibrium does not exist. It is generally accepted that some type of *maser action* is responsible for these inversions. The source of the radiation that is thus amplified is quite unknown. Some mechanism (and several have been proposed but none agreed upon) serves to greatly amplify the energy so as to "pump" the molecules into the appropriate excited state.

Some of the other radio molecules also show deviations from thermal equilibrium, although these are not necessarily all due to maser action. The most astounding case is that of formaldehyde, H_2CO, which appears to be *colder* than the 2.7 kelvins of the cosmic background radiation [see section 20–2(a)]. Apparently the formaldehyde molecules act as refrigerators!

17–3   THE ASSOCIATION OF DUST AND GAS

We have already seen from the discussion of interstellar molecules that gas and dust are closely associated in dense clouds. In fact, the formation and destruction of both molecules and grains are probably intimately linked. That the co-mingling of dust and gas is quite general is evident in representative photographs of some emission nebulae (see Figures 17–2, 17–3, 17–9, and 17–13), which show dark lanes or knots superimposed on the bright clouds. Moreover, many bright nebulae shine both by emission and by reflection, thus signifying the presence of both gas and dust. Theoretical calculations show that the gas will carry the dust along with it. The turbulent and ordered motions within clouds, as well as the orbital motions of individual clouds, are very complex. The intricate forms assumed by the nebulae give us some indication of this complexity. The density of H I clouds, found from the observed intensity of the 21-cm line, increases with the amount of interstellar extinction in the general galactic field. The proportion of dust to gas differs somewhat from region to region, but both are always present together. An approximate average figure for the ratio of gas to dust *by mass* is 100 to 1. In dense dust clouds, however, we observe little neutral atomic hydrogen, and it is surmised that the hydrogen is largely in the form of the molecule H_2.

The spatial concentration and physical conditions of the gas and dust vary widely throughout the Galaxy. The number of hydrogen atoms per cubic centimeter is about 0.1 to 0.5, when averaged over the plane of the Galaxy, with temperatures ranging from 100 K to over 1000 K. Within interstellar clouds, the gas density ranges from 10 atoms cm^{-3} to over 10^6 atoms cm^{-3}, and the temperature varies from 10 K to 200 K with considerable uncertainty at both extremes. When the dust and gas are very dense, conditions become favorable for the birth of stars.

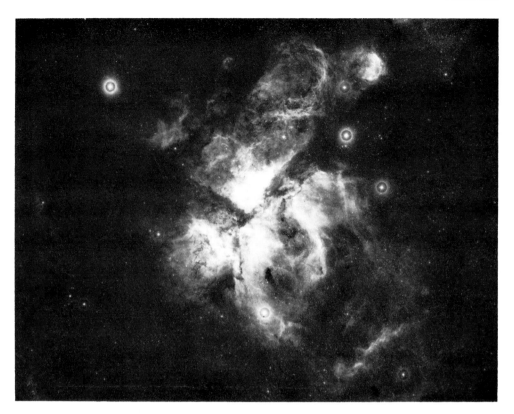

Figure 17–13. *Another H II Region* (Eta Carina). (Cerro Tololo Observatory)

Problems

17–1. An A0 V star has an apparent visual magnitude of $V = 12^m\!.5$ and an apparent blue magnitude of $B = 13^m\!.3$.
(a) What is the color excess for this star?
(b) What is the visual absorption A_v in front of this star?
(c) Calculate the distance to the star (in parsecs).
(d) What error would have been introduced if you had neglected interstellar absorption?
(Hint: Consult the figures and tables in Chapter 12 and in the Appendix.)

17–2. (a) How does interstellar reddening alter the Planck spectral energy curve of a star? Sketch approximate curves for an A star (10,000 K), in the visible part of the spectrum, both with and without reddening.
(b) What effect would interstellar reddening have on the color-magnitude (H-R) diagram of a star cluster? Draw a diagram to support your answer.

17–3. What are the observational clues to the nature of the interstellar dust grains, and how have these clues been interpreted in terms of models?

17–4. Several galactic nebulae have been photographed in color. Two different types occur, red nebulae and blue nebulae.
(a) Explain the physics of these two types of nebulae.

(b) Briefly describe the spectra which you would expect to observe for these two types of nebulae.

17–5.   Why are none of the hydrogen Balmer lines seen as interstellar absorption lines, even though hydrogen is the most abundant element in the Universe? (Hint: Remember the energy level diagram of hydrogen.)

17–6.   (a) Name the two factors which primarily determine the size of an H II region.

(b) Explain the physical basis for your answer to part (a), in terms of the Strömgren sphere.

(c) The *brightness* of an H II region depends only on the gas density of the nebula. Explain this phenomenon in terms of what you know about the hydrogen atom. (Hint: Remember the Saha equation!)

17–7.   Forbidden lines are observed both in the solar corona and in gaseous nebulae, but they are *not the same* forbidden lines.

(a) How do the lines differ in the two cases?

(b) Why are nebular forbidden lines not emitted by the solar corona? What about coronal forbidden lines from nebulae?

17–8.   Somewhere in our Galaxy resides a cloud of neutral hydrogen gas, with a radius of 10 pc. The gas density is 10 hydrogen atoms per cubic centimeter.

(a) How many 21-cm photons does the cloud emit every second?

(b) If the cloud is 100 pc from the Sun, what is the energy flux of this radiation (erg cm^{-2} sec^{-1}) at the Sun?

17–9.   What are the energies (in eV) of the photons which characterize:

(a) the Lyman continuum limit ($\lambda = 912$ Å),

(b) the nebular line of [O III] at $\lambda = 5007$ Å,

(c) the neutral hydrogen line ($\lambda = 21$ cm),

(d) the ammonia (NH_3) emission ($\lambda = 1$ cm),

(e) the hydrogen Balmer line, Hα ($\lambda = 6563$ Å)?

Reading List

Blaauw, A., and Schmidt, Maarten, (eds.): *Galactic Structure*. (See especially Chapters 7, 9, 10, 23.) Chicago, The University of Chicago Press, 1965.

Dufay, Jean: *Galactic Nebulae and Interstellar Matter*. New York, Dover Publications, 1968.

Heiles, Carl: "Physical Conditions and Chemical Constitution of Dark Clouds." *Annual Review of Astronomy and Astrophysics*, Vol. 9, p. 293 (1971).

Lynds, B. T., (ed.): *Dark Nebulae, Globules, and Protostars*. Tucson, Arizona, University of Arizona Press, 1971.

Lynds, Beverly T., and Wickramasinghe, N. C., "Interstellar Dust." *Annual Review of Astronomy and Astrophysics*, Vol. 6, p. 215 (1968).

Middlehurst, B. M., and Aller, L. H., (eds.): *Nebulae and Interstellar Matter*. Chicago, The University of Chicago Press, 1968.

Chapter 18
Our Galaxy:
A Summary

In Chapter 13 we began this part of the book with a brief overview of our Milky Way Galaxy; that discussion set the stage for the subsequent chapters in which we considered the physical characteristics, observational properties, and evolutionary behavior of those objects which constitute our Galaxy. In the present chapter we summarize the structural properties of the Galaxy, as evidenced by the *spatial distribution* of dust, gas, stars, and magnetic fields. An important component of the galactic medium, the *cosmic rays*, is described at the end of the chapter.

18–1 THE STRUCTURE OF OUR GALAXY FROM 21-CM STUDIES

(a) 21-Cm Data and the Spiral Structure

In Chapter 17, we elucidated the physical origin of the hyperfine-transition emitting radiation from neutral hydrogen (H I) at the radio wavelength of 21 cm. The radial velocity formula for differential galactic rotation was derived in Chapter 14. By utilizing both of these concepts simultaneously, we may deduce the *spiral-arm structure* in the galactic plane; if the galactic rotation curve is known, the distances to concentrations of neutral hydrogen may be found from the observed *Doppler-shifted 21-cm line profiles*. Implicit in these distance determinations are the following **443** assumptions: (a) differential galactic rotation, and (b) circular galactic orbits for the gas near the galactic plane.

Since interstellar absorption is insignificant at the 21-cm wavelength, the line emission is observable throughout our Galaxy; in other words, the neutral hydrogen may be detected across distances as large as 30 kpc. Hence, we can probe the galactic regions far beyond the galactic center. The 21-cm line profile for a given line-of-sight exhibits several Doppler-shifted peaks which are fairly narrow and well defined. The concentration of hydrogen into spiral arms produces a slight perturbation of the *velocities* and these perturbations produce the observed forms of the line profiles. The density as such is only a secondary effect, for the ratio of arm to interarm density may only be 3 to 1. In Figure 18–1, we display the line profiles seen near the galactic plane, at 5° intervals in galactic longitude (ℓ). Note that the Doppler peaks shift in a concerted fashion as we vary ℓ. Each peak characterizes a spiral arm intersected by the line-of-sight. If we interpret the Doppler shift in terms of the radial velocity of that section of the arm and apply the rotation formulae of Chapter 14, we immediately find the *distance* to the arm; since we have assumed circular orbits, the distance is uncertain to the extent that asymmetric motions occur. More realistic results are obtained by including modifications to circular motions.

By combining 21-cm data from both the northern and southern hemispheres, we can construct a schematic picture of the neutral hydrogen distribution in the spiral arms of our Galaxy (see Figure 18–2). As is evident in the figure, the spiral structure is poorly determined near galactic longitudes $\ell = 0°$ and $\ell = 180°$— that is, toward the galactic center and in the diametrically opposed (*anticenter*) direction. Circularly-orbiting hydrogen clouds in those directions should exhibit *no* radial velocity;* hence, we cannot determine the distances to such clouds, for the line profile will be a single peak at 21 cm. Observationally, there is some Doppler-shifted radiation from the galactic-center direction, but this is due to gas which moves radially outward from the center (as we shall discuss later). A *distance ambiguity* exists for hydrogen clouds which are closer to the galactic center than is the Sun. Since the maximum radial velocity of recession occurs when the line of sight passes closest to the galactic center (tangent point), a cloud closer to the Sun than the tangent point may have the *same* (lower) recession speed as a cloud beyond the tangent point. In general, the nearer cloud will subtend a larger angle in galactic latitude, so that latitude scans (see the following paragraphs) can usually resolve the ambiguity.

Let us illustrate these points by interpreting the exemplary 21-cm profile shown in Figure 18–3. This profile corresponds to the line-of-sight at $\ell = 48°$ (shown in the lower part of the figure), and it consists of three Doppler peaks at radial velocities of $+55$, $+15$, and -50 km sec^{-1}. If we denote the Sun's distance from the galactic center by R_0, and note that positive ($+$) radial velocities correspond to recession, then two of the hydrogen clouds are receding while one is approaching. From the rotation formulae of Chapter 14, we find that recession corresponds to $R < R_0$, while an approaching cloud must lie at $R > R_0$. Remembering the approximate nature of equation (14–36), we compute the maximum radial velocity as $+57$ km

---

* The reader should be cautioned that the term "cloud" in the context of this section does not refer to a discrete entity, but rather to a velocity perturbation and density maximum.

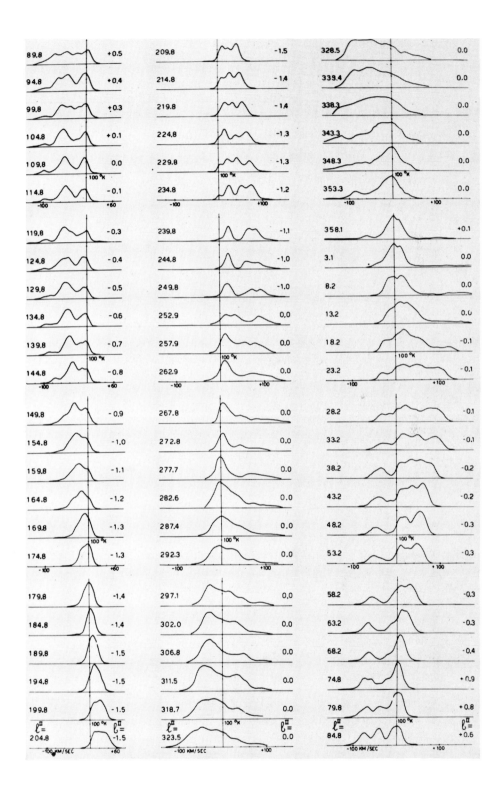

Figure 18–1. *21-cm Line Profiles in the Galactic Plane.* These profiles represent the region near the galactic plane ($|b| \leq 1.5°$), and span the entire galactic equator at intervals of approximately 5° in galactic longitude. (Courtesy of F. J. Kerr and G. Westerhout)

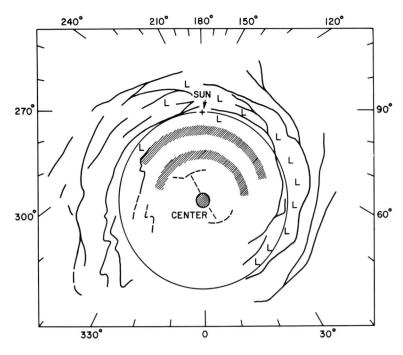

SKETCH OF GALACTIC HYDROGEN SPIRAL STRUCTURE

Figure 18–2. *The Spiral Structure of Our Galaxy from 21-cm Data.* This is a sketch of the neutral hydrogen distribution in the galactic plane, as deduced by Kerr, Hindman, and Henderson. The spiral structure is clearly visible (though complex). The paucity of structure near $\ell = 0°$ and $\ell = 180°$ is due to the large uncertainties in distances in these directions. The symbol L indicates regions of *low* hydrogen density. (Courtesy of F. J. Kerr)

$\sec^{-1}$; therefore, cloud A with a radial velocity of $+55$ km $\sec^{-1}$ must lie very close to the tangent point. Cloud D, which is approaching us (-50 km $\sec^{-1}$), lies at $R > R_0$, as indicated in the figure. There is a hint of double structure (clouds B and C) in the $+15$ km $\sec^{-1}$ peak; latitude scans imply that cloud B lies beyond the tangent point (near $R = R_0$), while cloud C is very close to the Sun.

(b) The Galactic Latitude Distribution of Hydrogen

Neutral hydrogen is concentrated to the galactic plane to a remarkable degree. If we define the *thickness* of the gas layer as the distance from the galactic plane to the half-density point (where the number density falls to *half* the value found at the galactic plane or $\ell = 0°$), then the thickness of the hydrogen layer is observed to range from 80 pc to greater than 220 pc. The smaller value refers to the region between the Sun and the galactic center. The thickness increases to 220 pc at the spiral arms near the Sun ($R = R_0$), and flares out to several hundred parsecs for $R > R_0$. The galactic latitude distribution of the neutral hydrogen has

Figure 18–3. *A Line Profile and the Neutral Hydrogen seen at ℓ = 48°.* **A,** An exemplary 21-cm hydrogen line profile at ℓ = 48°. **B,** The galactic geometry and line-of-sight corresponding to **A**.

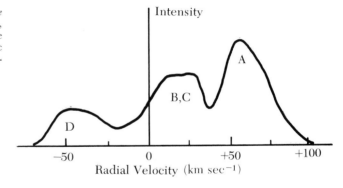

been studied extensively. As shown in Figure 18–4, the gas layer is observed to be very flat (near $\ell = 0°$) in the region $R < R_0$, while beyond $R = R_0$, the layer is bent or *distorted* in opposite directions (relative to $\ell = 0°$) at $\ell = 0°$ and $\ell = 180°$. The physical basis for this phenomenon is still obscure.

(c)   The Galactic Center Region

If the gas motion were perfectly circular, there would be no observed radial velocities (Doppler-shifted line profiles) near $\ell = 0°$. The distances to such gas clouds would be indeterminate, since the expected 21-cm line profile is a single intense peak at zero radial velocity. Nevertheless, the gas structure in this region may be approximated by extrapolating the features seen at longitudes flanking $\ell = 0°$.

The 21-cm line profiles actually observed within a few degrees of the galactic center are extremely complex, largely because of peculiar geometrical and velocity perturbations. The major feature seen is a sharp peak at -50 km sec^{-1}, which reveals material moving *outward* from the galactic center. Detailed 21-cm studies of the galactic center by G. W. Rougoor and J. H. Oort, and by F. J. Kerr, reveal

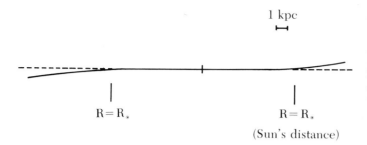

1 kpc

$R = R_*$

$R = R_*$

(Sun's distance)

Figure 18–4. *A Cross-Section of the Galactic Neutral Hydrogen Layer.* This is a schematic cross-section of the galactic plane, indicating the warping of the thin hydrogen layer at $R > R_\odot$ in the directions $\ell = 0°$ and $\ell = 180°$. (Courtesy of F. J. Kerr)

the neutral hydrogen structure depicted in Figure 18–5. There is an expanding arm or ring at $R \simeq 3$ to 4 kpc, which orbits the center at 200 km sec^{-1} and which is moving away from the center at about 50 km sec^{-1} along the direction toward the Sun. Other features of the observed 21-cm profiles are interpreted as (a) a rapidly-rotating central disk of gas about 1 kpc in diameter (this is sometimes conjectured to be an inner disk surrounded by a ring of gas), (b) individual clouds of hydrogen moving rapidly outward from the galactic center, and (c) possibly a high-velocity bar-like structure connecting the inner disk with the 3-kpc arm. There is also evidence of a ring of ionized hydrogen (H II) at $R \simeq 4$ to 5 kpc.

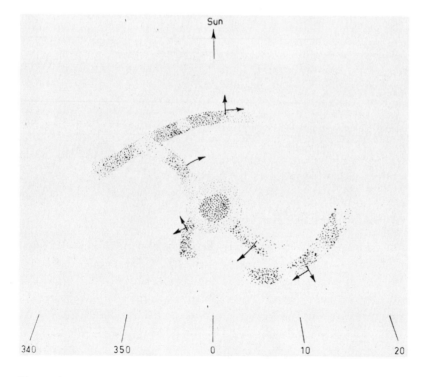

Sun

340   350   0   10   20

Figure 18–5. *Neutral Hydrogen Near the Galactic Center.* A model of the galactic-center region, showing the expanding 3-kpc spiral arm, the high-velocity barred structure, and the central gas disk. (Courtesy of F. J. Kerr)

(d)   High-Velocity Hydrogen Clouds

In Chapter 14, we saw that stars which move in highly eccentric orbits about the galactic center appear to us as high-velocity stars. Recently, 21-cm profiles at high galactic latitudes ($|b| \geq 10°$) have revealed previously-unsuspected hydrogen structures far from the galactic plane. This gas is distinctly distributed in the form of *discrete clouds* (real clouds in this case), which are usually divided into two velocity categories: (a) the *high-velocity clouds* with velocities (relative to the Sun) in excess of 70 km sec^{-1}, and (b) the *intermediate-velocity clouds* with velocities in the range 30 to 70 km sec^{-1}. Nearly all of these clouds exhibit radial velocities of *approach*, and the clouds appear to be concentrated near $\ell = 120°$ and $b = +40°$ (they occur in the region $\ell = 60$ to $200°$ and $b = +10$ to $+80°$).

These high-latitude clouds apparently falling into the Galaxy are relatively rare and far apart; their importance to the dynamics and evolution of our Galaxy is still unclear. Many conjectures have been advanced regarding their origin. Oort has suggested that the clouds represent intergalactic material being accreted by the Galaxy; others advocate that these are independent extragalactic objects, perhaps dwarf members of the Local Group of galaxies (see Chapter 19). Another idea is that the clouds originated in great explosions in the interior parts of our Galaxy, or at the galactic center, and that we observe matter "raining" back to the galactic plane. A recent suggestion proposes that these high-velocity clouds are distant, highly distorted spiral arms. Much observational and theoretical work remains to be done before we can understand the true nature of these mysterious hydrogen clouds.

18–2   THE DISTRIBUTION OF STARS AND GAS IN OUR GALAXY

(a)   The Spiral Arms: Extreme Population I Objects

Neutral hydrogen emission (the 21-cm line) enables astronomers to delineate the spiral-arm structure over a very large portion of our Galaxy, but spiral arms also include all other objects of *extreme Population I*. The objects belonging to the various populations, and their characteristics, are summarized in Table 18–1. Since gas and dust are inevitably found together, both occur in the spiral arms of our Galaxy. The newborn T Tauri stars (see Chapters 15 and 16) are usually surrounded by the gas and dust from which they have formed. These stars are not very bright, and even when they occur in groups, the *T-associations*, they cannot be seen to very great distances. We know that they are good examples of extreme Population I, but they are not good spiral-arm indicators because they are intrinsically too faint.

Massive stars evolve very rapidly, and they may be seen as main-sequence O or B stars; they too may still be enveloped by the gas and dust from which they formed. The ultraviolet radiation from such stars ionizes the surrounding gas, rendering it visible as an H II region (or emission nebula; see Chapter 17). The sudden heating of the gas, when O or B stars form, causes an expansion which may produce peculiar forms when the H II is braked by denser cold material; "elephant-trunk" structures are examples of such forms (see Figure 18–6). *H II regions* are usually intrinsically bright, for they characteristically include several bright hot stars, whose ultraviolet radiation is converted into visible light. H II regions and the O-associations exciting them are extreme Population I objects and are powerful indicators of spiral structure. *O-associations* are loosely-bound groups of O (and some early B) stars. Each group contains 10 to 100 such stars, and these associations are far less stable than ordinary open clusters. Their dimensions range from a few parsecs to some hundreds of parsecs. In many cases, one or more open clusters may be found within an association. An O-association may therefore contain many fainter main-sequence stars, but it is the O stars (O to B2) which primarily characterize these objects. In some cases, such as the ζ Persei group, the observed velocities of the member stars indicate that an association is rapidly dispersing; the formation of the member stars appears to have initiated this disruptive expansion.

The spiral-arm structure in the solar vicinity becomes apparent when the observed distribution of O-associations and their optical H II regions is projected onto the galactic plane. Although there is general agreement between the optical and the 21-cm data, some unexplained discrepancies still remain.

The radio continuum emission from H II regions cannot be used for finding their distances, but the recently-discovered *recombination lines* (see Chapter 17) allow us to determine their radial velocities. Such velocities can then be used in the same manner as the 21-cm line velocities to delineate the spiral structure, making possible a more complete comparison between the distributions of H I and H II regions. The spiral-arm patterns delineated by H I and H II are fairly similar, but a marked difference appears in the large-scale distribution, for the ionized hydrogen

Table 18–1. Characteristics of Stellar Populations

	Population Group				
	EXTREME POPULATION I	OLDER POPULATION I	DISK POPULATION II	INTERMEDIATE POPULATION II	HALO POPULATION II
Typical Objects	Interstellar dust and gas O and B stars Supergiants T Tauri stars Young open clusters Classical Cepheids O-associations H II regions	Sun Strong-line stars A stars Me dwarfs Giants Older open clusters	Weak-line stars Planetary nebulae Galactic bulge Novae RR Lyrae stars ($P < 0.4$ day)	High-velocity stars ($Z > 30$ km sec^{-1}) Long-period variables ($P < 250$ days)	Globular clusters Extreme metal-poor stars (subdwarfs) RR Lyrae stars ($P > 0.4$ day) Population II Cepheids
Characteristics					
$\langle |z| \rangle$, pc	120	160	400	700	2000
$\langle |Z| \rangle$, km sec^{-1}	8	10	17	25	75
Distribution	Extremely patchy in spiral arms	Patchy	Smooth	Smooth	Smooth
Age (10^9 yr)	<0.1	0.1 to 10	3 to 10	~ 10	$\gtrsim 10$
Brightest stars (M_{vis})	-8	-5	-3	-3	-3
Concentration to galactic center	None	Little	Considerable	Strong	Strong
Galactic orbits	Circular	Almost circular	Slightly eccentric	Eccentric	Highly eccentric

Adapted from A. Blaauw.

Figure 18–6.   *An "Elephant-Trunk" Structure.* (Lick Observatory)

attains its greatest concentration *closer* to the galactic center than does the neutral hydrogen; these peak concentrations occur in the regions 4 to 8 kpc and 7 to 15 kpc, respectively.

Since early-type stars are extreme Population I, it follows that the *young open clusters* which contain these stars must also be of this population. In these clusters, few stars have evolved away from the upper main sequence; NGC 2362 and *h* and χ Persei are good examples. Other clusters like NGC 2264 still have stars

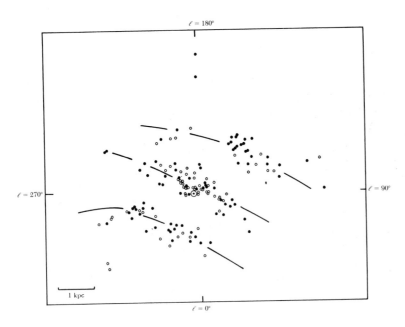

Figure 18–7. *OB Associations and H II Regions.* Here the spatial distribution of hot O and B stars in associations, and of H II regions, is shown projected on the galactic plane. The Sun is indicated by ⊙, and direction is given by galactic longitude ℓ. Note the three spiral-arm features (drawn-in lines). (After W. Becker and R. B. Fenkart)

evolving onto the lower main sequence, while the more-massive stars occupy the upper main sequence (see Figure 15–7). Many of these young clusters are surrounded by O- or T-associations.

(b) The Galactic Disk and Halo: Older Population I, Disk Population, and Population II Objects

When the positions of older, more evolved galactic clusters with well-determined color-magnitude diagrams are projected onto the galactic plane, the clear division into spiral arms is lost (see Figure 18–9). Such behavior is characteristic of *older Population I*. The inclusion of A-type main-sequence stars as Older Population I (see Table 18–1) means that clusters whose upper main-sequence stars have evolved to the giant and variable stage (the older open clusters) are prototypes of this population. Our Sun too belongs to Older Population I. Although they are not strictly confined to the spiral arms, Older Population I objects still lie fairly close to the galactic plane and have an inhomogeneous distribution throughout the Galaxy.

Disk and intermediate Population II stars constitute most of the total mass of our Galaxy, and they form a fairly thick layer or disk. The *Disk Population* is intermediate between Populations I and II. As we see from the $\langle |z| \rangle$ entry in

Figure 18–8. *A Young Open Cluster: h and χ Persei.* (Lick Observatory)

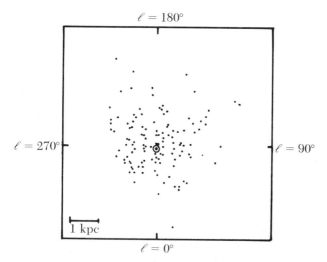

Figure 18–9. *Distribution of Old Open Clusters Near the Sun.* Spiral-arm structure is indiscernible when old open clusters are projected on the galactic plane. [From W. Becker and R. B. Fenkart: *In* I.A.U. Symposium No. 38, W. Becker and G. Contopoulos (eds.). Dordrecht, D. Reidel Publishing Co.]

Figure 18–10.   *An Older Open Cluster: Praesepe.* (Yerkes Observatory)

Figure 18–11.   *A Globular Cluster: M3.* (Lick Observatory)

Table 18–1, representative disk stars may lie quite far from the galactic plane. This quantity $\langle |z| \rangle$ is the mean stellar distance from the plane, while the related parameter $\langle |Z| \rangle$ is the mean of the stellar velocity components *perpendicular* to the galactic plane. The greater the perpendicular velocity component of a star, the greater will be the likelihood of that star being far from the galactic plane. Contrary to Population I, Disk Population objects are distributed fairly smoothly, and show no spiral structure.

Intermediate Population II stars may lie even farther from the galactic plane; their mean distance is 700 pc. They may also have highly eccentric orbits, since many of them are high-velocity stars. Disk Population objects show a slight concentration to the galactic center, but Population II stars are *strongly* concentrated to the center. We are dealing with highly-evolved stars, which are very old, when we speak of Population II. These stars probably formed at a place where heavy elements were scarce.

The *metal abundance* of stars is an indicator of the degree of heavy element enrichment of the gas out of which they formed. This metal abundance is *not* a smooth function of population, although this was thought to be the case earlier. Extreme halo Population II stars do have a low metal abundance, as do some intermediate II, but stars in the nucleus and some disk stars are metal rich, perhaps because of an accelerated birthrate of massive stars early in the life of the Galaxy.*

---

* "Metal rich" is a relative term. The metal abundance of ordinary Population II stars is about one-hundredth that of the Sun, while for metal rich Population II it may go up to one-tenth or somewhat higher.

Enveloping the disk and spiral arms is the *galactic halo*, which extends far above the plane but is still concentrated to the galactic center. Objects which lie in this domain are called *Halo Population II;* these include the globular clusters and RR Lyrae stars with periods greater than 0.4 days. The orbits of these objects are highly eccentric, with large velocity components perpendicular to the galactic plane.

(c)   The Central Bulge and Galactic Nucleus

Wide-angle pictures of our Galaxy (see Chapter 13) show that it has a *central bulge* similar to that of other spiral galaxies, such as the Andromeda galaxy. This bulge, which is about 2 kpc in radius, is not well defined, but is rather amorphous. Apparently the bulge contains a mixture of heavy-element-enriched stars, especially late-type M giants, ordinary Population I K giants, and metal-poor objects like the RR Lyrae stars.

At the very center of the bulge is the *galactic nucleus*, which is analogous to the stellar-like nucleus in M31. Spectra of the Andromeda galaxy nucleus show that it too consists of metal-rich giants as well as a large number of low-mass dwarfs.

Interstellar absorption makes it impossible to observe the nucleus of our Galaxy at visible wavelengths; in fact, observations of the central bulge are restricted to small "windows" relatively clear of obscuration. Much of what we know about the nucleus is deduced from *infrared* observations, for it is only in the far infrared that we can really observe the nucleus. At wavelengths from 1 to 5 μ and longer, we observe a fairly extended bright source (1°) with a sharp brightness peak near the expected position of the center of the Galaxy. This infrared radiation is attributed to a region with a very high density of stars which is obscured at visual wavelengths by 25 to 30 magnitudes. A part of the radiation at these and longer wavelengths may be stellar ultraviolet and optical radiation absorbed and *re-emitted* by dust that is also concentrated to the center. The possibility that some of the infrared radiation is nonthermal cannot be excluded.

The strong *radio source* Sagittarius A lies in the same direction as one of the small, strong infrared components. There are actually several components to the radio source originally called Sagittarius A, but only the strong central source is thought to be the nucleus; this main source lies close to the apparent center of symmetry of the Galaxy. Assuming a distance of 10 kpc, the nucleus has a diameter of only 10 pc, truly a semistellar object. The radio spectrum is characteristic of a *nonthermal* source, and the radiation is presumed to be synchrotron emission from relativistic electrons; such synchrotron radiation requires the presence of a magnetic field, which is estimated to have a strength of less than 10^{-5} Gauss. The relativistic gas must be mixed with the stars in the nucleus.

There are several *thermal* radio sources in the vicinity of the primary radio source in the galactic nucleus. Since one of the radio recombination lines has been observed from some of them, these thermal sources must be massive H II regions lying within 100 pc of the center. In addition, we note that molecular radio lines of OH, NH_3, and H_2CO (see Chapter 17) appear in absorption, superimposed on the continuum sources around Sagittarius A. There are about a dozen such molecular clouds which are very small and which do not completely coincide with the neutral and ionized hydrogen features. Apparently molecular material is strongly concentrated to the center of our Galaxy. The peculiar behavior of neutral hydrogen gas around the galactic center, such as the outward motions, the rotating disk, and the high-velocity clouds (see section 18–1), must also be explained by any theoretical model of the highly active galactic nucleus.

Gravitational waves from the center of the Galaxy appear to have been detected by J. Weber. Such waves are a consequence of Einstein's theory of relativity. If indeed gravitational waves of the intensity indicated by the observations originate from the galactic center, they may be due to a high rate of conversion of mass to energy, or to a rotating black hole. This then is still another indication of the active nature of the nucleus.

(d)   The Distribution of Mass in the Galaxy

In spite of the importance of both neutral and ionized hydrogen gas in the Galaxy, we should keep in mind that the gas represents only a small fraction of the total mass of the Galaxy. It is estimated that all the hydrogen gas in the Galaxy contributes only *five to ten per cent* of the total mass, and of this fraction, the major contribution is from neutral hydrogen. On the average, approximately one per cent of all the hydrogen is ionized, but the percentage varies with distance from the galactic center. For instance, there is a ring of ionized gas beyond the galactic bulge at about 4 kpc, and in this region the percentage of ionized hydrogen is about ten per cent. This ring is not uniform, but consists of many smaller H II regions.

The total mass of the Galaxy can be determined, to a first approximation, on the assumption that the Sun moves in a circular Keplerian orbit about a point mass; such calculations lead to a value of $1.45 \times 10^{11} M_\odot$. A more refined method (due to M. Schmidt) explains the observed rotation curve in terms of a "point mass" and a concentric series of highly-flattened ellipsoids. This model leads to a total mass for the Galaxy of $1.86 \times 10^{11} M_\odot$; approximately half of the mass lies outside an eccentric ellipsoid passing through the Sun's position. The radius of the Galaxy on this model is about 25 kpc. The concentration of mass to the center is still very high, with the central mass point containing slightly less than one-twentieth of all the mass.

(e)   The Theoretical Basis of the Spiral Structure

Why is it that galaxies such as ours exhibit a *spiral* structure? Another question which we would like answered is, how stable are spiral arms? Spiral structure may arise from an instability or perturbation, so that some density irregularity is pulled into a spiral form by the differential galactic rotation. The difficulty with this explanation is that such a feature is expected to persist for only a short time (about 5×10^8 yrs) before being pulled apart again by the differential rotation. An additional problem is that the initial spiral-like distortion would not extend over the entire galaxy, but would occupy only a small part of it. Yet we observe spiral structure throughout the entire plane of a galaxy, and such spiral galaxies are sufficiently common to suggest that they are in fact stable. A very promising approach to this problem has been worked out by C. C. Lin, F. H. Shu, and their associates. This is the *density-wave theory*, a modification and extension of the original work of B. Lindblad. The spiral structure of a galaxy is regarded as a *wave pattern* resulting from gravitational instabilities. The wave extends far beyond the initial localized perturbation. The presence of a density wave means that the distribution of mass is nonuniform; therefore, the gravitational potential is variable over the galactic disk. Stars and gas are concentrated to the regions where the gravitational potential is low, and these mass concentrations in turn influence the orbits of other stars and clouds of gas. The density wave is therefore a self-sustained phenomenon, and is *stable*. The spiral pattern produced by the density waves is not tied to the matter, but instead moves *through* the matter. According to this theory, the angular velocity of the pattern may differ appreciably (say a factor of 1/2) from that of the material. Hence, stars that are formed from the gas and dust concentrated to the arm eventually migrate out of the arms. Gas concentrated near the potential minimum largely defines the spiral arms; this is ever-changing gas, for some is consumed in star formation and some is ejected from stars by some mode of mass loss. Stars traveling in orbits which differ greatly from circular orbits come under rapidly varying gravitational attractions. Density waves certainly influence their motions, but not systematically, and therefore no structure will persist for these eccentric-orbit stars.

By studying the galactic distribution of matter, and the behavior of these density waves, Lin and his colleagues have made great strides in this challenging subject of spiral structure. Their ability to fit their theory to some of the observations is rather impressive. Nevertheless, much work remains to be done. The role of the galactic magnetic field in determining spiral structure has not yet been investigated properly; in addition, the actual origin of the density waves (and therefore the spiral arms) has not yet been elucidated.

18–3   COSMIC RAYS AND GALACTIC MAGNETIC FIELDS

(a)   Basic Definitions and Observational Data for Cosmic Rays

Before we can discuss the role of *cosmic rays* in our Galaxy, we must examine their characteristics. They are not "rays" at all, but rather *high-energy charged particles*. These particles may be nuclei of atoms whose electrons have been stripped away, or they may be electrons, or even positrons. They are called cosmic rays when they travel at essentially the speed of light; in general, they possess incredibly large kinetic energies. As we saw in Chapter 9, our Sun ejects low-energy (tens to hundreds of MeV) cosmic rays as part of the flare phenomenon. In addition to generating its own cosmic rays, the Sun also modulates those cosmic rays coming from outside the Solar System; the interplanetary magnetic field and the solar wind severely distort the orbits of particles with energies less than 10^9 electron Volts

(1 GeV). The amount of modulation varies with solar activity, and the modulation obscures the *intrinsic* properties of low-energy galactic cosmic rays. We do know, however, that the number of particles increases rapidly with decreasing energy. Most of the observational data on extra-Solar System cosmic rays pertain to particles having energies in excess of 10^9 eV. Energies as high as 10^{20} eV have been observed for single particles, but these particles are extremely rare.

The *chemical composition* of cosmic rays gives us information on both their source and their journey through space. As in stellar abundance determinations, hydrogen nuclei (protons) are by far the most abundant component, constituting about 90 per cent of all the cosmic ray nuclei. Another nine per cent are helium nuclei, while the remainder is divided among heavier elements. Much interest centers on the so-called *light nuclei*—lithium, beryllium, and boron. These nuclei are virtually absent in stellar atmospheres, where their abundance is 10^{-7} that of helium, while among cosmic rays the light-nuclei-to-helium ratio is about $1:100$. The much greater proportion of such nuclei among cosmic rays is attributed to the impact of original heavy cosmic rays upon interstellar matter. During such a collision the heavy nuclei break up into lighter ones; this process is termed *spallation*. Observations of the light, and therefore *secondary*, nuclei enable us to estimate the amount of material traversed by the original cosmic rays. The density of the interstellar medium and the lifetime of cosmic rays are inexorably linked together. Current data suggest an average density of interstellar matter from less than 0.1 to 1 atom cm^{-3} and a mean lifetime of cosmic rays from a few million to some tens of millions of years.

Another important datum concerns the *distribution* of cosmic rays. From what directions do they come? Observationally, they appear to come more or less uniformly from all directions; in other words, they are *isotropic*. This phenomenon is not necessarily due to a random spatial distribution of the cosmic ray sources, but probably arises because galactic magnetic fields deflect cosmic rays to such an extent that they cannot travel in straight paths. In fact, these high-energy particles travel along the magnetic lines of force in spiral (or helical) paths, whose size is determined by both the magnetic field strength and the particle energy.

(b)   The Source and Acceleration of Cosmic Rays

If cosmic rays arrive at the Solar System isotropically, how can we locate and identify their sources? One clue is the high energy of the particles; we should look for highly energetic phenomena. But we have more direct evidence concerning the sources, for one component of cosmic rays is the electrons, which represent about one per cent of all cosmic ray particles. Cosmic ray electrons have been observed directly. The fact that positrons (positive electrons) are only one-tenth as numerous indicates that these electrons are primary particles, and not secondaries like the light nuclei. This conclusion is based on the fact that more positrons than electrons are produced when primary cosmic rays collide with interstellar atoms.

We already know that high-energy electrons are observable from a distance, for they emit synchrotron radiation at radio frequencies when they move in a magnetic field. As we saw in Chapter 16, supernova remnants like the Crab nebula are strong sources of synchrotron radiation from relativistic electrons. Moreover, *supernovae* generate energy on a scale sufficient to produce cosmic rays. Other sources of cosmic rays may be rotating neutron stars, such as those observed as pulsars, and perhaps even rotating white dwarfs.

The lifetime of cosmic rays in the Galaxy is limited; some escape from the Galaxy altogether and others are consumed by interactions with the interstellar medium. It follows that the supply of cosmic rays must be steady and continuous if the energy density and the total energy in cosmic rays are to remain more or less constant. Supernovae can indeed maintain the supply, if supernova explosions occur every 30 to 50 years in the Galaxy. Statistics based on observations of supernovae in other galaxies suggest that this is a reasonable estimate of the frequency of supernova outbursts.

A possible additional source of cosmic rays is the *galactic nucleus*, which appears to be a strong source of synchrotron radiation. The highest-energy cosmic rays (over 10^{18} eV) may well come from outside our Galaxy—that is, from the extragalactic radio galaxies and quasars which are observed to generate such enormous amounts of energy (see Chapter 19).

Charged particles must be powerfully accelerated if they are to attain the energies or velocities observed for cosmic rays. Most of this acceleration occurs at the time of generation (e.g., at the time of the supernova explosion). The energy of gravitational collapse of the supernova is presumed to produce a *shock wave*, which then accelerates part of the stellar envelope to produce cosmic rays. Alternatively, *changing* magnetic fields in the outer parts of the supernovae or from the rotating magnetic fields of pulsar-neutron stars may accelerate high-energy particles. Some magnetic acceleration of particles may also occur in interstellar space, but this process is quite inefficient and is therefore not a major factor determining cosmic ray energies.

(c)   The Galactic Magnetic Field

In the preceding discussion, and in Chapter 17, we have repeatedly referred to the *galactic magnetic field*. The existence of a galactic magnetic field is proved by the observation of the Faraday rotation of radiation from radio sources; such radio sources include galactic sources such as pulsars (Chapter 16) and extragalactic sources. As discussed in Chapter 17, *Faraday rotation* is the rotation of the plane of polarization as linearly-polarized radiation passes through a magnetized plasma. Interstellar electrons are the most important source of Faraday rotation in our Galaxy. The amount of rotation depends on the frequency of the radiation; therefore, by studying the polarization position angle at different frequencies, the amount of rotation can be established. If the electron density and the distance traversed are known, we can solve for the mean magnetic field strength. The best current estimates of the galactic magnetic field strength suggest that it is 3 to 6 $\times$ 10^{-6} Gauss. Before the observations of Faraday rotation, a galactic magnetic field had already been surmised to exist, both on the basis of interstellar polarization observations (Chapter 17) and from cosmic ray data.

The orientation of the magnetic field with respect to the spiral structure is primarily *along* the axes of spiral arms. Superimposed on this mean field is a *local field* in our vicinity, which dominates most of the observations, making it difficult to ascertain the true nature of the general field. Locally, within 500 pc of the Sun, the field seems to have a turbulent, possibly helical form, with the mean magnetic axis along the spiral arm. Because cosmic rays travel in spiral paths about magnetic field lines, it is clear that they are closely tied to the galactic magnetic field. Produced by supernovae and pulsars(?) in the Galaxy, cosmic rays are, in all probability, largely contained within the Galaxy.

The importance of cosmic rays to galactic structure is best seen from the fact that their *mean energy density* is comparable to the mean energy densities of *each* of the following: (a) the galactic magnetic field, (b) the kinetic energy of interstellar gas, and (c) stellar radiation. The pressures exerted by all of these galactic constituents are therefore similar, and at times rival the gravitational forces! Therefore, galactic structure must be influenced to some extent by each of these components of the Galaxy. The relative importance of each factor remains to be ascertained. Hence, the theory of galactic structure is still in its formative stages, awaiting much hard, detailed work, and perhaps a flash of insight to carry it to fruition and maturity.

Problems

18–1. The figure shown below is a sketch of the distribution of the neutral hydrogen maxima in the spiral arms of our Galaxy; the position of our Sun is denoted by ☉. Draw the *21-cm line profiles* which you would expect to observe in the directions $\ell = 50°$, 110°, and 230°; label the points on these profiles corresponding to the spiral arms. Do not make any detailed calculations of radial velocity; just make certain that the signs of the velocities and the relative positions of the peaks are correct.

18–2. Name *three* important physical characteristics which distinguish Population I stars from Population II stars. Explain these differences in terms of the evolution of our Galaxy.

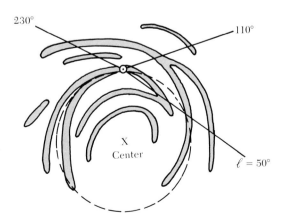

18–3.   The objects which constitute our Galaxy are usually divided into five major *population classes*.
(a) What are these classes?
(b) Give an example of an object from each class.
(c) Draw an edge-on view of our Galaxy, indicating the spatial distribution of each of the five classes. Label your diagram carefully.

18–4.   (a) What would be the apparent magnitude of a star like our Sun if it were at the distance of the galactic center (10 kpc) from the Sun?
(b) The Milky Way Galaxy contains 1.5×10^{11} stars. If we assume that all of these stars are like our Sun ($M_b = 4.7$), what is the absolute magnitude of the whole Galaxy? Compare this result with the *apparent* magnitude of the Sun.

18–5.   When a particular globular cluster is at its farthest point from the galactic center (apogalacticon), its distance from the center is 10^4 pc. What is its *period* of galactic revolution? What assumptions must you make to arrive at a unique answer; can you give any physical justification for *your* assumptions?
(Hint: 1 pc $= 2 \times 10^5$ AU. Assume that the mass of the Galaxy is $10^{11} M_\odot$.)

18–6.   A spider whose mass is 2 gm falls from a ceiling. Just before it hits the floor its velocity is 1 m sec^{-1}. Compute the *kinetic energy* of this spider, and compare your result with the kinetic energy of
(a) a 1 GeV cosmic ray proton,
(b) a 10^{20} eV cosmic ray particle.

18–7.   We can deduce the average time between *stellar collisions* by considering the figure below. If we have *identical* stars of radius R, scattered through space with a mean number density N (stars per unit volume), then a star moving with speed V will sweep out the volume $\pi R^2 V$ per unit time. The average number of stars in this volume is $\pi R^2 V N$, so that in the time $T = 1/(\pi R^2 V N)$ the star will encounter (i.e., *collide with*) one star (on the average)! The average distance between each collision is just $L = VT = 1/(\pi R^2 N)$; this is called the *mean free path*. In each of the following situations compute the mean collision time and the mean free path for:
(a) *The solar neighborhood*, where $V = 20$ km sec^{-1} and $N = 0.1$ pc^{-3}; consider stars of radius $R = R_\odot$.
(b) *A galactic nucleus*, where $V = 1000$ km sec^{-1} and where there are 10^9 stars (of radius $R = 10R_\odot$) within a sphere of radius 5 pc. Comment briefly upon your results.

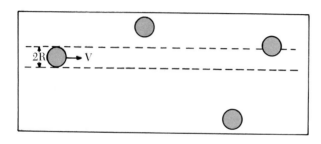

Reading List

See also Chapter 13 Reading List.

Atanasijević, I.: *Selected Exercises in Galactic Astronomy*. Dordrecht, Holland, D. Reidel Publishing Co., 1971.

Becker, W., and Contopoulos, G., (eds.): *The Spiral Structure of Our Galaxy*. I.A.U. Symposium No. 38. Dordrecht, Holland, D. Reidel Publishing Co., 1970.

Blaauw, A., and Schmidt, Maarten, (eds.): *Galactic Structure*. Chicago, The University of Chicago Press, 1965.

Bok, B. J., and Bok, P. F.: *The Milky Way*. Fourth Ed. Cambridge, Harvard University Press, 1973.

Kerr, Frank J.: "The Large-Scale Distribution of Hydrogen in the Galaxy." *Annual Review of Astronomy and Astrophysics*, Vol. 7, pp. 39–66 (1969).

King, I. R.: "Stellar Populations in Galaxies." *Publications of Astronomical Society of the Pacific*, Vol. 83, p. 377 (1971).

Mavridis, L. N., (ed.): *Structure and Evolution of the Galaxy*. Dordrecht, Holland, D. Reidel Publishing Co., 1971.

Mihalas, D.: *Galactic Astronomy*. San Francisco, W. H. Freeman & Co., 1968.

Ogorodnikov, K. F.: *Dynamics of Stellar Structures*. Oxford, Pergamon Press, 1965.

Part 5
The Universe

Chapter 19
Extragalactic
Objects

Let us now leave our own Milky Way Galaxy and consider the vast reaches of *extragalactic space*. In Chapter 13, we alluded to the existence of numerous *galaxies* filling this space; although these objects are strikingly beautiful when seen on telescopic photographs, they appear nebulous and unimpressive when viewed visually through a telescope. Hence, they were given the name *extragalactic nebulae* in the past; remember that "nebula" is the Latin word meaning cloud. In 1924, E. Hubble resolved Cepheid variable stars in the Andromeda nebula (M31; see Figure 13–3), thereby ending a long-standing debate between astronomers as to the nature of the extragalactic nebulae. Hubble's observation proved that these entities are vast assemblages of stars (i.e., galaxies) which lie beyond our Galaxy. Using the period-luminosity relation for Cepheids, which had been discovered in the Magellanic Clouds (two small companion galaxies to our Galaxy), Hubble determined an enormous distance for M31. Since many other galaxies resemble M31, these objects must also lie far beyond the Milky Way Galaxy.

In Chapter 13, we saw that galaxies are huge aggregates of stars and gas held together by gravitational attraction. The various forms of galaxies are determined by the distribution of mass within the galaxy, and by the relative proportions of stars and gas. Let us therefore begin our discussion of galaxies by considering the classification of galactic types.

467

19–1   THE CLASSIFICATION OF GALAXIES

(a)   The Classification Scheme

The present classification of galactic types is still based upon Hubble's original classification scheme for galaxies according to their *appearance*—his famous *tuning fork diagram* (see Figure 19–1). Most galaxies may be separated into three major categories: the *elliptical galaxies*, the *spiral galaxies*, and the *irregular galaxies*. Although we illustrate several galaxies here, the reader is urged to browse through *The Hubble Atlas of Galaxies* (see reading list) for more examples.

Elliptical galaxies (designated by E) have the shape of an oblate spheroid; their appearance in the sky is that of a luminous elliptical disk. The distribution of light is smooth, with the surface brightness decreasing outward from the center. Elliptical galaxies are subclassified according to the elongation of the apparent *projected image*; that is, if a and b are the major and minor axes of the apparent ellipse, then $10(a - b)/a$ expresses the observed ellipticity. The true ellipticity cannot be found because the orientation of a particular galaxy cannot be determined. Therefore, an E0 galaxy appears circular, while increasingly elliptical galaxies are given designations from E1 to E7 (the latter is the flattest to have been observed). Statistical studies lead to the conclusion that the true ellipticities of these galaxies are represented fairly uniformly from E0 to E7. Some clusters of galaxies contain exceptionally large elliptical galaxies, the giant ellipticals. An important subclass of these is the D giant or supergiant galaxies with elliptical nuclei and extended envelopes. Most, if not all, D-galaxies are strong radio sources.

The *spiral galaxies* (designated by S) may be subdivided into the *normal* and the *barred* spirals. Both types have spiral-shaped arms, with two arms generally placed symmetrically about the center. In the normal spirals, the arms emerge directly from the nucleus, whereas in the barred spirals there is a bar through the center and the arms originate from the ends of the bar. Both types are subclassified according to (a) how tightly the arms are wound, (b) the patchiness of the arms,

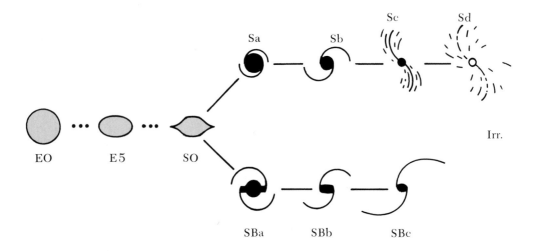

Figure 19–1.   *Hubble's Tuning Fork Diagram.* This classification scheme for galaxy types is discussed in detail in the text.

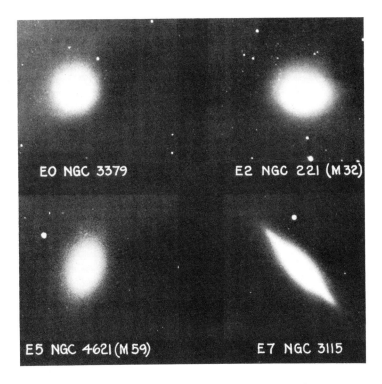

Figure 19–2. *The Elliptical Galaxy Sequence.* (Yerkes Observatory)

and (c) the relative size of the nucleus. Normal spirals of type Sa have smooth, ill-defined arms which are tightly wound about the nucleus; in fact, the arms form almost a circular pattern. The intermediate Sb galaxies have more open arms, which are often partly resolved into patches of H II regions and stellar associations. The nuclei in Sc galaxies are characteristically quite small, while the spiral arms are extended and well-resolved into knots or clumps of stars. Barred spirals exhibit a parallel sequence of types: SBa, SBb, and SBc. The degree to which the spiral arms are developed is related to the absolute luminosity of the galaxy; this is used in van den Bergh's classification to add another dimension to the Hubble scheme.

There is a peculiar type of galaxy which is still given the S classification of spirals, although spiral structure is not usually seen. These *S0 galaxies* are intermediate between the ellipticals, E7, and the true spirals, Sa. They are flatter than the E7s, and also differ from ellipticals in having a thin plane or disk as well as a lenticular (i.e., lense-shaped) nucleus. In many respects, S0 galaxies resemble the true spirals such as Sa or Sb, but without the Population I objects. This fact may be of great significance in determining how such objects came about, for it has been suggested that they are remnants left over after a collision of two spiral galaxies. In such a collision, the interstellar dust and gas would be swept out and left behind in intergalactic space, while the stars remained virtually unaffected. The fact that S0 galaxies are found predominantly in clusters of galaxies, where the probability of collisions is high, lends support to this hypothesis. Another possibility is that they represent an intermediate form to which protogalaxies collapse—namely, between ellipticals and spirals.

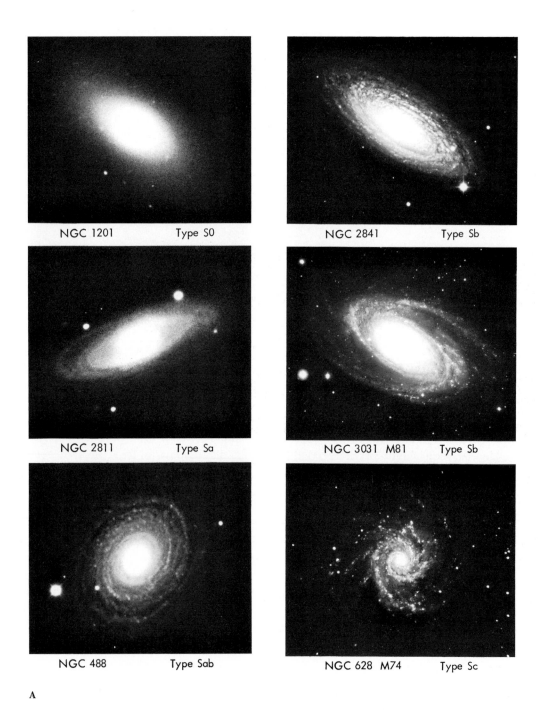

A

Figure 19–3. *The Spiral Galaxy Sequence.* **A**, Normal spirals. **B**, Barred spirals. Some of the barred spiral galaxies have rings, with the arms attached tangentially. These are subtyped (r) in contradistinction to the pure barred spirals (s), in which the arms extend from the ends of the bar. (Hale Observatories)

Illustration continued on opposite page.

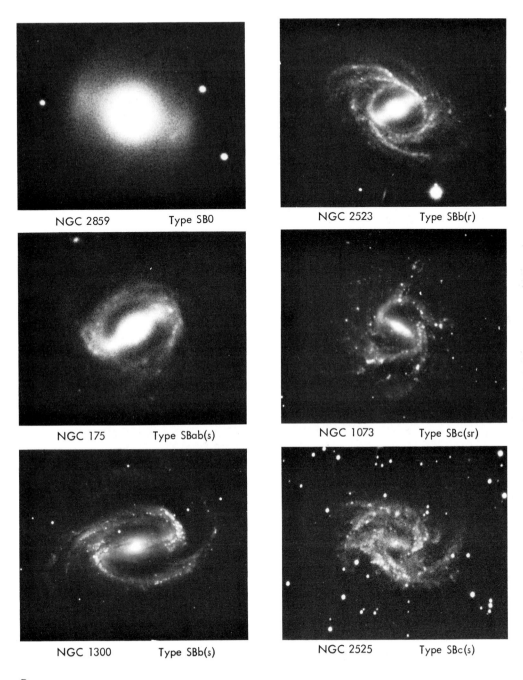

NGC 2859   Type SB0

NGC 2523   Type SBb(r)

NGC 175   Type SBab(s)

NGC 1073   Type SBc(sr)

NGC 1300   Type SBb(s)

NGC 2525   Type SBc(s)

B

There are other galaxies which are classified as *irregular*, since they show no symmetrical or regular structure; nevertheless, even these galaxies may be divided into two distinct groups. Those of type Irr I can be thought of as the extension of the spiral galaxies beyond Sc, for they resolve into O and B stars and H II regions and clearly have a large Population I component. Both Magellanic Clouds (see Figure 19–4) are examples of this type. In some Irr I galaxies, incipient spiral arms seem to occur. The classification Irr II is quite ambiguous, and may include galaxies that are simply peculiar. Primarily, however, these galaxies are amorphous and do not resolve into stars. This lack of stellar resolution implies that the brightest stars are fainter than $-4^{m}.0$. Such galaxies do show marked absorption by interstellar dust, and gaseous emission is also observed. The peculiar galaxy M82 has been classified as Irr II (See Figure 19–14); it is remarkable in that material has

Figure 19–4.   *An Irregular Galaxy of Type Irr I: The Large Magellanic Cloud.* (Lick Observatory)

been observed being ejected at speeds of 1000 km sec^{-1} [see section 19–5(b)]. Other Irr II galaxies are also difficult to interpret.

The central concentration of light is an important characteristic of galaxies and serves as a basis for W. W. Morgan's classification. It is closely correlated with the Hubble sequence and with the integrated spectral type (see section 19–2). Extreme concentration is seen in galaxies that do not fit into any of these classification schemes, but which represent some of the most intriguing extragalactic objects. These are *compact galaxies*, which may appear wholly stellar (quasi-stellar objects) or may have fainter envelopes surrounding brilliant nuclei (N-galaxies and Seyfert galaxies), or they may be just bright, small galaxies, often with peculiarities. In many of these galaxies, there is strong evidence that explosions have occurred in the nuclei. Many are also strong radio sources (see sections 19–5 and 19–6).

(b) The Implications of the Classification Scheme

The classification scheme is not to be thought of as an evolutionary sequence, although Hubble originally thought it might be. We still continue the convention of speaking of Sa spirals as "early" and Sc's as "late." It is likely that the form a galaxy takes depends significantly upon its *angular momentum*, so that the greater the angular momentum (i.e., spin) the more flattened the galaxy becomes. In elliptical galaxies, the condensation of gas into stars was efficient and therefore rapid, which led to a more or less spherical distribution of stars and a high concentration of stars at the nucleus. Conversely, in spiral galaxies, star formation occurred more slowly and a dual type of distribution resulted. The slowly rotating system contains stars distributed spherically, while the rapidly rotating part formed a flattened, disk-like system containing stars, dust, and gas. The dust and gas in the plane become subjected to the density waves that create spiral arms [see section 18–2(e)]. Turbulence and magnetic fields may also play important roles in controlling the final shape of a galaxy.

The total mass of the system *cannot* be an important factor in determining the type, for elliptical galaxies range from *dwarf systems* (with masses similar to that of a large globular cluster) to the giant systems that dominate the clusters of galaxies. These giant ellipticals are far larger and more massive than the largest spirals. Spiral galaxies exhibit a relatively small range of masses. The data which characterize the various galactic types are tabulated in Table 19–1.

Table 19–1. The Fundamental Characteristics of Galaxies

	Ellipticals	Spirals			Irregulars I
Mass ($M_\odot$)	10^5 to 10^{13}	10^9 to 4×10^{11}			10^8 to 3×10^{10}
Absolute magnitude	-9 to -23	-15 to -21			-13 to -18
Luminosity ($L_\odot$)	3×10^5 to 10^{10}	10^8 to 2×10^{10}			10^7 to 10^9
M/L ($M_\odot/L_\odot = 1$)	100	2 to 20			1
Diameter (kpc)	1 to 200	5 to 50			1 to 10
Population content	II and old I	I in arms, II and old I overall			I, some II
Presence of dust	almost none	yes			yes
		Sa	Sb	Sc, Sd	
M_{HI}/M_T (%)*	0	2 ± 2	5 ± 2	10 ± 2	22 ± 4
Spectral type	K	K	F to K	A to F	A to F

* From M.S. Roberts

It is difficult to estimate the relative proportions of various types of galaxies, because observational selection prevents us from obtaining a truly complete count of *small* galaxies for a given volume of space. From our own Local Group [see section 19–3(b)] it is clear that dwarf galaxies must constitute the great majority by number.

Dwarf galaxies may be either ellipticals or Irregular I. In either case they are small, often sparsely populated with stars; therefore, they are faint and of low surface brightness, and they are always hard to detect. Dwarf ellipticals represent the low mass end of the ellipticals, and differ from globular clusters only in having about ten times their diameter. Whereas giant ellipticals may contain metal-rich stars, the dwarf ellipticals are extreme Population II. In contrast, dwarf irregulars contain large amounts of neutral hydrogen, and their stellar content also represents Population I.

19–2   OTHER IMPORTANT CHARACTERISTICS OF GALAXIES

Many of the important characteristics of galaxies, such as their structure and their population characteristics, are inherent in their classifications. From the stellar population content we can conclude that elliptical galaxies contain primarily old stars, while spiral galaxies have a mixture of both old (in their disks and haloes) and young stars (in their spiral arms). The late-type spirals and Irr I galaxies contain an increasingly higher proportion of young stars.

(a)   Integrated Spectra

The *spectrum of a galaxy* is due to all of its constituent stars, with the radiation contribution from the brightest hot stars competing with the light from the fainter (but far more numerous) cool stars. Because of its close correlation with the composite spectrum, W. W. Morgan's classification is also a measure of the relative contributions of population types to the total light. Attempts have been made to use the integrated spectra, or even the related integrated colors, to estimate the luminosity functions of different stellar types. This is a very complex problem, which still has not been entirely solved. Spiral galaxies are particularly difficult because their bright nuclei consist of stars older than those found in the arms; therefore, we must differentiate between the central and the outer regions. Older stars do not mean extreme Population II. Metal-rich stars are characteristic of old Population I and dominate the centers of all galaxies except the dwarf ellipticals.

(b)   The Masses of Galaxies

Spectra can also be used to determine the *masses* of galaxies. This method involves finding the velocity or rotation curve as a function of the distance from the center of the galaxy; hence, the method applies only to galaxies which are near enough and bright enough to permit one to obtain spectra at several points. A model of the mass distribution is then fitted to the velocity curve, and the total mass is calculated.

Recently, 21-cm radio data have been used in a similar fashion to find the velocity curves and thence the masses of some of the nearer galaxies. The velocity determination is usually superior with the 21-cm line, but the position measurements are, of course, far inferior because of the poor resolution of radio telescopes. Nevertheless, by assuming uniform motions within the galaxy, good data may be obtained for the nearest galaxies.

Some galaxies are in binary systems; if the relative velocities of the two galaxies are found, we can apply Kepler's third law in the same manner as it is applied to binary stars. Since we cannot wait for even a small part of the orbit to be observed, it is common practice to assume that the measured difference in velocities is the projected circular velocity. Projection effects are taken into account through statistical handling of the data.

Still another method employs observations of clusters of galaxies, and determines masses from the motions of individual galaxies relative to the whole cluster. This technique was first developed and applied to star clusters, both open and globular clusters, in which case the motions of member stars are used. The *virial theorem* is invoked in either case. This theorem states that the magnitude of the total gravitational potential energy of a cluster equals twice the total kinetic energy, on the average:

$$2\langle KE \rangle = -\langle PE \rangle \quad \text{or} \quad \left\langle \frac{mv^2}{2} \right\rangle = \left\langle \frac{GMm}{2r} \right\rangle \tag{19-1}$$

Here m is the mass of an individual galaxy, v is its speed, r is the distance from the center of the cluster, and M is the mass of the whole cluster. The *mean velocity* can be estimated from the velocities of the individual members, while the *mean distance* (from the center of the cluster) is deduced by averaging the measured distances (with a correction for projection effects). We can then use average values in equation (19-1) to find the total mass, for the individual masses cancel out. The total mass can be divided by the number of known or estimated members to find the mean individual mass. This procedure yields considerably higher masses for galaxies than do the other methods. There are several possible explanations for this result: (a) the mass-to-luminosity ratio is higher than is usually estimated for most galaxies; perhaps some of the excess mass is in the form of black holes; (b) the number of dwarf galaxies has been grossly underestimated; it would take a great many to contribute the necessary mass; (c) the virial theorem may not be strictly applicable if the observed velocities are not gravitational equilibrium velocities; (d) much of the mass in clusters of galaxies is in the form of intergalactic matter. This last suggestion has gained considerable support from the X-ray satellite observation that several clusters of galaxies are bright and extensive X-ray sources. The X-ray emission may be explained as coming from intergalactic gas.

(c)   The Neutral Hydrogen Content

We have seen how the powerful tool of neutral hydrogen 21-cm radiation can delineate the spiral structure in our own Galaxy (Chapter 18). It is not surprising, therefore, that this same radiation gives us much information about some of the nearby galaxies. Because of the paucity of hydrogen gas in elliptical galaxies, only one such elliptical has been observed to emit 21-cm radiation. Galaxies selected for 21-cm observations have been primarily of the late spiral types or irregulars Irr I; this selection was based on the fact that these types are known to have a high percentage of neutral hydrogen.

The information which may be obtained from a study of the 21-cm radiation includes: (a) an estimate of the total amount of hydrogen contained within the galaxy, (b) the ratio of the mass of the hydrogen gas to the total mass of the galaxy, (c) the distribution of the neutral hydrogen within the system, (d) the rotation curves as a function of distance from the center of the particular galaxy and the derivation of its mass, as discussed earlier, and (e) the radial velocity of the galaxy as a whole. The most complete data, of course, are for galaxies which may be resolved by radio telescopes. As we might expect, the total amount of hydrogen contained in a galaxy is primarily a function of the size of the galaxy. On the other hand, the ratio of the hydrogen mass to the total mass of the galaxy (M_{HI}/M_T) depends on the galaxy type, as shown in Table 19–1. The percentage of the total mass that is in the form of neutral hydrogen is relevant to our ideas concerning evolution within galaxies. The less hydrogen there is relative to stars, the more original gas there was which must have already been condensed into stars.

The present rate of star formation depends on both the amount of hydrogen available and its density. Late-type spirals and the irregular galaxies are not necessarily younger than the early-type spirals, but their development has been different. We note, for instance, that Population II stars have been observed in irregular galaxies (like the Magellanic Clouds), even though these galaxies have high M_{HI}/M_T ratios—a case of retarded star formation perhaps.

As might be expected from the observed distribution of neutral hydrogen in our own Galaxy, the distribution in a system like the Andromeda galaxy shows fluctuations across the galaxy that are clearly related to the spiral arms.

For later-type spirals, Sc and Sd, and irregulars, the extent of the hydrogen in many cases is almost *double* that of the optical size of the galaxy. For example, studies of the Magellanic Clouds show that there is hydrogen between these two galaxies, both as a type of bridge and as a common envelope surrounding both galaxies.

In addition to revealing rotation curves, the 21-cm line Doppler shifts also give information about the radial velocity of the galaxy as a whole—the systematic radial velocity. In determining such velocities, 21-cm techniques permit high precision; this precision is considerably better than that obtained from low-dispersion spectra at optical wavelengths. On the other hand, the distances to which such measurements can be made are severely limited, for only nearby galaxies are sufficiently bright at radio wavelengths to be observed at 21 cm.

(d) Continuum Radio Emission from Normal Galaxies

In addition to the radiation due to neutral hydrogen at 21 cm, most spiral galaxies also radiate in the continuum at radio frequencies. The total energies radiated, however, are several orders of magnitude smaller than those from the strong radio sources called radio galaxies (see section 19–5). In some cases, such as M31, the continuum seems to come from a region which is more extensive than the optical galaxy; this region is apparently the galactic halo. In the case of some other spirals, the continuum radiation is far more restricted and indeed appears to arise from a source no larger than the nucleus of the galaxy.

This phenomenon may be related to the violent explosions that seem to occur in the centers of some (perhaps all!) galaxies at some time in their evolution. Normal elliptical galaxies do not display much radio emission, nor do the S0 galaxies. We do know that our own galactic center is a strong emitter at continuum radio wavelengths. Although part of this continuum emission is thermal in character (i.e., due to the free-free interactions of the electrons), the rest is nonthermal and is explained in terms of synchrotron emission. Similarly, some of the continuum radiation from other galaxies is also nonthermal. We postpone discussion of the strong extragalactic radio sources and their kin until after the section on the distance scale (section 19–4).

19–3   CLUSTERS OF GALAXIES

At least half of all galaxies, and perhaps most galaxies, are members of some type of cluster. The Milky Way Galaxy is one of the dominant members of the Local Group of galaxies, which in turn *may* be part of a local super-cluster that also includes the rich Virgo cluster.

(a)   The Types of Clusters

There are different ways of defining categories of clusters of galaxies, but the simplest is that of G. O. Abell, who separates clusters into *regulars* and *irregulars*. *Regular clusters* tend to be giant systems, with spherical symmetry and a high degree of central condensation; they frequently contain many thousands of member galaxies, of which perhaps a thousand are brighter than absolute magnitude -15. Almost all members of regular clusters are either elliptical or S0 galaxies, whereas *irregular clusters* contain a mixture of all types of galaxies. Among the irregular clusters of galaxies are included (a) small groups, such as our own Local Group, (b) loose aggregates of subgroups with several centers of condensation, and (c) fairly large but diffuse clusters. The *luminosity function for galaxies* in clusters (the number of galaxies per integrated magnitude interval) indicates that there is a great preponderance of *faint* galaxies. As noted earlier, several clusters of galaxies have been identified with strong and greatly extended X-ray sources.

(b)   The Local Group of Galaxies

Our Milky Way Galaxy and the Andromeda galaxy (M31) dominate the small group of galaxies, referred to as the Local Group, which contains at least 17 members (see Table 19–2). The other members are fainter and less massive, and to some extent seem to be concentrated around one or the other of the two large galaxies just mentioned. A wide range in galaxy types is included in the group,

Figure 19–5.   *A Cluster of Galaxies in Hercules.* (Hale Observatories)

from the two giant spirals (our Galaxy and M31) to the dwarf ellipticals and irregulars. Once again, the high proportion of dwarf galaxies is readily apparent.

The confines of the Local Group are somewhat uncertain. Some astronomers would include a considerably larger number of galaxies and a correspondingly larger volume. Of particular interest are two large galaxies discovered by P. Maffei and named after him. These lie in the direction of the constellation Perseus in the galactic plane; they escaped detection for a long time because of the heavy obscuration (about 5^m) from dust in that direction. They were in fact discovered on infrared plates. Their appearance in the infrared as well as other data (photometric, spectroscopic, and radio) indicate that Maffei 1, the brighter of the two, is a giant elliptical galaxy and Maffei 2 is a spiral. Present distance estimates place them at about 2.7 megaparsecs (Mpc = one million parsecs), well beyond the 1 Mpc usually considered as the boundary of the Local Group. On the other hand, it may be dynamically more correct to enlarge the group to include not only Maffei 1 and 2 but also other galaxies at similar distances. If one does include Maffei 1, the apparent lack of a giant elliptical in the Local Group is removed. Figure 19–6 illustrates the location of Local Group member galaxies, within the 1 Mpc limit, relative to the plane of our Galaxy. The galactic coordinates of Maffei 1 and 2 are $\ell \approx 136°$, $\ell \approx -0°\!.5$; therefore, they would lie on the far side of M31 and at almost four times its distance in such a projection onto the plane of the Galaxy.

Table 19–2. The Local Group of Galaxies

Name	R.A. (1950)	Dec.	l	b	Type	M_v	Distance (kpc)	$m - M$	Mass ($M_\odot$)	Rel. Vel. (km sec^{-1})
M31 = NGC 224	00^h 40^m	+41°0	121°	−20°	Sb	−21.1	690	24.6	3×10^{11}	−267
Galaxy	17 42	−29.0	0°	0°	Sb or Sc	−20?	9	14.7	2×10^{11}	—
M33 = NGC 598	01 31	+30.4	135°	−31°	Sc	−18.9	690	24.6	4×10^{10}	−190
LMC*	05 24	−69.8	280°	−33°	Irr I	−18.5	50	18.6	6×10^{9}	+275
SMC*	00 51	−73.2	303°	−45°	Irr I	−16.8	60	19.1	1.5×10^{9}	+163
NGC 205	00 38	+41.4	121°	−21°	E6p	−16.4	690	24.6	—	−239
M32 = NGC 221	00 40	+40.6	121°	−22°	E2	−16.4	690	24.6	2×10^{9}	−220
NGC 6822	19 42	−14.9	26°	−20°	Irr I	−15.7	460	24.2	1.4×10^{9}	−34
NGC 185	00 36	+48.1	121°	−14°	dE0	−15.1	690	24.5	—	−270
NGC 147	00 30	+48.2	120°	−14°	dE4	−14.8	690	24.5	—	—
IC 1613	01 02	+01.8	129°	−60°	Irr I	−14.8	740	24.5	4×10^{8}	−235
Fornax	02 37	−34.7	237°	−66°	dE3	−13.0	188	21.4	2×10^{7}	−73
Sculptor	00 57	−34.0	286°	−83°	dE3	−11.7	84	19.7	3×10^{6}	—
Leo I	10 06	+12.6	226°	+49°	dE3	−11.0	220	21.8	3×10^{6}	—
Leo II	11 11	+22.4	219°	+67°	dE0	−9.4	220	21.8	10^{6}	—
Ursa Minor	15 08	+67.3	103°	+45°	dE6	−8.8	67	19.5	10^{5}	—
Draco	17 19	+58.0	86°	+35°	dE3	−8.6	67	19.6	10^{5}	—

* LMC = Large Magellanic Cloud; SMC = Small Magellanic Cloud

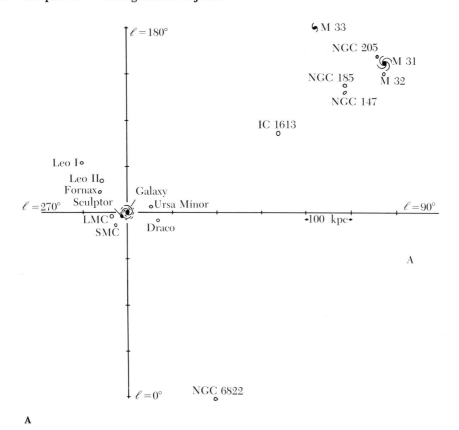

$\ell = 180°$

M 33

NGC 205

M 31

NGC 185

M 32

NGC 147

IC 1613

Leo I

Leo II

Fornax

$\ell = 270°$ Sculptor

Galaxy

Ursa Minor

$\ell = 90°$

LMC

Draco

100 kpc

SMC

A

$\ell = 0°$ NGC 6822

A

Figure 19–6. *The Local Group of Galaxies.* **A**, A top view of the Local Group projected onto the plane of the Galaxy. Note the distance scale and the galactic longitude coordinates (ℓ). *Illustration continued on opposite page.*

The importance of the Local Group lies not only in the fact that it is a representative cluster of galaxies, but also in the fact that the study of its individual galaxies enables us to learn a great deal about the characteristics of galaxies; we may then use this knowledge to extrapolate to the more distant galaxies. We have seen how the study of the Andromeda galaxy enabled Baade to first outline the differences between stellar populations (see Chapter 13). Recent studies of several of the local galaxies have greatly refined our ideas on populations and on the stellar content of galaxies, and consequently on the evolution of both stars and galaxies. Earlier, the study of Cepheids in the Magellanic Clouds led to the discovery of the period-luminosity law, and then Hubble's application of this law to M31 proved that galaxies are truly extragalactic. The importance of the Local Group members as calibrators for distance scales cannot be overemphasized.

19–4 THE DISTANCE SCALE

(a) The Distances to Galaxies

The sizes and integrated absolute magnitudes which we obtain for galaxies depend on our ability to ascertain their distances; in fact, questions regarding the

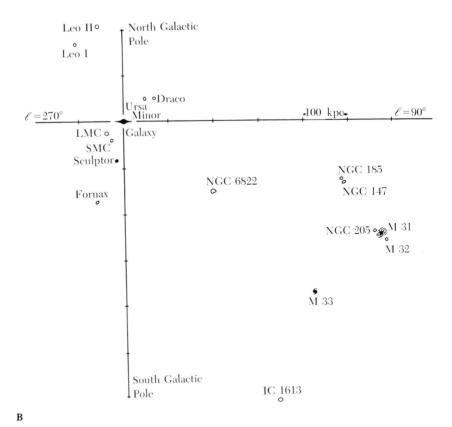

B, A side view of the Local Group, looking toward $\ell = 180°$. Note the two concentrations of galaxies, one near our Galaxy and one near M31.

Universe and cosmology involve the *distance scale*—the scale of astronomical distances. This is a good point to look back and to see how the distance scale is established in a step-wise fashion. Distances to even the nearest galaxies are far too great to be measured *directly* (i.e., by trigonometric parallax or other geometrical techniques; see Chapter 10), so we must use indirect methods related to the brightness of stars and the inverse-square law of radiation intensity (Chapter 10). Table 19–3 lists the most important *distance indicators* used within our own Galaxy and for other galaxies, while Figure 19–7 emphasizes the pyramidal nature of the distance scale. The reader is urged to study both Table 19–3 and Figure 19–7, and to assure himself that he fully understands them.

Most of the objects listed in Table 19–3, which are brighter than absolute magnitude 0^m, can be used as distance indicators for galaxies within the Local Group. Galactic cluster main-sequence fitting is not useful even for the nearby galaxies, but this method is basic to the *calibration* of bright stars, such as OB stars and classical Cepheids. Although the methods listed in the second part of the table permit one to find distances to galaxies with large distance moduli, these distance indicators are substantially less reliable than those in the first part of the table. Their power lies in the very great brightness of the objects and the subsequently large limiting distance moduli. The bright objects can give distances to galaxies well beyond the Local Group (i.e., to other larger clusters of galaxies, such

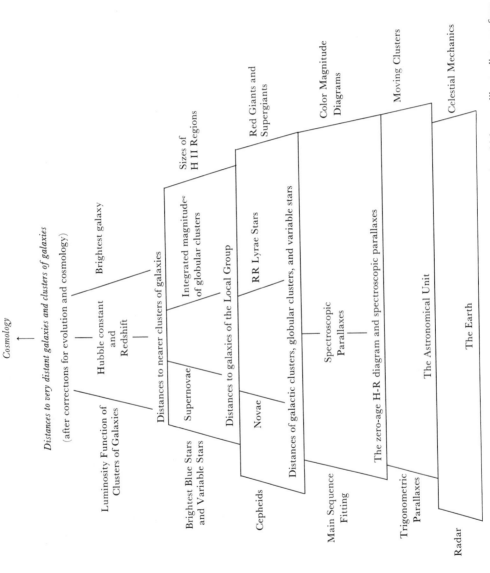

Figure 19–7. *The Distance Scale Pyramid.* This diagram illustrates the bootstrap process by which we calibrate distances from our Earth to the farthest reaches of the Universe. See also Table 19–3. (After P. W. Hodge)

Table 19–3.   Distance Indicators

Object	M_v	Population	Method	Basis of Calibration	Reference Section
Nearby stars	$\gtrsim 0^m$	Disk	Trigonometric parallax	Radar determination of AU	7–2
Galactic clusters	—	I	Main-sequence fitting	Trigonometric parallaxes	10–1
				Moving clusters	14–3
				Main sequence of nearby stars	12–3
Lower main-sequence stars (A–M)	$\gtrsim 0^m$	Disk	Spectroscopic parallax	Trigonometric parallaxes	10–1
O and B stars	0^m to -6^m	I	Spectroscopic parallax	Galactic cluster C-M diagrams	12–3
				Galactic cluster C-M diagrams	12–3, 15–3
				Statistical parallaxes	14–2
Supergiants	$-6^m, -7^m$	I	Spectroscopic parallax	Galactic cluster C-M diagrams	12–3
RR Lyrae stars	$+0^m\!.5$	II	Period from light curve	Statistical parallaxes	14–2
				Globular clusters	15–3
Classical Cepheids	$-0^m\!.5$ to -6^m	I	P-L law	Statistical parallaxes	14–2
				Galactic cluster C-M diagrams	15–3
W Virginis stars (Population II Cepheids)	0^m to -3^m	II	P-L law	Statistical parallaxes	14–2
				Globular cluster C-M diagrams	15–3
Globular clusters	-5^m to -9^m	II	Integrated magnitude	RR Lyrae stars	16–2
				C-M diagrams	15–3
Novae	-8^m	Disk	Maximum light	Expansion rate of shell	16–5
H II regions	-9^m	I	Angular size	Nearby galaxies	17–2
Supernovae	-16^m to -20^m	I and II	Maximum light	Nearby galaxies	16–5
Brightest galaxies in clusters of galaxies	-21^m	I and II	Integrated mag. (a) Brightest galaxy (b) Fifth brightest (c) Mean of ten brightest	Nearby galaxies	19–3, 19–4
Galaxies	—	I and II	Hubble constant of redshift (expansion of Universe)	Doppler shifts of nearest clusters of galaxies	19–4

C-M = Color-Magnitude (= HR) ; P-L = Period-Luminosity

The entries below the line are objects used to find distances to more distant galaxies and they are considerably less accurate than those above the line. The entries above the line also serve as distance indicators in our own Galaxy.

as the Virgo cluster). Statistics from these clusters in turn tell us the absolute magnitude of the *brightest galaxies* in such a cluster and the luminosity function within a cluster. Although there is some variation from cluster to cluster, the statistics are sufficiently good to give a first approximation of the distance, and they indicate a reasonable amount of homogeneity from one cluster to another. In fact, there appear to be distinct groupings at certain absolute magnitudes; for instance, the brightest galaxy in a large cluster is nearly always a giant elliptical!

Among the most important distance indicators for galaxies are the *novae* and *supernovae*. Although their outbursts are relatively rare, supernovae are so bright that they may rival the total brightness of the galaxy in which they occur. The type of supernova can be determined from its light-curve.

(b) Redshift as an Indicator of Distance

Galaxies are known to move relative to one another. The radial velocities of other galaxies (relative to our Galaxy) may be ascertained by studying *Doppler-shifted features* (see Chapter 8) in the spectra of these galaxies. At visible wavelengths, we may use Doppler-shifted spectral emission lines, while at radio wavelengths, the 21-cm emission line of neutral hydrogen is extremely useful.

In 1912, V. M. Slipher discovered that *most* of the galaxies he observed had *redshifted* spectral features; that is, most galaxies are receding from our Galaxy. E. Hubble and M. Humason later established that the amount of redshift increased with the faintness of the galaxy; this correlation, which we shall express quantitatively later in this section, is known as *Hubble's law*. Let us consider a given galaxy at a distance r from us. If this galaxy emits a spectral line of wavelength λ_0, and we detect this line at a (greater) wavelength λ, then the *redshift* z is defined as

$$z = \left(\frac{\lambda - \lambda_0}{\lambda_0}\right) = \frac{\Delta\lambda}{\lambda_0} \tag{19–2}$$

If this change in wavelength is interpreted as a *velocity* Doppler shift, then Chapter 8* tells us that the *speed of recession* of the observed galaxy is

$$v = (\Delta\lambda/\lambda_0)c = zc \tag{19–3}$$

where c is the speed of light. By using the apparent magnitude of the galaxy to measure its relative distance, Hubble discovered the correlation

$$\boxed{zc = Hr} \tag{19–4a}$$

---

* When $v \gtrsim c/10$, we must use the *relativistic* expression for the relation between Doppler shift $(\Delta\lambda/\lambda_0)$ and recession speed v, which is given in section 8–1(a), equation (8–14).

RELATION BETWEEN RED-SHIFT AND DISTANCE FOR EXTRAGALACTIC NEBULAE

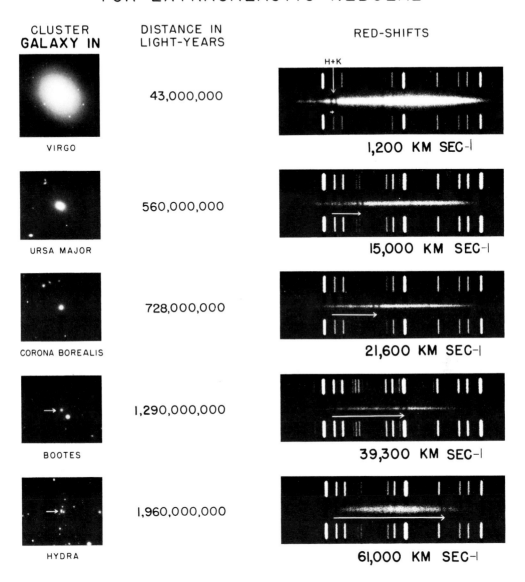

CLUSTER GALAXY IN	DISTANCE IN LIGHT-YEARS	RED-SHIFTS
VIRGO	43,000,000	1,200 KM SEC⁻¹
URSA MAJOR	560,000,000	15,000 KM SEC⁻¹
CORONA BOREALIS	728,000,000	21,600 KM SEC⁻¹
BOOTES	1,290,000,000	39,300 KM SEC⁻¹
HYDRA	1,960,000,000	61,000 KM SEC⁻¹

Figure 19–8. *Spectra Showing Redshifts of Galaxies.* Redshifts are expressed as velocities rather than z. Arrows indicate the shift for the calcium H and K lines. Distances are calculated on the basis of a Hubble constant of 100 km sec^{-1} Mpc^{-1}. (Hale Observatories)

where H is *Hubble's constant;* comparing equation (19–4a) with equation (19–3), we find an alternative form of Hubble's law as

$$v = Hr$$

(19–4b)

Since apparent magnitude goes as log *r*, we may illustrate Hubble's *linear* relationship between redshift and distance (equation (19–4a)) by plotting log *z* versus the apparent magnitude of the galaxy (see Figure 19–9).

If *H* is known, then it is a very straightforward matter to find the *distance* to a galaxy, once we know its Doppler shift. It is clear from the preceding equations that the dimensions of *H* are velocity per unit distance; the units currently used for *H* are km sec^{-1} Mpc^{-1}, where Mpc = megaparsec = 10^6 pc! To calibrate the Hubble constant, we must rely upon other methods for establishing distances to the galaxies; as we have seen, these methods are not very accurate. The distance scale of the Local Group plays a major role in this calibration, but for distant clusters of galaxies we must use integrated apparent magnitudes and the known magnitudes of the brightest galaxies in clusters. The current acceptable range for *H* is from less than 50 to 125 km sec^{-1} Mpc^{-1} (i.e., the value is uncertain by a factor of 2 or more); a current analysis by A. R. Sandage favors 55 ± 7 km sec^{-1} Mpc^{-1}, which we may round off to 50 km sec^{-1} Mpc^{-1}. Note that *H* may also be expressed in the units yr^{-1}, if we convert Mpc → km and sec → yr; therefore, the Hubble constant appears to have some bearing on the *age of the Universe*, since $H^{-1} \cong 10^{10}$ years for a value of $H = 100$ km sec^{-1} Mpc^{-1}, or 2×10^{10} years if $H = 50$ km sec^{-1} Mpc^{-1}.

As we shall discuss more fully in the next chapter, the fact that all galaxies beyond the Local Group are receding from us implies that the Universe of galaxies is expanding. If we apply equation (19–4b) to the Local Group, we find $v \lesssim 60$ km sec^{-1}; the fact that *nearby* galaxies do not seem to participate in the cosmic expansion is due to their *random motions* within the Local Group (see Table 19–2). It is clear, therefore, that Hubble's law does not become important until $r \gtrsim 1$ Mpc, so that the general expansion is only observable at large distances. Since we observe all other galaxies to be receding from *us*, Hubble's law appears to imply that we stand at the *center* of the Universe. This conclusion is patently incorrect, since

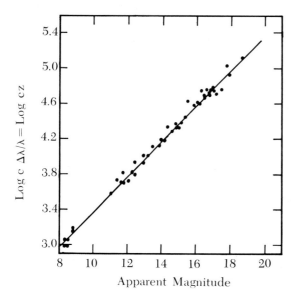

Figure 19–9. *Hubble's Law of Cosmic Expansion.* Here *z* is the redshift of spectral features, while the horizontal scale indicates the integrated apparent magnitude of the galaxy. The observational points are from data obtained by A. R. Sandage, and they illustrate Hubble's relation between velocity of galactic recession and distance.

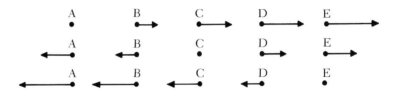

Figure 19–10.   *Receding Galaxies in an Expanding Universe.* We have reduced the motions to one dimension to clarify the effects of cosmic expansion. All galaxies are receding from all other galaxies at a speed proportional to their distance of separation (Hubble's Law). In this figure, we illustrate what an observer will see, if he is situated on galaxy A, C, or E. General recession is observed in *all* cases! (After S. P. Wyatt)

Figure 19–10 illustrates that, in a generally expanding Universe, *each* galaxy will see *all other* galaxies receding from it; therefore, there is *no* center of the Universe!

19–5   RADIO GALAXIES AND RELATED OBJECTS

(a)   Radio Galaxies

Whereas a majority of the spiral galaxies of apparent visual magnitude 11^m or brighter do radiate in the radio continuum, the amount of this radiation appears to peak at about 10^{40} erg sec^{-1}. There are a few *extragalactic radio sources* which are exceptionally strong radio emitters, with some of them generating energies in excess of 10^{45} erg sec^{-1}. When identified optically, these galaxies all seem to show certain peculiarities. Several radio galaxies also radiate strongly at X-ray frequencies. Henceforth, we will use the term strong radio source or *radio galaxy* for those galaxies with a radio luminosity greater than 10^{40} erg sec^{-1}. This is not to say that the "weak" sources are all normal, but rather that all normal galaxies are relatively weak sources. Some of the peculiar galaxies have not been observed as radio sources at all, and in some cases their energies correspond to those of weak sources; a few are strong sources. No spiral galaxies, peculiar or not, have been identified as strong radio galaxies.

Radio galaxies may be giant or supergiant ellipticals, particularly the D-galaxies; more rarely, they may be "dumb-bell" (db) galaxies, in which two nuclei are embedded in the same envelope. Alternatively, radio galaxies may be compact objects, such as the N-galaxies with their star-like nuclei surrounded by a far fainter envelope.

A few elliptical galaxies identified with extragalactic radio sources have certain peculiarities, such as absorption features or jets (i.e., unsymmetrical extensions of the galaxy).

The radio structure of the strong sources is usually *double* in nature. In most cases, the two components of such a double source are well separated from the optical galaxy and are much larger in size. As a rule, these sources lie on a line, with the galaxy at about the "center of gravity" of the radio emission pattern. An example

TYPES OF OPTICAL OBJECTS WHICH PRODUCE EXTRAGALACTIC RADIO SOURCES
The absolute linear scales have been made equal.

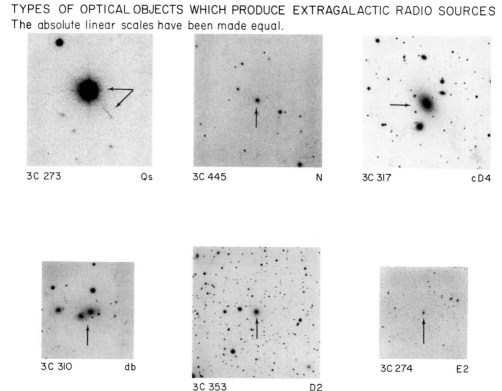

Figure 19–11. *Types of Optical Objects which Produce Extragalactic Radio Sources.* 3C 273 is a quasi-stellar object. See the text for a description of the characteristics of these types of objects. (Courtesy of T. A. Matthews; Hale Observatories photographs)

of such a double radio source is Fornax A (shown in Figure 19–12), while a somewhat special case is Centaurus A (Figure 19–13), which is a double-double source. For Centaurus A, one pair of radio regions is separated by 240 kpc, well away from the optical galaxy, and the other pair is only 10 kpc apart and is superimposed on the galaxy. Occasionally, the components themselves may be very small, with an angular diameter less than 0".001. A less common structure of radio emission is that of a core and halo.

The double sources comprise the large majority of all the resolved radio galaxies. In some cases the radio components are fairly near to the optical galaxy, while in other cases they are very far removed. The dimensions range from a separation of less than 15 kpc between the components to perhaps 500 kpc. A few radio galaxies, which have been unambiguously identified with optical galaxies, have radio components less than 4" in size. Most radio galaxies, however, do show finite structure when studied at radio wavelengths by interferometer techniques. We emphasize this point to show the contrast to the observations of quasi-stellar sources, which are discussed in the following section. In summary, strong radio sources usually have two radio components well displaced on either side of the optical galaxy, and the optical galaxy identified with the source is in nearly every

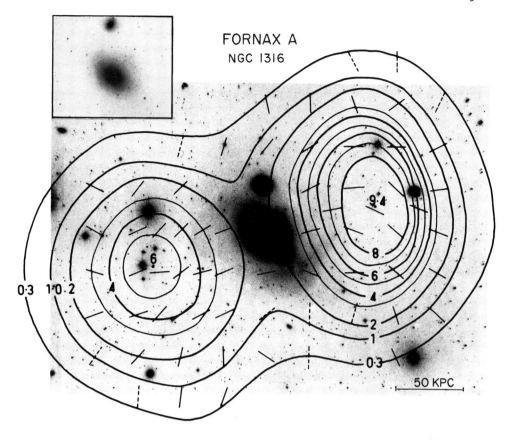

Figure 19–12. *A Double Radio Source: Fornax A.* Intensity contours of the radio components are superimposed on a negative photograph of the galaxy. The radio intensity falls off smoothly from two centers. The short lines are perpendicular to the magnetic lines of force. A shorter exposure of the galaxy appears in the insert. (Courtesy of T. A. Matthews; Hale Observatories photograph)

case either a giant galaxy, such as a D type, or a compact N-galaxy. This is true not only for the individual galaxies, but also for radio galaxies observed in clusters. The radio emission from clusters of galaxies is apparently due to individual giant or supergiant galaxies, and is not from the entire cluster. In some cases, the radio components of a radio galaxy may be separated by 500 kpc, but the *average* distance between optical galaxies in the cluster is only 100 kpc.

One of the first radio galaxies to be identified, Cygnus A, was originally believed to be two galaxies in collision, and at that time the hypothesis was put forth that most radio sources were in fact *colliding galaxies*. Centaurus A was thought to be another example. Since then it has been shown that the energy generated through such a collision of galaxies is quite insufficient to explain the observed energy output, and therefore, other mechanisms were investigated. The radio spectral characteristics—the (radio) energy flux as a function of wavelength— indicate that the emission from all nonthermal radio sources is *synchrotron radiation* arising from the acceleration of electrons in a magnetic field. The origin of the energy that accelerates the electrons is still largely a matter of speculation, although it is believed to be some kind of explosion.

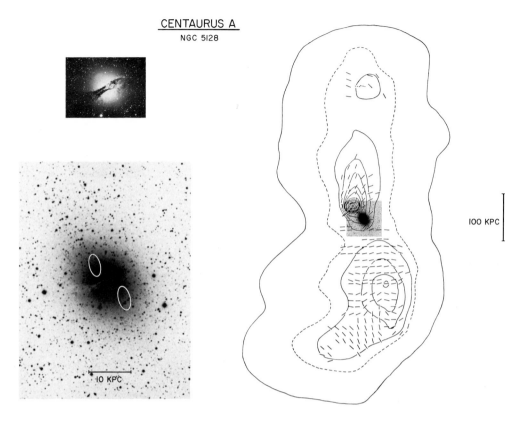

Figure 19–13.   *A Double-Double Radio Source: Centaurus A.* The positions of the central double radio source components, as superimposed on a negative photograph of the galaxy in red light, appear at the lower left. The outer radio intensity contours are shown on the right; note the difference in scale. The short lines are parallel to the magnetic lines of force. At the upper left is a positive photograph of NGC 5128 in blue light. (Composite courtesy of T. A. Matthews; Hale Observatories photographs)

(b)   The Evidence for Explosions in Galaxies

One of the most spectacular instances of a galaxy showing evidence of an *explosion* is M82, the irregular galaxy (Irr II) illustrated in Figure 19–14. This is a radio source, but not a strong one, since it radiates less than 10^{40} erg sec^{-1}. Spectral and polarization data at both optical and radio wavelengths again point to synchrotron emission, and therefore to the presence of relativistic electrons in a magnetic field. Photographs of this galaxy in Hα show filaments extending above the plane of the galaxy to a distance of 3 to 4 kpc, suggesting that the material was thrown out from the galactic plane. Spectra of this galaxy also show material expanding outward from the center at high velocities. The most distant material seems to be moving at a velocity of about 1000 km sec^{-1}. The combined data indicate that the material must have been ejected outward from the nucleus of the galaxy about a million and a half years ago.

Other galaxies with high expansion velocities are the so-called *Seyfert*

Figure 19–14.   *An Exploding Galaxy, M82.* **A,** This photograph was made in blue light. The galaxy is classified as an Irr II. **B,** This Hα photograph clearly shows the explosive nature of the galaxy. Note that the scales are not the same for the two photographs. The bright star above the galaxy and to the left in the Hα picture is identical to the very bright star almost embedded in the galaxy on the blue photograph. (Hale Observatories)

galaxies (named after their discoverer). These are galaxies with strong, very broad, emission lines and very bright, small nuclei. Some have spiral-like structure. Seyfert galaxies are closely related to, if not identical with, the N-type galaxies. It is possible that M82 is a Seyfert galaxy. Although a few Seyfert galaxies are strong radio sources, most are weak, and some are not radio sources at all. Many, however, have very high infrared luminosities for which the origin is not known, although their spectral characteristics indicate both thermal and nonthermal radiation.

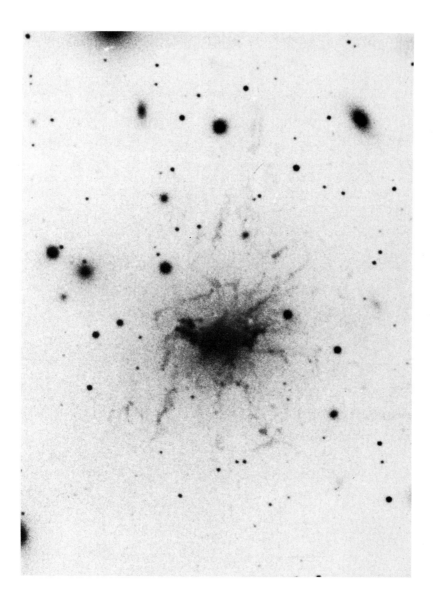

Figure 19–15.   *The Seyfert Galaxy NGC 1275.* This remarkable galaxy is a strong radio source and shows filamentary structure indicative of an explosive event. It should not be considered a typical Seyfert galaxy. The photograph was made with an interference filter centered on Hα at the radial velocity of the nucleus. (Courtesy of R. Lynds, Kitt Peak National Observatory)

Some strong radio sources, like Centaurus A, can also be interpreted in terms of explosions. The double nature of many sources could be due to clouds of electrons ejected from the nucleus of the galaxy. Since such clouds lie at distances of 250 kpc (about 800,000 light years) from the parent galaxy, they must have been ejected from the galaxy more than 800,000 years ago, for their velocities cannot exceed the speed of light. It is not known whether more than one type of violent event is responsible for these radio sources, nor is it known what mechanisms are responsible for the events. We do know, however, that the energy of such outbursts must be enormous. Calculations indicate that the total energy output ranges from 10^{55} to 10^{61} ergs, and the time scales over which such events take place may be as short as a thousand years and as long as a million years. These explosions have been compared with supernovae outbursts, but no concrete connection has been established between these phenomena. The strong X-radiation emitted by several radio galaxies points to high energies and activity of these objects.

The nucleus of our own Galaxy may well have been the seat of one or more explosions. The nonthermal radio and infrared radiations from the galactic center, the expanding 3-kpc arm, and the peculiar behavior of the high-velocity clouds all indicate highly energetic events.

That magnetic fields and synchrotron radiation play a major role in the radio emission is also clear from the observed *polarization* in these sources. Galactic magnetic fields may also control the ejection, so that the plasma clouds follow the magnetic field and remain on a line with the parent galaxy. Since double sources occur both superimposed on the optical galaxy and well separated from it, it seems likely that they represent different stages of the ejection. The plasma clouds expand as they move outward, becoming fainter with time until they fade away altogether.

19–6   QUASI-STELLAR OBJECTS

Quasi-stellar objects* are, as their name implies, stellar in appearance, but they are distinguished by some very nonstellar characteristics, such as *greatly* redshifted emission line spectra! We include them in this chapter on galaxies because external galaxies are the only other objects observed with redshifts even approaching those of the quasi-stellar objects (henceforth referred to as QSOs), and because other evidence indicates that they are in fact extragalactic. As we shall see, whether they are actually as distant as their redshifts would indicate on the basis of Hubble's law is still very much a matter of controversy. Some QSOs are strong radio sources; indeed, optical identification of such quasi-stellar sources (QSS) first drew attention to the QSOs. The majority of QSOs, however, are weak radio sources or altogether radio-quiet. These objects are usually first identified on the basis of their excessive ultraviolet radiation, compared to ordinary stars of comparable apparent magnitude, say 13^m and fainter. We then verify that we have true QSOs when their

---

* In analogy with the term "pulsar," many astronomers and astrophysicists refer to a quasi-stellar object as a *quasar*. (In actuality, the word "quasar" appeared before the word "pulsar.") This terminology is extremely convenient, but it is not yet universally accepted nor does it distinguish between QSO and QSS.

spectra show large redshifts. The total number of QSOs brighter than 19th apparent magnitude is estimated as over a million.

(a)   Spectral Characteristics

The first quasi-stellar object whose spectral redshift was identified was the radio source 3C 273 (number 273 in the Third Cambridge Catalogue of radio sources). Prior to that time, the spectra obtained of QSSs were a complete mystery, for they showed broad emission lines which did not seem to fit any known spectrum. The emission lines in the spectrum of 3C 273 showed a familiar regularity, reminiscent of the hydrogen Balmer lines, with the exception that they seemed to be greatly displaced to the red of the normal Balmer lines. Nevertheless, M. Schmidt did identify them as the Balmer lines Hβ through Hϵ, and he calculated a redshift of $z = 0.158$ (where z is $\Delta\lambda/\lambda$). With the acceptance of the concept of large z, other lines could be identified, not only in this particular object but also in many others. Most of the emission features, all of which are *broad* lines, are similar to those found in gaseous nebulae. In addition to the hydrogen Balmer lines, there are forbidden lines, such as [O II], [O III], [Ne III], and [Ne V], and permitted lines of Mg II, C III and C IV.

The spectra of most quasi-stellar objects show very broad emission lines similar to those found in the spectrum of 3C 273. The outstanding feature of these lines is the fact that they are all very greatly shifted to the red, so that the redshift corresponds to values of z from 0.06 to 2.88! For example, in several cases the Lyman α line, normally at 1216 Å, is shifted into the visible part of the spectrum. The presence of nebular lines (especially the forbidden lines) requires the presence of a *low-density gas*, which is an important factor in regard to the nature of these objects.

Some QSOs have been discovered with relatively sharp *absorption lines* (in addition to emission lines) in their spectra. These absorption lines frequently correspond to approximately the same z as the emission lines. In some cases, the

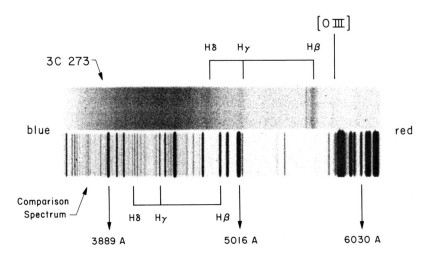

Figure 19–16.   *The Spectrum of Quasi-Stellar Object 3C 273.* (Courtesy of M. Schmidt, Hale Observatories Photograph)

absorption line redshift and the emission line redshift are not identical, but they generally satisfy the relation $z_{abs} \leq z_{em}$. Since the differences in redshift are not large, this implies that the absorption lines actually arise *in the objects* and not in the intergalactic region. This absorbing material is assumed to be matter ejected from the QSO. In several cases, the absorption lines are *multiple*, with redshifts corresponding to a wide range in z. Again, these could be due to multiple ejections of matter or to several intervening "clouds" in the intergalactic medium.

When we consider the observed continuum spectra (over the entire electromagnetic spectrum), we find that QSOs differ considerably from one another in the distribution of intensity as a function of wavelength. We have already noted that some QSOs are strong radio emitters, while others are not. Most of these objects, however, do share the characteristic of strong continuous radiation in both the near ultraviolet and the red. Several, such as 3C 273, emit far more infrared radiation than would be expected on the basis of the optical and radio emission; moreover, 3C 273 is also an X-ray source.

(b) Radio Properties

The QSOs that are radio sources include some of the strongest radio sources; it should be remembered, however, that some of the radio galaxies have comparable apparent radio luminosities. In fact, the radio properties of QSSs do not differ very greatly from those of radio galaxies. It is true that some QSSs have diameters of less than $0\overset{''}{.}001$ at radio wavelengths. But we have already seen that this is not a unique property of the QSS, for a small fraction of the radio sources of that dimension are Seyfert galaxies or N-galaxies. Moreover, several QSSs have radio components with sizes comparable to those of the radio galaxies, and a few are actually double sources with rather large separations between the components. Another characteristic which does seem to set many QSSs apart from radio galaxies is their *variation* in radio brightness, on a time-scale of months or a year. Although few radio galaxies have been observed to change their output at radio wavelengths, several Seyfert nuclei have variable radio emission.

(c) Optical Appearance

As suggested by their names, the *optical appearance* of QSSs is stellar; that is, their angular diameters are less than $1''$. Some, however, have faint nebulosities associated with them (see Figure 19–11). The optical variations of several QSSs have been well established; in some cases, changes of as much as 1 magnitude over a few months have been observed. Some of these variations are long term, with time scales of years; in other cases, the magnitude changes have been very rapid, with small changes (of the order of $0\overset{m}{.}04$) occurring in a matter of 15 minutes! Other changes occur more slowly, over a matter of several days or a month or two. One phenomenal case, 3C 446, changed its luminosity by a factor of 20 (or 3 magnitudes) within a year, and continued to show sizeable variations thereafter. QSSs which vary appreciably at radio frequencies are also optical variables; in addition, many nonvariable QSSs and radio-quiet QSOs also vary in the optical region. Therefore, it seems that the optical variations are not closely related to the radio characteristics, and the two types of variability may be due to different mechanisms. A strong correlation does exist, however, between optical variability and polarization. In fact, the brightness changes are often accompanied by a variation in polarization, indicating that it is the polarized component which is responsible for the light fluctuations.

Optical variations are of great importance in setting limits on the *size* of QSOs. It can be shown that if an object varies with a period τ, the size of the object must be equal to or smaller than τ times the speed of light—$R \lesssim c\tau$. Outbursts at different places within the source would average out to only slight overall changes. A total variation by any appreciable factor requires that the outbursts be *synchronized*; this means that the signal from one region must travel the distance to all other regions within the period of variation. Moreover, if an object larger than $c\tau$ varied as a whole, the light travel time would smooth out the time variations. If $c\tau$ is a light-month, but the object is a light-year in diameter, radiation from the farthest point would be delayed by a year compared to that from the side nearest the observer, thereby masking the monthly variations.

Since the variations in QSOs have been observed to take place with time scales of less than a month (even days), this suggests that the size of the object must be about a *light-day* or 10^{15} cm. (Recall that a light-year is about 10^{18} cm.) It is true that the entire QSO need not participate in the variation, but since variations by a factor of 2 and even more have been observed, it does mean that a very sizeable portion of the total radiation from the object participates in this variation; therefore, the region from which this radiation emanated is restricted by this size limitation.

The small size of the emitting region of QSOs, implied by the luminosity fluctuations, is unacceptable to many scientists. One way to increase the size of this region is to consider an expanding shell of gas which interacts with an external medium, thus producing radiation over a region of much larger scale than implied earlier. Other possibilities include relativistic effects. The only way to overcome this problem of size is to understand the true nature of the quasi-stellar phenomenon; this problem is still unsolved, but some possibilities are appearing on the horizon.

(d)   The Nature of Quasi-Stellar Objects

The restriction on the size of QSOs implies an enormous concentration of energy in a very small region, and it is natural to ask what can produce this energy. Several theories have been proposed, but none has been accepted very widely. They commonly invoke the synchrotron mechanism to produce the radio radiation; this requires a high magnetic field strength to accelerate the electrons. A source of energetic electrons is also needed. In their search for a powerful energy source, some astronomers and physicists have suggested that QSOs are similar to super-novae; either a *chain reaction* occurs with one supernova exploding after another, or else there is an enormously massive supernova which undergoes periodic explosions or outbursts. Several theories consider the QSOs to be massive superstars undergoing gravitational collapse (see Chapter 15). It has also been suggested that some form of interaction between matter and antimatter may be responsible for the QSOs, but none of these ideas has led to a definitive theory which is generally accepted.

One difficulty with the interpretation of QSOs as highly-massive objects is the appearance of forbidden lines in the spectrum. Forbidden lines can only occur when gas densities are very low; this state seems difficult to attain in a highly-massive condensed object. It is interesting to note, however, that the light variations observed in QSOs do not appear to affect these emission lines. This implies that the region responsible for the forbidden emission lines is separate, presumably outside the main object; the picture which comes to mind is a cloud of nebulosity surrounding the core region. In fact, some astronomers argue that the region producing the light variations is only a relatively small part of the total object.

Light and radio variations are not unique to QSOs, for Seyfert galaxies also exhibit radio and optical variability. Seyfert galaxies and QSOs both have bright, highly-condensed nuclei, and both exhibit small regions of radio emission. This latter characteristic applies to about half of the QSOs and to several Seyfert galaxies. These similarities have led some scientists to suggest that both classes of objects are related; the hypothesis advanced is that they may represent similar stages in the evolution of some galaxies, when enormous amounts of energy are released.

The question on which most of the controversy regarding quasi-stellar objects centers is the *origin of the large redshift*. We have seen that extragalactic redshifts are usually interpreted in terms of the expansion of the Universe. Can we apply this same interpretation to QSOs? If so, the recession velocities of those objects with the highest z are approximately $0.88c$, where c is the speed of light! By this interpretation, the QSOs, like galaxies, participate in the expansion of the Universe. This is sometimes referred to as the *cosmological interpretation* of quasi-stellar objects. Strong support for this interpretation comes from the observation of at least one case of a QSO superimposed on a cluster of galaxies and having a redshift identical to one of the member galaxies. The cluster and the QSO must lie at the same distance.

There are several problems related to this cosmological interpretation, however, for if we accept the distance as given by the Hubble law we find that QSOs radiate 10^2 to 10^3 times as much energy as do ordinary galaxies, yet they must be very much smaller according to the light variations! The total energy output does not, however, differ much from that of other radio galaxies. The problem lies in trying to explain that much energy from so small a volume. Such energies must be generated by large masses. Together with the limitations on size, this requires very high densities, much higher than are known for any other galaxies.

Some astronomers have argued that QSOs are not at cosmological distances, but are far more local. A limit on just how near they can be is set by the observed lack of proper motions; since no transverse motions have been observed, they must be at least 10^5 light-years distant. According to the *local interpretation*, the redshifts are still attributed to Doppler shifts but not to the cosmological expansion. The fact that no blueshifts (i.e., velocities of approach) have been observed implies that the center of the originating explosion must have been fairly nearby so that all the objects would now appear to be moving away from us. One suggestion (due to J. Terrell) is that an explosion may have occurred at the center of our own Galaxy, ejecting the QSOs at relativistic speeds; the argument would still be valid if the explosion had occurred sufficiently long ago in some nearby galaxy, so that any objects originally moving toward us have already passed us by and are now moving away from us. The redshifted multiple absorption lines observed in some QSO spectra also suggest the existence of explosive phenomena, if we assume that the lines arise in the source itself. The explosive nature of some galactic nuclei is used as an argument in favor of interpreting QSOs as massive objects ejected from the nuclei of galaxies. H. C. Arp also cites the groupings and configurations of QSOs in close proximity in the sky with peculiar fairly nearby galaxies. There are even cases of apparently interconnecting matter or "bridges." The number of such configurations seems too large to be explained in terms of chance coincidences, yet the redshifts of *these* QSOs greatly exceed those of the "companion" galaxies, in contrast to the case mentioned earlier. With this so-called *local hypothesis*, the energy requirements for QSOs are reduced considerably. While the total energy output of QSOs at cosmological distances is 10^{60} to 10^{61} ergs or even greater, the figure drops to about 10^{56} ergs for the local hypothesis. In both cases, these figures for the total energy output are based upon the total observed radiation and on calculations of the total radiation expected from the synchrotron mechanism. The chief argument against the local hypothesis is the fact that the ejection or explosion itself must have consumed an enormous amount of energy, and there is *no evidence* in our own Galaxy, or indeed in any nearby galaxy, that such an explosion occurred. Enormous amounts of energy would have been required to eject such bodies at the observed speeds. In addition, the energy density and inferred mass of the objects seem unreasonable.

A third suggestion is that a part or all of the redshift is not due to a velocity Doppler shift, but rather is gravitational in nature. A *gravitational redshift* will occur in the presence of a very massive object; from Einstein's theory of general relativity the redshift z is given by the relation $z = GM/Rc^2$. To explain all of the observed redshift in this manner, we find that we need masses greater than $10^{12}\ M_\odot$. Several models have been suggested, based on the idea of gravitational redshift; all of these models encounter difficulties in explaining the other observed phenomena, such as the emission of forbidden lines and the fact that no gravitational perturbations are observed for our own Galaxy or for the planets in our Solar System, as would be the case if the QSOs were within a megaparsec. The possibility that the redshift is actually a *combination* of gravitational and cosmological effects has been discussed but has not found favor, since the multiple-absorption-line velocities become difficult to understand. In this *combination theory*, there is no clear-cut way to determine the distance to a QSO.

(e) A Summary of the Properties of Quasi-Stellar Objects

In summary, we may list the following characteristics which seem to set quasi-stellar sources and quasi-stellar objects apart from other known objects. We must keep in mind, however, that many of these characteristics are shared by some galaxies.

Properties of Quasi-Stellars

1. Star-like objects subtending less than $1''$; identified as strong radio sources in the case of the QSS; some have a faint nebulosity associated with the stellar object. Starred characteristics are those shared with Seyfert galaxies.

*2. Radio diameters are often <0″.001; many are double sources.

*3. Many show variation in the optical continuum.

*4. Some show variation in radio luminosity.

*5. They radiate strongly in the ultraviolet region of the spectrum, as well as in the red and infrared.

*6. Spectra show broad emission lines which include the hydrogen lines, ionized magnesium lines, and some of the nebular lines.

7. A number of QSOs have absorption lines as well as emission lines in their spectra. In some cases, several sets of absorption lines having different redshifts are seen.

8. The spectral lines correspond to *large* redshifts.

9. A few have large and variable polarization.

The discovery and further investigation of these quasi-stellar objects and sources have presented a real challenge to astronomy and physics. At this point we are still far from an understanding of their nature, and indeed they present such a puzzle that some scientists have suggested that we need a *new physics* to explain them. We may have to look deeper into the fundamental behavior of matter and the nature of physical laws in the presence of large gravitational forces.

In closing this section on the quasi-stellars, it seems very appropriate to quote Hamlet:

> And therefore as a stranger give it welcome.
> There are more things in heaven and earth, Horatio,
> than are dreamt of in your philosophy.

Problems

19–1. From Figure 19–6, extract an approximate distance for the separation between our Galaxy and the Magellanic Clouds. What will be the observed apparent magnitude for the following stars in the Magellanic Clouds:
(a) an RR Lyrae variable with a period of 0.5 days,
(b) a classical Cepheid with a pulsation period of 100 days,
(c) a Population II Cepheid with a ten-day period?

19–2. Elliptical galaxies are designated by Ex, where x = $10(a - b)/a$; the quantities a and b are the major and minor axes (respectively) of the apparent elliptical disk of the galaxy. Draw a reasonably accurate picture of the observed shapes of E0, E2, E4, and E7 galaxies.

19–3. Our Galaxy and the Andromeda galaxy (M31) are by far the most massive members of the Local Group of galaxies. If these two giant galaxies form a *binary system*, and move about one another in circular orbits, then calculate:
(a) the distance to the center-of-mass of the system from our Galaxy,
(b) the orbital period.
Perform the same computations for the orbit of M32 about M31.
[Hint: Consult Table 19–2.]

19–4. Which methods of distance determination are most useful for finding the distance to:
(a) the Pleiades,
(b) a globular cluster in our Galaxy,
(c) the Large Magellanic Cloud,

(d) M31, the Andromeda galaxy,

(e) the Virgo cluster of galaxies,

(f) the Hercules cluster of galaxies,

(g) our Sun,

(h) the nucleus of our Galaxy?

19–5.   The special theory of relativity teaches us that no material object can move faster than the speed of light. Equation (19–3) in this chapter is based upon the classical Doppler formula, so that when the redshift z is greater than one, we have $v > c$ (which is impossible!).

(a) Referring back to Chapter 8, derive the *exact relativistic relation* between v (speed) and z (redshift). Express your result in the form $v =$ function of z.

(b) Make a table with three columns: column one for z, where you should enter five exemplary values of z from 0 to 3.0; the second column headed $(v/c)_{classical}$, where you compute the *nonrelativistic* result for your five values of z; and the last column headed $(v/c)_{rel}$, where you use your formula from part (a).

(c) Comment briefly upon your results.

19–6.   If the optical galaxy associated with the radio source Centaurus A is the same size as our Galaxy, draw a scale diagram of the radio and optical emission regions of Centaurus A. Clearly indicate the dimensions and the relative positions of the various components.

19–7.   The radio galaxy Cygnus A has an observed radio *flux density* of 2.18×10^{-23} watt m^{-2} Hz^{-1} at a frequency of 10^3 MHz. (Note that 1 watt $= 10^7$ erg sec^{-1}, and that the unit of bandwidth Δv is 1 Hz.) The observed redshift of the galaxy Cygnus A is $\Delta\lambda/\lambda_o = z = 0.170$.

(a) If the radiation is received at 10^3 MHz, at what (*rest*) frequency was it emitted by Cygnus A?

(b) What is the distance to Cygnus A? (Use a Hubble constant of H $= 50$ km sec^{-1} Mpc^{-1}.)

(c) What is the *radio luminosity* (erg sec^{-1} Hz^{-1}) of this radio source at 10^3 MHz?

(d) To find the total radio luminosity of Cygnus A, we must multiply the result of part (c) by the *bandwidth* Δv of our detector. Assume $\Delta v = 10^4$ Hz, and compute the energy radiated per second at radio frequencies.

(e) What is the minimum mass of hydrogen (in solar masses) that must be converted to helium during each second to provide this luminosity?

(f) If Cygnus A continues to radiate at this rate for 10^8 years, how many solar masses of H must be converted to He? Express this result in terms of the mass of our Galaxy ($\cong 10^{11}\ M_\odot$).

19–8.   3C9 is a quasi-stellar object which has a redshift of $z = 2.0$ and an apparent visual magnitude of $m_v = +18.2$. Answer the following questions *twice*, the first time using the cosmological interpretation of the redshift, and the second time assuming that *half* the redshift is *gravitational*:

(a) What is the speed of recession?

(b) What is the distance to 3C9?

(c) What is the intrinsic luminosity relative to that of our Galaxy?

(d) What is the maximum size of the emitting region if 3C9 exhibits luminosity variations on a time scale of two months?

Reading List

Abell, George O.: "Clustering of Galaxies." *Annual Review of Astronomy and Astrophysics*, Vol. 3, pp. 1–22 (1965).

Baade, W.: *Evolution of Stars and Galaxies.* Cambridge, Massachusetts, Harvard University Press, 1963.

van den Bergh, S.: "Galaxies of the Local Group." Communication from David Dunlap Observatory. No. 195, 1968.

Burbidge, E. M.: "Quasi-Stellar Objects." *Annual Review of Astronomy and Astrophysics*, Vol. 5, pp. 399–452 (1967).

Burbidge, G. R.: "The Nuclei of Galaxies." *Annual Review of Astronomy and Astrophysics*, Vol. 8, pp. 369–460 (1970).

Burbidge, G. R., and Burbidge, E. M.: *Quasi-Stellar Objects.* London, E. H. Freeman, Ltd., 1967.

Douglas, K. N., et al., (eds.): *Quasars and High-Energy Astronomy.* New York, Gordon and Breach Science Publishers, 1969.

Evans, D. E., (ed.): *External Galaxies and Quasi-Stellar Objects.* IAU Symposium No. 44. Dordrecht, Holland, D. Reidel Publishing Co., 1972.

Hodge, Paul W.: "Dwarf Galaxies." *Annual Review of Astronomy and Astrophysics*, Vol. 9, pp. 35–66 (1971).

Hodge, Paul W.: *Galaxies and Cosmology.* New York, McGraw-Hill Book Company, 1966.

Hubble, E.: *The Realm of the Nebulae.* New York, Dover Publications, 1958.

Kahn, F. D., and Palmer, H. P.: *Quasars.* Cambridge, Massachusetts, Harvard University Press, 1967.

King, Ivan R.: "The Dynamics of Galaxies." *Annual Review of Astronomy and Astrophysics*, Vol. 1, pp. 179–194 (1963).

Lequeux, J.: *Structure and Evolution of Galaxies.* New York, Gordon and Breach Science Publishers, 1969.

Maran, S. P., and Cameron, A. G. W., (eds.): *Physics of Non-Thermal Radio Sources.* Washington, D.C., National Aeronautics and Space Administration, 1964.

Moffet, Alan, T.: "The Structure of Radio Galaxies." *Annual Review of Astronomy and Astrophysics*, Vol. 4, pp. 145–170 (1966).

O'Connell, D. J. K., (ed.): *Nuclei of Galaxies.* Amsterdam, North Holland Publishers, 1971.

Peebles, P. J. E.: *Physical Cosmology.* Princeton, Princeton University Press, 1971.

Sandage, A. R.: *The Hubble Atlas of Galaxies.* Washington, D.C., Carnegie Institute, 1961.

Schmidt, Maarten: "Quasistellar Objects." *Annual Review of Astronomy and Astrophysics*, Vol. 7, pp. 527–552 (1969).

Sciama, D. W.: *Modern Cosmology.* Cambridge, Cambridge University Press, 1971.

Shapley, H.: *Galaxies.* Cambridge, Massachusetts, Harvard University Press, 1961.

Woltjer, L., (ed.): *Galaxies and the Universe.* New York, Columbia University Press, 1968.

Chapter 20
Our Universe
and Cosmology

20–1 INTRODUCTION

We have come a long way and have learned a great deal since the beginning of this book. We began our journey of exploration on our home planet, the Earth, whose structure we could sense and probe directly. Standing on the shoulders of a giant, Sir Isaac Newton, we felt our way first to our Moon, and then to the other bodies which comprise our Solar System. By extending our minds and senses with theories and instruments, we probed the very small (atomic structure) and then pushed outward to the very large, our Sun and the stars. The structure and evolution of stars were then found to be intimately connected with the domain of the extremely small (nuclear structure). Looking ever outward, we measured and categorized the content, structure, and motions of the Galaxy in which we reside—the Milky Way Galaxy. The next great leap carried us up beyond our own Galaxy, to the vastness of extragalactic space, wherein billions of other galaxies (alone and in clusters) dwell and evolve. Thus, we arrived finally at the domain of the *expanding Universe*, whose outer limits we will now probe.

By definition, our *Universe* is the sum total of all that which exists or has existed, both in space and in time. In this final chapter, we shall first study the observational characteristics of the Universe. We will then consider both past and present theories of our Universe and its evolution. We end the chapter by sketching the current consensus view of the evolution of the Universe.

20–2 THE FUNDAMENTAL OBSERVATIONAL DATA OF OUR UNIVERSE

(a) The Physical Content of the Universe

This entire book deals with the detailed characteristics of those entities which compose our Universe. On the largest scale, our Universe is filled with numerous *clusters of galaxies*, each of which contains many hundreds or even thousands of galaxies. Each cluster is actually a swarm of galaxies, bound together by mutual gravitational attraction. To the limiting magnitudes obtainable with the largest telescopes, we can "see" (i.e., photograph) about 10^9 galaxies in all! As we shall see later in the chapter, the average separation between visible galaxies is about 1 Mpc; since these large galaxies are about 100 kpc in diameter, the spatial distribution of galaxies is similar (in relative scale) to that of dinner plates scattered at random (except for the clusters) with mean separations of about 5 meters. Remember that there must be a great preponderance of *dwarf galaxies*, since they are vastly more difficult to detect than are the giant elliptical and spiral galaxies.

Since galaxies tend to be found in clusters of galaxies, where their mean separation is on the order of 200 kpc, their spatial distribution is quite *inhomogeneous*. If, however, we average volumes larger than 10^3 Mpc^3, we find that the distribution of clusters of galaxies is very *homogeneous* (i.e., uniform). In fact, the *luminous matter* (i.e., clusters of galaxies) which we see filling our Universe has the following four observed properties: (1) the mean number density of galaxies is uniform out to distances of at least 1000 Mpc—we term this property *spatial homogeneity;* (2) the mean angular distribution of galaxies (i.e., number of galaxies per square degree on the sky) is uniform to the same distance—we say that the distribution is *isotropic;* (3) the mean mass density of this luminous matter is approximately 3×10^{-31} gm cm^{-3}, which is equivalent to 2×10^{-7} hydrogen atoms cm^{-3}; and (4) the clusters of galaxies are moving away from one another in accordance with Hubble's law of cosmic expansion [see Chapter 19 and section 20–2(c)]. Even within the clusters, the orientations of the planes (and therefore the rotation axes) of spiral galaxies are observed to be random. In section 20–3, we shall see that these properties are very useful in helping us formulate theoretical models of our Universe. Some astronomers claim that clusters of galaxies tend to aggregate into even larger groupings, called *superclusters;* since refined statistical analyses do not support this claim, the consensus among astronomers is that superclustering does not exist.

As we discussed in Chapter 19, numerous *radio galaxies* exist. Radio telescopes are so sensitive that these radio sources can be detected and counted to enormous distances. Just as we used *number counts* of stars (Chapter 13) to measure their spatial distribution, so also may we use number counts of galaxies (as we did to verify their spatial homogeneity) and radio sources. In the radio region of the electromagnetic spectrum, this number count technique is extremely simple: we merely count the number (N) of radio sources from which we receive *more than* a given amount of *radio flux density* (S),* and then plot N versus S. If r denotes distance, and

---

* The standard radio astronomical unit of flux density is the *flux unit* (f.u.), where 1 f.u. $= 10^{-26}$ watt m^{-2} Hz^{-1} (see Problem 19–7). Note that the radio flux density received at a distance of 1 km from a standard AM radio station is about 10^{20} f.u.

radio sources are *uniformly* distributed, then (a) N increases as r^3 (since volume goes as $4\pi r^3/3$), and (b) S varies as r^{-2} (the inverse-square law for intensity). Therefore, uniformly distributed radio sources (of approximately the same intrinsic radio luminosity) should follow the relation

$$N \propto S^{-3/2} \qquad\qquad (20\text{--}1)$$

or taking logarithms of both sides

$$\log N = -(3/2) \log S + \text{constant} \qquad\qquad (20\text{--}2)$$

Hence, on a graph of $\log N$ versus $\log S$ (see Figure 20–1), equation (20–2) yields a straight line of slope $= -1.5$. G. G. Pooley and M. Ryle have carried out this procedure down to about 0.1 f.u.; their observations are shown in Figure 20–1. Although the procedure is simple, the results are open to many interpretations! The most reasonable interpretation is one of the following two: (1) the number density of radio sources *increases* at great distances; or (2) the number density remains constant, but there are *brighter* (i.e., less evolved) sources at great distances. Since 0.1 f.u. is believed to correspond to a redshift of $z \cong 3$ to 4, or a distance near 2000 Mpc, these data unequivocally imply that *something* happens at great distances.

Let us now discuss the physical content of *intergalactic space* (i.e., the space between the galaxies). Progress in probing these dark regions has been slow and spotty; many fascinating observations have been made, but their interpretation is either nonexistent or still speculative. Let us comment briefly upon the observations; the data (and conjectures) are listed in Table 20–1.

As we shall see in section 20–3, the nature of the observed expansion of the Universe places an *upper limit* to the average mass density (ρ) of the Universe: $\rho \lesssim 10^{-28}$ gm cm^{-3}. Since we shall be discussing *massless* entities (electromagnetic fields, photons, neutrinos, and gravitational waves) later in the chapter, we must

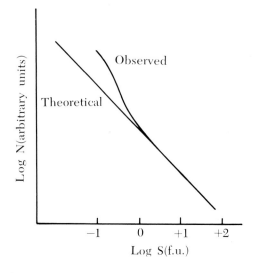

Figure 20–1. *The Log N-Log S Plot for Radio Sources.* N is the number of radio sources with a flux density greater than S. The units of S are 1 flux unit $= 10^{-26}$ watt m^{-2} Hz^{-1}. The theoretical (lower) line is based on a uniform distribution, so that $N \propto S^{-3/2}$, and the line has a slope of $-3/2$. The observations (upper curve) indicate evolution of great distances.

use Einstein's relation between energy and mass (i.e., $E = mc^2$) to convert energy densities u (erg cm^{-3}) to equivalent mass densities ρ:

$$\rho = u/c^2 \qquad\qquad (20\text{–}3)$$

where c is the speed of light.

In 1965, A. A. Penzias and R. W. Wilson discovered an isotropic flux of background radio radiation, which was apparently of extragalactic origin. Further investigations have shown that this (electromagnetic) radiation corresponds to *blackbody radiation* (see Chapter 8) at 2.7 K, and that it is isotropic to better than 0.1%. Evolutionary cosmologists believe that this radiation is a relic of the *primordial fireball* which existed at the beginning of our Universe; indeed, this is the only reasonable explanation which has been advanced to date. Since this blackbody radiation has an equivalent mass density of $\rho = 5 \times 10^{-34}$ gm cm^{-3}, it is not important to the dynamical evolution of the Universe today. Nevertheless, we shall see (section 20–4) that it dominated our Universe in the distant past.

The content of *hydrogen gas* (neutral = H I, or ionized = H II) in intergalactic space is extremely uncertain. Studies of 21-cm emission and Lyman α absorption show apparently conflicting results, but we can be certain that the present mass density of H I is less than 10^{-31} gm cm^{-3}. The first real hint of hot (10^7 K) intergalactic gas has come from recent satellite observations of X-ray emission from several clusters of galaxies. These X-ray sources are large and diffuse, encompassing the clusters entirely.

Intergalactic dust grains may exist to a mass density of around 10^{-28} gm cm^{-3}, and still be spread so thinly that we cannot detect them. It is not clear where and how intergalactic dust grains would originate. If present, these dust grains would imply a general intergalactic obscuration, as well as polarization (if an intergalactic magnetic field exists) and line absorption phenomena. Since these effects are *not* observed, there are two possible alternatives: (1) the dust is not present at all, and (2) the dust is concentrated into small dark clouds! Only future observations can resolve this uncertainty.

Other possible constituents of the intergalactic medium include dilute starlight (from the galaxies), cosmic rays (from supernovae and galactic-core explosions?), neutrinos, and gravitational waves. Of these, we are only certain that

Table 20–1.   The Physical Content of the Universe

Constituent	Equivalent Mass Density (gm cm^{-3})	
Luminous matter = galaxies	3×10^{-31}	
2.7 K background radiation	5×10^{-34}	
Intergalactic magnetic field	$\lesssim 10^{-40}$	
Neutral hydrogen (H I)	? $(<10^{-31})$	
Ionized hydrogen (H II)	?	
Dust grains	?	
Dilute starlight	?	(all $<10^{-28}$)
Cosmic rays	?	
Neutrinos	?	
Gravitational waves	?	

the starlight and *some* cosmic rays exist in intergalactic space. The *starlight* must exist because we can *see* external galaxies, but its equivalent mass density is negligible. The subject of *intergalactic cosmic rays* is still totally open ended, since we know only that those cosmic rays observed with energies in excess of 10^{18} eV must come from outside our Galaxy. The galactic magnetic field is too weak to contain such energetic cosmic rays; even if they were generated in the Galaxy, they would immediately escape.

Both neutrinos and gravitational waves are massless phenomena which are fantastically difficult to detect. Nevertheless, *neutrinos* have recently been produced and detected in the laboratory, and they are being measured as they come to us from the Sun; but no extragalactic neutrinos have ever been identified. Around 1970, J. Weber claimed to have detected *gravitational waves* and verified that they are coming to us from the direction of the nucleus of our Galaxy. If his measurements are even approximately correct, gravitational radiation may provide the dominant form of mass-energy density in our Universe today!

(b)   The Size and Distance Scale

In Chapter 19, we described the bootstrap process by which increasingly larger astronomical *distances* are measured. Once we know the angle subtended by a given object, and the distance to it, we can immediately find the *size* or extent of that object. In this subsection, we will attempt to give the reader some perspective on the typical dimensions found in our astronomical Universe; to aid us in this task, we have constructed Figure 20–2, which expresses these dimensions on a logarithmic scale.

We already have an intuitive feeling for how large the Earth is compared to a man, and for the relative dimensions of our Solar System (Chapter 5). Let us, therefore, begin with the stars. If stars are imagined to be the size of cherries, then we must place these cherries (10^{11} of them!) an average of 1 km apart to mimic the mean stellar distribution in our Galaxy. On this scale, our Galaxy would be about 100,000 km in diameter. Let us now shrink our Galaxy to the size of a flat doorknob (about 4 cm across). The Andromeda galaxy (M31) is a similar doorknob about 1 m away; this is essentially the size of the Local Group of galaxies. In addition, this scale represents the mean distribution of galaxies in our Universe, except that in clusters of galaxies there may be thousands of these doorknobs about 20 cm apart in a volume of diameter 10 m or so. On this same scale, the most distant optical galaxy is 1 km away, while the most distant quasar (on the cosmological interpretation) is about 2 km away. The greatest distance from which light can reach us now in our Universe (sometimes called "the edge") would be about 3 km away, since such light has been traveling to us for essentially the lifetime of the Universe. Therefore, we can begin to appreciate the fact that interplanetary and interstellar space are incredibly empty, while the extragalactic regions are fairly densely populated with galaxies. It is small wonder that we frequently have difficulty comprehending cosmic dimensions, when they are a factor of 10^{26} larger than we are.

(c)   Cosmic Expansion and the Age of the Universe

In Chapter 19, we discussed Hubble's discovery that all distant galaxies appear to be moving away from us with recession speeds (v) directly proportional

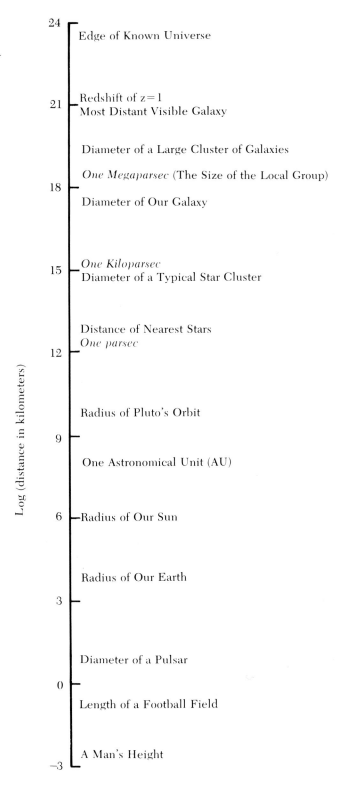

Figure 20–2. *Characteristic Dimensions in Our Universe.* Note the logarithmic scale; for every *unit* on this scale we multiply distances by a factor of 10! For example, 13 means 10^{13} km = 1 pc.

to their distances (r)—Hubble's law:

$$v = Hr \qquad\qquad\qquad (20\text{–}4)$$

where Hubble's constant is $H \cong 50$ km sec^{-1} Mpc^{-1}.* A more correct statement of this effect is that all *clusters* of galaxies recede in accordance with equation (20–4); since most individual galaxies are gravitationally bound to the clusters, these galaxies are not free to move apart at will—only the clusters can move apart in accordance with Hubble's law. Therefore, the entire Universe of clusters is expanding; this is why we frequently refer to the *expanding Universe*. Once we have obtained the recession speed of a galaxy (or cluster) by measuring its redshift (see Chapter 19), we can immediately find its distance from equation (20–4); hence, the distance scale of the Universe must be related to Hubble's constant H. The simplest way to obtain the approximate *size of the Universe* is to compute r [from equation (20–4)] when $v = c$; that is, the farthest galaxy is moving at the speed of light! Since $H^{-1} \cong 2 \times 10^{10}$ years, we quickly obtain $r_{\text{edge}} = cH^{-1} \cong 2 \times 10^{28}$ cm (note that this is the number we use in Figure 20–2).

It should be apparent that H^{-1} is related to the *age of our Universe*, since it has the dimension of years. To better understand this relationship, let us consider Figure 20–3. Here we have plotted the distance (r) to an arbitrary galaxy as a function of time (t). If the Universe has always expanded at a constant rate (i.e., $\tau = 1/H$), then equation (20–4) tells us that H^{-1} years ago $r = 0$; therefore, all the galaxies (and the Universe itself) must have originated in a titanic explosion about 20 billion years ago! On the other hand, every bit of matter (and energy) in the Universe gravitationally attracts every other bit; therefore, the expansion could not be free-coasting (the solid line in Figure 20–3), but rather there must be a *deceleration* (represented by the dashed curve). To better illustrate this point, consider a rock thrown vertically upward from the Earth's surface. This rock would move at a constant speed if there were *no* gravitational attraction from the Earth;

---

* In the discussions that follow we shall use $H = 50$ km sec^{-1} Mpc^{-1} whenever we need a specific value. The reader is cautioned, however, that current estimates of H range from 50 to over 100 km sec^{-1} Mpc^{-1}.

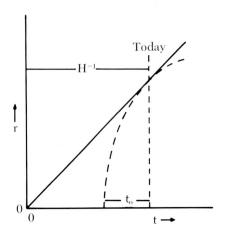

Figure 20–3. *The Age of the Universe.* This diagram illustrates the distance (r) to a certain galaxy as a function of time (t). If the galaxy has always moved at a *constant* speed (solid line), we have $r = 0$ at a time $H^{-1} \cong 2 \times 10^{10}$ years in the past; the gravitational self-attraction of the matter in the Universe *decelerates* the expansion (dashed curve), leading to an age $t_0 < H^{-1}$.

but we are well aware that the rock slows down and eventually falls back to the Earth. If we could give the rock parabolic or hyperbolic velocities (see Chapter 3), it would never return to the Earth; but it would slow down nonetheless. Hence, the actual age of the Universe must be $t_0 < H^{-1}$.

Since H^{-1} is apparently a good order-of-magnitude estimate of the age of the Universe (at least an *upper limit*), we should check whether all of the content of the Universe is less than 2×10^{10} years old. Using lead-uranium (Pb-U) dating, the greatest age on the Earth is found to be about 3.5×10^9 years. It is quite possible that erosion and internal (plutonic) processes have erased all traces of the primordial Earth's surface, so a better place to look is the dead surface of the Moon. One rock in the Apollo 12 lunar surface samples has been found to have an age of 4.6×10^9 years. This finding corroborates the dating of meteorites, the plurality of which have ages of about 4.6×10^9 years; apparently this figure represents the age of our Solar System. To find older objects, we must look to the stars. In particular, the ages of globular clusters may be deduced from theoretical considerations by using the main-sequence turnoff method (see Chapter 15). This is a rather imprecise method that indicates an age for the oldest stars of about 15×10^9 years with an uncertainty of 5×10^9 years. When we consider the factor-of-2 uncertainty in the value of H, it is clear that H^{-1} is a reasonable estimate of the age of our Universe.

20–3 COSMOLOGY: THEORIES OF THE UNIVERSE

(a) The History of Cosmology

Cosmology means literally "the science of the cosmos or Universe." We will not consider experimental cosmology here, for we have already discussed the observational data; rather let us concentrate upon the *theoretical* concepts, models, and theories of our Universe.

The ancient Hindus pictured their world as the shell of an immense tortoise, which stood upon the backs of four elephants. The early Greeks and other Eurasian peoples envisioned the Earth as a large island, set in the midst of a vast ocean; this ocean (or river, in some cases) was surrounded by either fire or regions of eternal cold. Later peoples spoke of the "bowl of the sky" or the "heavenly dome," which spread over the Earth and upon which the "fiery stars" or "eternal lights" were fixed. The Pythagoreans introduced the concept of nested polyhedra, upon which medieval scholars later based their image of the crystalline spheres. In this model, the Earth lies at the center surrounded by "crystalline spheres," placed concentric to the Earth at harmonious distances; each planet was attached to its own sphere, and the stars were set in the immobile outermost regions. As the spheres turned and the planets moved, there issued forth the heavenly "music of the spheres." Ptolemy's epicycle system eventually gave way to Copernicus' heliocentric theory of our Solar System, and this led to the development of celestial mechanics, from the time of Galileo and Newton to the end of the 19th century.

Attempts to understand the Universe in terms of Newton's theories of dynamics and gravitation met with little apparent success. To Newton and his successors, the Universe was characterized by absolute (immutable) space and time; their predilection was to assume that the Universe must be infinite, static, and unchanging. But the Newtonian gravitation theory implied that the Universe must either expand or collapse*; this was one reason for rejecting the theory. In 1826, H. W. M. Olbers recognized a physical phenomenon which utterly destroyed the static and infinite Newtonian cosmology. Olbers assumed that we have an infinite, static Universe filled with a uniform distribution of stars; he then proved that the night sky should be at least as bright as the surface of the Sun! Since the night sky is obviously dark, we have *Olbers' Paradox*. The proof is completely straightforward: (a) in an infinite Universe of stars, if we look in any direction, our line-of-sight will inevitably encounter the surface of some star (the *surface brightness* of the star, which is the energy flux per unit solid angle received at the Earth, does not depend upon that star's distance from us); or, alternatively, (b) even if interstellar space is filled with obscuring dust, so that we cannot see beyond a certain distance, the dust itself will heat up to the surface temperature of the stars (i.e., thermal equilibrium) and the sky will be just as bright. The only way out of Olbers' Paradox in the Newtonian theory is to assume a finite Universe, with ourselves situated approximately at the center, but this was philosophically unpalatable to the Newtonian cosmologists. Today we realize that the expansion of the Universe resolves Olbers' Paradox, since the light from increasingly distant galaxies is Doppler redshifted to increasingly lower energies and, hence, intensities.

It was only with the appearance of Einstein's special (1905) and general (1915) theories of relativity that theoretical cosmology became a true science. In the special theory, Einstein united space and time into an indivisible four-dimensional space-time continuum, and deduced the correct kinematical theory for all objects—even when they moved with a relative speed $v \lesssim c$. The general theory of relativity is an extension to include all gravitational phenomena; hence, it is a *theory of gravitation*. In this theory, the geometry of space-time is *curved*, and the trajectories of objects are determined by this curvature; the curvature, in turn, is produced by all of the material content of the Universe. We shall consider this theory in more detail in the next subsection.

Many other fascinating cosmological theories have been advanced, but those which have stood up under the observational data are either developments within Einstein's theory or extensions of it. Only two full-fledged alternative theories remain besides Einstein's; these are (a) the Steady-State theory of Bondi and Hoyle, and (b) the Scalar-Tensor theory of Brans and Dicke. Neither of these theories is presently faring very well, but we cannot yet reject them without further studies.

(b) The Current Theories

The theories of cosmology currently in contention are all relativistic in nature. For simplicity (*Occam's Razor* is the term used to denote elegance or simplicity in

---

*We know now that our Universe is expanding. Newton and his followers might have formulated a reasonable theory of an expanding Universe over 300 years ago, if they had believed strongly enough in their own theory!

a theory), each of these theories makes an assumption concerning the symmetry of the Universe; these assumptions are collectively referred to as the *cosmological principle*. This principle, in both Einstein's theory and the Brans-Dicke (Scalar-Tensor) theory, consists of these statements: (a) at a given cosmic time, every fundamental observer measures the same thing—*homogeneity*, (b) at a given cosmic time, every fundamental observer perceives the same view in every direction—*isotropy*, and (c) observables depend only on time itself. In the Bondi-Hoyle (Steady-State) theory, this principle is carried one step further; in addition to (a) and (b) above, all observables are constant in cosmic time (that is, independent of the time at which they are observed)—this we call the *perfect cosmological principle*.

(i) *The Bondi-Hoyle Steady-State Theory*

In 1952, H. Bondi and F. Hoyle announced their Steady-State theory of cosmology. This theory recognizes the Hubble expansion of the Universe (as it must!), but it also advocates that the *average* properties of the Universe are *stationary* (i.e., unchanging) in both space and time; this idea is called the *perfect cosmological principle*. The consequences of this one assumption are curious and far reaching. For instance, since the galaxies are moving away from one another, new matter must be *created out of nothing* to replace the old; this new matter (preferably hydrogen atoms) must also participate in the universal expansion. Although the principle of energy conservation is violated by this continual creation of matter, the violation is small, since the rate of creation is only 10^{-35} gm cm^{-3} sec^{-1} (or about one hydrogen atom per cubic centimeter every 10^4 years). Fortunately (for energy conservation), the theory makes other predictions which are apparently not borne out by the observations. Among these are: (a) the expansion of the Universe is *accelerating*; the observations of A. Sandage show that it is actually decelerating, (b) a log N-log S curve is predicted which does not agree at all with the observed relation for radio sources (see section 20–2) and (c) there seems to be no way to produce the observed 2.7 K background radiation in this model. Taken together, these results strongly imply that we should reject the Steady-State theory.

(ii) *The Brans-Dicke Scalar-Tensor Theory*

During the past decade, C. Brans and R. H. Dicke have advocated a variant of Einstein's theory; this is known as the Scalar-Tensor theory. In Einstein's theory, gravitation is represented by a tensor field*; in addition, Brans and Dicke have appended a small amount of scalar field. The main consequence is to radically alter the theoretical evolution of cosmological models, but other more readily testable consequences also appear. Since this theory is such a small variation on Einstein's theory, very accurate measurements are necessary to either prove or disprove it. The theory already disagrees with the observed orbital motions of the terrestrial planets (which Einstein's theory predicts correctly), and other experimental tests are currently underway. Although we will reserve judgment on the eventual fate of the Scalar-Tensor theory, let us leave it now for more fertile ground.

---

* The reader need not worry about the concept of scalar and tensor fields. Only the fact that they are different is important.

(iii)   *Einstein's Theory of General Relativity*

In 1905, Einstein had the insight to realize that space and time must be considered together as one harmonious entity; with his general theory of relativity in 1915, he replaced gravitation by a coupling between this geometrical space-time and the material content of the Universe. His initial predictions from the theory concerned the warped geometry produced by the mass of our Sun; these are the famous *perihelion precession* and *bending-of-light* effects (see Figure 20–4). The perihelion of Mercury's orbit is observed to precess (i.e., rotate) prograde by an amount of $43''$ century^{-1}; Einstein predicted a prograde precession rate of $43''.16$ century^{-1}. Starlight passing the limb of the Sun (observable during total solar eclipses) is expected to be deflected toward the Sun by $1''.75$; the observations agree with this prediction to within 15 per cent. Several other general relativistic effects have been verified to within 10 per cent; therefore, we seem to have a reasonable theory of gravitation, which can now be extrapolated to the cosmological domain.

Einstein's general theory of relativity yields equations of the form

$$\text{(space-time curvature)} \propto \text{(mass-energy density)}$$

Therefore, cosmological models may be constructed by making reasonable estimates of the type and amount of material content of the Universe; the equations then yield the *evolutionary* behavior of the model. Because he believed that the Universe should be static, Einstein's first cosmological model was a *faux pas;* he introduced

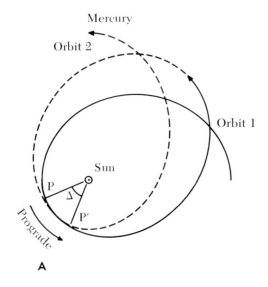

A

B

Figure 20–4. *Famous Tests of General Relativity.* **A,** General relativity causes the perihelion of Mercury's orbit to precess from P to P' (prograde) by $\Delta = 43''$ century^{-1}. **B,** The Sun's gravitational field deflects starlight (at the limb) by $\delta = 1''.75$.

complicating *ad hoc* assumptions into his equations to produce the type of Universe he wanted, and thereby failed to *predict* the expansion of the Universe. W. de Sitter (1917), G. E. LeMaitre (1931), and A. S. Eddington (1930) found other interesting, but nonphysical, cosmological models; de Sitter's model is of some interest because it contains *no* material at all—we therefore have an empty Universe which evolves!

The great turning point in relativistic cosmology came when A. Friedmann (1922, 1924), H. P. Robertson (1928, 1929), and A. G. Walker (1936) independently discovered the classic cosmological models designated by their names. Their assumptions were simple, and were based upon solid observational evidence:

(1)   There exists a "cosmic time" for every "observer" (i.e., galaxy).

(2)   At a given cosmic time all observers see the same thing (i.e., spatial homogeneity).

(3)   The Universe is *homogeneous* and *isotropic* everywhere.

(4)   The material content of the Universe is uniform "dust," where each dust mote is a cluster of galaxies.

Three generic types of *expanding* Universes follow from these simple assumptions and the unadulterated Einstein equations. We term these types the *closed, flat,* and *open* models, respectively. Each model is characterized by a function $R(t)$, which represents the "radius" of the Universe; since t is "cosmic time," the models evolve. In Figure 20–5 we display this evolutionary behavior.

All three models are presently *expanding* (t_0), in agreement with Hubble's observations in the late 1920s. They all suffer *deceleration*, with the *closed* model decelerating fastest and the *open* model decelerating the least; Sandage's observations do not yet permit us to single out any one model over the rest. All three models emerge (at $t = 0$) from an initial *creation*, where $R = 0$; since the volume of the Universe goes as R^3, the entire material content was crushed to *infinite* density (and therefore had a very high, perhaps infinite temperature) at the creation! This

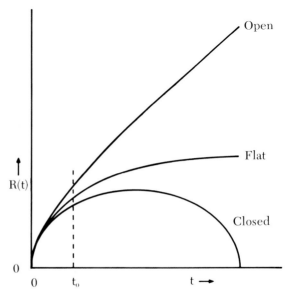

Figure 20–5.   *The Homogeneous and Isotropic Models of the Universe.* These general relativistic models were discovered by Friedmann, Robertson, and Walker. $R(t)$ is the "radius" of the Universe, and it evolves in time t. Note that all models emerge from *creation* at $t = 0$, but only the *closed* model collapses again to a point after a finite time. The present is approximately indicated by t_0.

high-density phase is believed to be the *primordial fireball* which produced the observed 2.7 K background radiation. Since the slope of each curve represents the Hubble rate H, we see that the present epoch (t_0) is greatest for the *open* model and smallest for the *closed* model. Note that only the *closed* model reaches a *maximum* size, and then re-collapses to zero volume in a finite time.

There is an extremely useful analogy, which should help us to understand these cosmological models better. Let us neglect air drag, and throw a stone into the sky. If we do not give the stone enough kinetic energy, it will fall back to the Earth. The *closed* cosmological model behaves similarly, for the density is sufficiently high so that gravitational deceleration eventually halts the expansion. If we give the stone *escape speed*, it moves in a parabolic orbit which carries it to infinity, where it comes to rest; the *flat* model is the analog here. Finally, with an excess of kinetic energy the stone will move in a hyperbolic orbit to infinity; this picture concurs with the *open* model. Let us now discuss each model in somewhat more detail.

The *flat* model marks the critical point between the *open* and *closed* models. Its properties are quite simple. The model evolves with $R(t) \propto t^{2/3}$; therefore, the Hubble rate goes as $H = 2/3t$. If we denote the present by the subscript 0, we find:

$$\left.\begin{array}{l} \text{Age of Universe} \equiv t_0 = (2/3)H_0{}^{-1} \cong 13 \times 10^9 \text{ yr} \\[2em] \text{Mass Density} \equiv \rho_0 \cong 10^{-29} \text{ gm cm}^{-3} \end{array}\right\} \quad \textbf{(20–5)}$$

Note that the critical values listed in equation (20–5) are uncertain by about a factor of 2, since H_0 is uncertain by that amount. We see that t_0 is in reasonable agreement with the observed ages of objects in our Universe; ρ_0 is the *critical mass density* which determines whether our Universe is *open* or *closed*, and it is the number responsible for the current frantic search to find the "missing mass density" in intergalactic space.

The *open* model is also simple in its consequences. We find that it has:

$$\left.\begin{array}{l} (2/3)H_0{}^{-1} \lesssim t_0 \lesssim H_0{}^{-1} \\[2em] \rho_0 \lesssim 10^{-29} \text{ gm cm}^{-3} \end{array}\right\} \quad \textbf{(20–6)}$$

It is clear that t_0 goes to its maximum value (i.e., 2×10^{10} yr), as $\rho_0 \to 0$. Since the age of the Universe is somewhat greater in this model, and since ρ_0 can be matched to the mass density of the observed *luminous* matter, many scientists are inclined to accept this model.

The *closed* model is the most complicated, and therefore the most interesting. As expected, its properties are given by:

$$\left.\begin{array}{l} t_0 \lesssim (2/3)H_0{}^{-1} \\[2em] \rho_0 \gtrsim 10^{-29} \text{ gm cm}^{-3} \end{array}\right\} \quad \textbf{(20–7)}$$

The value of ρ_0 cannot exceed 10^{-27} gm cm^{-3}, and is probably less than 10^{-28} gm cm^{-3}; otherwise, the age of the Universe becomes less than the age of our Earth!

To many people, it is philosophically satisfying that this model stops expanding and recontracts; this implies a finite lifetime for the Universe, and permits the possibility of a rebirth after $R \to 0$. Hence, the Universe may be *cyclic*, pulsating from one collapsed state to the next along repeating, identical curves. Note that the recollapse of the *closed* model implies that galaxies will approach us and appear *blueshifted* after we pass the maximum of expansion.

In recent years, many other fascinating cosmological models have been constructed using Einstein's theory. *Anisotropy* (i.e., things look different in different directions) and *magnetic fields* have been considered, but the observed isotropy of the 2.7 K background radiation (deviations from isotropy as less than 0.1%) implies that these effects will be practically negligible. Currently, realistic *rotating* cosmologies are being constructed, but again the background radiation implies that only an extremely small amount of cosmic rotation is permissible. Still in their formative stages are *inhomogeneous* or *nonuniform* cosmological models; if they are to be physical, these models must be extremely complicated. Some interesting effects may arise when these "clumpy" models are developed further.

20–4   THE EVOLUTION OF OUR UNIVERSE

In all reasonable cosmological models based upon Einstein's general theory of relativity, it is known that an initial *creation* event must have occurred about 2×10^{10} years before the present. Theoretical cosmologists refer to this event as the *initial singularity*, because at this point the volume of the Universe goes to zero while the density of mass-energy becomes infinite. From the equations which govern the *open*, *flat*, and *closed* models discussed earlier, and from Figure 20–5, we find that all three models behave in precisely the same manner *near* this singularity. As we move far from the singularity, the evolution of the three models changes, since each model is characterized by a different *spatial curvature*; as we look out at the Universe today, these curvature effects become observable only at redshifts of $z \gtrsim 1$. At present, we have insufficient observational data at these large distances to unambiguously select a single model for our Universe.

For simplicity, therefore, let us consider the *flat* cosmological model of our Universe, and follow the astrophysical consequences of its evolution from creation to the present epoch. This exemplary evolution will illustrate all of the important effects occurring in homogeneous, isotropic, cosmological models; at the end of this section we will speculate upon the future evolution of our Universe.

(a)   In the Beginning . . .

Our Universe emerged from a state of infinite density and exceedingly high temperature about 2×10^{10} years ago. We might envisage this initial state as the "primordial atom" of G. E. LeMaitre and G. Gamov, but at these temperatures all matter behaves like photons (since each particle moves at essentially the speed of light); hence, the initial state is actually a chaotic, gaseous holocaust or inferno of high-energy elementary particles and photons! For this reason, we speak of it as the *primordial fireball*. An abundance of relativistic particles and antiparticles exists in equilibrium at these temperatures. As the Universe expands, the temperature drops, and the heavier particles (hyperons and mesons) annihilate and decay into the less-massive "stable" particles (protons, neutrons, electrons, and neutrinos).* About *one second* after the creation, the temperature has dropped to 10^{10} K, and all that remains are these stable particles and an overwhelming number of photons.

(b)   The Radiation Era and Cosmic Nucleosynthesis

For $T \lesssim 10^{10}$ K, the Universe is dominated by photons—the *photon gas*. Since thermal equilibrium prevails, these photons are characterized by the Planck *blackbody* spectral energy distribution (see Chapter 8); it is this photon gas which eventually cools down to the observed 2.7 K cosmic blackbody radiation! It turns out that, in this model, *lengths* scale as $R(t)$, so that the wavelength λ of a photon is proportional to $R(t)$. Therefore, the energy of the photon goes as $h\nu \propto R^{-1}$, and we find that the Planck distribution maintains its *form* with the temperature dependence $T \propto R^{-1}$. In Chapter 8, we saw that the energy density (erg cm^{-3}) of blackbody radiation is proportional to T^4; converting to the equivalent mass density ρ_r, we find

$$\rho_r \propto R^{-4} \tag{20-8}$$

where the subscript r denotes radiation. From the Einstein equations we then find $R(t) \propto t^{1/2}$, and thence

$$T \cong 10^{10} t^{-1/2} \text{ K} \tag{20-9}$$

Since the mass density ρ_m of massive particles (i.e., hydrogen atoms or clusters of galaxies!) is inversely proportional to the volume in which they are contained, we have

$$\rho_m \propto R^{-3} \tag{20-10}$$

We may now demonstrate that the radiation will dominate the matter near the initial singularity. Clearly the mass density of the radiation increases faster than that of the matter as we go to smaller values of R. Using equations (20–8) and (20–10), and the present values $\rho_{m0} \cong 2 \times 10^{-29}$ gm cm^{-3} and $\rho_{r0} \cong 5 \times 10^{-34}$ gm cm^{-3}, we immediately find that $\rho_r \gtrsim \rho_m$ when $R \lesssim (3 \times 10^{-5})R_0$. This implies that the radiation was dominant until about 2000 years after the creation (we call this period the *radiation era*), when the matter took over.

Near the beginning of the radiation era, when the temperature was about $T \cong 10^9 - 10^8$ K, *cosmic nucleosynthesis* took place. Before this epoch, composite nuclei could not exist, for they would immediately dissociate into their constituent nucleons owing to the ferocity of the radiation. At $T \cong 10^9$ K, nucleosynthesis

---

* Of these particles, only the neutron decays (by β-decay; $n^0 \rightarrow {}_1H^1 + e^- + \bar{\nu}$), with a half-life of about 10^3 sec. This half-life is long compared to the characteristic expansion time of the Universe at this stage.

(i.e., the buildup of nuclei) began with the production of deuterium via the thermonuclear reaction

$$_1H^1 + n \rightarrow {_1H^2} + \gamma$$

The normal proton-proton reaction that occurs in stellar interiors (see Chapter 15) is far too slow to have been operative at this time. Moreover, there was an abundance of neutrons. The presence of deuterium then leads to other, even faster reactions in which deuterons combine to produce helium.

The decay of the original neutrons provided additional protons for this reaction, and at the end of about *30 minutes* the nucleosynthesis was complete. The end product of this phase of cosmic evolution was hydrogen ($_1H^1$), deuterium ($_1H^2$) and helium ($_2He^4$), in the relative *mass* fractions $0.75:0.01:0.25$; *no* heavier elements were formed at this time! Since helium is extremely difficult to destroy, we may test this prediction of the theory by observing the *minimum* mass fraction of helium in old (i.e., Population II) objects. Many observations to date appear to be consistent with this prediction, but there remain many ambiguities and unsolved problems.

(c)   **The Matter Era and Galaxy Formation**

After $t \cong 2000$ yr, when matter (i.e., hydrogen and helium) began to dominate the Universe and radiation became a secondary constituent, we had the *matter era*. The evolution of this phase is governed by the standard *flat* model relation $R \propto t^{2/3}$. The matter existed in *ionized* form until the temperature dropped to the point where recombination to the neutral form could predominate; the helium recombined to He I before the hydrogen finally predominated as H I at $T \cong 3000$ K (i.e., $t \cong 10^6$ years). At this stage, the matter and radiation decoupled, and both evolved independently thereafter. The radiation cooled to become the 2.7 K cosmic blackbody radiation observed today, but the gaseous matter underwent a critically important transformation.

Prior to the decoupling point, the radiation had maintained a reasonably uniform distribution of ionized matter. As the plasma recombined, very small inhomogeneities in the distribution of mass density were freed from the iron grip of the radiation, and this subtle clumpiness began to condense (as a result of self-gravitation) into discrete clouds of gas—these were the *protogalaxies*! In a manner analogous to the collapse of a dust-and-gas cloud to form a star, each protogalaxy evolved into a *galaxy*. Therefore, the decoupling point set the stage for *galaxy formation*. There is a great deal of uncertainty concerning the details of galaxy formation; in fact, this subject is currently a very active and exciting branch of astrophysics. J. H. Jeans first analyzed the gravitational instability of gas clouds; the minimum size of a cloud which will collapse is called the *Jeans length*. Today we know only that the protogalaxies (probably a misnomer) had a mass in the range 10^5 to 10^{15} $M_\odot$. The lower limit is the mass of a typical globular cluster, while the upper limit corresponds to a typical cluster of galaxies. These objects probably formed within 10^9 years of the creation, at redshifts in the range $3-30$. Astronomers are actively searching for the protogalaxies, but we cannot say much more on this subject until extensive theoretical and observational investigations have been carried to fruition.

(d) The Present State and the Probable Future

It is evident that the protogalaxies condensed into stars in the galaxies which we observe today. The present state of the Universe has been considered in detail throughout this book, so we need not repeat it here. To summarize the preceding discussion, and to bring us up to the present time, we have constructed Figure 20–6; therein we depict the characteristic stages of the evolution of our Universe, from the beginning to now.

Until we can determine which cosmological model most closely represents our Universe, the *future of cosmic evolution* is uncertain. Both the *open* and the *flat* models imply similar behavior, with continued expansion forever; stars and galaxies evolve and age, eventually burning out so that the Universe becomes a dark, dead entity filled with cold cinders of matter separated by enormous (eventually infinite) distances. On the other hand, if our Universe is *closed*, then a fascinating future lies in store for it. The Universe will reach a maximum size, and will begin to collapse to the second singularity (see Figure 20–5). Past the maximum, nearby galaxies begin to *approach* our Galaxy, and (if human observers still exist) *blueshifts* will be observed. Although stars and galaxies continue to evolve and age, their radiation is shifted to higher and higher frequencies as the Universe collapses faster and faster. Eventually, the galaxies will be brought together into a single gigantic collection of stars, and the sky will begin to brighten. Radiation will once again predominate, and matter will be dissociated to its most elementary forms; the final collapse to the second singularity will occur much as did the initial explosion from creation, but now time appears to be running backward! We can only speculate

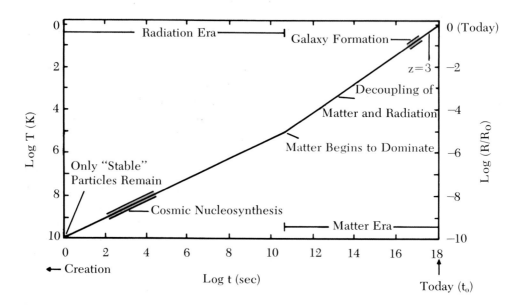

Figure 20–6. *The Probable Evolution of Our Universe.* The evolution of our Universe is represented schematically on this graph of size (R) and temperature (T) versus cosmic time (t). Characteristic events which take place during cosmic evolution are shown. The subscript 0 refers to the present; note the logarithmic scales.

about the evolution *after* this second singularity. Perhaps the Universe will "bounce" and a new Universe will be born; but perhaps time itself will end at the second singularity, so that we cannot even ask about the period "after."

Problems

20–1.   Assume that *dust grains*, whose characteristic size is 1 micron, are uniformly distributed throughout intergalactic space at a mean mass density of 10^{-29} gm cm^{-3} (the *critical* mass density).
(a) What is the number density (number cm^{-3}) of this dust, and what is the average separation between the grains?
(b) Compare your answers to (a) with the number density and separation of the interstellar dust grains in our Galaxy.
(c) Show that this hypothetical intergalactic dust will drastically *redden* the stars observed in the Andromeda galaxy (such reddening is *not* observed in practice).

20–2.   If intergalactic space is filled with H II at a temperature of 10^6 K (i.e., a *plasma*):
(a) What are the mean speed and mean kinetic energy (per particle) of these protons?
(b) To what wavelength of electromagnetic radiation does this individual kinetic energy correspond? Could such radiation be detected from the surface of the Earth?

20–3.   What is the approximate *volume* of our Galaxy (express your answer in kpc^3)? By what scale factor must the dimensions of our Universe shrink if there is to be no empty space between the galaxies? Is this stage of cosmic expansion a reasonable time for *galaxy formation?*

20–4.   Planck's law for the intensity of *blackbody radiation* (see Chapter 8) may be written as

$$I_\lambda = (2hc^2/\lambda^5)(e^{hc/\lambda kT} - 1)^{-1}$$

As the Universe expands with a scale factor (i.e., "radius") $R(t)$, the intensity varies as $I_\lambda \propto R^{-5}$, while the wavelength goes as $\lambda \propto R$.
(a) Show that $T \propto R^{-1}$, if the blackbody formula is to remain valid.
(b) At what wavelength (λ_{max}) does the blackbody curve reach a maximum for the observed 2.7 K background radiation?

20–5.   If the *Hubble constant* is observed to be $H_0 = 75 \pm 30$ km sec^{-1} Mpc^{-1}, what are the permissible ranges for the "age" of the Universe ($t_0 \cong H_0^{-1}$), the "size" of the Universe ($r_{edge} \cong cH_0^{-1}$), and the critical mass density ($\rho_0 \propto H_0^2$)?

20–6.   A spectral line is emitted at a wavelength λ_{em} from a galaxy whose *redshift* is z.
(a) What is the observed wavelength (λ_{obs}) of this line?
(b) If $z \to \infty$ as we look to the "edge" of our Universe, explain (in your own words) how *Olbers' Paradox* is resolved in this expanding Universe filled with galaxies.

20–7.   The diagram below illustrates the famous *expanding balloon* analogy for our Universe. All of space is represented by the surface of the spherical balloon, and clusters of galaxies are represented by spots painted on this surface; the radius of the balloon corresponds to $R(t)$—the "radius" of the Universe.

(a) As the balloon expands, the spots remain at constant angular separations (θ) from one another. Let the balloon expand at a *constant* rate, and verify that

$$\left(\frac{\Delta s}{\Delta t}\right) = \left(\frac{1}{R}\frac{\Delta R}{\Delta t}\right)s$$

where s is the separation between any two spots (on the surface) and $(\Delta s/\Delta t)$ is the speed of recession of one spot from another. (Note that this is Hubble's law!)

(b) The *photons* from distant galaxies may be represented by ants, which crawl along the balloon's surface at speed 1! Show that for uniform cosmic expansion [i.e., $(\Delta R/\Delta t) =$ constant] there is a distance (s) from beyond which these ants cannot ever reach our Galaxy (this distance is called the *horizon*)!

(c) Discuss the effects which take place if the balloon's expansion is decelerated [i.e., the increase of $R(t)$ is slowed down].

20–8.   Consider an expanding (gravitating) gaseous sphere of uniform mass density ρ, with a total mass M and a radius $R(t)$. A gas particle at the surface of this sphere will move radially outward in accordance with the *Vis-Viva equation* (Chapter 3):

$$\frac{v^2}{2} = \frac{GM}{R} + \text{(constant)}$$

where $v = (\Delta R/\Delta t)$ is the radial speed.

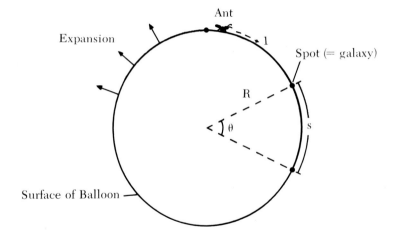

(a) Show that the Vis-Viva equation may be written in the form:

$$\left(\frac{1}{R}\frac{\Delta R}{\Delta t}\right)^2 = \frac{8\pi G\rho}{3} + \frac{2(\text{constant})}{R^2}$$

Note that this is the equation which governs the *expansion of our Universe*, and which leads to the three cosmological models discussed in this chapter!

(b) From your knowledge of the Vis-Viva equation, show that the (constant) can be either positive, zero, or negative; illustrate the evolution of $R(t)$ in each case by drawing an approximate graph of R versus t. Comment upon your results in detail.

Reading List

Alfvén, Hannes: *Worlds and Antiworlds*. San Francisco, W. H. Freeman and Company, 1966.

Bondi, H.: *Cosmology*. New York, Cambridge University Press, 1960.

Bonnor, William: *The Mystery of the Expanding Universe*. New York, The Macmillan Company, 1963.

Klein, H. Arthur: *The New Gravitation*. Philadelphia, J. B. Lippincott Company, 1971.

McVittie, G. C.: *Fact and Theory in Cosmology*. New York, The Macmillan Company, 1961.

North, J. D.: *The Measure of the Universe*. Oxford, Clarendon Press, 1965.

Novikov, Igor D., and Zel'dovich, Ya. B.: "Cosmology." *Annual Review of Astronomy and Astrophysics*, Vol. 5, pp. 627–648 (1967).

Peebles, P. J. E.: *Physical Cosmology*. Princeton, Princeton University Press, 1971.

Ryle, Martin: "The Counts of Radio Sources." *Annual Review of Astronomy and Astrophysics*, Vol. 6, pp. 249–266 (1968).

Sciama, D. W.: *Modern Cosmology*. New York, Cambridge University Press, 1972.

Singh, Jagit.: *Great Ideas and Theories of Modern Cosmology*. New York, Dover Publications, 1961.

Whitrow, G. J.: *The Structure and Evolution of the Universe*. New York, Harper and Row, 1959.

Woltjer, L., (ed.): *Galaxies and the Universe*. New York, Columbia University Press, 1968.

Appendix 1
Bibliography

A1-1 GENERAL TEXTS IN ASTRONOMY

Abell, George: *Exploration of the Universe*. Second Ed. New York, Holt, Rinehart and Winston, 1969.

Aller, Lawrence H.: *Atoms, Stars, and Nebulae*. Cambridge, Massachusetts, Harvard University Press, 1971.

Baker, R. H., and Fredrick, L. W.: *An Introduction to Astronomy*. Ninth Ed. Princeton, New Jersey, D. Van Nostrand Company, Inc., 1968.

Bok, B. J., and Bok, P. F.: *The Milky Way*. Fourth Ed. Cambridge, Massachusetts, Harvard University Press, 1973.

Brandt, J. C., and Maran, Stephen P.: *New Horizons in Astronomy*. San Francisco, W. H. Freeman and Company, 1972.

Dufay, Jean: *Introduction to Astrophysics: The Stars*. New York, Dover Publications, Inc., 1964.

Gingerich, Owen: *Frontiers in Astronomy*. Readings from *Scientific American*. San Francisco, W. H. Freeman and Company, 1970.

Jastrow, Robert, and Thompson, M. H.: *Astronomy: Fundamentals and Frontiers*. New York, John Wiley and Sons, Inc., 1972.

Menzel, Donald H., Whipple, Fred L., and de Vaucouleurs, Gerard: *Survey of the Universe*. Englewood Cliffs, New Jersey, Prentice-Hall, Inc., 1970.

Minnaert, M. G. J.: *Practical Work in Elementary Astronomy*. New York, Springer-Verlag New York Inc., 1969.

Motz, Lloyd, and Duveen, Anneta: *Essentials of Astronomy*. Belmont, California, Wadsworth Publishing Company, Inc., 1966.

Rose, W. K.: *Introduction to Astrophysics*. New York, Holt, Rinehart and Winston, 1973.

Russell, H. N., Dugan, R. S., and Stewart, J. Q.: *Astronomy*. Boston, Ginn and Company, 1926. (A classic)

Swihart, Thomas L.: *Astrophysics and Stellar Astronomy*. New York, John Wiley and Sons, Inc., 1968.

Unsold, Albrecht: *The New Cosmos*. New York, Springer-Verlag New York Inc., 1969.

Wyatt, Stanley: *Principles of Astronomy*. Second Ed. Boston, Allyn and Bacon, Inc., 1971.

A1–2 JOURNALS AND PERIODICALS

(a) General Astronomy (often suitable for the interested layman)

Sky and Telescope, monthly. Sky Publishing Corporation, 49 Bay State Road, Cambridge, Massachusetts.

Scientific American, monthly. 415 Madison Avenue, New York, New York.

Mercury, bimonthly. Astronomical Society of the Pacific, California Academy of the Sciences, Golden Gate Park, San Francisco, California.

Science, weekly. American Association for the Advancement of Science, 1515 Massachusetts Avenue, N.W., Washington, D.C.

Physics Today, monthly. American Institute of Physics, 335 E. 45th Street, New York, New York.

Publications of the Astronomical Society of the Pacific, bimonthly. California Academy of the Sciences, Golden Gate Park, San Francisco, California.

Journal of the Royal Astronomical Society of Canada, bimonthly. 252 College Street, Toronto, Canada.

The Observatory, monthly. Royal Greenwich Observatory, Herstmonceux Castle, Hailsham, Sussex, England.

Proceedings of the Astronomical Society of Australia. Sydney University Press, Sydney, Australia.

Journal of the British Astronomical Association, bimonthly. 303 Bath Road, Hounslow West, Middlesex, England.

The Irish Astronomical Journal, quarterly. Observatories of Armagh and Dunsink, Armagh, Northern Ireland.

(b) Publications of a More Technical Nature

The Astrophysical Journal, semimonthly. University of Chicago Press, 5750 Ellis Avenue, Chicago, Illinois.

The Astronomical Journal, monthly. American Institute of Physics, 335 E. 45th Street, New York, New York.

Astronomy and Astrophysics, monthly. Springer-Verlag New York, Inc., 175 Fifth Avenue, New York, New York.

Monthly Notices of the Royal Astronomical Society, monthly. Blackwell Scientific Publications, Ltd., 5 Alfred Street, Oxford, England.

Icarus, International Journal of Solar System Studies, bimonthly. Academic Press, 111 Fifth Avenue, New York, New York.

Space Science Reviews, monthly. D. Reidel Publishing Company, Singel 419–421, Dordrecht, Holland.

Nature, weekly. Macmillan, Basingstoke, Hampshire, England.

(c) Symposia and Continuing Series Publications

International Astronomical Union Symposia (and Colloquia). D. Reidel Publishing Company, Singel 419–421, Dordrecht, Holland.

Annual Review of Astronomy and Astrophysics. Annual Review, Inc., 231 Grant Avenue, Palo Alto, California.

Advances in Astronomy and Astrophysics. Academic Press, 111 Fifth Avenue, New York, New York.

Astrophysics and Space Science Library. D. Reidel Publishing Company, Singel 419–421, Dordrecht, Holland.

A1–3 THE OBSERVATIONAL TOOLS OF ASTRONOMY—A READING LIST

Instrumentation does not fit well into a book of this type and in fact is well covered in other texts. The following list directs the interested reader to some useful sources on the subject.

---

* Comprehensive review articles on astronomical subjects often appear in most of the above journals. Starred publications are especially recommended.

The standard material on telescopes and their accessories is well covered in many of the general texts in Astronomy, in particular:

Abell, George: *Exploration of the Universe*. Second Ed. New York, Holt, Rinehart and Winston, 1969.

Ingalls, Albert G., (ed.): *Amateur Telescope Making*. Books One, Two and Three. New York, Scientific American Inc., 1964, 1965, 1966.

Intermediate-Level Books Devoted to the Subject

Miczaika, G. R., and Sinton, W. M.: *Tools of the Astronomer*. Cambridge, Massachusetts, Harvard University Press, 1961.

Motz, Lloyd, and Duveen, Anneta: *Essentials of Astronomy*. Belmont, California, Wadsworth Publishing Company, Inc., 1966.

Steinberg, J. L., and Lequeux, J., (Bracewell, R. N., translator): *Radio Astronomy*. New York, McGraw-Hill Book Company, 1963.

Wyatt, Stanley: *Principles of Astronomy*. Second Ed. Boston, Allyn and Bacon, Inc., 1971.

More Advanced References

Hiltner, W. A., (ed.): *Astronomical Techniques*. Chicago, University of Chicago Press, 1962.

Kuiper, G. A., and Middlehurst, Barbara M., (eds.): *Telescopes*. Chicago, University of Chicago Press, 1960.

Strand, K. Aa., (ed.): *Basic Astronomical Data*. Chicago, University of Chicago Press, 1963.

Several of the articles in *Frontiers of Astronomy* (Edited by Owen Gingerich, W. H. Freeman and Company) make reference to some of the new techniques (e.g., X-ray, ultraviolet and infrared instrumentation).

Appendix 2
Tables

Table A2–1. The Constellations

Constellation Name	Genitive Form	Abbreviation	EQUATORIAL COORDINATES α	EQUATORIAL COORDINATES δ	GALACTIC COORDINATES ℓ	GALACTIC COORDINATES b
			h	°	°	°
Andromeda	Andromedae	And	1	+40	135	−25
Antlia	Antliae	Ant	10	−35	270	+15
Apus	Apodis	Aps	16	−75	315	−15
Aquarius	Aquarii	Aqr	23	−15	50	−60
Aquila	Aquilae	Aql	20	+5	45	−15
Ara	Arae	Ara	17	−55	335	−10
Aries	Arietis	Ari	3	+20	160	−35
Auriga	Aurigae	Aur	6	+40	175	+10
Boötes	Boötis	Boo	15	+30	45	+65
Caelum	Caeli	Cae	5	−40	245	−35
Camelopardalis	Camelopardalis	Cam	6	+70	145	+20
Cancer	Cancri	Cnc	9	+20	210	+35
Canes Venatici	Canum Venaticorum	CVn	13	+40	110	+80
Canis Major	Canis Majoris	CMa	7	−20	230	−10
Canis Minor	Canis Minoris	CMi	8	+5	215	+20
Capricornus	Capricorni	Cap	21	−20	30	−40
Carina	Carinae	Car	9	−60	270	−10
Cassiopeia	Cassiopeiae	Cas	1	+60	125	−5
Centaurus	Centauri	Cen	13	−50	305	+10
Cepheus	Cephei	Cep	22	+70	110	+10
Cetus	Ceti	Cet	2	−10	170	−65
Chamaeleon	Chamaeleontis	Cha	11	−80	300	−20
Circinus	Circini	Cir	15	−60	320	0
Columba	Columbae	Col	6	−35	240	−25
Coma Berenices	Comae Berenices	Com	13	+20	320	+85
Corona Australis	Coronae Australis	CrA	19	−40	355	−20
Corona Borealis	Coronae Borealis	CrB	16	+30	50	+50
Corvus	Corvi	Crv	12	−20	290	+40

Table A2–1   (*Continued*)

Constellation Name	Genitive Form	Abbreviation	EQUATORIAL COORDINATES		GALACTIC COORDINATES	
			α	δ	ℓ	b
Crater	Crateris	Crt	11	-15	270	$+40$
Crux	Crucis	Cru	12	-60	295	0
Cygnus	Cygni	Cyg	21	$+40$	85	-5
Delphinus	Delphini	Del	21	$+10$	60	-25
Dorado	Doradus	Dor	5	-65	275	-35
Draco	Draconis	Dra	17	$+65$	95	$+35$
Equuleus	Equulei	Equ	21	$+5$	55	-30
Eridanus	Eridani	Eri	3	-20	205	-60
Fornax	Fornacis	For	3	-30	225	-60
Gemini	Geminorum	Gem	7	$+20$	195	$+10$
Grus	Gruis	Gru	22	-45	355	-55
Hercules	Herculis	Her	17	$+30$	50	$+35$
Horologium	Horologii	Hor	3	-60	380	-50
Hydra	Hydrae	Hya	10	-20	260	$+25$
Hydrus	Hydri	Hyi	2	-75	300	-40
Indus	Indi	Ind	21	-55	340	-40
Lacerta	Lacertae	Lac	22	$+45$	100	0
Leo	Leonis	Leo	11	$+15$	230	$+60$
Leo Minor	Leonis Minoris	LMi	10	$+35$	190	$+55$
Lepus	Leporis	Lep	6	-20	225	-20
Libra	Librae	Lib	15	-15	345	$+35$
Lupus	Lupi	Lup	15	-45	325	$+10$
Lynx	Lyncis	Lyn	8	$+45$	175	$+30$
Lyra	Lyrae	Lyr	19	$+40$	70	$+15$
Mensa	Mensae	Men	5	-80	290	-30
Microscopium	Microscopii	Mic	21	-35	10	-40
Monoceros	Monocerotis	Mon	7	-5	210	0
Musca	Muscae	Mus	12	-70	300	-10
Norma	Normae	Nor	16	-50	330	0
Octans	Octantis	Oct	22	-85	305	-30
Ophiuchus	Ophiuchi	Oph	17	0	30	$+15$
Orion	Orionis	Ori	5	$+5$	195	-15
Pavo	Pavonis	Pav	20	-65	330	-30
Pegasus	Pegasi	Peg	22	$+20$	80	-25
Perseus	Persei	Per	3	$+45$	145	-10
Phoenix	Phoenicis	Phe	1	-50	300	-70
Pictor	Pictoris	Pic	6	-55	260	-30
Pisces	Piscium	Psc	1	$+15$	125	-45
Piscis Austrinus	Piscis Austrini	PsA	22	-30	20	-50
Puppis	Puppis	Pup	8	-40	255	-5
Pyxis	Pyxidis	Pyx	9	-30	255	$+10$
Reticulum	Reticuli	Ret	4	-60	270	-45
Sagitta	Sagittae	Sge	20	$+10$	50	0
Sagittarius	Sagittarii	Sgr	19	-25	10	-15
Scorpius	Scorpii	Sco	17	-40	345	0
Sculptor	Sculptoris	Scl	0	-30	10	-80
Scutum	Scuti	Sct	19	-10	25	-5
Serpens	Serpentis	Ser	17	0	20	$+5$
Sextans	Sextantis	Sex	10	0	240	$+40$
Taurus	Tauri	Tau	4	$+15$	180	-30
Telescopium	Telescopii	Tel	19	-50	350	-20
Triangulum	Trianguli	Tri	2	$+30$	140	-30
Triangulum Australe	Trianguli Australis	TrA	16	-65	320	-10
Tucana	Tucanae	Tuc	0	-65	310	-50
Ursa Major	Ursae Majoris	UMa	11	$+50$	160	$+60$
Ursa Minor	Ursae Minoris	UMi	15	$+70$	110	$+45$
Vela	Velorum	Vel	9	-50	260	0
Virgo	Virginis	Vir	13	0	310	$+65$
Volans	Volantis	Vol	8	-70	280	-20
Vulpecula	Vulpeculae	Vul	20	$+25$	65	-5

Table A2–2. Planetary Orbits

Planet	Symbol	Synodic Period DAYS	Sidereal Period TROPICAL YEARS	Sidereal Period DAYS	Semi-Major Axis AU	Semi-Major Axis 10^6 KM	Eccentricity	Inclination to Ecliptic
Mercury	☿	115.9	0.241	87.96	0.387	57.9	0.206	7°00
Venus	♀	583.9	0.615	224.70	0.723	108.2	0.007	3.39
Earth	⊕	——	1.000	365.26	1.000	149.6	0.017	0.00
Mars	♂	779.9	1.88	686.98	1.524	228.0	0.093	1.85
Jupiter	♃	398.9	11.86	——	5.203	778.3	0.048	1.31
Saturn	♄	378.1	29.46	——	9.54	1427	0.056	2.49
Uranus	♅	369.7	84.01	——	19.18	2869	0.047	0.77
Neptune	♆	367.5	164.79	——	30.07	4498	0.008	1.78
Pluto	♇	366.7	248.43	——	39.44	5900	0.249	17.17

Planetary Rotation

Planet	Sidereal Rotation Period	Oblateness	Obliquity*
Mercury	58.6 days	?	<7°
Venus	243 days	?	179°
Earth	$23^h 56^m 4\overset{s}{.}1$	0.0034	23°5
(Moon)	27.3 days	0.0006	6°7
Mars	$24^h 37^m 22\overset{s}{.}6$	0.0052	25°0
Jupiter	$9^h 50^m - 9^h 55^m$	0.062	3°1
Saturn	$10^h 14^m - 10^h 38^m$	0.096	26°7
Uranus	$10^h 49^m$	0.06	98°
Neptune	16^h	0.02	29°
Pluto	6.4 days	?	?

* Obliquity is defined as the inclination of equator to orbit plane. Obliquities greater than 90° imply retrograde rotation.

Table A2–3. Planetary Physical Data

| Planet | Mass | | Equatorial Radius | | Average Density | Surface Gravity | Albedo | Escape Speed | Temperature (K) | | |
	10^{24} KG	$\oplus = 1$	KM	$\oplus = 1$	GM CM^{-3}	$\oplus = 1$		KM SEC^{-1}	EQUILIBRIUM BLACKBODY	OBSERVED	SUBSOLAR BLACKBODY
TERRESTRIAL											
Mercury	0.33	0.055	2,432	0.38	5.5	0.38	0.06	4.2	445	100–700	633
Venus	4.87	0.815	6,050	0.95	5.2	0.91	0.76	10.3	325	700	464
Earth	5.98	1.000	6,378	1.00	5.52	1.00	0.3–0.5	11.2	277	250–300	395
Moon	0.07	0.012	1,738	0.27	3.34	0.16	0.07	2.4	277	120–390	395
Mars	0.64	0.107	3,394	0.53	3.9	0.39	0.16	5.1	225	210–300	319
Pluto	0.66	0.1	≤2,900	<0.46	~6?	~0.5	?	?	44	?	63
JOVIAN											
Jupiter	1900	318	68,700	10.77	1.40	2.74	0.51	61	122	110–150	173
Saturn	569	95	57,550	9.01	0.71	1.17	0.50	36	90	95	127
Uranus	87	14.5	25,050	3.93	1.32	0.94	0.66	22	63	(130–220)	90
Neptune	103	17.2	24,700	3.87	1.63	1.15	0.62	23	50	(180)	72

Note: The data for Pluto (and the data in parentheses) are extremely uncertain.

Table A2–4. Planetary Satellites

Planet	Moon	Distance from Planet 10³ KM	Sidereal Period DAYS	Orbital Eccentricity	Orbital Inclination* DEGREES	Radius† KM	Mass PLANET MASS / MOON MASS	Mass 10²⁰ KG	Apparent Magnitude at Opposition
Earth	Moon	384	27.32	0.055	(5.1)	1738	81	735	−12.7
Mars	Phobos	9	0.32	0.021	1.1	10	—	—	+11.5
	Deimos	23	1.26	0.003	1.6	6	—	—	+12.5
Jupiter I	Io	422	1.77	<0.01	0.03	1750	24×10^3	800	+5.5
II	Europa	671	3.55	<0.01	0.5	1550	38×10^3	500	+5.5
III	Ganymede	1,071	7.16	<0.01	0.2	2500	12×10^3	1650	+5.1
IV	Callisto	1,883	16.69	<0.01	0.3	2450	19×10^3	1020	+6.3
V		181	0.50	0.003	0.4	(85)			+13
VI		11,470	250.6	0.158	28	(65)			+14
VII		11,740	259.8	0.206	26	(22)			+18
VIII		23,500	737	0.40	33R	(6)			+18
IX		23,700	758	0.27	25R	(7)			+19
X		11,850	255	0.13	28	(7)			+19
XI		22,560	692	0.21	16R	(8)			+19
XII		21,200	631	0.16	33R	(6)			+19
Saturn	Janus	158	(0.75)	~0	~0	(185)	—		+14
	Mimas	186	0.94	0.020	1.5	(450)	15×10^6	0.4	+12
	Enceladus	238	1.37	0.004	0.0	(275)	8×10^6	0.7	+12
	Tethys	295	1.89	0.000	1.1	600	87×10^4	6.5	+11
	Dione	377	2.74	0.002	0.0	(410)	55×10^4	10.3	+11
	Rhea	527	4.52	0.001	0.3	650	4×10^5	15	+10
	Titan	1,222	15.95	0.029	0.3	2425	4150	1370	+8
	Hyperion	1,481	21.28	0.104	0.5	(175)	2×10^6	(3.1)	+14
	Iapetus	3,560	79.33	0.028	15	575	4×10^5	15	+11
	Phoebe	12,950	550.5	0.163	30R	(130)			+14
Uranus	Miranda	128	1.41	~0	—	(225)	10^6	(1)	+17
	Ariel	192	2.52	0.003	0	(735)	7×10^4	(12)	+14
	Umbriel	267	4.14	0.004	0	(480)	17×10^4	(5)	+15
	Titania	438	8.71	0.002	0	(880)	2×10^4	(40)	+14
	Oberon	586	13.46	0.001	0	(800)	3×10^4	(26)	+14
Neptune	Triton	353	5.88	0.00	20R	(1900)	750	1400	+14
	Nereid	5,600	360	0.75	28	(270)	3×10^6	0.3	+20

* With respect to ecliptic for Moon; with respect to planet's equator in all other cases; R means retrograde motion.
† The radii in parentheses are approximate, being determined by the apparent brightness of the satellite.

Table A2–5. The Nearest Stars (within 5 parsecs)

Star	α (1975)	δ (1975)	Distance* (parsecs)	Proper† Motion (″/yr)	A m_v	A M_v	A Spec	B m_v	B M_v	B Spec	C m_v	C M_v	C Spec
α Centauri	14^h 38^m.0	−60°44′	1.31	3.68	0.01	+4.4	G2 V	+1.4	+5.8	K5 V	10.7	+15	M5 eV
Barnard's Star	17 56.6	+4 37	1.83	10.34	9.54	13.2	M5 V			unseen companion			
Wolf 359	10 55.3	+7 10	2.32	4.71	13.66	16.8	M6 eV						
BD +36° 2147	11 2.0	+36 8	2.49	4.78	7.47	10.5	M2 V			unseen companion			
Sirius	6 44.0	−16 41	2.65	1.32	−1.47	1.4	A1 V	8.7	11.5	wd			
Luyten 726-8	1 37.7	−18 5	2.74	3.35	12.5	15.4	M5.5 eV	12.9	15.8	M6 eV			
Ross 154	18 48.3	−23 52	2.90	0.72	10.6	13.3	M4.5 eV						
Ross 248	23 40.7	+44 3	3.16	1.60	12.24	14.7	M5.5 eV						
ε Eridani	3 31.8	−9 33	3.28	0.97	3.73	6.1	K2 V						
Luyten 789-6	22 37.2	−15 27	3.31	3.25	12.58	14.9	M5.5 eV						
Ross 128	11 46.4	+0 57	3.32	1.40	11.13	13.5	M5 V						
61 Cygni	21 5.8	+38 37	3.43	5.22	5.19	7.5	K5 V	6.02	8.3	K7 V			
ε Indi	22 1.4	−56 53	3.44	4.69	4.73	7.0	K5 V						unseen companion
Procyon	7 38.0	+5 17	3.48	1.25	0.34	2.7	F5 IV–V	10.7	13.0	wd			
BD +59° 1915	18 42.5	+59 35	3.52	2.29	8.90	11.1	M4 V	9.69	11.9	M5 V			
BD +43° 44	0 16.9	+43 53	3.55	2.91	8.07	10.3	M2.5 eV	11.04	13.2	M4 eV			
CD −36° 15693	23 4.2	−36 0	3.59	6.90	7.39	9.6	M2 V						
τ Ceti	1 42.9	−16 4	3.67	1.92	3.50	5.7	G8 Vp						
BD +5° 1668	7 26.1	+5 18	3.76	3.73	9.82	11.9	M4 V			unseen companion			
CD −39° 14192	21 15.8	−38 58	3.85	3.46	6.72	8.7	M0 V						
CD −45° 1841	5 10.6	−45 0	3.91	8.72	8.81	10.8	M0						
Kruger 60	22 27.1	+57 34	3.94	0.87	9.77	11.8	M3 V	11.43	13.4	M4.5 eV			
Ross 614	6 28.1	−2 48	4.02	1.00	11.13	13.1	M4.5 eV	14.8	16.8	?			
BD −12° 4523	16 28.9	−12 36	4.02	1.18	10.13	12.0	M4.5 V						
v. Maanen's Star	0 47.9	+5 16	4.28	2.98	12.36	14.3	wd						
Wolf 424	12 32.1	+9 10	4.37	1.78	12.7	14.4	M5.5 eV	12.7	14.4	M6 eV			
CD −37° 15492	0 3.9	−37 29	4.45	6.11	8.59	10.3	M3 V						
BD +50° 1725	10 9.9	+49 35	4.61	1.45	6.59	8.3	M0 V						
CD −46° 11540	17 26.8	−46 52	4.63	1.06	9.34	11.3	M4 V						
CD −49° 13515	21 31.9	−49 7	4.67	0.81	9	11	M eV						
CD −44° 11909	17 35.3	−44 18	4.69	1.14	11.2	12.8	M5 V						
Luyten 1159-16	1 58.7	+12 58	4.72	2.08	12.3	13.9	M7?						
BD +15° 2620	13 44.5	+15 2	4.80	2.30	8.6	10.2	M2 V						
BD +68° 946	17 36.6	+68 22	4.83	1.31	9.1	10.7	M3 V			unseen companion			
Luyten 145-141	11 44.2	−64 42	4.85	2.69	11	12.5	wd						
Ross 780	22 51.9	−14 24	4.85	1.12	10.2	11.8	M5 V						
o² Eridani	4 14.2	−7 42	4.87	4.08	4.5	6.0	K0 V	9.2	10.7	wd	11.0	12.5	M5 eV
BD +20° 2465	10 18.2	+20 0	4.95	0.49	9.4	10.9	M4.5 V			unseen companion			

* Distances are from P. van de Kamp
† Proper Motions from W. Gliese

BD refers to Bonner Durchmusterung
CD refers to Cordoba Durchmusterung

Table A2–6.   The Twenty-Five Brightest Stars

Star	α (1975)	δ (1975)	m_v	Distance (parsecs)	Proper Motion ($''$/yr)	Spec	M_v
Sirius, α CMa	$6^h\ 44^m0$	$-16°41'$	$-1.5*$	2.7	1.32	A1 V	$+1.4$
Canopus, α Car	6 23.6	-52 41	-0.7	55	0.02	F0 Ib	-3.1
α Centauri	14 38.0	-60 44	$-0.3*$	1.3	3.68	G2 V	$+4.4$
Arcturus, α Boo	14 14.5	$+19$ 19	-0.1	11	2.28	K2 III	-0.3
Vega, α Lyr	18 36.0	$+38$ 46	0.0	8.1	0.34	A0 V	$+0.5$
Capella, α Aur	5 14.8	$+45$ 52	$0.0*$	14	0.44	G2 III	-0.7
Rigel, β Ori	5 13.3	-8 14	$0.1*$	250	0.00	B8 Ia	-6.8
Procyon. α CMi	7 38.0	$+5$ 17	$0.3*$	3.5	1.25	F5 IV–V	$+2.7$
Achernar, α Eri	1 37.8	-57 22	0.5	20	0.10	B5 V	-1.0
β Centauri	14 02.1	-60 15	$0.6*$	90	0.04	B1 III	-4.1
Altair, α Aql	19 49.5	$+8$ 48	0.8	5.1	0.66	A7 IV–V	$+2.2$
Betelgeuse, α Ori	5 53.8	$+7$ 24	$0.8†$	150	0.03	M2 Iab	-5.5
Aldebaran, α Tau	4 34.0	$+16$ 28	$0.9*$	16	0.20	K2 III	-0.2
α Crucis	12 25.2	-63 00	$0.9*$	120	0.04	B1 IV	-4.0
Spica, α Vir	13 23.9	-11 01	$1.0†$	80	0.05	B1 V	-3.6
Antares, α Sco	16 27.8	-26 22	$1.0*†$	120	0.03	M1 Ib	-4.5
Pollux, β Gem	7 43.8	$+28$ 05	1.2	12	0.62	K0 III	$+0.8$
Fomalhaut, α PsA	22 56.2	-29 45	1.2	7	0.37	A3 V	$+2.0$
Deneb, α Cyg	20 40.6	$+45$ 11	1.3	430	0.00	A2 Ia	-6.9
β Crucis	12 46.2	-59 33	1.3	150	0.05	B0.5 IV	-4.6
Regulus, α Leo	10 7.0	$+12$ 5	$1.4*$	26	0.25	B7 V	-0.6
ϵ Canis Majoris	6 57.7	-28 56	1.5	240	0.00	B2 II	-5.4
Castor, α Gem	7 33.0	$+31$ 56	1.6	14	0.20	A1 V	$+0.9$
λ Scorpii	17 31.8	-37 5	1.6	96	0.03	B2 IV	-3.3
Bellatrix, γ Ori	5 23.8	$+6$ 20	1.6	210	0.02	B2 III	-3.6

* Multiple star apparent magnitude is integrated magnitude, other data are brightest component.
† The star is a variable.
Distances for more distant stars are from spectroscopic parallaxes.

Table A2–7. Stellar Characteristics by Spectral Type and Luminosity Class

Spectral Type	M_v V	M_v III	M_v Ib*	(B–V) V	(B–V) III	(B–V) I	T_{eff}(K) V	T_{eff}(K) III	T_{eff}(K) I	BC V	(R/R_☉) V	(R/R_☉) III	(R/R_☉) I	(M/M_☉) V	(M/M_☉) III	(M/M_☉) I
O5	−6.0			−0.32	−0.32	−0.32	50000			−4.30	18			40		100
B0	−4.1	−5.0	−6.2	−0.30	−0.30	−0.24	27000			−3.17	7.6	16	20	17		50
B5	−1.1	−2.2	−5.7	−0.16	−0.16	−0.09	16000			−1.39	4.0	10	32	7		25
A0	+0.6	−0.6	−4.9	0.00	0.00	+0.01	10400			−0.40	2.6	6.3	40	3.6		16
A5	+2.1	+0.3	−4.5	+0.15	+0.15	+0.07	8200			−0.15	1.8		50	2.2		13
F0	+2.6	+0.6	−4.5	+0.30	+0.30	+0.24	7200			−0.08	1.3		63	1.8		13
F5	+3.4	+0.7	−4.5	+0.45	+0.45	+0.45	6700	6500	6200	−0.04	1.2	4.0	80	1.4		10
G0	+4.4	+0.6	−4.5	+0.60	+0.65	+0.76	6000	5500	5050	−0.06	1.04	6.3	100	1.1	2.5	10
G5	+5.2	+0.3	−4.5	+0.65	+0.86	+1.06	5500	4800	4500	−0.10	0.93	10	126	0.9	3	13
K0	+5.9	+0.2	−4.5	+0.81	+1.01	+1.42	5100	4400	4100	−0.19	0.85	16	200	0.8	4	13
K5	+8.0	−0.3	−4.5	+1.18	+1.52	+1.71	4300	3700	3500	−0.71	0.74	25	400	0.7	5	16
M0	+9.2	−0.4	−4.5	+1.39	+1.65	+1.94	3700	3500	3300	−1.20	0.63		500	0.5	6	16
M5	+12.3	−0.5	−4.5	+1.69	+1.85	+2.15	3000	2700		−2.10	0.32			0.2		

* All class Ia stars have an absolute visual magnitude of $−7^{m}.0$.

Appendix 3
Atomic Elements

Element	Symbol	Atomic Number	Atomic Weight*	Element	Symbol	Atomic Number	Atomic Weight*
hydrogen	H	1	1.008	chromium	Cr	24	52.0
helium	He	2	4.003	manganese	Mn	25	54.9
lithium	Li	3	6.9	iron	Fe	26	55.9
beryllium	Be	4	9.0	cobalt	Co	27	58.9
boron	B	5	10.8	nickel	Ni	28	58.7
carbon	C	6	12.0	copper	Cu	29	63.5
nitrogen	N	7	14.0	zinc	Zn	30	65.4
oxygen	O	8	16.0	gallium	Ga	31	69.7
fluorine	F	9	19.0	germanium	Ge	32	72.6
neon	Ne	10	20.2	arsenic	As	33	74.9
sodium	Na	11	23.0	selenium	Se	34	79.0
magnesium	Mg	12	24.3	bromine	Br	35	79.9
aluminum	Al	13	27.0	krypton	Kr	36	83.8
silicon	Si	14	28.1	rubidium	Rb	37	85.5
phosphorus	P	15	31.0	strontium	Sr	38	87.6
sulfur	S	16	32.1	yttrium	Y	39	88.9
chlorine	Cl	17	35.5	zirconium	Zr	40	91.2
argon	A	18	39.9	niobium	Nb	41	92.9
potassium	K	19	39.1	molybdenum	Mo	42	96.0
calcium	Ca	20	40.1	technetium	Tc	43	(99)
scandium	Sc	21	45.0	ruthenium	Ru	44	101.1
titanium	Ti	22	47.9	rhodium	Rh	45	102.9
vanadium	V	23	51.0	palladium	Pd	46	106.4

* Where mean atomic weights have not been well determined, the atomic mass numbers of the most stable isotopes are given in parentheses.

Element	Symbol	Atomic Number	Atomic Weight*	Element	Symbol	Atomic Number	Atomic Weight*
silver	Ag	47	107.9	iridium	Ir	77	192.2
cadmium	Cd	48	112.4	platinum	Pt	78	195.1
indium	In	49	114.8	gold	Au	79	197.0
tin	Sn	50	118.7	mercury	Hg	80	200.6
antimony	Sb	51	121.8	thallium	Tl	81	204.4
tellurium	Te	52	127.6	lead	Pb	82	207.2
iodine	I	53	126.9	bismuth	Bi	83	209.0
xenon	Xe	54	131.3	polonium	Po	84	(209)
cesium	Cs	55	132.9	astatine	At	85	(210)
barium	Ba	56	137.4	radon	Rn	86	(222)
lanthanum	La	57	138.9	francium	Fr	87	(223)
cerium	Ce	58	140.1	radium	Ra	88	226.1
praseodymium	Pr	59	140.9	actinium	Ac	89	(227)
neodymium	Nd	60	144.3	thorium	Th	90	232.1
promethium	Pm	61	(147)	protoactinium	Pa	91	(231)
samarium	Sm	62	150.4	uranium	U	92	238.1
europium	Eu	63	152.0	neptunium	Np	93	(237)
gadolinium	Gd	64	157.3	plutonium	Pu	94	(244)
terbium	Tb	65	158.9	americium	Am	95	(243)
dysprosium	Dy	66	162.5	curium	Cm	96	(248)
holmium	Ho	67	164.9	berkelium	Bk	97	(247)
erbium	Er	68	167.3	californium	Cf	98	(251)
thulium	Tm	69	168.9	einsteinium	E	99	(254)
ytterbium	Yb	70	173.0	fermium	Fm	100	(253)
lutetium	Lu	71	175.0	mendeleevium	Md	101	(256)
hafnium	Hf	72	178.5	nobelium	No	102	(253)
tantalum	Ta	73	181.0	lawrencium	Lw	103	(256)
tungsten	W	74	183.9	rutherfordium	Rf	104	(261)
rhenium	Re	75	186.2	hahnium	Ha	105	(260)
osmium	Os	76	190.2				

Appendix 4
Elementary
Particles

Class	Name	Symbol	Mass (in electron masses M_e)	Charge (in units e)	Mass (in atomic units)
Electromagnetic	Photon	γ	0	0	0
Leptons	Electron	e^-	1	-1	1/1836
	Positron	e^+	1	$+1$	1/1836
	Neutrino	ν	0	0	0
	Anti-Neutrino	$\bar{\nu}$	0	0	0
Nucleons	Proton	p	1836.1	$+1$	1
	Neutron	n	1838.7	0	1

Appendix 5
Constants and Units

A5–1. Astronomical Constants:

Astronomical unit	$AU = 1.496 \times 10^{13}$ cm
Parsec	$pc = 206{,}265$ AU
	$= 3.26$ ly
	$= 3.086 \times 10^{18}$ cm
Light year	$ly = 6.324 \times 10^{4}$ AU
	$= 0.307$ pc
	$= 9.46 \times 10^{17}$ cm
Sidereal year	$1 \text{ yr} = 365.26$ days
	$= 3.16 \times 10^{7}$ sec
Mass of Earth	$M_\oplus = 5.98 \times 10^{27}$ gm
Radius of Earth at equator	$R_\oplus = 6378$ km
Orbital velocity of Earth	$V_\oplus = 30$ km sec^{-1}
Mass of Sun	$M_\odot = 1.99 \times 10^{33}$ gm
Radius of Sun	$R_\odot = 6.96 \times 10^{5}$ km
Luminosity of Sun	$L_\odot = 3.90 \times 10^{33}$ erg sec^{-1}
Effective temperature of Sun	$T_{\text{eff}} = 5800$ K
Mass of Moon	$M_{\leftmoon} = 7.3 \times 10^{25}$ gm $= 0.0123 M_\oplus$
Radius of Moon	$R_{\leftmoon} = 1738$ km $= 0.273 R_\oplus$
Radius of Moon's orbit	$d_{\leftmoon} = 3.84 \times 10^{5}$ km
Sidereal month	$P_{\leftmoon} = 27.3$ days
Synodic month	$= 29.5$ days
Distance of Sun from center of Galaxy	$R_\odot = 10$ kpc
Velocity of Sun about galactic center	$V_\odot = 250$ km sec^{-1}
Diameter of Galaxy	$= 60$ kpc
Mass of Galaxy	$M = 1.5 \times 10^{11} M_\odot$

A5–2. Physical and Mathematical Constants:

Velocity of light	$c = 3.00 \times 10^{10}$ cm sec^{-1}
Constant of gravitation	$G = 6.67 \times 10^{-8}$ dyne cm^2 gm^{-2}
Planck constant	$h = 6.625 \times 10^{-27}$ erg sec
Boltzmann's constant	$k = 1.38 \times 10^{-16}$ erg K^{-1}
Rydberg constant	$R = 1.097 \times 10^5$ cm^{-1}
Stefan-Boltzmann constant	$\sigma = 5.67 \times 10^{-5}$ erg cm^{-2} sec^{-1} K^{-4}
Wien's law constant	$\lambda_{max} T = 2.898 \times 10^7$ Å K
Mass of hydrogen atom	$m_H = 1.67 \times 10^{-24}$ gm
Mass of electron	$m_e = 9.11 \times 10^{-28}$ gm
Charge of electron	$e = 4.80 \times 10^{-10}$ esu
Electron Volt	1 eV $= 1.602 \times 10^{-12}$ erg
Wavelength equivalence of eV	1 eV $\rightarrow 1.24 \times 10^4$ Å
Ångstrom unit	1 Å $= 10^{-8}$ cm (exact)
Gram-calorie	1 cal $= 4.2 \times 10^7$ erg
Terrestrial atmospheric pressure	1 atm $= 760$ mm Hg
	$= 1.013 \times 10^6$ dyne cm^{-2}
	$= 1.013 \times 10^6$ microbar

$\pi = 3.1416$

$e = 2.7183$; $\log_{10} e = 0.4343$

A5–3. Units and Conversions:

Metric $\rightleftarrows$ English units

1 cm = 0.3937 inch; 1 inch = 2.54 cm

1 cm = 10 mm = 10,000 microns $(\mu) = 10^8$ Å

  $= 10^{-2}$ m $= 10^{-5}$ km

 giga $= 10^9$

mega $= 10^6$

 kilo $= 10^3$

centi $= 10^{-2}$

 milli $= 10^{-3}$

micro $= 10^{-6}$

 nano $= 10^{-9}$

 pico $= 10^{-12}$

1 km = 0.6214 mile; 1 mile = 1.6093 km

 1 kg = 2.205 lbs;   1 lb = 0.454 kg

 1 gm = 0.035 oz;   1 oz = 28.35 gm

Temperature

K $= °C + 273°$

$°F = (5/9)°C + 32°$

Angular measure; degrees and time

$360° = 24^h = 2\pi$ radians; 1 radian $= 57°17'45'' = 206{,}264''.8$

  $1° = 60' = 3600''$; $1° = 0.01745$ rad

   $= 4^{min}$ $1'' = 4.848 \times 10^{-6}$ rad

  $15° = 1^{hr}$ Solid angle: sphere $= 4\pi$ steradians

Appendix 6
The Greek
Alphabet

Alpha	A	α	Nu	N	ν	
Beta	B	β	Xi	Ξ	ξ	
Gamma	Γ	γ	Omicron	O	o	
Delta	Δ	δ	Pi	Π	π, ϖ	
Epsilon	E	ϵ	Rho	P	ρ	
Zeta	Z	ζ	Sigma	Σ	σ	
Eta	H	η	Tau	T	τ	
Theta	Θ	θ	Upsilon	Y	υ	
Iota	I	ι	Phi	Φ	ϕ, φ	
Kappa	K	$\kappa, \varkappa$	Chi	X	χ	
Lambda	Λ	λ	Psi	Ψ	ψ	
Mu	M	μ	Omega	Ω	ω	

Appendix 7
Mathematical Appendix

In the following seven sections we briefly review the basic mathematical methods of astronomy and astrophysics: trigonometry, exponent notation, analytical geometry, vector analysis, series, the calculus, and mensuration formulae. The most useful results are placed in boxes and in tables for handy reference.

A7–1 TRIGONOMETRY

(a) Angular Measure

Figure A–1 depicts a circle of unit radius. The angular measure θ may be specified in three ways. The most ancient and familiar procedure is to divide the circle's circumference into 360 equal parts and to term that θ corresponding to one of these parts an *arc-degree* (°). Each arc-degree is further subdivided into 60 *arc-minutes* ('), and each arc-minute into 60 *arc-seconds* ("). Hence, there are $360 \times 60 \times 60 = 1,296,000"$ in the full circle.

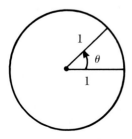

Astronomically, one rotation of the Earth requires 24 *hours* (^{h}) of time; we are accustomed to dividing the hour into 60 *minutes* (^{m}) and each minute into 60 *seconds* (^{s}). Hence, there are $24 \times 60 \times 60 = 86,400^{s}$ per rotation. But a complete rotation (24^{h}) corresponds to the full circle ($360°$), so that we may say: $1^{h} = 15°$, $1^{m} = 15'$, and $1^{s} = 15''$.

Finally, we may define a *radian* (rad) as that angle θ corresponding to a unit distance along the circumference of our unit circle. Since the entire circumference is 2π units in length ($\pi \equiv 3.141593 \ldots$), there are 2π radians in the full $360°$. Therefore: $1 \text{ rad} = 360°/2\pi = 57°.2958 = 206,264''.81$. [Radian measure is extended to *angular areas* by noting that the surface area of a sphere of unit radius is 4π square units, that is 4π *steradians* (sr). Since a steradian is one square radian, there are 41,252.96 square arc-degrees on the sphere.]

(b) The Right Triangle

The triangle *OHA* in Figure A–2 is a *right triangle*, since the angle at vertex *H* is $90°$. With respect to the angle θ, the three sides of this triangle are labelled *a* (adjacent), *o* (opposite), and *h* (hypotenuse). The fundamental trigonometric functions, *sine* (sin) and *cosine* (cos), are defined via

$$\sin \theta = o/h, \qquad \cos \theta = a/h$$

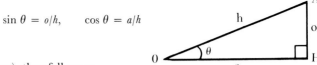

A dependent function, the *tangent* (tan), then follows as

Figure A–2.

$$\tan \theta = o/a = (o/h)/(a/h) = \sin \theta/\cos \theta$$

Also occasionally encountered are the three reciprocal functions:

$$cosecant \quad \rightarrow \quad \csc \theta = h/o = 1/\sin \theta$$

$$secant \quad \rightarrow \quad \sec \theta = h/a = 1/\cos \theta$$

$$cotangent \quad \rightarrow \quad \cot \theta = a/o = 1/\tan \theta$$

The following trigonometric identities are extremely useful:

Pythagorean	$\{ \sin^2 \theta + \cos^2 \theta = 1 \qquad 1 + \tan^2 \theta = \sec^2 \theta$
Sum-and-Difference	$\begin{cases} \sin (\theta \pm \phi) = \sin \theta \cos \phi \pm \cos \theta \sin \phi \\ \cos (\theta \pm \phi) = \cos \theta \cos \phi \mp \sin \theta \sin \phi \end{cases}$
Double-Angle	$\begin{cases} \sin 2\theta = 2 \sin \theta \cos \theta \\ \sin^2 \theta = (1/2)(1 - \cos 2\theta) \quad \cos^2 \theta = (1/2)(1 + \cos 2\theta) \end{cases}$

The trigonometric functions may be extended to the full circle ($0° \leq \theta \leq 360°$) by using the signs ($\pm$) given in Table A–1, the special values listed in Table A–2, and the values for every arc-

Table A–1

Region	sin	cos	tan
$0°-90°$	+	+	+
$90°-180°$	+	−	−
$180°-270°$	−	−	+
$270°-360°$	−	+	−

Table A–2

Angle		sin	cos	tan	cot
ARC-DEGREES	RADIANS				
0	0	0	1	0	∞
30	$\pi/6$	$1/2$	$\sqrt{3}/2$	$\sqrt{3}/3$	$\sqrt{3}$
45	$\pi/4$	$\sqrt{2}/2$	$\sqrt{2}/2$	1	1
60	$\pi/3$	$\sqrt{3}/2$	$1/2$	$\sqrt{3}$	$\sqrt{3}/3$
90	$\pi/2$	1	0	∞	0

degree from $0°$ to $90°$ tabulated in Table A–3. The following practical identities are needed in this extension:

$$\sin \theta = +\cos(\theta - 90°) = -\sin(\theta - 180°) = -\cos(\theta - 270°)$$

$$\cos \theta = -\sin(\theta - 90°) = -\cos(\theta - 180°) = +\sin(\theta - 270°)$$

$$\tan \theta = -\cot(\theta - 90°) = +\tan(\theta - 180°) = -\cot(\theta - 270°)$$

(c) The Plane Triangle

Figure A–3 illustrates the general *plane triangle ABC*, with vertex angles A, B, C and corresponding opposite sides a, b, c. For any such triangle, the following formulae obtain:

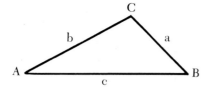

Figure A–3.

$$\text{Area} = \sqrt{s(s-a)(s-b)(s-c)}, \quad \text{where} \quad s = (1/2)(a+b+c)$$

Law of Sines $\quad \left\{ \dfrac{a}{\sin A} = \dfrac{b}{\sin B} = \dfrac{c}{\sin C} \right.$

Law of Cosines $\quad \left\{ \begin{array}{l} a^2 = b^2 + c^2 - 2bc \cos A \\ b^2 = c^2 + a^2 - 2ca \cos B \\ c^2 = a^2 + b^2 - 2ab \cos C \end{array} \right.$

A7–2 EXPONENT NOTATION

(a) Powers and Roots

When a positive number (a) is multiplied against itself an integer (m) number of times, the result is the *mth power of a*:

$$a \times a \times a \times \cdots (m \text{ times}) \cdots \times a = a^m$$

When several powers of the same number are multiplied together, their *exponents* add: $a^m a^n = a^{m+n}$. If we define $a^0 \equiv 1$, then negative exponents are admitted and are called *reciprocals*: $a^{-m} = 1/a^m \Rightarrow a^m a^{-m} = a^{m-m} = a^0 = 1$.

Similarly, we term $a^{1/m}$ the *mth root of a*, since we recover a when its root is raised to the *m*th power: $(a^{1/m})^m = a^{m/m} = a^1 = a$. Note that, when a power or root is raised to a power, the two exponents involved multiply.

Table A–3*

Angle (°) ↓	sin	cos	tan	cot	
0 →	0.0000	1.000	0.0000	∞	← 90
1	0.0175	0.9999	0.0175	57.29	89
2	0.0349	0.9994	0.0349	28.64	88
3	0.0523	0.9986	0.0524	19.08	87
4	0.0698	0.9976	0.0699	14.30	86
5 →	0.0872	0.9962	0.0875	11.43	← 85
6	0.1045	0.9945	0.1051	9.514	84
7	0.1219	0.9926	0.1228	8.144	83
8	0.1392	0.9903	0.1405	7.115	82
9	0.1564	0.9877	0.1584	6.314	81
10 →	0.1737	0.9848	0.1763	5.671	← 80
11	0.1908	0.9816	0.1944	5.145	79
12	0.2079	0.9782	0.2126	4.705	78
13	0.2250	0.9744	0.2309	4.332	77
14	0.2419	0.9703	0.2493	4.011	76
15 →	0.2588	0.9659	0.2680	3.732	← 75
16	0.2756	0.9613	0.2868	3.487	74
17	0.2924	0.9563	0.3057	3.271	73
18	0.3090	0.9511	0.3249	3.078	72
19	0.3256	0.9455	0.3443	2.904	71
20 →	0.3420	0.9397	0.3640	2.748	← 70
21	0.3584	0.9336	0.3839	2.605	69
22	0.3746	0.9272	0.4040	2.475	68
23	0.3907	0.9205	0.4245	2.356	67
24	0.4067	0.9136	0.4452	2.246	66
25 →	0.4226	0.9063	0.4663	2.145	← 65
26	0.4384	0.8988	0.4877	2.050	64
27	0.4540	0.8910	0.5095	1.963	63
28	0.4695	0.8830	0.5317	1.881	62
29	0.4848	0.8746	0.5543	1.804	61
30 →	0.5000	0.8660	0.5774	1.732	← 60
31	0.5150	0.8572	0.6009	1.664	59
32	0.5299	0.8481	0.6249	1.600	58
33	0.5446	0.8387	0.6494	1.540	57
34	0.5592	0.8290	0.6745	1.483	56
35 →	0.5736	0.8192	0.7002	1.428	← 55
36	0.5878	0.8090	0.7265	1.376	54
37	0.6018	0.7986	0.7536	1.327	53
38	0.6157	0.7880	0.7813	1.280	52
39	0.6293	0.7772	0.8098	1.235	51
40 →	0.6428	0.7660	0.8391	1.192	← 50
41	0.6561	0.7547	0.8693	1.150	49
42	0.6691	0.7431	0.9004	1.111	48
43	0.6820	0.7314	0.9325	1.072	47
44	0.6947	0.7193	0.9657	1.036	46
45 →	0.7071	0.7071	1.000	1.000	← 45 ↑
	cos	sin	cot	tan	Angle (°)

* For angles from 0° to 45°, use the left-most column and the top headings; for angles from 45° to 90°, use the right-most column and the bottom headings.

These results are readily generalized to *any real exponent* (not necessarily an integer or a rational fraction) by the following formulae:

$$a^0 = 1 \qquad a^{-m} = 1/a^m$$

$$(ab)^m = a^m b^m$$

$$a^m a^n = a^{m+n} \qquad (a^m)^n = a^{mn}$$

We define the *factorial* of an integer (n) as the product of n with all smaller integers (down to 1): $n! \equiv n(n-1)(n-2)\cdots(3)(2)(1)$. It is conventional to also define $0! \equiv 1$.

The following simple examples illustrate these manipulations:

$$3^4 = 3 \times 3 \times 3 \times 3 = 81$$

$$2^{-3} = 1/2^3 = 1/(2 \times 2 \times 2) = 1/8$$

$$15^2 = (3 \times 5)^2 = 3^2 \times 5^2 = 9 \times 25 = 225$$

$$6^2 \times 6^3 = 6^{2+3} = 6^5$$

$$(\sqrt{2})^3 = (2^{1/2})^3 = 2^{3/2}$$

$$4! = 4 \times 3 \times 2 \times 1 = 24$$

(b)   Exponentials and Logarithms

When the *base a* is given, the *exponential* formula

$$y = a^x = \text{``the base } a \text{ to the power } x\text{''}$$

yields a value of y for every value of x (*exponent*) which we choose. However, if we know both a and y and desire to learn x, we must invert this relationship to obtain the *logarithmic* formula

$$x = \log_a y = \text{``the exponent of } a \text{ which yields } y\text{''}$$

For example, given $8 = 2^x$, we know that $x = 3$, since $2^3 = 2 \times 2 \times 2 = 8$; hence, $\log_2 8 = 3$.

The general properties of powers and roots (above) lead at once to the following useful relations for logarithms:

Product	$\log_a (xy) = \log_a x + \log_a y$
Quotient	$\log_a (x/y) = \log_a x - \log_a y$
Power	$\log_a (y^n) = n \log_a y$
Change-of-Base	$\log_a y = (\log_a b)(\log_b y)$

In this text, we most frequently encounter the *decimal* base, $a = 10$; logarithms with respect to this base are termed *common* logarithms (written log). Every common logarithm consists of two parts: an integer (the *characteristic*) and an "endless" decimal (the *mantissa*). For example,

$$\log 33.7 = \log 10^{1.5276} = 1.\underbrace{5276}$$

$$\underset{\text{characteristic}}{} \qquad \underset{\text{mantissa}}{}$$

Table A–4*

N	0	0.1	0.2	0.3	0.4	0.5	0.6	0.7	0.8	0.9
1	0.000	0.041	0.079	0.114	0.146	0.176	0.204	0.230	0.255	0.279
2	0.301	0.322	0.342	0.362	0.380	0.398	0.415	0.431	0.447	0.462
3	0.477	0.491	0.505	0.519	0.532	0.544	0.556	0.568	0.580	0.591
4	0.602	0.613	0.623	0.634	0.644	0.653	0.663	0.672	0.681	0.690
5	0.699	0.708	0.716	0.724	0.732	0.740	0.748	0.756	0.763	0.771
6	0.778	0.785	0.792	0.799	0.806	0.813	0.820	0.826	0.833	0.839
7	0.845	0.851	0.857	0.863	0.869	0.875	0.881	0.887	0.892	0.898
8	0.903	0.909	0.914	0.919	0.924	0.929	0.935	0.940	0.945	0.949
9	0.954	0.959	0.964	0.969	0.973	0.978	0.982	0.987	0.991	0.996

* An example will illustrate the use of this table. Consider log 540. Since $540 = 5.4 \times 10^2$, log $540 = 2 +$ log 5.4. The characteristic is 2. The mantissa is 0.732 (log 5.4), found by going down the N-column to 5, then across to the column headed 0.4. Therefore, log $540 = 2.732$.

When we use the "powers-of-ten" notation, $33.7 = 3.37 \times 10^1$, the characteristic 1 is immediately evident. This fact is used in constructing Table A–4 (table of common logarithms), where only the mantissa is given.

Important in the calculus (see section A7–6), although infrequently encountered in this text, are exponentials to the base $e \equiv 2.71828 \ldots$. The associated *natural* or *Naperian* logarithms are denoted ln. In all practical computations, we will make a change of base to the decimal system (common logarithms) and employ Table A–4 via the relations:

$$e^x = 10^{0.4343x}$$

$$\ln x = (2.3026) \log x$$

A7–3 ANALYTICAL GEOMETRY

(a) Coordinate Systems

There are three common coordinate systems used to locate a point in three-dimensional space. The most familiar system is *rectangular Cartesian* coordinates (x, y, z). Beginning at the *origin O* $(x = 0, y = 0, z = 0)$, we move out the x-axis x units, across parallel to the y-axis y units, and up parallel to the z-axis z units (see Figure A–4).

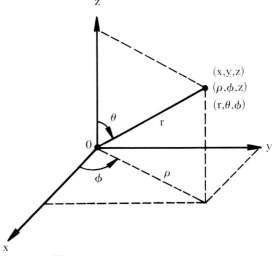

Figure A–4.

In *cylindrical polar* coordinates (ρ, ϕ, z), the point is found by moving out from the origin in the *x-y* plane the distance ρ at the angle ϕ to the *x*-axis, then up z units parallel to the *z*-axis (see Figure A–4). These coordinates are clearly related to Cartesian coordinates by:

$$x = \rho \cos \phi, \qquad y = \rho \sin \phi, \qquad z = z$$

Finally, in *spherical* coordinates (r, θ, ϕ), we move the distance r from the origin at the angle θ to the *z*-axis; the projection of this motion on the *x-y* plane is inclined the angle ϕ to the *x*-axis and has the length $\rho = r \sin \theta$ (see Figure A–4). The connection to Cartesian coordinates is therefore given by:

$$x = r \sin \theta \cos \phi, \qquad y = r \sin \theta \sin \phi, \qquad z = r \cos \theta$$

(b) Graphs

We define "*y* as a *function* of *x*" by the algebraic equation, $y = y(x)$. Therefore, for each value of *x*, the function yields a value of *y*; we have an (x, y) pair. To better illustrate the properties of the function, let us *graph* every (x, y) pair as a point in a two-dimensional Cartesian coordinate system; the result is a *curve*.

Consider the *linear* equation, $y = mx + b$, where *m* and *b* are constants. When $x = 0, y = +b$. When $x = -b/m, y = 0$. And for every unit increase of *x*, *y* "increases" by *m* units; we say that the *slope* is *m*. The graph of this function is the *straight line* shown in Figure A–5.

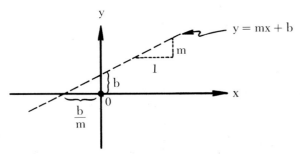

Figure A–5.

Now consider the *quadratic* equation, $y = ax^2 + bx + c$, where *a*, *b*, and *c* are constants. When $x = 0, y = +c$; the two "zeros" of the equation (where $y = 0$) are given by the *quadratic formula*:

$$x = \frac{-b \pm \sqrt{b^2 - 4ac}}{2a}$$

The graph of this function (shown schematically in Figure A–6) is a *parabolic* curve.

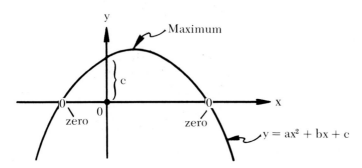

Figure A–6.

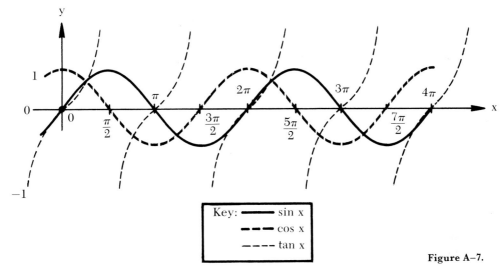

Figure A–7.

The usefulness of graphs is most evident when we consider more complicated functions. Figure A–7 shows the trigonometric functions: sin x, cos x, and tan x. Figure A–8 depicts the exponential function, $y = a^x$; the logarithmic function, $x = \log_a y$, may be seen by rotating the diagram 90° counterclockwise.

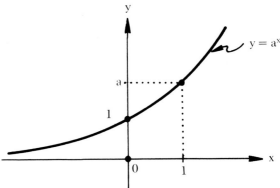

Figure A–8.

(c)   The Conic Functions

In plane polar coordinates (ρ, ϕ), all gravitational orbits may be described by the single equation, $\rho = d(1 + e)/(1 + e \cos \phi)$, where $\rho = d$ is the distance of *closest approach* to the origin (at $\phi = 0°$). The graph of this function yields a variety of curves, called the *conic sections* (see Figure A–9).

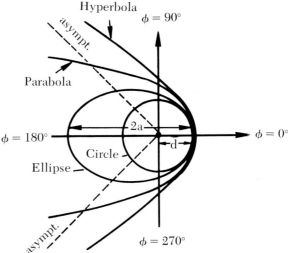

Figure A–9.

When $e = 0$, we have a *circle* of radius d. When e lies in the range $0 < e < 1$, the curve is an *ellipse;* we usually write $d = a(1 - e)$, so that the *major axis* (longest dimension) of the ellipse is $2a$. When $e = 1$, the curve is a *parabola*, which is "open" to the left at $\phi = 180°$. Finally, we speak of a *hyperbola* when $e > 1$; this curve exhibits $\rho \to \infty$ at the two angles where $\cos \phi = -1/e$ (along lines called *asymptotes*).

A7–4   VECTOR ANALYSIS

(a)   Vectors

A *vector* is like an *arrow*, for it has both a *magnitude* (length) and a *direction*. The magnitude is a *scalar*, a simple number without a direction (like temperature or mass). We denote a vector by **c**, and its magnitude by c.

Two vectors are added, $\mathbf{c} = \mathbf{a} + \mathbf{b}$, using the *parallelogram rule of vector addition* illustrated in Figure A–10. Conversely, a vector may always be decomposed into two *component* vectors. For convenience, we decompose along the coordinate axes, and write the vector as $\mathbf{c} = (c_x, c_y)$. Now the rule of vector addition may be stated in terms of components as:

$$c_x = a_x + b_x, \qquad c_y = a_y + b_y$$

From the Pythagorean theorem and Figure A–10, it is clear that the magnitude of **c** is just: $c = (c_x^2 + c_y^2)^{1/2}$. In terms of the angle α between **c** and the x-axis, the direction of **c** is given by: $\tan \alpha = c_y/c_x$. Finally, as a consequence of vector addition, the magnitude of **c** may be written as

$$c = \left[c_x^2 + c_y^2\right]^{1/2}$$

$$= \left[(a_x + b_x)^2 + (a_y + b_y)^2\right]^{1/2}$$

$$= \left[(a_x^2 + a_y^2) + (b_x^2 + b_y^2) + 2(a_xb_x + a_yb_y)\right]^{1/2}$$

$$= \left[a^2 + b^2 + 2(\mathbf{a} \cdot \mathbf{b})\right]^{1/2}$$

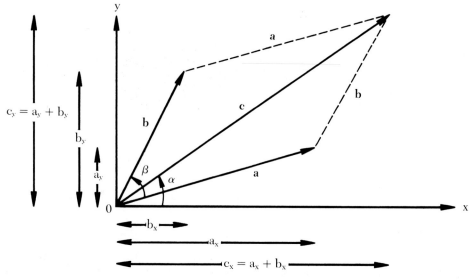

Figure A–10.

(see the vector dot product in the following subsection), or from the law of cosines

$$c^2 = a^2 + b^2 + 2ab \cos \beta$$

where β is the smallest angle between **a** and **b**.

As an example, consider the vectors, $\mathbf{a} = (1, 1)$ and $\mathbf{b} = (3, -4)$. Their magnitudes are:
$a = (a_x^2 + a_y^2)^{1/2} = (1^2 + 1^2)^{1/2} = (1 + 1)^{1/2} = (2)^{1/2} = \sqrt{2}$, and

$$b = (3^2 + 4^2)^{1/2} = (9 + 16)^{1/2} = 5$$

Their vector sum is $\mathbf{c} = \mathbf{a} + \mathbf{b} = (a_x + b_x, a_y + b_y) = (1 + 3, 1 - 4) = (4, -3) = (c_x, c_y)$, with the magnitude $c = (4^2 + 3^2)^{1/2} = 5$. [Construct a diagram like Figure A–10 for these vectors, and show that **a**, **b**, **c** form an *isoceles* triangle!]

(b)   Dot Product

In three-dimensional Cartesian coordinates, the *vector dot product* of **a** and **b** is defined as the scalar:

$$\mathbf{a} \cdot \mathbf{b} \equiv a_x b_x + a_y b_y + a_z b_z$$

If ψ is the smallest angle between **a** and **b**, then we can easily show that

$$\mathbf{a} \cdot \mathbf{b} \equiv ab \cos \psi$$

Therefore, the dot product is a measure of the component of **a** *in the direction of* **b** (or vice versa), and $\mathbf{a} \cdot \mathbf{b} = 0$ when the two vectors are perpendicular ($\psi = 90°$). (See Figure A–11.)

Consider the example from the previous section, $\mathbf{a} = (1, 1)$ and $\mathbf{b} = (3, -4)$. Now, $\mathbf{a} \cdot \mathbf{b} = a_x b_x + a_y b_y = (1)(3) + (1)(-4) = 3 - 4 = -1$. The angle ψ satisfies:

$$\cos \psi = \mathbf{a} \cdot \mathbf{b}/ab = (-1)/(\sqrt{2})(5) = -\sqrt{2}/10 = -0.1414$$

so that Tables A–1 and A–3 imply: $\psi \cong 98°$.

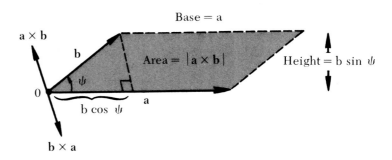

Figure A–11.

(c)   Cross Product

The *vector cross product* of **a** and **b**, denoted **a** × **b**, is *another vector* which is perpendicular to both **a** and **b**. The direction of the resultant vector is given by the *right hand rule*: "Align the fingers of your right hand along **a**, then rotate this hand through the smallest angle (ψ) between **a** and **b** toward **b**;

your thumb will point in the direction of the cross-product vector." In terms of components, the cross product is defined by

$$\mathbf{a} \times \mathbf{b} \equiv (a_y b_z - a_z b_y,\ a_z b_x - a_x b_z,\ a_x b_y - a_y b_x)$$

The cross product is essentially a measure of the component of **a** *perpendicular to* **b** (or vice versa), and it is therefore also given by

$$|\mathbf{a} \times \mathbf{b}| \equiv ab \sin \psi$$

Note that when **a** and **b** are parallel (or anti-parallel), $\mathbf{a} \times \mathbf{b} = 0$; also, it is true that $\mathbf{b} \times \mathbf{a} = -\mathbf{a} \times \mathbf{b}$ (check this using the right hand rule and the component definition of cross product).

Figure A–11 illustrates some of the properties of the dot product and the cross product of two vectors, **a** and **b**.

We conclude with the calculation of the cross product of $\mathbf{a} = (1, 1)$ and $\mathbf{b} = (3, -4)$—see the preceding subsections:

$$\mathbf{a} \times \mathbf{b} = (0, 0, -4 - 3) = (0, 0, -7), \quad \text{since } a_z = 0 = b_z$$

Therefore, $\mathbf{a} \times \mathbf{b}$ is directed in the negative z direction (perpendicular to both **a** and **b**, which lie in the x-y plane), and the area of the parallelogram indicated in Figure A–11 is $|\mathbf{a} \times \mathbf{b}| = 7$. An alternate method for finding $\mathbf{a} \times \mathbf{b}$ is the following: First, discover its direction by using the right hand rule. Second, find its magnitude via $|\mathbf{a} \times \mathbf{b}| = ab \sin \psi$:

$$|\mathbf{a} \times \mathbf{b}| = (\sqrt{2})(5) \sin 98° = (1.414)(5)(0.99) = 6.999 = 7$$

A7–5   SERIES

In the functional relation, $y = y(x)$, we term x the *argument*. In many practical applications of astronomy and astrophysics (and particularly in the calculus; see section A7–6), we need to know the behavior of certain functions for very small values of the argument ($0 \lesssim x \ll 1$). Hence, we expand the function in a *series* of powers of x; useful series expansions are listed below (together with the precise range of applicable x values):

Binomial	$\begin{cases} (1 \pm x)^n = 1 \pm nx + (1/2)n(n-1)x^2 \pm (1/6)n(n-1)(n-2)x^3 + \cdots \\ \\ \end{cases}$	$(x^2 < 1;\ \text{all } n)$
Trigonometric	$\begin{cases} \sin x = x - (1/6)x^3 + (1/120)x^5 - \cdots \\ \cos x = 1 - (1/2)x^2 + (1/24)x^4 - \cdots \\ \tan x = x + (1/3)x^3 + (2/15)x^5 + \cdots \end{cases}$	$(x^2 < 1)$ $(x^2 < 1)$ $(x^2 < \pi^2/4)$
Exponential	$\{\quad e^x = 1 + x + (1/2)x^2 + (1/6)x^3 + \cdots$	$(x^2 < 1)$
Logarithmic	$\{\ \ln(1 + x) = x - (1/2)x^2 + (1/3)x^3 - (1/4)x^4 + \cdots$	$(x^2 < 1)$

Three simple examples will serve to illustrate the use of these series. First, let us evaluate $\sqrt{e}$. Then we have (approximately):

$$e^{1/2} = 1 + (1/2) + (1/2)(1/2)^2 + (1/6)(1/2)^3 + \cdots$$

$$= 1 + 1/2 + 1/8 + 1/48 + \cdots = 79/48 + \cdots \cong 1.65$$

Secondly, consider the very narrow triangle used in stellar parallax (see Chapter 4), with short side = 1 AU, adjacent side = d(AU), and included angle at the star = π(rad) $\ll$ 1. Then we compute,

$$1 \text{ AU}/d(\text{AU}) = \tan \pi(\text{rad}) \approx \pi(\text{rad}) \Rightarrow d(\text{pc}) \cong 1/\pi''$$

since there are 206,265″ per radian and 1 pc = 206,265 AU. Finally, when computing tidal accelerations (see Chapter 4), we seek the very small difference between two large quantities: $GM/[r \pm (d/2)]^2$. Extract the r^2 in the denominator, and use the binomial series on the remaining denominator (since $d \ll r \Rightarrow x = d/2r \ll 1$):

$$\left(1 \pm \frac{d}{2r}\right)^{-2} = 1 \mp 2\left(\frac{d}{2r}\right) + 3\left(\frac{d}{2r}\right)^2 \quad \cdots$$

$$\approx 1 \mp \frac{d}{r} + \cdots$$

Therefore, we quickly find:

$$\frac{GM}{\left(r - \dfrac{d}{2}\right)^2} - \frac{GM}{\left(r + \dfrac{d}{2}\right)^2} = \frac{GM}{r^2}\left[\left(1 - \frac{d}{2r}\right)^{-2} - \left(1 + \frac{d}{2r}\right)^{-2}\right]$$

$$= \frac{GM}{r^2}\left[\left(1 + \frac{d}{r} + \cdots\right) - \left(1 - \frac{d}{r} + \cdots\right)\right]$$

$$\cong 2GMd/r^3$$

A7–6   THE CALCULUS

(a)   Derivatives

We seek the *derivative* (or *instantaneous slope*) of the function $y(x)$ at the point x. As illustrated in Figure A–12, we select a nearby point, $x + \Delta x$, evaluate $y(x + \Delta x)$, and in the limit as Δx becomes infinitesimally small $\left(\lim\limits_{\Delta x \to 0}\right)$ we *define* the derivative as

$$\boxed{\frac{dy}{dx} \equiv \lim_{\Delta x \to 0} \frac{y(x + \Delta x) - y(x)}{\Delta x}}$$

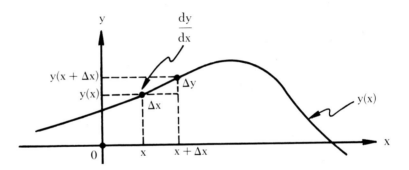

Figure A–12.

Let us use this definition to derive two simple derivatives. Consider $y(x) = x^2$; then,

$$y(x + \Delta x) = (x + \Delta x)^2 = x^2 + 2x(\Delta x) + (\Delta x)^2$$

Therefore,

$$\frac{dy}{dx} = \lim_{\Delta x \to 0} \frac{x^2 + 2x(\Delta x) + (\Delta x)^2 - x^2}{\Delta x} = \lim_{\Delta x \to 0} 2x + \Delta x = 2x$$

hence, the derivative of x^2 at x is $2x$. Secondly, consider $y(x) = \sin x$. Then (using the sum-of-angles identity): $y(x + \Delta x) = \sin (x + \Delta x) = \sin x \cos \Delta x + \cos x \sin \Delta x$. But since Δx will become very small, we may use the series expansions for $\sin \Delta x \approx \Delta x$ and $\cos \Delta x \approx 1$. Therefore,

$$\frac{dy}{dx} = \lim_{\Delta x \to 0} \frac{\sin x + (\Delta x) \cos x - \sin x}{\Delta x} = \cos x$$

the sought-for derivative of $\sin x$.

Proceeding in just this fashion, one may easily verify the following useful formulae for derivatives [where a and n are constants, while $u = u(x)$ and $v = v(x)$]:

Definitions	$\{ \quad da/dx = 0 \qquad dx/dx = 1$
Linearity	$\begin{cases} \quad d(au)/dx = a(du/dx) \\ \\ d(u + v)/dx = (du/dx) + (dv/dx) \end{cases}$
"Chain Rule"	$\{ \qquad d(uv)/dx = u(dv/dx) + v(du/dx)$
Powers	$\{ \qquad d(u^n)/dx = nu^{n-1}(du/dx)$
Trigonometric	$\begin{cases} d(\sin u)/dx = \cos u(du/dx) \\ d(\cos u)/dx = -\sin u(du/dx) \\ d(\tan u)/dx = \sec^2 u(du/dx) \end{cases}$
Exponential	$\begin{cases} \quad d(a^u)/dx = a^u(\ln a)(du/dx) \\ \\ d(e^u)/dx = e^u(du/dx) \end{cases}$
Logarithmic	$\{ \quad d(\ln u)/dx = (1/u)(du/dx)$

For example, here are the steps in finding $d(x \sin x)^2/dx$:

$$\frac{d}{dx}(x \sin x)^2 = 2(x \sin x) \frac{d}{dx}(x \sin x) \qquad\qquad \leftarrow \quad d(u^n)/dx$$

$$= 2(x \sin x)\left[x \frac{d(\sin x)}{dx} + \sin x \left(\frac{dx}{dx}\right)\right] \quad \leftarrow \quad \text{chain rule}$$

$$= 2(x \sin x)(x \cos x + \sin x) \qquad\qquad \leftarrow \quad \frac{dx}{dx} = 1 \quad \text{and} \quad \frac{d(\sin x)}{dx} = \cos x$$

$$= 2x^2 \sin x \cos x + 2x \sin^2 x \qquad\qquad \leftarrow \quad \text{expanding}$$

[Note that the derivative of a vector is defined in terms of the derivatives of its components as: $d\mathbf{a}/dx = (da_x/dx, da_y/dx, da_z/dx)$.]

(b) Integrals

The *integral* of the function $y(x)$ may be either indefinite or definite. The *indefinite integral*, denoted by $\int y(x)\, dx$, is to be thought of as "that function of x whose derivative is $y(x)$." Hence, it is clear that $\int \cos x\, dx = \sin x$, since $d(\sin x)/dx = \cos x$. Therefore, the indefinite integral is the *inverse* of the derivative, in the sense that $\int [dy(x)/dx]\, dx = y(x)$.

The *definite integral*, denoted by $\int_a^b y(x)\, dx$, is the net area under the curve $y(x)$ between $x = a$ and $x = b$ (see Figure A–13). If we have $y(x) = df(x)/dx$, then by definition it follows that:

$$\int_a^b y(x)\, dx = \int_a^b (df/dx)\, dx = f(x)\Big]_a^b = f(b) - f(a)$$

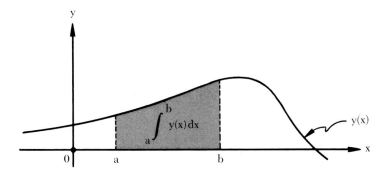

Figure A–13.

In general, indefinite integrals are found by trial and error, but we can tabulate here some useful known results (see the list of derivatives above):

Linearity	$\begin{cases} \int ay(x)\, dx = a \int y(x)\, dx \\[2mm] \int (u + v)\, dx = \int u\, dx + \int v\, dx \end{cases}$
"By Parts"	$\{ \quad \int u\, dv = uv - \int v\, du$
Powers	$\{ \quad \int x^n\, dx = x^{n+1}/(n+1) \qquad$ (except for $n = -1$)
Trigonometric	$\begin{cases} \int \sin x\, dx = -\cos x \\[2mm] \int \cos x\, dx = \sin x \\[2mm] \int \sec^2 x\, dx = \tan x \end{cases}$
Exponential	$\{ \quad \int e^{ax}\, dx = e^{ax}/a$
Logarithmic	$\begin{cases} \int [(dy/dx)/y(x)]\, dx = \ln y(x) \\[2mm] \int (1/x)\, dx = \ln x \end{cases}$

For a vastly more extensive tabulation, the student is referred to any standard *table of integrals*.

We may illustrate the usefulness of this brief table of integrals by considering $\int (\sin^2 x) \cos x \, dx$. Let $u = \sin x$, then $du = \cos x \, dx$, so that our integral is $\int u^2 \, du = u^3/3$. Substituting back in for $u = \sin x$, we have the answer: $(1/3) \sin^3 x$. If this had been the definite integral, $\int_0^{\pi/2} (\sin^2 x) \cos x \, dx$, we would find:

$$\int_0^{\pi/2} (\sin^2 x) \cos x \, dx = \left[(1/3) \sin^3 x \right]_0^{\pi/2}$$

$$= (1/3) \left[\sin^3 (\pi/2) - \sin^3 (0) \right]$$

$$= (1/3)(1^3 - 0^3) = 1/3$$

[Note that the integral of a vector is *another vector*, defined in terms of the components or: $\int \mathbf{a} \, dx = \int (a_x(x), a_y(x), a_z(x)) \, dx = (\int a_x \, dx, \int a_y \, dx, \int a_z \, dx)$.]

A7–7   MENSURATION FORMULAE

Such things as lengths, areas, and volumes are given by *mensuration formulae*; a typical example is the area of a circle of radius R, $A = \pi R^2$. In section A7–7(a) we show how to derive these formulae using the integral calculus; those interested only in the answers should proceed at once to section A7–7(b).

(a)   Multiple Integrals

At a given point in a coordinate system, we make infinitesimal changes in the three coordinates and define (a) infinitesimal lengths, (b) infinitesimal surface areas, and (c) infinitesimal volumes. By appropriately summing (i.e., integrating) these, we obtain finite lengths, areas, and volumes. In general, we will be dealing with *multiple integrals*.

In rectangular Cartesian coordinates (x, y, z), the infinitesimal extensions are (dx, dy, dz). The distance along the x-axis from $x = 0$ to $x = L$ is then just $\int_0^L dx = x]_0^L = L$. The infinitesimal surface areas are $dx \, dy$ [in the x-y plane at (x, y, z)], $dy \, dz$, and $dz \, dx$. Therefore, the area in the x-y plane bounded between $0 \le x \le L$ and $0 \le y \le W$ is $\int_0^L dx \int_0^W dy = x]_0^L y]_0^W = LW$. Finally, at (x, y, z) the infinitesimal volume is $dx \, dy \, dz$. The volume of a rectangular parallelopiped of dimensions $L \times W \times H$ is clearly $\int_0^L dx \int_0^W dy \int_0^H dz = LWH$.

In cylindrical polar coordinates, the elementary lengths are $(d\rho, \rho \, d\phi, dz)$, the elementary areas are $(\rho \, d\rho \, d\phi, \rho \, d\phi \, dz, \text{ and } d\rho \, dz)$, and the elementary volume is $\rho \, d\rho \, d\phi \, dz$. Therefore, the *circumference* of a circle of radius $\rho = R$ is $\int_0^{2\pi} R \, d\phi = R \int_0^{2\pi} d\phi = R\phi]_0^{2\pi} = 2\pi R$; the *area* of this circle is $\int_0^R \rho \, d\rho \int_0^{2\pi} d\phi = (1/2)\rho^2]_0^R \phi]_0^{2\pi} = (R^2/2)(2\pi) = \pi R^2$; and the *volume* of a right cylinder (of radius $\rho = R$ and height $z = H$) is $\int_0^R \rho \, d\rho \int_0^{2\pi} d\phi \int_0^H dz = \pi R^2 \int_0^H dz = \pi R^2 H$.

In spherical coordinates, the basic lengths are $(dr, r \, d\theta, r \sin \theta \, d\phi)$, the basic areas are $(r \, dr \, d\theta, r^2 \sin \theta \, d\theta \, d\phi, \text{ and } r \, dr \sin \theta \, d\phi)$, and the basic volume element is $r^2 \, dr \sin \theta \, d\theta \, d\phi$. Therefore, the *surface area* of a sphere of radius $r = R$ is $\int_0^\pi R^2 \sin \theta \, d\theta \int_0^{2\pi} d\phi = 2\pi R^2 \int_0^\pi \sin \theta \, d\theta = 2\pi R^2 [-\cos \theta]_0^\pi = 4\pi R^2$, while the volume of the sphere is $\int_0^R r^2 \, dr \int_0^\pi \sin \theta \, d\theta \int_0^{2\pi} d\phi = [(1/3)r^3]_0^R [-\cos \theta]_0^\pi [\phi]_0^{2\pi} = 4\pi R^3/3$.

A final example will illustrate how these techniques are extended to more complex cases. Suppose we want to know the surface area of a sphere of radius $r = R$ in the range $0 \le \theta \le \theta_0$. The appropriate multiple integral is $R^2 \int_0^{\theta_0} \sin \theta \, d\theta \int_0^{2\pi} d\phi = 2\pi R^2 [-\cos \theta]_0^{\theta_0} = 2\pi R^2 (1 - \cos \theta_0)$. Note that the area is $2\pi R^2$ when $\theta_0 = \pi/2$ (half the sphere's surface) and $4\pi R^2$ when $\theta_0 = \pi$ (the entire surface), as should be the case!

(b)   Useful Mensuration Formulae

Using methods similar to those shown in section A7–7(a), it is relatively straightforward to verify the following handy formulae:

Planar
  Arbitrary Triangle
    Area $= (1/2)$(base length) $\times$ (vertical height)
    $= \sqrt{s(s-a)(s-b)(s-c)}$ $\begin{cases} \text{where } s = (1/2)(a+b+c) \text{ and} \\ \text{the sides have lengths } a, b, c. \end{cases}$

  Parallelogram and Rhombus
    Area $=$ (base length) $\times$ (vertical height)
  Trapezoid
    Area $= (1/2)(a+b) \times$ (vertical height) $\begin{cases} \text{where } a \text{ and } b \text{ are the} \\ \text{lengths of top and bottom.} \end{cases}$

  Circle
    Circumference $= 2\pi$(radius) $= \pi$(diameter)
    Area $= \pi$(radius)$^2 = (\pi/4)$(diameter)^2
    Area of Segment $= (1/2)$(radius)$^2(\theta - \sin\theta)$ $\begin{cases} \text{where } \theta \text{ is the central} \\ \text{angle in radians.} \end{cases}$
    Area of Sector $= (1/2)$(radius)$^2\theta$
    Area of Thin Annulus $= 2\pi R(\Delta R)$ $\begin{cases} \text{where } R \text{ is the radius, and} \\ \Delta R \text{ the radial thickness.} \end{cases}$

  Ellipse
    Area $= \pi ab$ ($a =$ semi-major axis, $b =$ semi-minor axis)
Solid
  Rectangular Parallelopiped
    Volume $= abc$ (the sides have lengths a, b, c)
  Pyramid and Cone
    Volume $= (1/3)$(base area) $\times$ (vertical height)
  Right Cylinder
    Volume $= \pi R^2 H$ ($R =$ radius, $H =$ height)
  Sphere
    Surface Area $= 4\pi$(radius)$^2 = \pi$(diameter)^2
    Volume $= (4\pi/3)$(radius)$^3 = (\pi/6)$(diameter)^3
    Surface Area of Segment $= 2\pi$(radius) $\times$ (height of segment)
    Volume of Segment $= (\pi/3)$(height)$^2 \times (3\text{ radius} - \text{height})$
  Ellipsoid
    Volume $= (4\pi/3)abc$ $\begin{cases} \text{where } a, b, c \text{ are the lengths} \\ \text{of the three semi-axes.} \end{cases}$

Index

Key: **Boldfaced** type signifies key references or definitions of words. *Italicized* type indicates references to illustrations or tables. Roman type denotes all other references.